DIFFERENTIAL ANALYSIS

T. M. FLETT

DIFFERENTIAL ANALYSIS

Differentiation, differential equations and differential inequalities

CAMBRIDGE UNIVERSITY PRESS

CAMBRIDGE

LONDON   NEW YORK   NEW ROCHELLE

MELBOURNE   SYDNEY

CAMBRIDGE UNIVERSITY PRESS
Cambridge, New York, Melbourne, Madrid, Cape Town, Singapore, São Paulo, Delhi

Cambridge University Press
The Edinburgh Building, Cambridge CB2 8RU, UK

Published in the United States of America by Cambridge University Press, New York

www.cambridge.org
Information on this title: www.cambridge.org/9780521224208

First published 1980
This digitally printed version 2008

A catalogue record for this publication is available from the British Library

Library of Congress Cataloguing in Publication data
Flett, TM
Differential analysis.

Bibliography: p.
Includes index.
1. Calculus, Differential. 2. Differential equations.
3. Differential inequalities. I. Title.
QA304.F57   515'.3   78-67303

ISBN 978-0-521-22420-8 hardback
ISBN 978-0-521-09030-8 paperback

CONTENTS

PREFACE

On 13 February 1976 Professor T. M. Flett died at the early age of 52. At that time he had almost completed the manuscript of the present book. In order that so much effort should not be lost, I undertook the task of finishing the work. My guiding principle has been that the book is still Professor Flett's: although I have no doubt that he would have made alterations in arriving at his own final version, I am sure that the reader would wish to hear his voice, even imperfectly, rather than mine. The text he left has been altered only where there were clear indications that he himself had intended to do this or on the rare occasions when errors had crept in. A few parts of the book which he clearly proposed to include did not exist even in manuscript. The reader will wish to know that I am solely responsible for §2.13, for the historical note on differentials (§4.7) and for the notes on chapters 3 and 4. In connection with the last two I probably owe apologies to many mathematicians. I do not have Professor Flett's encyclopaedic knowledge of the literature in this field and my attribution of theorems to their originators is not as detailed as his corresponding work for the earlier chapters; I fear that some names which should have appeared will have been omitted.

Professor Flett would have wished to thank Mr P. Baxendale who made valuable comments on a draft of Chapter 1. I myself am deeply indebted to Dr F. Smithies whose meticulous and painstaking examination of the whole manuscript saved me from many errors. Both Professor Flett and I owe much to Mrs E. Bennett who made our jobs easier by producing excellent typescripts of many drafts of the work.

J. S. Pym

INTRODUCTION

The title *Differential Analysis* indicates clearly the content of this book. It is concerned with those parts of analysis in which the idea of differentiation, derivative or differential plays a central role. It is true that the functions to be discussed all take values in normed spaces and, for a large part of the book, they also have subsets of normed spaces as domains, but the basic aim is the generalization and subsequent application of the fundamental theorems of the differential calculus of functions of one real variable.

To be sure, this generalization, simple as it is in essence, extends the range of applications widely. One differential equation involving functions of one variable but with values in a normed space can be equivalent to an infinite system of equations involving scalar-valued functions. The calculus of variations becomes a part of the 'ordinary' theory of maxima and minima. Very general questions about constrained maxima and minima become problems about inclusions between tangent spaces. When set in Banach space, the Newton method, giving an iterative procedure for finding (approximate) solutions of equations, becomes applicable to integral and differential equations.

Although the theory has such a wide import, it remains basically elementary. The reader who glances through the appendix (listing results required in the text) will disbelieve this assertion: some deep theorems are given there. This is indeed true, but except in one or two sections, appeals to these results are rare. For example, the Ascoli–Arzelà theorem is used (on a few occasions) to obtain the existence of solutions to the fundamental equation $y' = f(t, y)$, but the theory then proceeds without further recourse to sophisticated theorems. The book has the flavour of the calculus, not the flavour of functional analysis.

For functions of one real variable with values in a normed space, the definition of derivative is straightforward enough (though the formulation and proof of results often requires ingenuity). This forms the subject of Chapter 1, while the applications to the equation $y' = f(t, y)$ are dealt with in Chapter 2. For functions whose domains are subsets of normed spaces, the 'correct' definition of differential is less clear, and there are several possibilities. One of the best established, and the strongest, is the Fréchet differential, which is discussed in Chapter 3. Two of the other candidates are presented in Chapter 4. The Gâteaux (or directional) differential is the weakest: it can be defined in any vector space. The Hadamard differential lies between the other two; it coincides with the Fréchet differential in finite dimensions, but is more general in infinite-dimensional spaces. It turns out to be exactly what is required for certain considerations in tangent spaces and in aspects of the theory of differential equations (for example, 'differentiation along the curve').

The above discussion has mentioned functions of a real variable. In fact, many of the theorems in the book are true for functions of a complex variable, or when the field of scalars of the normed space is complex. However, to leave the matter there would be misleading. Just as in the case of one variable, there are significant differences between the real and complex theories. No attempt has been made to obtain the stronger conclusions valid for the latter case, although the results have been stated for both real and complex scalars where this was possible.

There will also be found in the book three long historical notes. These are intended to put the modern theory presented in an historical context. They do not, of course, amount to a complete history of the subject, but it is hoped they will be seen not merely as curiosities, but as a real aid in understanding the material.

Each result in the book is assigned a triple of numbers, for example, (3.6.2); this means the second numbered result of §3.6, the sixth section of Chapter 3. Within some sections certain formulae are numbered on the right, for example, (6); a reference to (6) means the sixth numbered formula in the section in which the reference occurs, unless other instructions are given.

1

Differentiation of functions of one real variable

Throughout this chapter, Y denotes a normed vector space, and, except where noted, Y may be over either the real or complex field. The completeness of Y rarely enters our arguments, and we therefore allow Y to be incomplete unless otherwise stated.

1.1 The derivative of a real- or vector-valued function of a real variable

Let $A \subseteq \mathbf{R}$, let $\phi : A \to Y$, and let t_0 be a non-isolated point of A, i.e. $t_0 \in A$ and for each $\delta > 0$ there exists $t \in A$ satisfying $0 < |t - t_0| < \delta$.† The function ϕ is said to have a derivative at t_0 if the limit ‡

$$\lim_{t \to t_0} \frac{\phi(t) - \phi(t_0)}{t - t_0} = \lim_{h \to 0} \frac{\phi(t_0 + h) - \phi(t_0)}{h}$$

exists in Y, and the value of this limit is then called the *derivative of ϕ* at t_0, and is denoted by $\phi'(t_0)$. The *derivative ϕ'* of ϕ is the function $t \mapsto \phi'(t)$ whose domain is the set of non-isolated points t of A at which $\phi'(t)$ exists (ϕ' may, of course, be the empty function).

In the sequel, when we say that a function ϕ has a derivative at a point t_0, we take it as understood that t_0 is a non-isolated point of its domain. The most important cases are those where the domain of ϕ is an interval (when every point of the domain is non-isolated) and where t_0 is an interior point of the domain.*

---

† t_0 is a point of A that is also a point of accumulation of A.

‡ Here, by 'the limit as t tends to t_0', we mean that t is to tend to t_0 through the points of the set $A \backslash \{t_0\}$, so that the only values of t involved are ones for which the quotient $(\phi(t) - \phi(t_0))/(t - t_0)$ is defined. This convention is followed throughout the book.

* It is probably more usual to define the derivative of a function ϕ only at an interior point of its domain. However, in the theory of differential equations we frequently encounter functions from a compact interval J into Y possessing a derivative at each point of J. The endpoints of J can, of course, be dealt with by the use of left-hand and right-hand derivatives (see below), but this leads to a clumsy notation, and the wider definition of derivative adopted here seems preferable.

A function ϕ which possesses a derivative $\phi'(t)$ at a point t is sometimes said to be *differentiable at t*. Similarly, if $\phi'(t)$ exists at every point t of an interval I, then ϕ is said to be *differentiable on I*; if in addition ϕ' is continuous on I then ϕ is *continuously differentiable on I*.

When the function ϕ is given by a specific formula, the derivative $\phi'(t_0)$ of ϕ at t_0 is sometimes denoted by $[d\phi(t)/dt]_{t_0}$, so that, for example, $[dt^2/dt]_{t_0}$ denotes the derivative of the function $t \mapsto t^2$ at t_0. We also write $d\phi(t)/dt$ for the value of the derivative of ϕ at the point t.

A number of other notations for the derivative are in use; the most common, dy/dt, employs a 'dependent variable' y, which stands either for the function ϕ or for the value of ϕ at t, so that dy/dt can be interpreted as either the function ϕ' or its value $\phi'(t)$ according to context. We use this 'dependent variable' notation in a purely formal way in Chapter 2, and otherwise do not use any of these further notations.

Example (a)

If $\phi : A \to Y$ is constant, then $\phi'(t)$ exists and is equal to 0 at each non-isolated point t of A.

Example (b)

If $\phi : \mathbf{R} \to Y$ is given by
$$\phi(t) = c_0 + tc_1 + t^2 c_2 + \ldots + t^m c_m,$$
where $c_0, \ldots, c_m \in Y$, then for all $t \in \mathbf{R}$
$$\phi'(t) = c_1 + 2tc_2 + \ldots + mt^{m-1} c_m.$$

Example (c)

Let $\phi : A \to \mathbf{R}^m$, where $A \subseteq \mathbf{R}$, let $\phi_1, \ldots, \phi_m : A \to \mathbf{R}$ be the components of ϕ given by
$$\phi(t) = (\phi_1(t), \ldots, \phi_m(t)) \qquad (t \in A), \tag{1}$$
and let t_0 be a non-isolated point of A. Then $\phi'(t_0)$ exists and is equal to $d = (d_1, \ldots, d_m)$ if and only if, for $i = 1, \ldots, m$, $\phi_i'(t_0)$ exists and is equal to d_i. This follows immediately from the identity
$$\left\| \frac{\phi(t) - \phi(t_0)}{t - t_0} - d \right\| = \left(\sum_{i=1}^m \left| \frac{\phi_i(t) - \phi_i(t_0)}{t - t_0} - d_i \right|^2 \right)^{1/2}.$$
For example, if $\phi : \mathbf{R} \to \mathbf{R}^3$ is given by $\phi(t) = (\cos t, \sin t, t)$, then, for all $t \in \mathbf{R}$, $\phi'(t) = (-\sin t, \cos t, 1)$.

More generally, we may replace $\mathbf{R}^m$ by the product of m normed vector spaces $Y_1 \times Y_2 \times \dots \times Y_m$, the components $\phi_i : A \to Y_i$ ($i = 1, \dots, m$) being defined by (1) as before.

Example (d)

Let Y be a complex Hilbert space, let $\phi : A \to Y$, where $A \subseteq \mathbf{R}$, and let $\psi(t) = \| \phi(t) \|^2$. Then $\psi'(t) = 2 \operatorname{Re} \langle \phi'(t), \phi(t) \rangle$ whenever $\phi'(t)$ exists. In fact

$$\psi(t + h) - \psi(t) = \| \phi(t + h) \|^2 - \| \phi(t) \|^2$$
$$= \langle \phi(t + h), \phi(t + h) \rangle - \langle \phi(t), \phi(t) \rangle$$
$$= \langle \phi(t + h) - \phi(t), \phi(t + h) \rangle + \langle \phi(t), \phi(t + h) - \phi(t) \rangle,$$

whence

$$\psi'(t) = \langle \phi'(t), \phi(t) \rangle + \langle \phi(t), \phi'(t) \rangle$$
$$= \langle \phi'(t), \phi(t) \rangle + \overline{\langle \phi'(t), \phi(t) \rangle}$$
$$= 2 \operatorname{Re} \langle \phi'(t), \phi(t) \rangle.$$

If Y is a real Hilbert space, then $\psi'(t) = 2 \langle \phi'(t), \phi(t) \rangle$.

Example (e)

Let c_0 denote the Banach space of sequences $y = (y_n)_{n \geq 1}$ of real numbers that converge to the limit 0, with the sup norm $\| y \| = \sup_n |y_n|$. Let also ϕ be a function from a set $A \subseteq \mathbf{R}$ into c_0, and for $n = 1, 2, \dots$ define $\phi_n : A \to \mathbf{R}$ by $\phi(t) = (\phi_n(t))$ (so that the sequence $(\phi_n(t))$ is convergent to the limit 0 for each $t \in A$). If $\phi'(t_0)$ exists for some non-isolated point t_0 of A, then $\phi_n'(t_0)$ exists for each n and $\phi'(t_0) = (\phi_n'(t_0))$, so that also $(\phi_n'(t_0)) \in c_0$. (For, if $\phi'(t_0) = (d_n)$, then for all $n \geq 1$ and all $t \in A$

$$\left| \frac{\phi_n(t) - \phi_n(t_0)}{t - t_0} - d_n \right| \leq \left\| \frac{\phi(t) - \phi(t_0)}{t - t_0} - \phi'(t_0) \right\| .)$$

The converse is false, i.e. there is a function ϕ such that $\phi'(t_0)$ does not exist, even though $\phi_n'(t_0)$ exists for each n and $(\phi_n'(t_0)) \in c_0$. For example, let $A = \mathbf{R}$, let $\phi_n(t) = n^{-1} \log(1 + n^2 t^2)$ ($n \geq 1$), and let $\phi(t) = (\phi_n(t))$. If $\phi'(0)$ exists, then, by the preceding argument, it must be equal to $(\phi_n'(0)) = (0)$. However, for all $t \neq 0$

$$\left\| \frac{\phi(t) - \phi(0)}{t - 0} - (0) \right\| = \left\| \frac{\phi(t)}{t} \right\| = \sup_n (|nt|^{-1} \log(1 + n^2 t^2)),$$

and the expression under the supremum sign on the right here is greater than or equal to log 2 when t is the reciprocal of an integer, so that $\phi'(0)$ does not exist.

Similar results hold for other sequence spaces.

Example (f)

A derivative need not be continuous, even if it exists at each point of an interval. Consider for instance the function $\phi : \mathbf{R} \to \mathbf{R}$ given by

$$\phi(t) = t^2 \sin(\pi/t^2) \quad (t \neq 0), \quad \phi(0) = 0.$$

Here $\phi'(0) = 0$, and for $t \neq 0$

$$\phi'(t) = 2t \sin(\pi/t^2) - (2\pi/t) \cos(\pi/t^2),$$

so that $\phi'(1/n^{1/2}) = \pm 2\pi n^{1/2}$ according as n is an odd or even integer (thus the derivative is in fact unbounded on each compact interval containing 0).†

The right-hand and left-hand derivatives $\phi'_+(t_0)$ and $\phi'_-(t_0)$ are defined in a manner similar to that for $\phi'(t_0)$. If t_0 is a point of A such that for each $\delta > 0$ there exists $t \in A$ satisfying $t_0 < t < t_0 + \delta$, then the (*strong*) *right-hand derivative* $\phi'_+(t_0)$ of ϕ at t_0 is the limit

$$\lim_{t \to t_0+} \frac{\phi(t) - \phi(t_0)}{t - t_0} = \lim_{h \to 0+} \frac{\phi(t_0 + h) - \phi(t_0)}{h}$$

whenever this limit exists in Y. The *left-hand derivative* $\phi'_-(t_0)$ is defined similarly.

Example (g)

Let I be an interval in $\mathbf{R}$, and let $\phi : I \to \mathbf{R}$ be convex. Then $\phi'_+(t)$ and $\phi'_-(t)$ exist and satisfy $\phi'_-(t) \leq \phi'_+(t)$ for each interior point t of I (so that also ϕ is continuous on I°), and $\phi'_-(t) = \phi'_+(t)$ for all except a countable set of $t \in]a,b[$. Further, ϕ'_- and ϕ'_+ are both increasing on I°.

Let s, t, u be points of I° such that $s < u < t$. We see from (A.3.6) that

$$\frac{\phi(u) - \phi(s)}{u - s} \leq \frac{\phi(t) - \phi(s)}{t - s} \leq \frac{\phi(t) - \phi(u)}{t - u}.$$

The left-hand inequality, with s, u, t replaced first by $s, s+h, s+k$, and then by $s, s+h, t$, implies that the function $h \mapsto (\phi(s+h) - \phi(s))/h$ is increasing on $]0, t - s]$ and is bounded above there by $(\phi(t) - \phi(s))/(t - s)$.

† The expressions $\cos(\pi/t^m)$ and $\sin(\pi/t^m)$ are frequently useful in the construction of counter-examples, and various instances will be encountered later.

Similarly, if $s - k \in I^\circ$, where $k > 0$, then the right-hand inequality, with s, u, t replaced by $s - k, s, s + h$, implies that $h \mapsto (\phi(s + h) - \phi(s))/h$ is bounded below by $(\phi(s) - \phi(s - k))/k$, and hence $\phi'_+(s)$ exists and satisfies

$$\frac{\phi(s - k) - \phi(s)}{k} \le \phi'_+(s) \le \frac{\phi(t) - \phi(s)}{t - s}.$$

In the same way, we show that $\phi'_-(t)$ exists and satisfies

$$\frac{\phi(t) - \phi(s)}{t - s} \le \phi'_-(t) \le \frac{\phi(t + l) - \phi(t)}{l},$$

where $l > 0$ and $t + l \in I^\circ$, and on combining these results we obtain that

$$\phi'_-(s) \le \phi'_+(s) \le \frac{\phi(t) - \phi(s)}{t - s} \le \phi'_-(t) \le \phi'_+(t).$$

Finally, let E be the set of $t \in I^\circ$ for which $\phi'_-(t) < \phi'_+(t)$, and for each $t \in E$ let $J_t =]\phi'_-(t), \phi'_+(t)[$. By the central pair of inequalities above, $J_s \cap J_t = \varnothing$ for all $s, t \in E$. It therefore follows by a standard argument that the set of intervals J_t is countable, whence so also is E.

In the theorems of this section we state the results for the ordinary derivative ϕ', and take for granted their extensions to $\phi'_\pm$.

The following result gives some equivalent formulations of the definition of the derivative; the proof is immediate.

(1.1.1) *Let $\phi : A \to Y$, where $A \subseteq \mathbf{R}$, and let t_0 be a non-isolated point of A. Then the following statements are equivalent:*
  (i) *ϕ has a derivative at t_0 equal to d;*
  (ii) *the function $\lambda : A \to Y$ given by*

$$\lambda(t) = \frac{\phi(t) - \phi(t_0)}{t - t_0} \quad (t \in A, t \ne t_0), \quad \lambda(t_0) = d,$$

is continuous at t_0;
  (iii) *for all $t \in A$*

$$\phi(t) = \phi(t_0) + (t - t_0)d + (t - t_0)v(t),$$

where $v : A \to Y$ is continuous at t_0 and has the value 0 there;
  (iv)   *$(\phi(t_0 + h) - \phi(t_0) - hd)/|h| \to 0$ in Y as $h \to 0$.*

The existence of $\phi'(t_0)$ obviously implies that ϕ is continuous at t_0, and indeed we have a stronger result:

(1.1.2) *If $\phi : A \to Y$ has a derivative at t_0, then for each $\varepsilon > 0$ there exists a neighbourhood U of t_0 in $\mathbf{R}$ such that for all $t \in U \cap A$*

$$\|\phi(t) - \phi(t_0)\| \le (\|\phi'(t_0)\| + \varepsilon)|t - t_0|.$$

This is an easy consequence of (1.1.1) (ii).

(1.1.3) (i) *If* $\phi : A \to Y$ *has a derivative at* t_0, *and* α *is a scalar, then* $\alpha\phi$ *has a derivative at* t_0 *equal to* $\alpha\phi'(t_0)$.
(ii) *If* $\phi : A \to Y$, $\psi : B \to Y$ *have derivatives at* t_0, *and* t_0 *is a non-isolated point of* $A \cap B$, *then* $\phi + \psi$ *has a derivative at* t_0, *equal to* $\phi'(t_0) + \psi'(t_0)$.
This too is immediate. We note that if t_0 is an interior point of A and B, then it is also an interior point of the domain $A \cap B$ of $\phi + \psi$.

This result implies in particular that if $A \subseteq \mathbf{R}$ and t_0 is a given non-isolated point of A, then the class of functions $\phi : A \to Y$ such that $\phi'(t_0)$ exists is a vector space under the operations of addition of functions and multiplication of functions by scalars. Moreover, the function $\phi \mapsto \phi'(t_0)$ is a linear functional on this space.

It should be noted that in general $(\phi + \psi)' \ne \phi' + \psi'$, since the domain of $\phi' + \psi'$ may be a proper subset of that of $(\phi + \psi)'$ (take, for example, $\phi(t) = -\psi(t) = |t|$).

(1.1.4) (Differentiation of a product and a quotient) *Let* Y *be a real normed space, and let* $\phi : A \to \mathbf{R}$, $\psi : B \to Y$ *be given functions, where* $A, B \subseteq \mathbf{R}$. *If* t_0 *is a non-isolated point of* $A \cap B$, *and* $\phi'(t_0)$, $\psi'(t_0)$ *exist, then the function* $t \mapsto \phi(t)\psi(t)$ *has a derivative at* t_0 *equal to* $\phi(t_0)\psi'(t_0) + \phi'(t_0)\psi(t_0)$. *If in addition* $\phi(t_0) \ne 0$, *then also the function* $t \mapsto \psi(t)/\phi(t)$ *has a derivative at* t_0 *equal to* $(\phi(t_0)\psi'(t_0) - \phi'(t_0)\psi(t_0))/(\phi(t_0))^2$. *The same result holds if* Y *is a complex normed space and* ϕ *maps* A *into* $\mathbf{C}$.
Here again the proofs are elementary and we omit them.

The next result is a generalization of the rule for differentiation of a product.

(1.1.5) *Let* $Y_1, \ldots, Y_m, Z$ *be normed vector spaces, all real or all complex, and let* $u : Y_1 \times \ldots \times Y_m \to Z$ *be a continuous multilinear function. Let also* $\phi_i : A_i \to Y_i$ $(i = 1, \ldots, m)$ *be given functions, where each* $A_i \subseteq \mathbf{R}$, *let* $A = \bigcap_{i=1}^m A_i$, *and define* $\phi : A \to Z$ *by* $\phi(t) = u(\phi_1(t), \ldots, \phi_m(t))$. *If* t_0 *is a non-isolated point of* A, *and* $\phi_i'(t_0)$ *exists for each i, then* $\phi'(t_0)$ *exists and is equal to*

$$\sum_{i=1}^m u(\phi_1(t_0), \ldots, \phi_{i-1}(t_0), \phi_i'(t_0), \phi_{i+1}(t_0), \ldots, \phi_m(t_0)).$$

Let $a_i = \phi_i(t_0), b_i = \phi_i(t)$ $(i = 1, \ldots, m; t \in A, t \neq t_0)$. Then

$$\phi(t) - \phi(t_0) = u(b_1, \ldots, b_m) - u(a_1, \ldots, a_m)$$

$$= \sum_{i=1}^{m} u(b_1, \ldots, b_{i-1}, b_i - a_i, a_{i+1}, \ldots, a_m).$$

Hence

$$\frac{\phi(t) - \phi(t_0)}{t - t_0} = \sum_{i=1}^{m} u(b_1, \ldots, b_{i-1}, \frac{b_i - a_i}{t - t_0}, a_{i+1}, \ldots, a_m),$$

and on making $t \to t_0$ in A we obtain the result.

The 'product rule' in (1.1.4) is the case where u is $(\alpha, y) \mapsto \alpha y$.

(1.1.6) (The chain rule) *Let* $A, B \subseteq \mathbf{R}$, *and suppose that* $\phi : A \to \mathbf{R}$ *has a derivative at* t_0, *that* $\psi : B \to Y$ *has a derivative at* s_0, *where* $\phi(t_0) = s_0$, *and that* t_0 *is a non-isolated point of the domain of* $\psi \circ \phi$. *Then* $\psi \circ \phi$ *has a derivative at* t_0, *equal to* $\phi'(t_0)\psi'(s_0)$.

(We recall that the domain $\mathcal{D}(\psi \circ \phi)$ of $\psi \circ \phi$ is the set of $t \in A$ for which $\phi(t) \in B$, i.e. the set $\phi^{-1}(B)$.)

By (1.1.1) (ii), we can write

$$\phi(t) = \phi(t_0) + (t - t_0)\lambda(t), \tag{2}$$

$$\psi(s) = \psi(s_0) + (s - s_0)\mu(s), \tag{3}$$

where the functions $\lambda : A \to \mathbf{R}, \mu : B \to Y$ are continuous at t_0, s_0, respectively, and $\lambda(t_0) = \phi'(t_0), \mu(s_0) = \psi'(s_0)$. Let $t \in \mathcal{D}(\psi \circ \phi)$, so that $\phi(t) \in B$. Then, from (3) with $s = \phi(t), s_0 = \phi(t_0)$, and (2),

$$\psi(\phi(t)) = \psi(\phi(t_0)) + (t - t_0)\lambda(t)\mu(\phi(t)).$$

Since $t \mapsto \lambda(t)\mu(\phi(t))$ is continuous at t_0 and has there the value

$$\lambda(t_0)\mu(\phi(t_0)) = \phi'(t_0)\psi'(s_0),$$

the result follows from a further application of (1.1.1) (ii).

We note that if t_0 is an interior point of A and s_0 is an interior point of B, then t_0 is an interior point of $\mathcal{D}(\psi \circ \phi)$. For B is then a neighbourhood of s_0 in $\mathbf{R}$, and since $\phi(t_0) = s_0$ and ϕ is continuous at t_0, it follows that $\phi^{-1}(B)$ is a neighbourhood of t_0 in A. Further, A is a neighbourhood of t_0 in $\mathbf{R}$, whence $\phi^{-1}(B)$ is a neighbourhood of t_0 in $\mathbf{R}$, i.e. t_0 is an interior point of $\mathcal{D}(\psi \circ \phi)$.

(1.1.7) *Let* ϕ *be real-valued, strictly monotone, and continuous on an interval* I, *and let* t_0 *be a point of* I *at which* ϕ *has a derivative* $\phi'(t_0) \neq 0$.

Then the inverse function of ϕ has a derivative at the point $s_0 = \phi(t_0)$, equal to $1/\phi'(t_0)$.

Obviously ϕ is one-to-one, so that ϕ^{-1} is well-defined. Since, in addition, ϕ is continuous and therefore maps compact sets to compact sets, ϕ^{-1} is continuous. The function λ of (1.1.1) (ii) is continuous at t_0, and since $\lambda(t_0) \neq 0$, $1/\lambda$ is continuous at t_0. This is equivalent to the assertion.

Exercises 1.1

1 Let ϕ be a mapping of the interval $[a, b]$ into the normed space Y, and let $a < t < b$. Prove that
   (i) if $\phi'(t)$ exists then $(\phi(v) - \phi(u))/(v - u)$ tends to $\phi'(t)$ as u, v tend to t in such a manner that $u < t < v$,
   (ii) if ϕ is continuous at t and $(\phi(v) - \phi(u))/(v - u)$ tends to l as u, v tend to t in such a manner that $u < t < v$, then $\phi'(t)$ exists and is equal to l.
2 Let p be a continuous convex function on Y, and let ϕ be a function from a subset of $\mathbf{R}$ into Y possessing a right-hand derivative at t. Prove that $p \circ \phi$ has a right-hand derivative at t.
   [Hint. Use the fact that if $y, z \in Y$ then $h \mapsto p(y + hz)$ is convex on $\mathbf{R}$.]
3 Let U, V be functions from a set $A \subseteq \mathbf{R}$ into the space $\mathscr{L}(Y, Y)$ of continuous linear maps of Y into itself. Prove that if U, V possess derivatives at a point $t_0 \in A$, then the function $t \mapsto U(t) \circ V(t)$ has a derivative at t_0 equal to $U'(t_0) \circ V(t_0) + U(t_0) \circ V'(t_0)$.
4 Let $A, B \subseteq \mathbf{R}$, and let $\phi : A \to \mathbf{R}$ and $\psi : B \to Y$ be functions such that ϕ has a right-hand derivative at t_0 and ψ has a derivative at an interior point s_0 of B, where $\phi(t_0) = s_0$. Prove that $\psi \circ \phi$ has a right-hand derivative at t_0 equal to $\phi'_+(t_0)\psi'(s_0)$.

1.2 Tangents to paths

It is familiar that the existence of the derivative of a real-valued function at a point t is equivalent to the existence of a tangent to the graph of the function at t. We prove here the analogous result for vector-valued functions, and we consider also the more general notion of a tangent to a path.

A *path* in a normed vector space Y is a continuous function ψ from a compact interval $[a, b]$ into Y. For each $t \in [a, b]$ the point $\psi(t)$ in Y is called *the point t of the path*, and the points $\psi(a), \psi(b)$ are called the *endpoints* of the path. The path is said to be *simple* if ψ is one-to-one; in this case ψ^{-1} is also continuous (since $[a, b]$ is compact), so that ψ is a homeomorphism of $[a, b]$ into Y.

The term 'path' has an obvious geometrical connotation, and intuitively we think of the point $\psi(t)$ as tracing out a continuous path in Y as t runs from a to b. If the path is simple, it does not cross itself.

To define the notion of a tangent to a path it is convenient to use the idea of a ray. For given distinct points $y_0, y \in Y$, the *ray from y_0 through y* is the set $\{y_0 + \lambda(y - y_0) : \lambda \geq 0\}$, and we define the *direction* of this ray to be the unit vector $(y - y_0)/\|y - y_0\|$. For a given ray, this direction is obviously independent of the choice of y on the ray.

Now let $\psi : [a,b] \to Y$ be a path, let $t_0 \in]a,b[$, and for each $t \in [a,b]$ for which $\psi(t) \neq \psi(t_0)$ let $d(t_0, t)$ be the direction of the ray from $\psi(t_0)$ through $\psi(t)$, i.e.

$$d(t_0, t) = \frac{\psi(t) - \psi(t_0)}{\|\psi(t) - \psi(t_0)\|}.$$

We say that the path ψ *has a tangent at t_0* if there exists a neighbourhood U of t_0 such that $\psi(t) \neq \psi(t_0)$ whenever $t \in U$ and $t \neq t_0$ (so that $d(t_0, t)$ exists for all $t \in U \setminus \{t_0\}$), and there exists a (unit) vector $q(t_0)$ in Y such that $d(t_0, t) \to q(t_0)$ as $t \to t_0 +$ and that $d(t_0, t) \to -q(t_0)$ as $t \to t_0 -$. The vector $q(t_0)$ is then called the *direction of the tangent* to the path ψ at t_0, and the set $\{\psi(t_0) + \lambda q(t_0) : \lambda \in \mathbf{R}\}$ is called the *tangent* to the path at t_0.

The graph of a continuous function $\phi : [a,b] \to Y$ can be regarded as a (simple) path in $\mathbf{R} \times Y$, namely the path $t \mapsto (t, \phi(t))$. We can therefore speak of a tangent to the graph, and here we have the following result.

(1.2.1) *Let $\phi : [a,b] \to Y$ be continuous, and let $t_0 \in]a,b[$. Then $\phi'(t_0)$ exists if and only if the graph of ϕ has a tangent at the point t_0 not parallel to the y-axis, i.e. with direction $(l,m) \in \mathbf{R} \times Y$ such that $l \neq 0$.*

It is obvious that if $t \neq t_0$ then $(t, \phi(t)) \neq (t_0, \phi(t_0))$, and that the direction of the ray in $\mathbf{R} \times Y$ from $(t_0, \phi(t_0))$ through $(t, \phi(t))$ is

$$D(t_0, t) = \frac{(t - t_0, \phi(t) - \phi(t_0))}{r(t_0, t)},$$

where

$$r(t_0, t) = \|(t, \phi(t)) - (t_0, \phi(t_0))\| = ((t - t_0)^2 + \|\phi(t) - \phi(t_0)\|^2)^{1/2}.$$

If $D(t_0, t) \to \pm(l, m)$ as $t \to t_0 \pm$, respectively, then

$$\frac{t - t_0}{r(t_0, t)} \to \pm l \quad \text{and} \quad \frac{\phi(t) - \phi(t_0)}{r(t_0, t)} \to \pm m$$

as $t \to t_0 \pm$. Hence, if $l \neq 0$, then $(\phi(t) - \phi(t_0))/(t - t_0) \to m/l$ as $t \to t_0$, so that $\phi'(t_0)$ exists and is equal to m/l. Conversely, if $\phi'(t_0)$ exists, then clearly

$$D(t_0, t) \to \pm \frac{(1, \phi'(t_0))}{(1 + \|\phi'(t_0)\|^2)^{1/2}} \quad \text{as} \quad t \to t_0 \pm,$$

so that the graph of ϕ has a tangent at t_0 not parallel to the y-axis.

Here the tangent is obviously the line in $\mathbf{R} \times Y$ with equation

$$y = \phi(t_0) + (t - t_0)\phi'(t_0).$$

The situation with a general path $\psi : [a,b] \to Y$ is more complicated, for the existence of the tangent to the path at the point t_0 neither implies nor is implied by the existence of the derivative of ψ at t_0. For example, if the path $\psi : [-1,1] \to \mathbf{R}^2$ is given by $\psi(t) = (t^{1/3}, t^{2/3})$ (so that the range of ψ is an arc of a parabola), then ψ has no derivative at 0, but $d(0,t) \to \pm(1,0)$ as $t \to 0\pm$, so that ψ has a tangent at 0 with direction $(1,0)$. Again, if the path $\psi : [-1,1] \to \mathbf{R}^2$ is given by $\psi(t) = (t|t|, t^2)$ (so that the range of ψ consists of two segments joined at right angles at the origin), then $\psi'(t)$ exists for all $t \in [-1,1]$, but

$$d(0,t) = \frac{(t|t|, t^2)}{t^2 \sqrt{2}} \to (\pm 1/\sqrt{2}, 1/\sqrt{2}) \quad \text{as} \quad t \to 0\pm,$$

so that ψ has no tangent at 0.

It is significant that $\psi'(0) = 0$ in the second example here, for in the case where ψ has a non-zero derivative we have the following result, which is an immediate consequence of the definition of a derivative.

(1.2.2) *Let $\psi : [a,b] \to Y$ be a path in Y, and let $t_0 \in\,]a,b[$. If $\psi'(t_0)$ exists and is not 0, then ψ has a tangent at t_0 with direction $\psi'(t_0)/\|\psi'(t_0)\|$.*

If Y is the plane $\mathbf{R}^2$, and ψ_1, ψ_2 are the components of ψ then $\psi'(t_0) = (\psi'_1(t_0), \psi'_2(t_0))$, and the condition that $\psi'(t_0) \neq 0$ is equivalent to the condition that $\psi'_1(t_0)$ and $\psi'_2(t_0)$ are not both 0. If this condition is satisfied, then the direction of the tangent at t_0 is

$$\frac{\psi'(t_0)}{\|\psi'(t_0)\|} = \frac{(\psi'_1(t_0), \psi'_2(t_0))}{((\psi'_1(t_0))^2 + (\psi'_2(t_0))^2)^{1/2}},$$

so that the tangent is the line through the point $(\psi_1(t_0), \psi_2(t_0))$ with slope $\psi'_2(t_0)/\psi'_1(t_0)$. If $\psi'_1(t_0) = 0$, this is to be interpreted as meaning that the tangent is vertical.

Exercise 1.2

1 Prove that a path ψ has a tangent at a point t_0 if the path $t \mapsto (t, \psi(t))$ has a tangent at t_0 with direction (l, m) such that $m \neq 0$.

1.3 The mean value theorems of Rolle, Lagrange and Cauchy

In the sequel we consider a number of generalizations of the classical mean value theorems associated with the names of Rolle, Lagrange and

Cauchy, and as a preliminary we recall the standard present-day versions of these theorems,† and some of their consequences. At the same time we consider some allied geometrical results concerning tangents.

We begin with a simple lemma, the proof of which is immediate.

(1.3.1) Lemma. *If a real-valued function ϕ has a positive derivative at a point t_0, there exists $\delta > 0$ such that $\phi(s) < \phi(t_0) < \phi(t)$ whenever $t_0 - \delta < s < t_0 < t < t_0 + \delta$ and s and t belong to the domain of ϕ. Similarly, if ϕ has a negative derivative at t_0, there exists $\delta > 0$ such that $\phi(s) > \phi(t_0) > \phi(t)$ whenever $t_0 - \delta < s < t_0 < t < t_0 + \delta$ and s and t belong to the domain of ϕ.‡*

(1.3.2) (Rolle's theorem) *Let $\phi : [a,b] \to \mathbf{R}$ be a continuous function such that $\phi'(t)$ exists for all $t \in]a,b[$ and that $\phi(a) = \phi(b)$. Then there is at least one $\xi \in]a,b[$ such that $\phi'(\xi) = 0$.*

If ϕ is constant, then $\phi'(\xi) = 0$ for all $\xi \in]a,b[$. If ϕ is not constant, then (since $\phi(a) = \phi(b)$) ϕ attains either its supremum or its infimum at an interior point ξ of $[a,b]$, and at such a point ξ the derivative $\phi'(\xi)$ is 0, by (1.3.1).

We note in passing that we can specify a solution ξ of $\phi'(\xi) = 0$ in a perfectly definite manner; for instance, if sup $\phi > \phi(a)$, we can take ξ to be the least element of $[a,b]$ at which ϕ attains its supremum, while if sup $\phi = \phi(a) > \inf \phi$ we can take ξ to be the least element of $[a,b]$ at which ϕ attains its infimum. On the other hand, the infimum and supremum of such solutions may not be a solution as is easily seen by considering the example $\phi : [-2,2] \to \mathbf{R}$ given by

$$\phi(t) = t \quad (-2 \le t \le 0), \quad \phi(t) = t - t^2 \sin(\pi/t) \quad (0 < t \le 2).$$

In geometrical terms, Rolle's theorem asserts that, if the chord from the point a of the graph of ϕ to the point b is horizontal (i.e. parallel to the t-axis), there is at least one intermediate point of the graph at which the direction of the tangent is the same as that of the chord.

Rolle's theorem contains the essence of the mean value theorem, but the name 'mean value theorem' is usually given to the following more

---

† See also §1.11.

‡ It should be noted that if ϕ is real-valued and $\phi'(t_0) > 0$ for a single point t_0, then ϕ is not necessarily increasing on some neighbourhood of t_0. For instance, if

$$\phi(t) = t - t^2 \cos(\pi/t) \quad (t \ne 0), \quad \phi(0) = 0,$$

then $\phi'(0) = 1 > 0$, while $\phi(1/(2n)) - \phi(1/(2n+1)) < 0$ for $n = 1,2,\dots$. It is, of course, obvious that if ϕ is increasing on a neighbourhood of t_0, then $\phi'(t_0) \ge 0$ whenever it exists.

general form in which the chord joining the endpoints of the graph need not be horizontal.

(1.3.3) (Lagrange's mean value theorem) *Let $\phi : [a,b] \to \mathbf{R}$ be a continuous function such that $\phi'(t)$ exists for all $t \in]a,b[$. Then there is at least one $\xi \in]a,b[$ such that*

$$\phi(b) - \phi(a) = (b - a)\phi'(\xi). \tag{1}$$

This follows from Rolle's theorem applied to the function $t \mapsto \phi(t) - \lambda t$, where $\lambda = (\phi(b) - \phi(a))/(b - a)$.

We note three immediate consequences of the mean value theorem (1.3.3). Generalizations of these results will be given in later sections.

(1.3.4) *If $\phi : [a,b] \to \mathbf{R}$ is a continuous function such that $\phi'(t)$ exists and is non-negative for all $t \in]a,b[$, then ϕ is increasing. If in addition $\phi'(t) > 0$ for all $t \in]a,b[$, then ϕ is strictly increasing.*

(1.3.5) *If $\phi : [a,b] \to \mathbf{R}$ is a continuous function such that $\phi'(t)$ exists and is equal to 0 for all $t \in]a,b[$, then ϕ is constant.*

(1.3.6) *Let $M \geq 0$, and let $\phi : [a,b] \to \mathbf{R}$ be a continuous function such that $\phi'(t)$ exists and satisfies $|\phi'(t)| \leq M$ for all $t \in]a,b[$. Then*

$$|\phi(b) - \phi(a)| \leq M(b - a).$$

The results of Rolle's theorem (1.3.2) and the mean value theorem (1.3.3) do not extend to vector-valued functions, i.e. if ϕ is a continuous function from an interval $[a,b]$ into a normed vector space Y such that $\phi'(t)$ exists for all $t \in]a,b[$, there may be no $\xi \in]a,b[$ for which (1) holds. For example, if $\phi : [0,2\pi] \to \mathbf{R}^2$ is given by $\phi(t) = (\cos t, \sin t)$, then ϕ is continuous, and $\phi(2\pi) - \phi(0) = (0,0)$. Also $\phi'(\xi) = (-\sin \xi, \cos \xi)$ for all $\xi \in [0,2\pi]$, so that $\| \phi'(\xi) \| = 1$, and hence in this case there is no ξ satisfying $\phi(2\pi) - \phi(0) = 2\pi\phi'(\xi)$.

Although the mean value theorem does not extend in the form (1), we have the following simple substitute.

(1.3.7) *Let Y be a real normed space, and let u be a continuous linear functional on Y. Let also $\phi : [a,b] \to Y$ be a continuous function such that $\phi'(t)$ exists for all $t \in]a,b[$, and let $d = (\phi(b) - \phi(a))/(b - a)$. Then there is at least one $\xi \in]a,b[$ such that*

$$u(d) = u(\phi'(\xi)).$$

This result follows easily from the mean value theorem (1.3.3) applied to the function $\psi = u \circ \phi$, for ψ is clearly continuous, and if $t \in]a,b[$ then

$$\frac{\psi(t+h) - \psi(t)}{h} = \frac{u(\phi(t+h)) - u(\phi(t))}{h} = u\left(\frac{\phi(t+h) - \phi(t)}{h}\right) \to u(\phi'(t))$$

as $h \to 0$, so that $\psi'(t)$ exists and is equal to $u(\phi'(t))$.

This result leads in turn to the following mean value inequality for vector-valued functions.

(1.3.8) *Let Y be a normed space, let $\phi : [a,b] \to Y$ be a continuous function such that $\phi'(t)$ exists for all $t \in]a,b[$, and let $d = (\phi(b) - \phi(a))/(b - a)$. Then there is at least one $\xi \in]a,b[$ such that*

$$\|d\| \le \|\phi'(\xi)\|. \tag{2}$$

We may obviously suppose here that Y is over the real field. If $d = 0$, the result is trivial. If $d \ne 0$, then, by the Hahn–Banach theorem (A.4.1 Corollary 1), we can find a continuous linear functional u on Y such that $\|u\| = 1$ and that $u(d) = \|d\|$. By (1.3.7), we can then find $\xi \in]a,b[$ such that $u(d) = u(\phi'(\xi))$, and hence

$$\|d\| = u(d) = u(\phi'(\xi)) \le \|u\| \|\phi'(\xi)\| = \|\phi'(\xi)\|.$$

The inequality (2) has a simple kinematic interpretation. If we think of $\phi(t)$ as the position of a moving particle at time t, then the difference quotient d represents the average velocity of the particle over the interval of time from a to b, while the derivative $\phi'(t)$ represents the instantaneous velocity of the particle at time t. The inequality (2) therefore shows that if the particle is to travel from one point to another in a certain time, it must at some instant move at least as fast as it would have to do if it travelled with constant velocity.

The use of the Hahn–Banach theorem in the proof of (1.3.8) can be avoided, and in §1.6 we obtain a sharper result by purely real-variable arguments.

Theorem (1.3.8) clearly implies the extension of (1.3.5) to vector-valued ϕ.

We return now to real-valued functions, and we consider next a simple consequence of Rolle's theorem which shows that, although derivatives need not be continuous, they nevertheless have the same intermediate value property as continuous functions.

(1.3.9) (Darboux's theorem) *If I is an interval, and $\phi : I \to \mathbf{R}$ is a function whose derivative $\phi'(t)$ exists for all $t \in I$, then the range A of ϕ' is an interval.*

It is enough to prove that if $y_1, y_2 \in A$ and $y_1 < y_0 < y_2$, then $y_0 \in A$. Moreover, by replacing ϕ by the function $t \mapsto \phi(t) - y_0 t$, we may suppose that $y_0 = 0$. But, if $0 \notin A$, then, by Rolle's theorem, ϕ is one-to-one. Since ϕ is continuous, it is therefore strictly monotone, and hence either $\phi'(t) \geq 0$ for all $t \in I$ or $\phi'(t) \leq 0$ for all $t \in I$. Since $y_1, y_2 \in A$ and $y_1 < 0 < y_2$, we obtain a contradiction, whence $0 \in A$.

Alternatively, we may argue as follows. Let D be the set of difference quotients $d(s,t) = (\phi(s) - \phi(t))/(s - t)$, where $s, t \in I$ and $s < t$. The mean value theorem (1.3.3) asserts that $D \subseteq A$, and obviously $A \subseteq \bar{D}$. It is therefore enough to show that D is an interval, or, equivalently, that D is connected. This, however, is obvious, since the function $(s,t) \mapsto d(s,t)$ from the triangle $\{(s,t) \in I \times I : s < t\}$ into $\mathbf{R}$ is clearly continuous, and D is its range.

The next result, which involves a pair of functions, leads directly to the standard version of Cauchy's mean value theorem.

(1.3.10) *Let $\psi_1, \psi_2 : [a,b] \to \mathbf{R}$ be continuous functions such that $\psi'_1(t)$ and $\psi'_2(t)$ exist for all $t \in {]}a,b{[}$. Then there is at least one $\xi \in {]}a,b{[}$ such that*

$$(\psi_1(b) - \psi_1(a))\psi'_2(\xi) = (\psi_2(b) - \psi_2(a))\psi'_1(\xi).$$

Here we consider the function $\phi : [a,b] \to \mathbf{R}$ given by

$$\phi(t) = (\psi_1(b) - \psi_1(a))(\psi_2(t) - \psi_2(a)) - (\psi_2(b) - \psi_2(a))(\psi_1(t) - \psi_1(a)).$$

Obviously $\phi(a) = \phi(b) = 0$, whence there is at least one $\xi \in {]}a,b{[}$ for which $\phi'(\xi) = 0$, and this gives the result.

(1.3.10. Corollary 1) (Cauchy's mean value theorem) *Let $\psi_1, \psi_2 : [a,b] \to \mathbf{R}$ be continuous functions such that, for each $t \in {]}a,b{[}$, $\psi'_1(t)$ and $\psi'_2(t)$ exist and $\psi'_1(t) \neq 0$. Then there is at least one $\xi \in {]}a,b{[}$ such that*

$$\frac{\psi_2(b) - \psi_2(a)}{\psi_1(b) - \psi_1(a)} = \frac{\psi'_2(\xi)}{\psi'_1(\xi)}. \tag{3}$$

This follows from the theorem and Rolle's theorem, since the condition that $\psi'_1(t) \neq 0$ implies that $\psi_1(b) - \psi_1(a) \neq 0$.

We remark that the extra generality of (3) in comparison with (1) is more apparent than real, for (3) is essentially a particular case of (1). As in the proof of (1.3.9), the condition that $\psi'_1(t) \neq 0$ implies that ψ_1 is strictly monotone, and therefore, by replacing ψ_1 by $\alpha \psi_1 + \beta$ with suitable α, β, we may suppose that $\psi_1(a) = a$ and $\psi_1(b) = b$. If we now apply (1) to the function $\psi_2 \circ \psi_1^{-1}$, and note that the derivative of this function at the point t of $]a,b[$ is $\psi_2'(s)/\psi_1'(s)$, where $s = \psi_1^{-1}(t)$, we obtain (3).

Exactly as in (1.3.7, 8), we can obtain analogues of Cauchy's mean value theorem for vector-valued functions, but we omit these here, since more general results will be obtained in §1.6.

The geometric interpretation of Cauchy's mean value theorem (more precisely, of (1.3.10)) is less obvious than the interpretations of Rolle's and Lagrange's theorems, and is as follows.

(1.3.10. Corollary 2) *Let* $\psi:[a,b] \to \mathbf{R}^2$ *be a path in* $\mathbf{R}^2$ *such that for each* $t\in[a,b[$ *the derivative* $\psi'(t)$ *exists and is not* $(0,0)$. *Suppose further that the end points* $A = \psi(a)$, $B = \psi(b)$ *of the path are distinct. Then there is at least one* $\xi\in]a,b[$ *such that the tangent to the path at* ξ *is parallel to the chord* AB.

It should be noted that the path can cross itself an arbitrary number of times; all that we ask is that the endpoints A,B are distinct, or, equivalently, that if ψ_1,ψ_2 are the components of ψ, then $\psi_1(b) - \psi_1(a)$ and $\psi_2(b) - \psi_2(a)$ are not both 0.

By (1.2.2), the path ψ has a tangent at each point other than A,B, and the slope of the tangent at the point ξ is $\psi_2'(\xi)/\psi_1'(\xi)$, the tangent being vertical if $\psi_1'(\xi) = 0$.

Next, by the main theorem, we can find $\xi\in]a,b[$ for which

$$(\psi_1(b) - \psi_1(a))\psi_2'(\xi) = (\psi_2(b) - \psi_2(a))\psi_1'(\xi),$$

and, by the hypothesis, at least one of $\psi_1'(\xi), \psi_2'(\xi)$ is not 0. If $\psi_1'(\xi) = 0$, then $\psi_1(b) - \psi_1(a) = 0$, so that both the chord AB and the tangent at ξ are vertical. If $\psi_1'(\xi) \neq 0$, then also $\psi_1(b) - \psi_1(a) \neq 0$ (for otherwise both $\psi_1(b) - \psi_1(a)$ and $\psi_2(b) - \psi_2(a)$ are 0, and $A = B$). Hence

$$\frac{\psi_2(b) - \psi_2(a)}{\psi_1(b) - \psi_1(a)} = \frac{\psi_2'(\xi)}{\psi_1'(\xi)},$$

i.e. the chord and the tangent at ξ have the same slope.

In view of this corollary, the proof of the main theorem (1.3.10) becomes less artificial if we observe that $2\phi(t)$ is the area of the triangle with vertices at the points A,B, and the point t of the path ψ. It is intuitive that, when the area of this triangle is a maximum, the tangent at the point t will be parallel to the chord AB, and we therefore look at those t for which $\phi'(t) = 0$.

It should be noted that in (1.3.10. Corollary 2) there may be no point

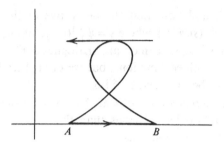

Figure 1.1

ξ of the path where the tangent *has the same direction* as the chord. This is evident from the path shown in Fig. 1.1; in this case there is exactly one point where the tangent is parallel to the chord, and there the tangent is pointing in the opposite direction to the chord.

We may ask whether the result of (1.3.10. Corollary 2) continues to hold if we replace the condition that $\psi'(t)$ exists and is not $(0,0)$ for each $t \in]a,b[$ by the weaker condition that the path ψ has a tangent at each point (cf.(1.2.2)). The answer is affirmative, and in the following theorem we put the result in a more general setting analogous to that of (1.3.7).

(1.3.11) *Let Y be a real normed space, let u be a non-zero continuous linear functional on Y, and let $\psi : [a,b] \to Y$ be a path in Y such that $u(\psi(a)) = u(\psi(b))$. Suppose further that the path ψ has a tangent at each point of $]a,b[$, the direction of the tangent at t being $q(t)$. Then there is at least one $\xi \in]a,b[$ such that $u(q(\xi)) = 0$.*

In geometrical language, the theorem asserts that if the chord joining the endpoints of the path ψ lies in the hyperplane with equation $u(y) = c$, there is at least one point ξ of the path such that the tangent at ξ is parallel to this hyperplane (the endpoints $\psi(a), \psi(b)$ need not be distinct).

If $u(\psi(t)) = u(\psi(a))$ for all $t \in [a,b]$, then for all $\xi \in]a,b[$

$$u(q(\xi)) = u\left(\lim_{t \to \xi +} \frac{\psi(t) - \psi(\xi)}{\|\psi(t) - \psi(\xi)\|} \right) = \lim_{t \to \xi +} \frac{u(\psi(t)) - u(\psi(\xi))}{\|\psi(t) - \psi(\xi)\|} = 0.$$

We may therefore suppose that $u(\psi(t))$ is somewhere positive or somewhere negative, say the former. Let ξ be a point at which the (continuous) function $t \mapsto u(\psi(t))$ attains its supremum on $[a,b]$. Then clearly $\xi \in]a,b[$. If $u(q(\xi)) > 0$, then for all t greater than and sufficiently near to ξ we have

$$u\left(\frac{\psi(t) - \psi(\xi)}{\|\psi(t) - \psi(\xi)\|} \right) = \frac{u(\psi(t)) - u(\psi(\xi))}{\|\psi(t) - \psi(\xi)\|} > 0, \tag{4}$$

whence $u(\psi(t)) > u(\psi(\xi))$ for such t, contrary to our choice of ξ. Similarly, if $u(q(\xi)) < 0$, then (4) holds for all t less than and sufficiently near to ξ, and again we obtain a contradiction. Hence $u(q(\xi)) = 0$.

In the example of a plane path shown in Fig. 1.1, where no tangent has the same direction as the chord AB, the path crosses itself, and it is natural to enquire whether the extra condition that the path is simple ensures that there is a point where the tangent has the same direction as the chord. The following theorem shows that the answer is again affirmative, though the proof is on a different level from that of (1.3.11).

(1.3.12) *Let* $\psi : [a,b] \to \mathbf{R}^2$ *be a simple path in* $\mathbf{R}^2$ *having a tangent at each point of* $]a,b[$, *and let* $A = \psi(a), B = \psi(b)$. *Then there is at least one* $\xi \in]a,b[$ *such that the tangent to the path at* ξ *has the same direction as the chord from A to B.*

Let ψ_1, ψ_2 be the components of ψ. We may evidently suppose that the chord AB lies on the x-axis and runs from left to right (so that $\psi_2(a) = \psi_2(b) = 0$ and $\psi_1(a) < \psi_1(b)$). Moreover, we may suppose that the path does not meet the x-axis to the right of B, for if there exist $t \in [a,b[$ for which $\psi(t)$ lies on the x-axis to the right of B, then there is a least such t, say $t = b_1$, and the restriction of ψ to $[a,b_1]$ is a path which does not meet the x-axis to the right of its endpoint $\psi(b_1)$.

We observe next that since $\psi_2(a) = \psi_2(b)$, either there exist ξ, η satisfying $a \le \eta < \xi < b$ such that $\psi_2(\xi) = \sup \psi_2$ and $\psi_2(\eta) = \inf \psi_2$, or there exist ξ, η satisfying $a \le \eta < \xi < b$ such that $\psi_2(\xi) = \inf \psi_2$ and $\psi_2(\eta) = \sup \psi_2$. Further, it is enough to prove the result when the first of these alternatives holds, for if ψ satisfies the second alternative, then the path $\bar{\psi}$ with components $\psi_1, -\psi_2$ satisfies the first alternative, and the result for $\bar{\psi}$ implies that for ψ.

Suppose then that AB lies on the x-axis and runs from left to right, that the path does not meet the x-axis to the right of B, and that ξ, η are points satisfying $a \le \eta < \xi < b$ such that $\psi_2(\xi) = \sup \psi_2$ and $\psi_2(\eta) = \inf \psi_2$. We show that the tangent to the path at ξ has the same direction as AB, thus proving the theorem.

The argument of (1.3.11) shows that the tangent at ξ is horizontal, so that it is enough to show that this tangent points to the right. Let C be the set in $\mathbf{R}^2$ consisting of a half-line drawn vertically upwards ending at $\psi(\eta)$, the arc of the path from η to ξ (i.e. the set $\psi([\eta,\xi])$), and a half-line drawn vertically upwards starting from $\psi(\xi)$. It follows easily from the

Jordan curve theorem for the plane $\mathbf{R}^2$ closed by the addition of a point at infinity† that the set $\mathbf{R}^2 \backslash C$ has two components, each with frontier C. Further, since C lies in the strip $\{(x,y): \inf \psi_1 \leq x \leq \sup \psi_1\}$, one of these components, R say, contains all points (x,y) with $x > \sup \psi_1$, and the other, L, contains all points (x,y) with $x < \inf \psi_1$. It is also obvious that the set $\psi(]\xi, b])$ is contained in R, since each point of this set can be joined via B to every point of the x-axis to the right of B by a path not meeting C.

Suppose now that the tangent at ξ points to the left, and for $\delta > 0$, $\varepsilon > 0$, let

$$W_+ = \{(\psi_1(\xi) + x, \psi_2(\xi) + y) : 0 < x < \delta, |y| < x\},$$

$$W_- = \{(\psi_1(\xi) - x, \psi_2(\xi) + y) : 0 < x < \varepsilon, |y| < x\}.$$

Then each of W_+, W_- is an open wedge-shaped region with vertex at the point ξ of the path and with the tangent at ξ as axis of symmetry. It is evident that W_- contains all points t of the path for which $t > \xi$ and t is sufficiently near ξ, and that W_+ contains all points t of the path for which $t < \xi$ and t is sufficiently near ξ. Choose $\zeta \in]\eta, \xi[$ so that $\psi(]\zeta, \xi[) \subseteq W_+$. Since the set $\psi([\eta, \zeta])$ is at a positive distance from the point $\psi(\xi)$, we can now choose ε so small that $W_- \cap \psi(]\eta, \zeta[) = \varnothing$, and then (with this choice of ε) W_- does not meet C. Since W_- contains points of $\psi(]\xi, b])$, we infer that $W_- \subseteq R$, and this gives a contradiction, since any point of W_- above its axis of symmetry can be joined to points of L by a horizontal segment which does not meet C. Hence the tangent at ξ points to the right.

Exercises 1.3

1 Prove that if $\phi : \mathbf{R} \to \mathbf{R}$ is given by $\phi(t) = t^3 \sin^2(\pi/t^2)$ $(t \neq 0), \phi(0) = 0$, then ϕ' has domain $\mathbf{R}$, and is bounded below on $[0,1]$, but there is no $\xi \in [0,1]$ for which $\phi'(\xi) = \inf_{[0,1]} \phi'$. [Thus in respect of its bounds a derivative does not behave like a continuous function; cf. (1.3.9).]

2 Let $m > 1$, and let $\phi : [-1,1] \to \mathbf{R}^m$ be given by

$\phi(t) = (0,0,\ldots,0)$ $(-1 \leq t \leq 0)$, $\phi(t) = (t^2 \cos(1/t),\ t^2 \sin(1/t), 0, \ldots, 0)$, $(0 < t \leq 1)$. Prove that $\phi'(t)$ exists for $-1 \leq t \leq 1$, and that the range of ϕ' is not connected. [Hence the analogue of Darboux's theorem (1.3.9) for functions with values in $\mathbf{R}^m$ with $m > 1$ is false.]

3 The function $\phi : [a,b] \to \mathbf{C}$ is continuous and one-to-one, and has a derivative at each point of $]a,b[$. Prove that there is at least one $\xi \in]a,b[$ for which

$$\arg(\phi(b) - \phi(a)) = \arg \phi'(\xi).$$

4 The path $\psi : [a,b] \to \mathbf{R}^2$ has a tangent at each point $t \in]a,b[$, and E is the set of

† See for instance, Newman (1951), Chapter 5, Theorem 10.2.

the directions of these tangents. Prove that if ψ is simple, then E is connected. Prove also that this result fails if ψ is not simple.

[Hint. For the first part, argue as in the second proof of Darboux's theorem (1.3.9), using (1.3.12) in place of (1.3.3).]

5 Let $m \geq 3$, and let $\psi : [-1,1] \to \mathbf{R}^m$ be given by $\psi(0) = (0,0,\ldots,0)$, $\psi(t) = (t, t^2 \cos(1/t), t^2 \sin(1/t), 0, \ldots, 0)$ $(0 < |t| \leq 1)$. Prove that ψ is a simple path with a tangent at each point, and that the set of the directions of these tangents is not connected.

1.4 Monotonicity theorems and an increment inequality

In this section we consider some generalizations of the condition of (1.3.4) for a function to be increasing, in which the derivative $\phi'(t)$ is replaced by one of the Dini derivatives of ϕ.

The four *Dini derivatives* of a given function $\phi : [a,b] \to \mathbf{R}$ at a point t_0, denoted by $D^+\phi(t_0), D_+\phi(t_0), D^-\phi(t_0), D_-\phi(t_0)$, are defined by the following relations:

$$D^+\phi(t_0) = \limsup_{t \to t_0+} \frac{\phi(t) - \phi(t_0)}{t - t_0} \qquad (a \leq t_0 < b);$$

$$D_+\phi(t_0) = \liminf_{t \to t_0+} \frac{\phi(t) - \phi(t_0)}{t - t_0} \qquad (a \leq t_0 < b);$$

$$D^-\phi(t_0) = \limsup_{t \to t_0-} \frac{\phi(t) - \phi(t_0)}{t - t_0} \qquad (a < t_0 \leq b);$$

$$D_-\phi(t_0) = \liminf_{t \to t_0-} \frac{\phi(t) - \phi(t_0)}{t - t_0} \qquad (a < t_0 \leq b).$$

The values ∞ and $-\infty$ are allowed. Obviously $D_+\phi(t_0) \leq D^+\phi(t_0)$ and $D_-\phi(t_0) \leq D^-\phi(t_0)$. It is also obvious that $\phi'_+(t_0)$ exists if and only if $D^+\phi(t_0)$ and $D_+\phi(t_0)$ are finite and equal, and similarly for $\phi'_-(t_0)$.

Here and later we will be concerned with properties that hold for all $t \in [a,b]$ except those t in a countable subset of this interval. We say that a property $P(t)$ holds *nearly everywhere* in a set E, or *for nearly all* $t \in E$, if there exists a countable subset H of E such that $P(t)$ holds for all $t \in E \setminus H$.†

The theorems of this section are less general than can be achieved, but they suffice for most applications, and their proofs are both elegant and simple. In §1.10 we give some further generalizations of these theorems, but more sophisticated arguments will be required there.

We begin with two lemmas, which lead easily to the main theorem of the section, (1.4.3).

---

† The term 'nearly everywhere in E' must be distinguished from the measure-theoretic term 'almost everywhere in E', i.e. except in a subset of E of Lebesgue measure zero.

(1.4.1) Lemma. *If $\phi : [a,b] \to \mathbf{R}$ is a continuous function such that $\phi(a) > \phi(b)$, and $A = \{t \in [a,b[: D^+\phi(t) \le 0\}$, then $\phi(A) \supseteq]\phi(b), \phi(a)[$.*

Let $c \in]\phi(b), \phi(a)[$. By the intermediate value theorem, there is a greatest element $t_c \in]a,b[$ such that $\phi(t_c) = c$, and for $t_c < t \le b$ we have $\phi(t) < c$, whence also $\phi(t) - \phi(t_c) < c - c = 0$. Hence $D^+\phi(t) \le 0$, i.e. $t_c \in A$, so that $c = \phi(t_c) \in \phi(A)$.

(1.4.2) Lemma. *In (1.4.1) the condition that ϕ is continuous can be replaced by the condition that $\limsup_{s \to t^-} \phi(s) \le \phi(t)$ for all $t \in]a,b]$ and $\phi(t) \le \limsup_{s \to t^+} \phi(s)$ for all $t \in [a,b[$.*

In the preceding argument, the continuity of ϕ is used only to establish the existence of the point t_c and the fact that $\phi(t) < c$ for $t_c < t \le b$. In fact, if t_c is the supremum of the set $B = \{t \in [a,b] : \phi(t) \ge c\}$, then the first lim sup condition implies that $\phi(t_c) \ge c$, whence $t_c < b$. The second lim sup condition then shows that $\phi(t_c) \le c$, so that $\phi(t_c) = c$.

(1.4.3) *If $\phi : [a,b] \to \mathbf{R}$ is continuous (or, more generally, if $\limsup_{s \to t^-} \phi(s) \le \phi(t)$ for all $t \in]a,b]$ and $\phi(t) \le \limsup_{s \to t^+} \phi(s)$ for all $t \in [a,b[$), and $D^+\phi(t) \ge 0$ nearly everywhere in $[a,b[$, then ϕ is increasing.*

It is obviously enough to prove that $\phi(a) \le \phi(b)$. Let $\varepsilon > 0$, and let $\chi(t) = \phi(t) + \varepsilon t$. Then χ satisfies the condition of (1.4.2). Moreover, $D^+\chi(t) = D^+\phi(t) + \varepsilon > 0$ nearly everywhere in $[a,b[$, so that the set where $D^+\chi(t) \le 0$ is countable. Hence the image of this set by χ is countable, and therefore contains no non-empty open interval. Using (1.4.1,2) we conclude that $\chi(a) \le \chi(b)$, and since ε is arbitrary this gives the result.

We have also a similar result for the upper left-hand derivative D^-, namely:

(1.4.3. Corollary) *If $\phi : [a,b] \to \mathbf{R}$ is continuous (or, more generally, if $\liminf_{s \to t^-} \phi(s) \le \phi(t)$ for all $t \in]a,b]$ and $\phi(t) \le \liminf_{s \to t^+} \phi(s)$ for all $t \in [a,b[$), and $D^-\phi(t) \ge 0$ nearly everywhere in $]a,b]$, then ϕ is increasing.*

This follows from the main theorem applied to the function $s \mapsto -\phi(a+b-s)$ on $[a,b]$.

The monotonicity theorems above imply inequalities for the increment $\phi(b) - \phi(a)$. For simplicity we confine ourselves here to continuous functions, although the condition of continuity can be weakened as in (1.4.3) and its corollary.

(1.4.4) *Let* $M \in \mathbf{R}$, *and let* $\phi : [a, b] \to \mathbf{R}$ *be a continuous function such that* $D\phi(t) \le M$ *for nearly all* $t \in [a, b]$, *where* D *denotes a fixed one of the four Dini derivatives. Then*

$$\phi(b) - \phi(a) \le M(b - a), \tag{1}$$

with equality if and only if $\phi(t) = \phi(a) + M(t - a)$ *for all* $t \in [a, b]$ *(so that if* $D\phi(t) < M$ *for at least one t, then* $\phi(b) - \phi(a) < M(b - a)$*).*

Let $\psi(t) = -(\phi(t) - \phi(a) - M(t - a))$ $(a \le t \le b)$. If D is D_+, (1.4.3) shows that ψ is increasing (for $D^+\psi(t) = M - D_+\phi(t)$). Since $\psi(a) = 0$, it follows that $\psi(b) \ge 0$, with equality if and only if ψ is identically 0, and this gives (1), together with the conditions for equality.

The case where D is D_- follows in a similar manner from (1.4.3. Corollary), and since the results for D_+ and D_- imply those for D^+ and D^-, this completes the proof.

In (1.4.4) the increasing property of the function ψ obviously implies that $M \ge D^+\phi(t)$ $(\ge D_+\phi(t))$ for *all* $t \in [a, b[$, and that $M \ge D^-\phi(t)$ $(\ge D_-\phi(t))$ for *all* $t \in]a, b]$. The next theorem generalizes this result.

(1.4.5) *Let* D *denote a fixed one of the four Dini derivatives, and let* ϕ, $\omega : [a, b] \to \mathbf{R}$ *be continuous functions such that* $D\phi(t) \le \omega(t)$ *for nearly all* $t \in [a, b]$. *Then* $D^+\phi(t) \le \omega(t)$ *for all* $t \in [a, b[$ *and* $D^-\phi(t) \le \omega(t)$ *for all* $t \in]a, b]$. *If in addition* $D\phi(t) = \omega(t)$ *for nearly all* $t \in [a, b]$, *then* $\phi'(t)$ *exists and is equal to* $\omega(t)$ *for all* $t \in [a, b]$.

Let $\varepsilon > 0$, and let $t \in [a, b[$. Since ω is continuous at t, we can find $\delta > 0$ such that $\omega(s) \le \omega(t) + \varepsilon$ for all $s \in [t, t + \delta[$, and then $D\phi(s) \le \omega(t) + \varepsilon$ for nearly all $s \in [t, t + \delta[$. By (1.4.4), $\phi(s) - \phi(t) \le (\omega(t) + \varepsilon)(s - t)$ for all $s \in [t, t + \delta[$, and therefore $D^+\phi(t) \le \omega(t)$, as required. A similar argument shows that $D^-\phi(t) \le \omega(t)$ for all $t \in]a, b]$, and for the last part we apply the two inequalities together to ϕ, ω and $-\phi, -\omega$.

Exercises 1.4

1 Let $\psi : [a, b] \to \mathbf{R}$ be continuous and strictly increasing. The upper right-hand Dini derivative of a function $\phi : [a, b] \to \mathbf{R}$ at the point t_0 relative to ψ, denoted by $D_\psi^+ \phi(t_0)$, is defined by

$$D_\psi^+ \phi(t_0) = \limsup_{t \to t_0+} \frac{\phi(t) - \phi(t_0)}{\psi(t) - \psi(t_0)},$$

the values $\infty, -\infty$ being allowed. The other relative Dini derivatives are defined similarly.

Prove that if $\limsup_{s \to t-} \phi(s) \le \phi(t)$ for all $t \in]a, b]$ and $\phi(t) \le \limsup_{s \to t+} \phi(s)$ for all $t \in [a, b[$, and $D_\psi^+ \phi(t) \ge 0$ nearly everywhere in $[a, b[$, then ϕ is increasing.

1.5 Applications of the increment inequality to differential equations and to a differential inequality

The theorems of this section can be regarded as generalizations of the increment inequality (1.4.3), and the proof of the key result (1.5.1) provides an interesting application of the ideas of §1.4. The theorems are intimately connected with the theory of ordinary differential equations, and our arguments give not only the required generalizations of the increment inequality, but also existence theorems for the solution of a differential equation for a function of one real variable. Other proofs of these results, together with various applications, will be given in Chapter 2.

Let E be a set in $\mathbf{R}^2$, and let $f : E \to \mathbf{R}$ be a given function. A function ϕ is said to be a *solution of the differential equation*†

$$y' = f(t, y) \tag{1}$$

on an interval $I \subseteq \mathbf{R}$ if ϕ is a continuous function from I into $\mathbf{R}$ such that, for each $t \in I$, the point $(t, \phi(t)) \in E$ (i.e. the graph of ϕ lies in E) and the derivative $\phi'(t)$ exists and satisfies

$$\phi'(t) = f(t, \phi(t)).$$

We consider in particular the solutions ϕ of the equation (1) that satisfy a condition of the form $\phi(t_0) = y_0$, where (t_0, y_0) is a given point of E. Such a condition is called an *initial condition*, and t_0, y_0 are called the *initial point* and *initial value* respectively (the word 'initial' is traditional in this context; t_0 may be any point of I).

There may be more than one solution of the equation satisfying the given initial condition. To illustrate this, consider the case where f has domain $\mathbf{R}^2$ and is given by $f(t, y) = 3y^{2/3}, t_0$ is any real number, $y_0 = 0$, and $I = \mathbf{R}$. The equation is then

$$y' = 3y^{2/3}, \tag{2}$$

and we seek solutions on $\mathbf{R}$ satisfying the initial condition that $y = 0$ when $t = t_0$.

It is obvious that the zero function on $\mathbf{R}$ is such a solution, and that, for all real u, v satisfying $u \le t_0 \le v$, so also are the functions ϕ given by

(i) $\phi(t) = 0$ $(t \le v)$, $\phi(t) = (t - v)^3$ $(t > v)$,

(ii) $\phi(t) = (t - u)^3$ $(t < u)$, $\phi(t) = 0$ $(t \ge u)$,

(iii) $\phi(t) = (t - u)^3$ $(t < u)$, $\phi(t) = 0$ $(u \le t \le v)$, $\phi(t) = (t - v)^3$ $(t > v)$.

Conversely, every solution ϕ on $\mathbf{R}$ satisfying $\phi(t_0) = 0$ is of one of these types. To prove this, let ϕ be a solution of (2) on $\mathbf{R}$, and consider first the behaviour of ϕ on $[t_0, \infty[$. Since $3y^{2/3} \ge 0$ for all real y, ϕ is increasing,

† See also §2.1.

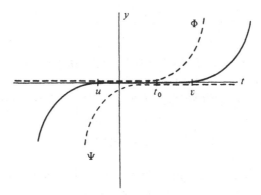

Figure 1.2

and hence either $\phi(t) = 0$ for all $t \geq t_0$, or there exists $v \geq t_0$ such that $\phi(t) = 0$ for $t_0 \leq t \leq v$ and $\phi(t) > 0$ for all $t > v$. In the latter case let $\psi = \phi^{1/3}$. Then $\psi(v) = 0$, and $\psi'(t) = \frac{1}{3}(\phi(t))^{-2/3}\phi'(t) = 1$ for $t > v$, whence for $t \geq v$ we have $\psi(t) = t - v$, and therefore $\phi(t) = (t - v)^3$. A similar argument applies for $t \leq t_0$.[†]

We note in particular that there are *maximal* and *minimal* solutions Φ, Ψ of the equation (2), satisfying the given initial condition, with the property that for every solution ϕ of (2) on $\mathbf{R}$ with $\phi(t_0) = 0$ we have $\Psi(t) \leq \phi(t) \leq \Phi(t)$ for all $t \in \mathbf{R}$. In fact Φ is the function of type (i) with $v = t_0$, and Ψ is the function of type (ii) with $u = t_0$ (Fig. 1.2).

A similar situation occurs with the general equation $y' = f(t, y)$, where f is a continuous mapping of the open set E in $\mathbf{R}^2$ into $\mathbf{R}$. However, the proof of this for an arbitrary open set E would take us too far into the theory of differential equations for the present, and we confine ourselves here to the proof of the key result, where E is a strip parallel to the y-axis and f is continuous and bounded, and defer the general case to §2.4.[‡]

Throughout we use J to denote the closed interval $[t_0, t_0 + \alpha]$, where $\alpha > 0$.

(1.5.1) *Let* $J = [t_0, t_0 + \alpha]$, *let* $f : J \times \mathbf{R} \to \mathbf{R}$ *be bounded and continuous,*

[†] We shall see in §2.7, p. 102, that for the differential equation (2) the set of points $(t_0, y_0) \in \mathbf{R}^2$ for which there exists more than one solution ϕ on $\mathbf{R}$ satisfying $\phi(t_0) = y_0$ is exactly the t-axis. However, examples of equations of the form (1) are known in which this set is the entire plane $\mathbf{R}^2$ (see Hartman, 1964, p. 18).

[‡] The additional condition that f is bounded does not ensure the uniqueness of the solutions satisfying a given initial condition. This may be seen by considering the case where $f : [0,1] \times \mathbf{R} \to \mathbf{R}$ is given by $f(t,y) = 3y^{2/3}$ ($|y| \leq 1$), $f(t,y) = 3$ ($y > 1$), $f(t,y) = -3$ ($y < -1$).

and let $y_0 \in \mathbf{R}$. Then the differential equation

$$y' = f(t, y) \qquad (3)$$

has maximal and minimal solutions Φ, Ψ on J satisfying the initial conditions that $\Phi(t_0) = \Psi(t_0) = y_0$. Further,

(i) *if $\phi : J \to \mathbf{R}$ is a continuous function such that $\phi(t_0) \leq y_0$ and that*

$$D\phi(t) \leq f(t, \phi(t)) \qquad (4)$$

for nearly all $t \in J$, where D is some fixed Dini derivative, then $\phi(t) \leq \Phi(t)$ for all $t \in J$,

(ii) *if $\psi : J \to \mathbf{R}$ is a continuous function such that $\psi(t_0) \geq y_0$ and that*

$$D\psi(t) \geq f(t, \psi(t))$$

for nearly all $t \in J$, where D is some fixed Dini derivative, then $\psi(t) \geq \Psi(t)$ for all $t \in J$.

First, the existence of the maximal solution Φ implies that of the minimal solution Ψ, for if Φ^* is the maximal solution of the equation $y' = -f(t, -y)$, then $-\Phi^*$ is the minimal solution of the equation (3). Similarly, the result of (i) implies that of (ii). Moreover, from (1.4.5), in proving (i), we may suppose that the inequality (4) holds with $D = D^+$ for all $t \in J_0$, where J_0 denotes the interval $[t_0, t_0 + \alpha[$.

Let Γ be the set of continuous $\phi : J \to \mathbf{R}$ such that $\phi(t_0) \leq y_0$ and that $D^+ \phi(t) \leq f(t, \phi(t))$ for all $t \in J_0$, and let $\Phi = \sup_\Gamma \phi$. It is enough to show that Φ is well-defined and finite on J and is a solution of (3) on J satisfying $\Phi(t_0) = y_0$. For, if ϕ is any solution of (3) on J satisfying $\phi(t_0) = y_0$, then $\phi \in \Gamma$, so that $\phi \leq \Phi$, whence Φ is the required maximal solution of (3).

Let $K = \sup |f|$. Then clearly the function $t \mapsto y_0 - K(t - t_0)$ belongs to Γ, and, by (1.4.4), each $\phi \in \Gamma$ is bounded above by the function $t \mapsto y_0 + K(t - t_0)$, whence Φ is well-defined and finite on J and $\Phi(t_0) = y_0$.

We show next that Φ is continuous. Let $t_0 \leq t_1 \leq t_2 \leq t_0 + \alpha$. By (1.4.4), we have $\phi(t_2) - \phi(t_1) \leq K(t_2 - t_1)$ for each $\phi \in \Gamma$, and this implies that

$$\Phi(t_2) - \Phi(t_1) \leq K(t_2 - t_1). \qquad (5)$$

Also, if $\phi \in \Gamma$, and ψ is the continuous function which agrees with ϕ on $[t_0, t_1]$ and whose graph on $[t_1, t_0 + \alpha]$ is a line segment of slope $-K$, then $\psi \in \Gamma$. Hence

$$\Phi(t_2) \geq \psi(t_2) = \phi(t_1) - K(t_2 - t_1),$$

and therefore

$$\Phi(t_2) \geq \Phi(t_1) - K(t_2 - t_1). \qquad (6)$$

The inequalities (5) and (6) together establish the continuity of Φ.

Next, for $\phi, \psi : J \to \mathbf{R}$ let $\phi \vee \psi = \max\{\phi, \psi\}$. We observe that if

ϕ, ψ are continuous, and $\phi(t) > \psi(t)$, then $D^+(\phi \vee \psi)(t) = D^+\phi(t)$, while if $\phi(t) = \psi(t)$, then $D^+(\phi \vee \psi)(t) = \max\{D^+\phi(t), D^+\psi(t)\}$. It follows that $\phi \vee \psi \in \Gamma$ whenever $\phi, \psi \in \Gamma$, and therefore $\Phi = \sup\{\phi \in \Gamma : \phi \geq \psi\}$ for each $\psi \in \Gamma$.

We prove now that $D^+\Phi(t) = f(t, \Phi(t))$ for all $t \in J_0$, and as a first step we show that $D^+\Phi(t) \leq f(t, \Phi(t))$, so that $\Phi \in \Gamma$. Let t be a fixed point of J_0, let $\varepsilon > 0$, and choose $\delta > 0$ such that $|f(u, y) - f(t, \Phi(t))| \leq \varepsilon$ whenever $u \in J$, $|u - t| \leq \delta$ and $|y - \Phi(t)| \leq \delta$. Choose also $\psi \in \Gamma$ such that $\Phi(t) - \psi(t) \leq \frac{1}{2}\delta$. Since Φ and Ψ are continuous, we can find a positive $\delta' \leq \delta$ such that

$$|\Phi(t+h) - \Phi(t)| \leq \delta \quad \text{and} \quad |\psi(t+h) - \psi(t)| \leq \tfrac{1}{2}\delta$$

whenever $0 \leq h \leq \delta'$. Then if ϕ is an element of Γ such that $\phi \geq \psi$, and $0 \leq h \leq \delta'$, we have

$$\Phi(t) - \delta \leq \psi(t) - \tfrac{1}{2}\delta \leq \psi(t+h) \leq \phi(t+h) \leq \Phi(t+h) \leq \Phi(t) + \delta,$$

and therefore also

$$D^+\phi(t+h) \leq f(t+h, \phi(t+h)) \leq f(t, \Phi(t)) + \varepsilon.$$

By (1.4.4), it follows that for such ϕ and h

$$\phi(t+h) - \phi(t) \leq (f(t, \Phi(t)) + \varepsilon)h,$$

and on taking suprema we deduce that

$$\Phi(t+h) - \Phi(t) \leq (f(t, \Phi(t)) + \varepsilon)h.$$

Hence $D^+\Phi(t) \leq f(t, \Phi(t)) + \varepsilon$, and since ε is arbitrary, $D^+\Phi(t) \leq f(t, \Phi(t))$.

Suppose now that there exists $t_1 \in J_0$ for which $D^+\Phi(t_1) < f(t_1, \Phi(t_1))$, and choose $\lambda \ (> -K)$ so that $D^+\Phi(t_1) < \lambda < f(t_1, \Phi(t_1))$. For some t_2 (which will be chosen later), let ϕ be the continuous function which agrees with Φ on $[t_0, t_1]$, whose graph on $[t_1, t_2]$ is a line segment of slope λ, and whose graph on the rest of J is a line segment of slope $-K$. Since f and ϕ are continuous, $\phi \in \Gamma$ if $t_2 - t_1$ is small enough. But then also $\phi \vee \Phi \in \Gamma$, and $\phi \vee \Phi \neq \Phi$, since $D^+(\phi \vee \Phi)(t_1) = \lambda \neq D^+\Phi(t_1)$. Since $\phi \vee \Phi \geq \Phi$, this contradicts the definition of Φ, and therefore $D^+\Phi(t) = f(t, \Phi(t))$ for all $t \in J_0$. By the second part of (1.4.5), it follows that $\Phi'(t)$ exists and satisfies $\Phi'(t) = f(t, \Phi(t))$ for all $t \in J$, and this completes the proof.

We end this section with a sharper version of the inequality of (1.5.1) in which we require only that the condition

$$D\phi(t) \leq f(t, \phi(t)) \tag{7}$$

should hold in the set of points t where the graph of ϕ breaks into a thin skin lying on top of the graph of Φ. This result is known as the 'epidermis property'.

(1.5.2) *Let* $f : J \times \mathbf{R} \to \mathbf{R}$ *be bounded and continuous, let* $y_0 \in \mathbf{R}$, *and let* Φ *be the maximal solution on* J *of the equation* $y' = f(t, y)$ *satisfying the initial condition that* $\Phi(t_0) = y_0$. *Let also* $\phi : J \to \mathbf{R}$ *be a continuous function such that* $\phi(t) \le y_0$, *let* ε_0 *be a given positive number, let* $S = \{t \in J : \Phi(t) < \phi(t) < \Phi(t) + \varepsilon_0\}$, *and suppose that* (7) *holds for nearly all* $t \in S$, *where* D *is some fixed Dini derivative. Then* $\phi(t) \le \Phi(t)$ *for all* $t \in J$ *(and so* S *is empty).*

Suppose on the contrary that there exists $\tau \in J$ for which $\phi(\tau) > \Phi(\tau)$, and let σ be the greatest element of $[t_0, \tau]$ such that $\phi(\sigma) = \Phi(\sigma)$ (so that $\phi(t) > \Phi(t)$ for all $t \in]\sigma, \tau]$). Since $\phi - \Phi$ is continuous, there exists $\eta \in]\sigma, \tau]$ such that $\Phi(t) < \phi(t) < \Phi(t) + \varepsilon_0$ for all $t \in]\sigma, \eta]$, and hence (7) holds for nearly all $t \in]\sigma, \eta]$. Now let ϕ_1 be the continuous function on J that agrees with Φ on $[t_0, \sigma]$ and with ϕ on $[\sigma, \eta]$, and whose graph on $[\eta, t_0 + \alpha]$ consists of a segment of slope inf f. Then evidently $D\phi_1(t) \le f(t, \phi_1(t))$ for nearly all $t \in J$, and hence, by (1.5.1), $\phi_1(t) \le \Phi(t)$ for all $t \in J$. In particular, $\phi(\eta) = \phi_1(\eta) \le \Phi(\eta)$, giving a contradiction.

Exercises 1.5

1 Let $J^* = [t_0 - \alpha, t_0]$, let $f : J^* \times \mathbf{R} \to \mathbf{R}$ be bounded and continuous, and let $y_0 \in \mathbf{R}$. Prove that the equation $y' = f(t, y)$ has maximal and minimal solutions Φ, Ψ on J^* satisfying the initial conditions that $\Phi(t_0) = \Psi(t_0) = y_0$. Prove further that
(i) if $\phi : J^* \to \mathbf{R}$ is a continuous function such that $\phi(t_0) \le y_0$ and that $D\phi(t) \ge f(t, \phi(t))$ for nearly all $t \in J^*$, where D is a fixed Dini derivative, then $\phi(t) \le \Phi(t)$ for all $t \in J^*$,
(ii) if $\psi : J^* \to \mathbf{R}$ is a continuous function such that $\psi(t_0) \ge y_0$ and that $D\psi(t) \le f(t, \psi(t))$ for nearly all $t \in J^*$, where D is a fixed Dini derivative, then $\psi(t) \ge \Psi(t)$ for all $t \in J^*$.
2 Let $J = [t_0, t_0 + \alpha]$, let $f : J \times \mathbf{R} \to \mathbf{R}$ be bounded and continuous, and let $\psi : J \to \mathbf{R}$ be a continuous function such that $D^+\psi(t) < f(t, \psi(t))$ for all $t \in [t_0, t_0 + \alpha[$. Prove that $\psi(t) \le \Psi(t)$ for all $t \in J$, where Ψ is the minimal solution of the equation $y' = f(t, y)$ on J such that $\Psi(t_0) = \psi(t_0)$.
    Show that this result fails if the condition that $D^+\psi(t) < f(t, \psi(t))$ for all $t \in [t_0, t_0 + \alpha[$ is replaced by the condition that $D^+\psi(t) < f(t, \psi(t))$ for all $t \in]t_0, t_0 + \alpha[$.

1.6 Increment and mean value inequalities for vector-valued functions

In §1.3 we combined the classical mean value theorem with the Hahn–Banach theorem to obtain a mean value inequality for vector-valued functions. In this section we use the increment inequality (1.4.4) to obtain stronger results of this type. In the statements of our theorems we employ

consistently the right-hand derivative ϕ'_+. Exactly similar results hold
for the left-hand derivative ϕ'_-, and obviously also for the ordinary
derivative ϕ'.

The following lemma provides the necessary connection with the
increment inequality (1.4.4).

(1.6.1) Lemma. *Let ϕ be a function from a set $A \subseteq \mathbf{R}$ into Y.*
(i) *If p is a continuous sublinear functional on Y, then*

$$- p(- \phi'_+(t)) \leq D_+ p(\phi(t)) \leq D^+ p(\phi(t)) \leq p(\phi'_+(t))$$

at each point t for which $\phi'_+(t)$ exists.
(ii) *If u is a continuous linear functional on Y then the function $t \mapsto u(\phi(t))$
has a right-hand derivative equal to $u(\phi'_+(t))$ at each point t for which $\phi'_+(t)$
exists.*

We recall (§A.3) that if p is sublinear then for all $y, z \in Y$ and $\lambda \geq 0$

$$p(y + z) \leq p(y) + p(z), \quad p(\lambda y) = \lambda p(y), \tag{1}$$

and hence

$$- p(z - y) \leq p(y) - p(z) \leq p(y - z). \tag{2}$$

From (2) and the second relation in (1) we obtain that if $s, t \in A$ and $s > t$
then

$$- p\left(- \frac{\phi(s) - \phi(t)}{s - t} \right) \leq \frac{p(\phi(s)) - p(\phi(t))}{s - t} \leq p\left(\frac{\phi(s) - \phi(t)}{s - t} \right),$$

and (i) follows from these inequalities and the continuity of p. The proof
of (ii) is immediate.

In fact, more is true than is asserted in (1.6.1) (i), for if p is sublinear then p
is convex, and hence the composite $\chi = p \circ \phi$ has a right-hand derivative
whenever $\phi'_+(t)$ exists (Exercise 1.1.2). Moreover, since $\chi'_+(t) = D^+ \chi(t) =
D^+ p(\phi(t))$, we have

$$- p(- \phi'_+(t)) \leq \chi'_+(t) \leq p(\phi'_+(t)).$$

However, we rarely need to use this stronger result.

The theorem below gives an increment inequality which is to some
extent an analogue of (1.3.10) for vector-valued functions. We shall
see later that the conditions for equality in this theorem provide useful
weapons of argument.

(1.6.2) *Let p be a continuous sublinear functional on Y, and let $\phi : [a, b] \to Y$
and $\psi : [a, b] \to \mathbf{R}$ be continuous functions whose right-hand derivatives*

$\phi'_+(t), \psi'_+(t)$ *exist and satisfy* $p(\phi'_+(t)) \le \psi'_+(t)$ *for nearly all* $t \in [a,b]$. *Then*

$$p(\phi(b) - \phi(a)) \le \psi(b) - \psi(a). \tag{3}$$

Moreover, there is equality in (3) *if and only if*

$$p(\phi(t) - \phi(a)) = \psi(t) - \psi(a)$$

for all $t \in [a,b]$, *and then also* $p(\phi'_+(t)) \ge \psi'_+(t)$ *whenever* $\phi'_+(t), \psi'_+(t)$ *exist (so that* $p(\phi'_+(t)) = \psi'_+(t)$ *nearly everywhere). Further, if* Y *is real and* p *is linear, and equality holds in* (3), *then* $p(\phi'_+(t)) = \psi'_+(t)$ *whenever* $\phi'_+(t), \psi'_+(t)$ *exist.*

Let $\sigma(t) = p(\phi(t) - \phi(a)) - (\psi(t) - \psi(a))$ ($a \le t \le b$). Obviously σ is continuous, and, by (1.6.1),

$$D^+\sigma(t) \le p(\phi'_+(t)) - \psi'_+(t) \tag{4}$$

whenever $\phi'_+(t), \psi'_+(t)$ exist, so that $D^+\sigma(t) \le 0$ nearly everywhere in $[a,b]$.† By (1.4.4), $\sigma(b) = \sigma(b) - \sigma(a) \le 0$, and this gives (3). Moreover, $\sigma(b) = 0$ if and only if $\sigma(t) = 0$ for all $t \in [a,b]$, and then also $D^+\sigma(t) = 0$ for all $t \in [a,b[$, and these results imply the statements concerning equality in the general case. Finally, if Y is real and p is linear, there is equality in (4), and hence if equality holds in (3) then $p(\phi'_+(t)) = \psi'_+(t)$ whenever $\phi'_+(t), \psi'_+(t)$ exist.

We note explicitly the case of (1.6.2) in which p is the norm function on Y and $\psi(t) = Mt$; this provides an extension of (1.3.6) to vector-valued functions.

(1.6.2. Corollary 1) *Let* $M \ge 0$, *and let* $\phi : [a,b] \to Y$ *be a continuous function whose right-hand derivative* $\phi'_+(t)$ *exists and satisfies* $\|\phi'_+(t)\| \le M$ *for nearly all* $t \in [a,b]$. *Then*

$$\|\phi(b) - \phi(a)\| \le M(b-a), \tag{5}$$

and there is strict inequality in (5) *if* $\|\phi'_+(t)\| < M$ *for at least one t.*

In particular, when $M = 0$ we obtain the following generalization of (1.3.5).

(1.6.2. Corollary 2) *If* $\phi : [a,b] \to Y$ *is a continuous function such that* $\phi'_+(t)$ *exists and is equal to* 0 *nearly everywhere, then* ϕ *is constant.*

---

† The derivatives $\phi'_+(t), \psi'_+(t)$ may possibly exist at a countable set of points for which $p(\phi'_+(t)) > \psi'_+(t)$. All that we require is that the set where $\phi'_+(t), \psi'_+(t)$ either both exist and satisfy $p(\phi'_+(t)) > \psi'_+(t)$, or do not both exist, should be countable (so that $D^+\sigma(t) \le 0$ nearly everywhere). Similar points occur in later theorems.

The next theorem provides a closer analogue of (1.3.10), where the equality of (1.3.10) is replaced by an inequality.

(1.6.3) *Let p be a continuous sublinear functional on Y, let $\phi : [a, b] \to Y$ and $\psi : [a, b] \to \mathbf{R}$ be continuous functions whose right-hand derivatives $\phi'_+(t)$, $\psi'_+(t)$ exist for nearly all $t \in [a, b[$, and let $\psi(a) < \psi(b)$. Then there exist uncountably many $\xi \in [a, b[$ for which*

$$p(\phi(b) - \phi(a))\psi'_+(\xi) \leq (\psi(b) - \psi(a))p(\phi'_+(\xi)). \tag{6}$$

Moreover, either (6) holds with strict inequality for uncountably many ξ, or (6) holds whenever $\phi'_+(\xi), \psi'_+(\xi)$ exist (and with equality nearly everywhere), and in this latter case

$$p(\phi(b) - \phi(a))(\psi(t) - \psi(a)) = (\psi(b) - \psi(a))p(\phi(t) - \phi(a))$$

for all $t \in [a, b]$.

For $a \leq t \leq b$ let

$$\sigma(t) = (\psi(b) - \psi(a))p(\phi(t) - \phi(a)) - p(\phi(b) - \phi(a))(\psi(t) - \psi(a)).$$

Clearly σ is continuous, and $\sigma(a) = \sigma(b) = 0$. Also, if t is a point of $[a, b[$ for which $\phi'_+(t), \psi'_+(t)$ exist, then, by (1.6.1),

$$D^+\sigma(t) \leq (\psi(b) - \psi(a))p(\phi'_+(t)) - p(\phi(b) - \phi(a))\psi'_+(t). \tag{7}$$

Now let A, B be respectively the sets of points $\xi \in [a, b[$ for which

$$p(\phi(b) - \phi(a))\psi'_+(\xi) < (\psi(b) - \psi(a))p(\phi'_+(\xi))$$

and

$$p(\phi(b) - \phi(a))\psi'_+(\xi) \geq (\psi(b) - \psi(a))p(\phi'_+(\xi)).$$

If A is uncountable, the first alternative of the theorem holds. If A is countable, so is $[a, b]\backslash B$. Also, by (7),

$$D^+\sigma(\xi) \leq (\psi(b) - \psi(a))p(\phi'_+(\xi)) - p(\phi(b) - \phi(a))\psi'_+(\xi) \leq 0$$

for all $\xi \in B$, and therefore nearly everywhere in $[a, b]$. Since $\sigma(b) - \sigma(a) = 0$, it follows from (1.4.4) that $\sigma(t) = 0$ for all $t \in [a, b]$, and that $D^+\sigma(t) = 0$ for all $t \in [a, b[$. Hence (6) holds whenever $\phi'_+(\xi), \psi'_+(\xi)$ both exist, and with equality for all $\xi \in B$, i.e. nearly everywhere, so that the second alternative of the theorem holds.

When p is the norm function on Y and $\psi(t) = t$, (1.6.3) gives the following result (cf. (1.3.8) and (1.6.2. Corollary 1)).

(1.6.3. Corollary) *Let $\phi : [a, b] \to Y$ be a continuous function whose right-hand derivative $\phi'_+(t)$ exists for nearly all $t \in [a, b[$. Then there exist uncountably many $\xi \in [a, b[$ for which*

$$\| \phi(b) - \phi(a) \| \le (b - a) \| \phi'_+(\xi) \|. \tag{8}$$

Moreover, either (8) holds with strict inequality for uncountably many ξ, or (8) holds whenever $\phi'_+(\xi)$ exists (and with equality nearly everywhere).

The result of (1.6.2. Corollary 1) shows that, if E is a closed ball in Y with centre 0, and for nearly all t the derivative $\phi'_+(t)$ belongs to E, then the difference quotient $(\phi(b) - \phi(a))/(b - a)$ also belongs to E. The next theorem shows that a similar result holds for any closed convex set E in Y. With later applications in mind, we give the form of the result corresponding to Cauchy's mean value theorem.

(1.6.4) *Let E be a closed convex set in Y, let $\phi : [a,b] \to Y$ and $\psi : [a,b] \to \mathbf{R}$ be continuous functions possessing right-hand derivatives nearly everywhere in $[a,b[$, let $\delta = (\phi(b) - \phi(a))/(\psi(b) - \psi(a))$, and suppose that, for nearly all $t \in [a,b[, \psi'_+(t) > 0$ and $\phi'_+(t)/\psi'_+(t) \in E$. Then $\delta \in E$.*

In proving this, we may obviously suppose that Y is over the real field. Further, since a closed convex set in Y is the intersection of the closed half-spaces that contain it (A.4.3), we may suppose that E is a closed half-space in Y. But then E is of the form $\{y \in Y : u(y) \le \alpha\}$, where u is a continuous linear functional on Y and $\alpha \in \mathbf{R}$, and the result for this E follows from (1.6.2) with $p = u$ and $\alpha\psi$ in place of ψ.

We note that the result of (1.6.4), applied to subintervals of $[a,b]$, shows that under the same hypotheses $\phi'_+(t)/\psi'_+(t) \in E$ whenever both derivatives exist and $\psi'_+(t) > 0$.

(1.6.5) *Suppose that the hypotheses of (1.6.4) are satisfied, and that in addition Y is over the real field, and E has a non-empty interior. Then either $\delta \in E°$, or there exists a closed hyperplane X in Y, contained in $Y \backslash E°$, such that $\phi'_+(t)/\psi'_+(t) \in X$ whenever both derivatives exist and $\psi'_+(t) > 0$ (and then also $\delta \in X$).*

By (1.6.4), $\delta \in E$. If δ is a boundary point of E, then E has a supporting closed hyperplane at δ (A.4.4). If X is such a hyperplane, we can choose its equation, $v(y) = \beta$ (where v is a continuous linear functional on Y), such that $E \subseteq \{y \in Y : v(y) \le \beta\}$, and then also $E° \subseteq \{y \in Y : v(y) < \beta\}$. By hypothesis, $v(\phi'_+(t)) \le \beta\psi'_+(t)$ for nearly all $t \in [a,b[$, and since $v(\delta) = \beta$, it follows from the conditions of equality in (1.6.2) (with $p = v$ and $\beta\psi$ in place of ψ) that $v(\phi'_+(t)) = \beta\psi'_+(t)$ whenever both derivatives exist, and this gives the second alternative of the theorem. This proves all but the parenthesis, which follows from (1.6.4) with $E = X$.

(1.6.5. Corollary) *The result of* (1.6.4) *continues to hold when E is an open convex set in Y.*

Since we may obviously suppose that Y is over the real field, and since the closure of an open convex set is convex, this is immediate.

By using this corollary, we obtain the following companion to (1.6.3) in which the conditions on p are weakened but with stronger conditions imposed on ψ.

(1.6.6) *Let $q : Y \to \mathbf{R}$ be a continuous quasi-convex function, let $\phi : [a,b] \to Y$ and $\psi : [a,b] \to \mathbf{R}$ be continuous functions possessing right-hand derivatives nearly everywhere in $[a,b[$, and let $\psi'_+(t) > 0$ for nearly all $t \in [a,b[$. Then there exist uncountably many $\xi \in [a,b[$ for which*

$$q\left(\frac{\phi(b) - \phi(a)}{\psi(b) - \psi(a)}\right) \le q\left(\frac{\phi'_+(\xi)}{\psi'_+(\xi)}\right). \tag{9}$$

Let $\delta = (\phi(b) - \phi(a))/(\psi(b) - \psi(a))$, and let $E = \{y \in Y : q(y) < q(\delta)\}$. Since q is continuous and quasi-convex, E is open and convex, by (A.3.7). If the result is false, then $\phi'_+(t)/\psi'_+(t) \in E$ for nearly all $t \in [a,b[$, and, by (1.6.5. Corollary), this implies that $\delta \in E$, which is impossible.

We note that there may be no ξ for which (9) holds with strict inequality (take, for example, $Y = \mathbf{R}$, $a = 0$, $b = 1, \phi(t) = t(1 - t), \psi(t) = t, q(y) = 0$ $(y < 0), q(y) = -y\ (y \ge 0)$).

We remark also that the general case of (1.6.6) can be reduced to the case where $\psi(t) = t$ by the device mentioned in the discussion of Cauchy's theorem in §1.3. In fact, by replacing ϕ, ψ by $\alpha\phi, \alpha\phi + \beta$ with suitable α, β, we may suppose that $\psi(b) = \psi(a)$, and then it is enough to apply the special case to $\phi \circ \psi^{-1}$.

In the final theorem of this section we consider the convex hull of the set of difference quotients of a function.

(1.6.7) *Let $\phi : [a,b] \to Y$ be a continuous function such that $\phi'_+(t)$ exists for nearly all $t \in [a,b]$, let A be the set of values of ϕ'_+, and let D be the set of difference quotients $(\phi(s) - \phi(t))/(s - t)$ for all distinct points $s, t \in [a,b]$. Then $\overline{\mathrm{co}}\, D = \overline{\mathrm{co}}\, A$. Further, if the dimension of Y is finite, then $D \subseteq \mathrm{co}\, A$.*

By (1.6.4), $D \subseteq \overline{\mathrm{co}}\, A$, and therefore $\overline{\mathrm{co}}\, D \subseteq \overline{\mathrm{co}}\, A$. On the other hand, it is obvious that $\bar{D} \supseteq A$, and therefore $\overline{\mathrm{co}}\, D = \overline{\mathrm{co}}\, \bar{D} \supseteq \overline{\mathrm{co}}\, A$.

Now let Y be finite-dimensional. We wish to prove that $D \subseteq \mathrm{co}\, A$, and we may obviously suppose that Y is over the real field. Further, it is enough to prove that $d = (\phi(b) - \phi(a))/(b - a) \in \mathrm{co}\, A$, or, equivalently, that $0 \in \mathrm{co}\, B$, where $B = A - d$.

Let $\sigma(t) = \phi(t) - \phi(a) - (t - a)d$ $(a \le t \le b)$. Then σ is continuous, and $\sigma(a) = \sigma(b) = 0$. Also B is the set of values of $\sigma'_+(t)$, and therefore, by (1.6.4), $\overline{\text{co}}\,B$ contains the set of difference quotients of σ. Hence $\sigma(t)/(t - a) = (\sigma(t) - \sigma(a))/(t - a) \in \overline{\text{co}}\, B$ for all $t \in \,]a, b]$, and, in particular, $0 \in \overline{\text{co}}B$.

Let X be the minimal closed linear subspace of Y containing $\overline{\text{co}}\,B$. From the preceding remark we see that X also contains the range of σ. Moreover, since $0 \in \overline{\text{co}}\,B$, X is also the minimal closed linear variety of Y containing $\overline{\text{co}}\,B$, and therefore $\overline{\text{co}}\,B$ has an interior point relative to X (A.3.2).

We now consider σ as a mapping into X. Since there is no proper subspace of X containing B, (1.6.5) shows that 0 is an interior point of $\overline{\text{co}}\,B$ relative to X. Since a convex set with a non-empty interior has the same interior as its closure, it follows that $0 \in \text{co}\,B$.

(1.6.7. Corollary) *Let Y be finite-dimensional, and let $\phi : [a, b] \to Y$ be a function whose derivative ϕ' is continuous on $[a, b]$. Then the extremal points of $\overline{\text{co}}\,D$ belong to A.*

Here the range A of ϕ' is compact, whence so also is co A. The extremal points of co A therefore belong to A, and since $\overline{\text{co}}\,D = \text{co}\,A$, this gives the result.

It should be noted that the sets A and D above can be disjoint, as also can be $\bar{A}$ and D. For example, if $\phi : [0, 2\pi] \to \mathbf{R}^2$ is given by $\phi(t) = (\cos t, \sin t)$, then $\phi'(t) = (-\sin t, \cos t)$ for all $t \in [0, 2\pi]$, so that the range A of ϕ' is the unit circle $\{(x, y) \in \mathbf{R}^2 : x^2 + y^2 = 1\}$. Also, if $0 \le t < s \le 2\pi$, and $u = \frac{1}{2}(s + t), v = \frac{1}{2}(s - t)$, then

$$d = \frac{\phi(s) - \phi(t)}{s - t} = \frac{\sin v}{v}(\cos(u + \tfrac{1}{2}\pi), \sin(u + \tfrac{1}{2}\pi)),$$

and therefore $\|d\| < 1$, since $0 \le v^{-1} \sin v < 1$ when $0 < v \le \pi$. Hence the set D of difference quotients of ϕ is contained in the open unit disc in $\mathbf{R}^2$ and therefore does not meet A (the set D is the shaded region in Fig. 1.3).

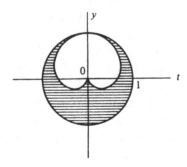

Figure 1.3

This example shows that some device such as taking the convex hull of A is necessary if we are to obtain information about D from a knowledge of A.

The results of this section can be generalized further by the use of the weak derivative. If $A \subseteq \mathbf{R}$, then a function $\phi : A \to Y$ is said to *have a weak derivative*† at a non-isolated point t_0 of A if there exists $l \in Y$ such that for every continuous linear functional u on Y

$$u\left(\frac{\phi(t) - \phi(t_0)}{t - t_0} \right) \to u(l)$$

as $t \to t_0$. This l is then called the *weak derivative of* ϕ at t_0, and we denote it by $\phi'_w(t_0)$; the function ϕ'_w thus defined is, of course, the *weak derivative of* ϕ. Similarly, we can define the weak right-hand and left-hand derivatives of ϕ, which we denote by ϕ'_{w+}, ϕ'_{w-}.

It is easy to verify that the result of (1.6.1) (ii) continues to hold if the right-hand derivative $\phi'_+(t)$ is replaced by the weak right-hand derivative $\phi'_{w+}(t)$. It therefore follows that if in (1.6.2) the space Y is over the real field and p is linear, then in this theorem we can replace $\phi'_+(t)$ throughout by $\phi'_{w+}(t)$. Since the proofs of (1.6.4–7) depend only on this special case of (1.6.2), these theorems too therefore hold for the weak derivative ϕ'_{w+}.

Exercises 1.6

1 Let I be an interval in $\mathbf{R}$, let $\phi : I \to Y$ be a continuous function such that ϕ' is defined nearly everywhere. Prove that if K is the domain of ϕ' and H is any countable subset of K then

$$\sup_{t \in K \setminus H} \| \phi'(t) \| = \sup_{t \in K} \| \phi'(t) \|,$$

infinite values being allowed.

2 Let $I =]t_0, t_0 + \alpha[$, and let $\phi : I \to Y$ be a continuous function such that for nearly all $t \in I$ the derivative $\phi'_+(t)$ exists and satisfies

$$\| \phi'_+(t) \| \leq \| \phi(t) \| / (t - t_0).$$

Prove that the function $t \mapsto \| \phi(t) \| (t - t_0)$ is increasing on I and that $t \mapsto \| \phi(t) \| / (t - t_0)$ is decreasing on I (so that $\phi(t)$ is either everywhere non-zero on I or identically zero on I).

3 Let $q : Y \to \mathbf{R}$ be a continuous convex function, let $\phi : [a, b] \to Y$ and $\psi : [a, b] \to \mathbf{R}$ be continuous functions possessing right-hand derivatives nearly everywhere in $[a, b[$, let $\psi'_+(t) > 0$ for nearly all $t \in [a, b[$, and let $\delta = (\phi(b) - \phi(a))/(\psi(b) - \psi(a))$. Prove that either $q(\delta) < q(\phi'_+(\xi)/\psi'_+(\xi))$ for uncountably many $\xi \in [a, b[$, or $q(\delta) \leq q(\phi'_+(\xi)/\psi'_+(\xi))$ whenever the derivatives exist and $\psi'_+(\xi) > 0$ (and with equality almost everywhere). Show also that in this result 'convex' can be replaced by 'strictly quasi-convex'.

---

† The derivative defined in §1.1 is then sometimes distinguished by calling it the *strong derivative*.

4 Extend the result of Exercise 3 to the weak derivative.

5 Let E be a closed convex set in Y with a non-empty interior, let $\phi : [a,b] \to Y$ be a continuous function whose right-hand derivative $\phi'_+(t)$ exists and belongs to E for nearly all $t \in [a,b[$, and let $d = (\phi(b) - \phi(a))/(b-a)$. By using the Minkowski functional of E, prove that $d \in E$ and that either $d \in E°$, or $d \in E \setminus E°$ and $\phi'_+(t) \in Y \setminus E°$ whenever $\phi'_+(t)$ exists (and $\phi'_+(t) \in E \setminus E°$ nearly everywhere).

6 Let C be an open convex cone with vertex 0 in the normed space Y, and let $\psi : [a,b] \to Y$ be a (continuous) path in Y with the property that for nearly all $t \in [a,b[$ there exist a unit vector $q(t) \in C$ and a sequence (t_n) of points of $]t,b]$ with the limit t such that $\psi(t_n) \neq \psi(t)$ $(n = 1,2,\ldots)$ and that the sequence of chord directions $(\psi(t_n) - \psi(t))/\|\psi(t_n) - \psi(t)\|$ tends to $q(t)$. Prove that $\psi(b) - \psi(a) \in C$.

[Hint. Since C is the intersection of all open half-spaces in Y with 0 as a boundary point and containing C, it is enough to prove the result when C is such an open half-space, i.e. when C is of the form $\{y \in Y : u(y) > 0\}$, where u is a continuous linear functional on Y. Further, for this C it is enough to prove that $u(\psi(b) - \psi(a)) \geq 0$. For suppose that this result holds. Then if t is a point for which $q(t) \in C$, we have $u(q(t)) > 0$ and therefore $u(\psi(t_n) - \psi(t)) > 0$ for all sufficiently large n. Since the result with '$\geq$' applied to ψ on $[a,t]$ and $[t_n,b]$ gives $u(\psi(t) - \psi(a)) \geq 0$ and $u(\psi(b) - \psi(t_n)) \geq 0$, we deduce that $u(\psi(b) - \psi(a)) > 0$. Finally, to prove that $u(\psi(b) - \psi(a)) \geq 0$, assume the contrary, so that $u(\psi(b)) < u(\psi(a))$. Then an argument similar to that of (1.4.1) shows that the set T of $t \in [a,b[$ for which there is no sequence of chord directions with limit $q(t)$ satisfying $u(q(t)) > 0$ is uncountable.]

7 Let Y be a real normed space, and let E be a closed convex set in Y. Let also $\psi : [a,b] \to \mathbf{R}$ be strictly increasing and continuous, and let $\phi : [a,b] \to Y$ be a function with the properties that, for each continuous linear functional u on Y,
(i) $u \circ \phi$ is continuous,
(ii) for nearly all $t \in [a,b[$ there exist a sequence (t_n) of points of $]t,b]$ with the limit t and a sequence (y_n) of points of E such that
$$u\left(\frac{\phi(t_n) - \phi(t)}{\psi(t_n) - \psi(t)} - y_n\right) \to 0 \text{ as } n \to \infty.$$
Prove that $(\phi(b) - \phi(a))/(\psi(b) - \psi(a)) \in E$.

[This result is a drastic generalization of the 'Cauchy' version of (1.6.6), and applies without change in a locally convex topological space.

Hint. Reduce the result to the case in which E is the closed half-space $\{y \in Y : u(y) \geq \alpha\}$, where u is a continuous linear functional on Y, and then use Exercise 1.4.1 to show that for this u the function $t \mapsto u(\phi(t)) - \alpha\psi(t)$ is increasing.]

1.7 Applications of the increment and mean value theorems

Theorems (1.6.2,3) and their corollaries have many important applications. In fact (1.6.2) and (1.6.2. Corollary 1) will suffice for almost all of these, but sometimes (1.6.3. Corollary) gives the more elegant proof, and we use here whichever result seems most appropriate.

Our first application deals with term-by-term differentiation of a sequence of functions.

(1.7.1) *Let (ϕ_n) be a sequence of continuous functions from the interval $I = [a, b]$ into a Banach space Y each possessing a derivative nearly everywhere on I. Suppose further that*
 (i) *there exists $c \in I$ such that the sequence $(\phi_n(c))$ converges,*
 (ii) *there exists a countable set $H \subseteq I$ such that the sequence (ϕ'_n) converges uniformly on $I \backslash H$.*
Then the sequence (ϕ_n) converges uniformly on I, and its limit-function ϕ is continuous on I and has a derivative $\phi'(t)$ equal to $\lim_{n \to \infty} \phi'_n(t)$ for all $t \in I \backslash H$.

Let $\varepsilon > 0$. Then we can find an integer q such that

$$\| \phi'_n(t) - \phi'_m(t) \| \leq \varepsilon \quad \text{whenever } n \geq m \geq q \text{ and } t \in I \backslash H. \tag{1}$$

Let $t, t_0 \in I$, and let $n \geq m \geq q$. By (1.6.2. Corollary 1) applied to the function $\phi_n - \phi_m$ on the closed interval with endpoints $t, t_0 \in I$, we obtain that

$$\| (\phi_n(t) - \phi_m(t)) - (\phi_n(t_0) - \phi_m(t_0)) \| \leq \varepsilon | t - t_0 |. \tag{2}$$

In particular, taking $t_0 = c$ we deduce that for $n \geq m \geq q$ and $t \in I$

$$\| \phi_n(t) - \phi_m(t) \| \leq \| \phi_n(c) - \phi_m(c) \| + \varepsilon(b - a).$$

Since $(\phi_n(c))$ is convergent, it follows that (ϕ_n) is uniformly convergent on I, to a function ϕ say, and ϕ is continuous on I.

Next, let $t_0 \in I \backslash H$. By (1.1.1)(ii), for each n and all $t \in I$ we can write

$$\phi_n(t) = \phi_n(t_0) + (t - t_0)\lambda_n(t), \tag{3}$$

where the function $\lambda_n : I \to Y$ is continuous at t_0 and $\lambda_n(t_0) = \phi'_n(t_0)$. By (2),

$$\| \lambda_n(t) - \lambda_m(t) \| \leq \varepsilon \tag{4}$$

whenever $n \geq m \geq q$ and $t \in I \backslash \{t_0\}$, and, by (1), this inequality (4) holds also for $t = t_0$. Hence the sequence (λ_n) converges uniformly on I to a function λ, and, by (3),

$$\phi(t) = \phi(t_0) + (t - t_0)\lambda(t)$$

for all $t \in I$. Since each λ_n is continuous at t_0, so is λ, whence $\phi'(t_0)$ exists and is equal to

$$\lambda(t_0) = \lim_{n \to \infty} \lambda_n(t_0) = \lim_{n \to \infty} \phi'_n(t_0).$$

(1.7.1. Corollary) *Let $I = [a, b]$ be a given compact interval, let Y be a Banach space, and let $C^1(I, Y)$ be the set of all functions $\phi : I \to Y$ whose derivative ϕ' is continuous on I. Then $C^1(I, Y)$ is a Banach space under the norm*

$$\| \phi \| = \sup_I \| \phi(t) \| + \sup_I \| \phi'(t) \|.$$

This is immediate. An equivalent norm is given by

$$\|\phi\|^* = \|\phi(t_0)\| + \sup_I \|\phi'(t)\|,$$

where t_0 is any fixed point of I.

The remaining applications of (1.6.2, 3) that we consider here deal with limit theorems of various types. The first result requires the completeness of Y, but the other results apply to any normed space.

(1.7.2) *Let ϕ be a continuous function from a bounded interval $[a, b[$ into a Banach space Y such that $\phi'_+(t)$ exists nearly everywhere in $[a, b[$, and suppose that there exist positive M, δ such that $\|\phi'_+(t)\| \leq M$ for nearly all $t \in [b - \delta, b[$. Then $\phi(t)$ tends to a limit (in Y) as $t \to b -$.*

This follows directly from (1.6.2. Corollary 1) and the completeness of Y.

The next result is more subtle.

(1.7.3) *Let ϕ be a continuous function from a bounded interval $[a, b[$ into a normed space Y such that $\phi_+(t)$ exists nearly everywhere in $[a, b[$, and suppose that there exists a sequence (t_n) of points of $[a, b[$ converging to b such that $(\phi(t_n))$ tends to a limit c in Y as $n \to \infty$. Suppose also that there exist positive M, δ such that $\|\phi'_+(t)\| \leq M$ for nearly all $t \in [b - \delta, b[$ for which $\|\phi(t) - c\| \leq \delta$. Then $\phi(t) \to c$ as $t \to b -$.*

If the result is false, we can find ε satisfying $0 < \varepsilon \leq \delta$ and a sequence (u_n) of points of $[a, b[$ converging to b such that $\|\phi(u_n) - c\| > \varepsilon$ for all n. Since $\phi(t_n) \to c$ and ϕ is continuous, we can therefore also find an increasing sequence (v_n) converging to b, with the points v_{2n} selected from the t_n, such that, for all n, $\|\phi(v_{2n}) - c\| \leq \frac{1}{2}\varepsilon$, $\|\phi(t) - c\| \leq \varepsilon$ for $v_{2n} \leq t \leq v_{2n+1}$, and $\|\phi(v_{2n+1}) - c\| = \varepsilon$. Then $\|\phi(v_{2n+1}) - \phi(v_{2n})\| \geq \frac{1}{2}\varepsilon$ for all n. On the other hand, $\|\phi'_+(t)\| \leq M$ for nearly all t satisfying $v_{2n} \leq t \leq v_{2n+1}$ and all sufficiently large n, so that

$$\|\phi(v_{2n+1}) - \phi(v_{2n})\| \leq M(v_{2n+1} - v_{2n}) \to 0 \quad \text{as} \quad n \to \infty,$$

giving a contradiction.

The next theorem shows that a derivative cannot have a jump discontinuity.

(1.7.4) *Let $\phi : [a, b] \to Y$ be a continuous function such that $\phi'_+(t)$ exists nearly everywhere in $[a, b]$ and that $\lim_{t \to b -} \phi'_+(t) = l$. Then $\phi'(b)$ exists and is equal to l.*

Here we apply (1.6.3. Corollary) to the function $t \mapsto \phi(t) - tl$ on the interval $[u,b]$. We deduce that

$$\left\| \frac{\phi(u) - \phi(b)}{u - b} - l \right\| \leq \left\| \phi'_+(\xi) - l \right\|$$

for uncountably many $\xi \in [u, b]$, and we now make $u \to b -$.

The argument of (1.7.4) can easily be generalized, using (1.6.3), to prove the following version of L'Hospital's theorem for vector-valued functions.

(1.7.5) *Let $a < b \leq \infty$, and let $\phi : [a, b[\to Y$ and $\psi : [a, b[\to]0, \infty[$ be continuous functions such that $\phi'_+(t), \psi'_+(t)$ exist and $\psi'_+(t) < 0$ nearly everywhere in $[a, b[$ and that $\psi(t) \to 0$ and $\phi'_+(t)/\psi'_+(t) \to l$ as $t \to b -$. Then $\phi(t) \to 0$ and $\phi(t)/\psi(t) \to l$ as $t \to b -$.†*

There is also a still stronger result, on the lines of (1.7.3), which we leave as an exercise (Exercise 1.7.4).

Exercises 1.7

1 Let J be a compact interval in $\mathbf{R}$, and let D be the set of continuous functions $\phi : J \to Y$ such that ϕ' is defined nearly everywhere on J and is bounded. Prove that
   (i) D is a linear space with the usual operations for function spaces,
   (ii) if for each $\phi \in D$

$$\| \phi \| = \sup_{t \in J} \| \phi(t) \| + \sup_{t \in K(\phi)} \| \phi'(t) \|,$$

where $K(\phi)$ is the domain of ϕ', then $\| \ \|$ is a norm on D, and D is complete if Y is complete.

2 Let J be a compact interval in $\mathbf{R}$, let $\phi : J \to Y$ be a function whose derivative $\phi'(t)$ exists for all $t \in J$, and let $\psi : J \times J \to Y$ be given by

$$\psi(s, t) = \frac{\phi(s) - \phi(t)}{s - t} \quad (s \neq t), \quad \psi(t, t) = \phi'(t).$$

Prove that ϕ' is continuous if and only if ψ is continuous at (t, t) for each $t \in J$ (or, equivalently, since ϕ is continuous, if and only if ψ is continuous).

3 Prove that if I is an interval in $\mathbf{R}$ and $\phi : I \to \mathbf{R}$ is convex, then ϕ'_+ and ϕ'_- are continuous at each interior point t of I for which $\phi'_+(t) = \phi'_-(t)$.

4 Let $a < b \leq \infty$, let $l \in Y$, and let $\phi : [a, b[\to Y$ and $\psi : [a, b[\to]0, \infty[$ be continuous functions with the properties that
   (i) $\psi(t) \to 0$ as $t \to b -$, and there exists a sequence (t_n) of points of $[a, b[$ with the limit b for which $\phi(t_n) \to 0$,
   (ii) $\phi'_+(t), \psi'_+(t)$ exist and $\psi'_+(t) < 0$ for nearly all $t \in [a, b[$,
   (iii) $\phi'_+(t)/\psi'_+(t) \to l$ as $t \to b -$ and $\phi(t) \to 0$, i.e. for each $\varepsilon > 0$ there exist $c \in [a, b[$

† When $b = \infty$, $t \to b -$ means $t \to \infty$.

and $\delta > 0$ such that $\| \phi'_+(t)/\psi'_+(t) - l \| \le \varepsilon$ for nearly all $t \in [c, b[$ for which $\| \phi(t) \| \le \delta$. Prove that

$$\lim_{t \to b-} \frac{\phi(t)}{\psi(t)} = l, \ \lim_{t \to b-} \phi(t) = 0, \text{ and } \lim_{t \to b-} \frac{\phi'_+(t)}{\psi'_+(t)} = l.$$

5 (Rectifiable paths) A *partition* P of the interval $[a, b]$ is a finite subset of $[a, b]$ which includes a and b. We assume without further reference that the points $t_0, t_1, \ldots, t_m$ of a partition P are enumerated in ascending order, so that $a = t_0 < \ldots < t_m = b$.

If $\psi : [a, b] \to Y$ is a path in Y (i.e. if ψ is continuous) and $P = \{t_0, t_1, \ldots, t_m\}$ is a partition of $[a, b]$, the sum

$$S(P) = \sum_{j=1}^{m} \| \psi(t_j) - \psi(t_{j-1}) \|$$

is the length of a polygon in Y with successive vertices at the points $\psi(t_0)$, $\psi(t_1), \ldots, \psi(t_m)$ of the path. It is a simple consequence of the triangle inequality that if $P \subseteq P'$ then $\| \psi(b) - \psi(a) \| \le S(P) \le S(P')$.

The *length* $l = l(a, b)$ of the path ψ is defined to be the supremum of the set of numbers $S(P)$ for all partitions P of $[a, b]$, the value ∞ being allowed. If l is finite, the path ψ is said to be *rectifiable*. Clearly $l \ge \psi(b) - \psi(a)$.

(i) Prove that if $M \ge 0$ and $\psi : [a, b] \to Y$ is a continuous function such that $\psi'_+(t)$ exists and satisfies $\| \psi'_+(t) \| \le M$ for nearly all $t \in [a, b[$, then ψ is a rectifiable path and its length does not exceed $M(b - a)$.

(ii) Prove that the path $\psi : [0, 1] \to \mathbf{R}^2$ given by $\psi(t) = (t, t \cos \pi/t)$ $(0 < t \le 1)$, $\psi(0) = (0, 0)$, is not rectifiable.

(iii) Let $\psi : [a, b] \to Y$ be a rectifiable path. The *length* $l(c, d)$ *of the arc of* ψ *from c to d*, where $a \le c < d \le b$, is the length of the path obtained by restricting ψ to $[c, d]$. Prove that

(a) if $a < c < b$ then $l(a, b) = l(a, c) + l(c, b)$,

(b) if $\sigma(t) = l(a, t)$ $(a \le t \le b)$, then σ is increasing and continuous on $[a, b]$,

(c) if $t_0 \in [a, b]$ and ψ' is defined nearly everywhere on some neighbourhood U of t_0 in $[a, b]$ and is continuous at t_0, then $\sigma'(t_0)$ exists and is equal to $\| \psi'(t_0) \|$,

(d) if the conditions of (c) are satisfied and $\psi'(t_0) \ne 0$, then $(\sigma(t) - \sigma(t_0))/\| \psi(t) - \psi(t_0) \|$ tends to ± 1 as $t \to t_0 \pm$ (i.e. the ratio of arc length to chord length tends to 1).

(e) By considering the example $\psi : [-1, 1] \to \mathbf{R}^2$ given by $\psi(0) = (0, 0)$, $\psi(t) = (t^3, t^3 \cos \pi/t)$ $(0 < |t| \le 1)$, show that the result of (d) fails if $\psi'(t_0) = 0$.

6 Let Y be a real normed space, and let $\psi : [a, b] \to Y$ be a continuous function with the property (∗) that for nearly all $t \in [a, b[$ there exists a sequence (t_n) of points of $]t, b]$ with the limit t such that $\psi(t_n) \ne \psi(t)$ $(n = 1, 2, \ldots)$ and that the sequence of chord directions $(\psi(t_n) - \psi(t))/\| \psi(t_n) - \psi(t) \|$ tends to a limit $q(t)$. Prove that if $q(t)$ belongs to the open half-space $\{y \in Y : u(y) > \alpha\}$ for nearly all $t \in [a, b[$, where u is a continuous linear functional on Y and $\alpha > 0$, then ψ is a rectifiable path with length not exceeding $u(\psi(b) - \psi(a))/\alpha$.

[Hint. Use Exercise 1.6.6 to show that if $a \le s < t \le b$, then $u(\psi(t) - \psi(s)) > \alpha \| \psi(t) - \psi(s) \|$.]

7 Let y be a unit vector in Y, let $0 < \mu < 1$, and let $\psi : [a, b] \to Y$ be a continuous function with the property (∗). Prove that if $\| y - q(t) \| < \mu$ for nearly all $t \in [a, b[$, then ψ is a rectifiable path with length not exceeding $\| \psi(b) - \psi(a) \|/(1 - \mu)$.

[Hint. Use Exercise 6 together with the Hahn–Banach theorem.]

8 Let $\psi : [a,b] \to Y$ be a path in the normed space Y having a tangent at every point, the direction of the tangent at t being $q(t)$. Suppose further that the function $t \mapsto q(t)$ is continuous on $[a,b]$. Prove that

(i) ψ is a rectifiable path,

(ii) if $\sigma(t)$ is the length of the arc of ψ from a to t and $\rho(s,t) = (\sigma(t) - \sigma(s))/ \| \psi(t) - \psi(s) \|$, then $\rho(s,t) \to 1$ as $t \to s+$ for each $s \in [a,b[$ and $\rho(s,t) \to -1$ as $s \to t-$ for each $t \in]a,b]$ (i.e. the ratio of arc length to chord length tends to 1),

(iii) if $\phi(s) = \psi(\sigma^{-1}(s))$ $(0 \le s \le \sigma(b))$, then $\phi'(s)$ exists and is equal to $q(\sigma^{-1}(s))$ for each $s \in [0, \sigma(b)]$ (i.e. if we reparametrize the path by taking arc length as the new parameter, then the reparametrized path is regular, i.e. it has a non-vanishing continuous derivative).

[In elementary differential geometry, it is customary to restrict the discussion of paths to regular arcs; this property implies in particular that the path has a continuously turning tangent. The exercise shows that, conversely, a path with a continuously turning tangent can be reparametrized to make it regular.

Hint. For (i) and (ii) use Exercise 7.]

1.8 Derivatives of second and higher orders; Taylor's theorem

Let ϕ be a function from a set $A \subseteq \mathbf{R}$ into Y, and let ϕ' be the first derivative of ϕ. We define the *second derivative* of ϕ, denoted by either ϕ'' or $\phi^{(2)}$, to be the first derivative of ϕ' (ϕ'' may be the empty function, and is certainly so if ϕ' is the empty function). Further, if the function ϕ'' is defined at the point t_0, we say that ϕ has a *second derivative at* t_0, equal to $\phi''(t_0)$. In other words, ϕ has a second derivative at t_0, equal to $\phi''(t_0)$, if and only if t_0 is a non-isolated point of the domain of ϕ', and the limit

$$\lim_{t \to t_0} \frac{\phi'(t) - \phi'(t_0)}{t - t_0}$$

exists and is equal to $\phi''(t_0)$.

Generally, the *r*th *derivative* $\phi^{(r)}$ of ϕ $(r = 2, 3, \ldots)$ is defined inductively as the first derivative of $\phi^{(r-1)}$. Further, if $\phi^{(r)}$ is defined at the point t_0, we say that ϕ has *a derivative of order* r *at* t_0, equal to $\phi^{(r)}(t_0)$. When ϕ is given by a specific formula, the value of the rth derivative of ϕ at t_0 is sometimes denoted by $[d^r\phi(t)/dt^r]_{t_0}$, and we also write $d^r\phi(t)/dt^r$ for the value of $\phi^{(r)}$ at the point t. In the theory of differential equations in Chapter 2 we also use the corresponding 'dependent variable' notation $d^r y/dt^r$, where $y = \phi(t)$.

It is obvious that if t_0 is a point at which ϕ has a derivative of some (positive) order, then either ϕ has derivatives of all orders at t_0, or there exists a positive integer p such that ϕ has derivatives of orders $1, \ldots, p$ at t_0 and has no derivative of order greater than p at t_0. It is also obvious that, for any positive integers r, s, the sth derivative of $\phi^{(r)}$ is $\phi^{(r+s)}$.

A function ϕ that possesses a derivative $\phi^{(r)}(t)$ of order r at a point t is sometimes said to be *r-times differentiable at t*. Similarly, if $\phi^{(r)}(t)$ exists at every point of an interval I, then ϕ is said to be *r-times differentiable on I;* if in addition $\phi^{(r)}$ is continuous on I, then ϕ is *r-times continuously differentiable on I.*

The most important results concerning rth derivatives are the various forms of Taylor's theorem, and we give three of these here.

(1.8.1) *Let q be a continuous quasi-convex function on Y, let r be a positive integer, and let $\phi : [a,b] \to Y$ be a function whose $(r-1)$th derivative $\phi^{(r-1)}(t)$ exists for all $t\in[a,b]$ and whose rth derivative $\phi^{(r)}(a)$ exists. Then there exist uncountably many $\xi\in\,]a,b[$ for which*

$$q(\phi(b) - \phi(a) - (b-a)\phi'(a) - \ldots - \frac{(b-a)^r}{r!}\phi^{(r)}(a))$$
$$\le q\left(\frac{(b-a)^r}{r!}\left(\frac{\phi^{(r-1)}(\xi) - \phi^{(r-1)}(a)}{\xi - a} - \phi^{(r)}(a)\right)\right).\dagger$$

This is obtained by successive applications of (1.6.6) to the pair of functions

$$t \mapsto (\phi(t) - \phi(a) - (t-a)\phi'(a) - \ldots - \frac{(t-a)^r}{r!}\phi^{(r)}(a))(b-a)^r$$

and $t \mapsto (t-a)^r$ and their derivatives of orders $1,2,\ldots,r-2$.

By taking q to be the norm function on Y, we deduce immediately:

(1.8.2) *Let r be a positive integer, and let $\phi:[a,b] \to Y$ be a function whose $(r-1)$th derivative $\phi^{(r-1)}(t)$ exists for all $t\in[a,b]$ and whose rth derivative $\phi^{(r)}(a)$ exists. Then*

$$h^{-r}\left(\phi(a+h) - \phi(a) - h\phi'(a) - \ldots - \frac{h^r}{r!}\phi^{(r)}(a)\right) \to 0 \quad \text{as} \quad h \to 0+.$$

A similar result holds to the left of b.

(1.8.3) *Let q be a continuous quasi-convex function on Y, let r be a positive integer, let $s > 0$, and let $\phi:[a,b] \to Y$ be a function whose $(r-1)$th derivative $\phi^{(r-1)}(t)$ exists for all $t\in[a,b]$ and whose rth derivative $\phi^{(r)}(t)$ exists nearly everywhere in $[a,b]$. Then there exist uncountably many $\xi\in\,]a,b[$ such that*

† If q is sublinear (in particular, if q is the norm function on Y), then the factor $(b-a)^r/r!$ on the right can be moved outside the 'q'. A similar remark applies in (1.8.3).

$$q(\phi(b) - \phi(a) - (b-a)\phi(a) - \dots - \frac{(b-a)^{r-1}}{(r-1)!}\phi^{(r-1)}(a))$$

$$\leq q\left(\frac{(b-\xi)^{r-s}(b-a)^s}{s((r-1)!)}\phi^{(r)}(\xi)\right). \qquad (1)$$

Here we apply (1.6.6) to the pair of functions

$$t \mapsto (-\phi(b) + \phi(t) + (b-t)\phi'(t) + \dots + \frac{(b-t)^{r-1}}{(r-1)!}\phi^{(r-1)}(t))(b-a)^s$$

and $t \mapsto -(b-t)^s$.

When $Y = R$, we can use (1.3.3) in place of (1.6.6) to obtain the familiar versions of Taylor's theorem for real-valued functions of a real variable. For example, the result of (1.8.1) can then be replaced by the assertion that there exists $\xi \in]a, b[$ for which

$$\phi(b) - \phi(a) - (b-a)\phi'(a) - \dots - \frac{(b-a)^r}{r!}\phi^{(r)}(a)$$

$$= \frac{(b-a)^r}{r!}\left(\frac{\phi^{(r-1)}(\xi) - \phi^{(r-1)}(a)}{\xi - a} - \phi^{(r)}(a)\right).$$

Similarly, the case $s = r$ of (1.8.3) corresponds to Lagrange's form of Taylor's theorem, while the case $s = 1$ corresponds to Cauchy's form.

The following limit theorem provides a simple application of (1.8.3).

(1.8.4) *If* $\phi :]a, \infty[\to Y$ *has a bounded second derivative on* $]a, \infty[$, *and* $\phi(t)$ *tends to a limit as* $t \to \infty$, *then* $\phi'(t) \to 0$ *as* $t \to \infty$.

From (1.8.3) we deduce that for $t > a$ and $h > 0$ we have

$$\|\phi'(t)\| \leq \|\phi(t+h) - \phi(t)\|/h + h\|\phi''(\xi)\|/2$$

for some $\xi \in]t, t + h[$. Since ϕ'' is bounded, given $\varepsilon > 0$ we can choose h so small that the second term on the right is less than ε. Since $\phi(t)$ tends to a limit as $t \to \infty$, we can find K such that $\|\phi(t+h) - \phi(t)\| < h\varepsilon$ whenever $t > K$. Hence $\|\phi'(t)\| < 2\varepsilon$ for all $t > K$, so that $\phi'(t) \to 0$.

We mention finally the following generalization of (1.7.1. Corollary).

(1.8.5) *Let* Y *be a Banach space, let* I *be a compact interval in* **R** *and let* r *be a positive integer. Then the class* $C^r(I, Y)$ *of functions* $\phi : I \to Y$ *such that* $\phi^{(r)}$ *is defined and continuous on* I *is a Banach space under the norm*

$$\|\phi\| = \sup_I \|\phi(t)\| + \sum_{n=1}^r \sup_I \|\phi^{(n)}(t)\|.$$

This is an easy consequence of (1.7.1).

1.9 Regulated functions and integration

Let $I = [a, b]$ be a compact interval in **R**. A function $\phi : I \to Y$ is said to be *regulated* if the right-hand limit $\phi(t+)$ exists for all $t \in [a, b[$ and the left-hand limit $\phi(t-)$ exists for all $t \in]a, b]$. Obviously any continuous function from I into Y is regulated, and, if $Y = \mathbf{R}$, then any monotone function is regulated.

A function $\phi : I \to Y$ is called a *step-function* if there exist a positive integer m, and points $\tau_0, \ldots, \tau_m \in I$ and $c_1, \ldots, c_m \in Y$, such that $a = \tau_0 < \tau_1 < \ldots < \tau_m = b$ and that

$$\phi(t) = c_j \quad \text{for} \quad \tau_{j-1} < t < \tau_j, \quad j = 1, 2, \ldots, m. \tag{1}$$

Any step-function is obviously regulated.

(1.9.1) *Let (ϕ_n) be a sequence of regulated functions from the interval $I = [a, b]$ into a Banach space Y converging uniformly on I to a function $\phi : I \to Y$. Then ϕ is regulated.*

Let $\varepsilon > 0$, and let $t_0 \in [a, b[$. We can find an integer n such that $\| \phi(t) - \phi_n(t) \| \leq \varepsilon$ for all $t \in I$. Moreover, since $\phi_n(t_0+)$ exists, we can find $\delta > 0$ such that $\| \phi_n(s) - \phi_n(t) \| \leq \varepsilon$ whenever $t_0 < s < t < t_0 + \delta$. Hence

$$\| \phi(s) - \phi(t) \| \leq \| \phi(s) - \phi_n(s) \| + \| \phi_n(s) - \phi_n(t) \| + \| \phi_n(t) - \phi(t) \| \leq 3\varepsilon$$

whenever $t_0 < s < t < t_0 + \delta$, and therefore the limit $\phi(t_0+)$ exists (since Y is complete). A similar argument applies to the left-hand limit of ϕ.

(1.9.2) *Let ϕ be a function from the interval $I = [a, b]$ into a Banach space Y. Then ϕ is regulated if and only if it is the uniform limit of a sequence of step-functions $\phi_n : I \to Y$.*

The 'if' follows from (1.9.1). To prove the 'only if', let ϕ be regulated, and let n be a positive integer. It is obviously enough to prove that there exists a step-function $\phi_n : I \to Y$ such that $\| \phi(t) - \phi_n(t) \| \leq 1/n$ for all $t \in I$.

Since ϕ has both left-hand and right-hand limits, for each $t \in I$ we can find $\lambda(t), \mu(t)$ satisfying $\lambda(t) < t < \mu(t)$ such that $\| \phi(s) - \phi(s') \| \leq 1/n$ whenever $s, s' \in]\lambda(t), t[\cap I$ and also whenever $s, s' \in]t, \mu(t)[\cap I$. Let $V(t)$ be the interval $]\lambda(t), \mu(t)[$. A finite number of intervals $V(t)$, say $V(t_1), \ldots, V(t_k)$, cover I, and we denote by $\tau_0, \ldots, \tau_m$ the points $a, b, t_i, \lambda(t_i), \mu(t_i)$ $(i = 1, \ldots, k)$ arranged in increasing order. For each j the point τ_j belongs to some $V(t_i)$, and then the points τ_j, τ_{j+1} either both belong to $[\lambda(t_i), t_i]$ or both belong to $[t_i, \mu(t_i)]$. In either case $\| \phi(s) - \phi(s') \| \leq 1/n$ whenever $s, s' \in]\tau_j, \tau_{j+1}[$. If now ϕ_n is the function that takes the same values as ϕ at the points τ_j, and takes the constant value $\phi(\frac{1}{2}(\tau_j + \tau_{j+1}))$ on each interval $]\tau_j, \tau_{j+1}[$, then ϕ_n clearly has the required property.

In particular, it follows from (1.9.2) that any regulated function is bounded, and is continuous nearly everywhere.

Let ϕ be a function from an interval $I \subseteq \mathbf{R}$ into Y. A function $\Phi : I \to Y$ is said to be a *primitive of ϕ on I* if Φ is continuous, and $\Phi'(t)$ exists and is equal to $\phi(t)$ nearly everywhere in I.

(1.9.3) *Let ϕ be a regulated function from the interval $I = [a,b]$ into a Banach space Y. Then ϕ has a primitive on I. Moreover, any two primitives of ϕ on I differ by a constant.*

If ϕ is a step-function, say of the form (1), let Φ be the function from I into Y such that $\Phi(a) = 0$ and that

$$\Phi(t) = c_j(t - \tau_{j-1}) + \sum_{r=1}^{j-1} c_r(\tau_r - \tau_{r-1}) \qquad (\tau_{j-1} < t \leq \tau_j; j = 1,\ldots,m).$$

Then clearly Φ is continuous, and $\Phi'(t) = \phi(t)$ except at the points τ_j, so that Φ is a primitive of ϕ on I.

Next, any regulated ϕ is the uniform limit of a sequence of step-functions Φ_n, and for each n we define Φ_n as above. Since the set of endpoints of all the intervals of constancy of all the functions ϕ_n is countable, it follows from (1.7.1) that the sequence (Φ_n) converges uniformly to a function Φ that is a primitive of ϕ on I.

Finally, the last part of the theorem is an immediate consequence of (1.6.2. Corollary 2).

(1.9.4) *Let ϕ be a continuous function from the interval $I = [a,b]$ into a Banach space Y, and let Φ be a primitive of ϕ on I. Then $\Phi'(t)$ exists and is equal to $\phi(t)$ for all $t \in I$.*

Let $\varepsilon > 0$ and let $t_0 \in I$. Then we can find $\delta > 0$ such that $\|\phi(t) - \phi(t_0)\| \leq \varepsilon$ whenever $t \in]t_0 - \delta, t_0 + \delta[\cap I$, and hence $\|\Phi'(t) - \phi(t_0)\| \leq \varepsilon$ for nearly all such t. By (1.6.2. Corollary 1) applied to the function $t \mapsto \Phi(t) - t\phi(t_0)$, we deduce that

$$\|\Phi(t) - \Phi(t_0) - (t - t_0)\phi(t_0)\| \leq \varepsilon |t - t_0|$$

whenever $t \in]t_0 - \delta, t_0 + \delta[\cap I$, and therefore $\Phi'(t_0)$ exists and is equal to $\phi(t_0)$.

Let $\phi : I \to Y$ be a regulated function, where Y is again a Banach space. For any $c, d \in I$ the *integral of ϕ between c,d*, denoted by $\int_c^d \phi(t)dt$, is defined by

$$\int_c^d \phi(t)dt = \Phi(d) - \Phi(c),$$

where Φ is any primitive of ϕ on I. By (1.9.3), the integral exists and is independent of the particular primitive Φ that is considered.

Obviously $\int_c^d \phi(t)dt = -\int_d^c \phi(t)dt$, and for all $c,d,u \in I$

$$\int_c^u \phi(t)dt + \int_u^d \phi(t)dt = \int_c^d \phi(t)dt.$$

To obtain further properties of the integral, we have only to translate results concerning derivatives into results concerning primitives. For instance, (1.1.3) shows that if $\phi, \psi : I \to Y$ are regulated functions and α, β are scalars, then for all $c, d \in I$

$$\int_c^d (\alpha\phi(t) + \beta\psi(t))dt = \alpha \int_c^d \phi(t)dt + \beta \int_c^d \psi(t)dt$$

(note that $\alpha\phi + \beta\psi$ is regulated). Hence the function $\phi \mapsto \int_c^d \phi(t)dt$ is a linear functional on the vector space of regulated functions from I into Y. Similarly the standard formulae for integration by parts and substitution follow from the rule for differentiation of a product (1.1.4) and the chain rule (1.1.6) (see Exercises 1.9.1, 2).

Again, the increment inequality (1.6.2) shows that if $\phi : I \to Y$ is regulated, $c < d$, and $\|\phi(t)\| \le M$ for nearly all $t \in [c,d]$, then

$$\left\| \int_c^d \phi(t)dt \right\| \le \int_c^d \|\phi(t)\| \, dt \le M(d-c). \tag{2}$$

In fact, if Φ is a primitive of ϕ, and Ψ is a primitive of $\|\phi\|$ (note that $\|\phi\|$ is regulated), then $\|\Phi'(t)\| = \|\phi(t)\| = \Psi'(t)$ nearly everywhere, whence, by (1.6.2),

$$\left\| \int_c^d \phi(t)dt \right\| = \|\Phi(d) - \Phi(c)\| \le \Psi(d) - \Psi(c) = \int_c^d \|\phi(t)\| \, dt.$$

The second inequality in (2) follows similarly from (1.6.2. Corollary 1).

We note also that (1.9.4) shows that if $\phi : I \to Y$ is continuous and $c \in I$, then for all $u \in I$

$$\frac{d}{du}\left(\int_c^u \phi(t)dt \right) = \phi(u), \quad \text{and} \quad \frac{d}{du}\left(\int_u^c \phi(t)dt \right) = -\phi(u).$$

We have also the following consequence of (1.9.1) and (1.7.1).

(1.9.5) *If* (ϕ_n) *is a sequence of regulated functions from* I *into a Banach space* Y *converging uniformly on* I *to a function* ϕ, *then* (ϕ *is regulated and*) *for all* $c, d \in I$

$$\lim_{n \to \infty} \int_c^d \phi_n(t)dt = \int_c^d \phi(t)dt.$$

The definition of the integral can be further extended by requiring that the integrand ϕ is defined nearly everywhere on I and is equal nearly everywhere on I to a regulated function ϕ^*. Under these circumstances, for any $c, d \in I$ we define the integral of ϕ between c, d by the relation

$$\int_c^d \phi(t)dt = \int_c^d \phi^*(t)dt,$$

where the integral on the right is defined as above. Moreover, any primitive of ϕ^* is also called a primitive of ϕ. In particular, *if $\psi : I \to Y$ is a continuous function such that ψ' is defined nearly everywhere on I and agrees nearly everywhere on I with a regulated function, then ψ is a primitive of ψ' so that for all $c, d \in I$*

$$\int_c^d \psi'(t)dt = \psi(d) - \psi(c).$$

Also, if Φ is a primitive of a regulated function, then Φ is a primitive of Φ'.

We consider finally the definition of integrals of functions over half-open or open intervals.

Let ϕ be a function defined everywhere or nearly everywhere on the interval $[a, b[$ and with values in a Banach space Y, and suppose that, for each $c \in [a, b[, \phi$ agrees nearly everywhere on $[a, c]$ with a regulated function on $[a, c]$. It follows easily from (1.9.3) that ϕ has a primitive on $[a, b[$, and that any two primitives of ϕ on $[a, b[$ differ by a constant. If for any primitive Φ of ϕ on $[a, b[$ the limit $\Phi(b-)$ exists,† we say that ϕ *has an integral over* $[a, b[$, and this integral is given by

$$\int_a^b \phi(t)dt = \Phi(b-) - \Phi(a) = \lim_{c \to b-} (\Phi(c) - \Phi(a)) = \lim_{c \to b-} \int_a^c \phi(t)dt.$$

In this case we say also that the integral $\int_a^b \phi(t)dt$ *converges*. If the limit $\Phi(b-)$ does not exist, we say that the integral $\int_a^b \phi(t)dt$ *diverges*, or *does not exist*. If $\Phi(b-) = \infty$, we say also that $\int_a^b \phi(t)dt = \infty$, or, since this property is independent of a, $\int^b \phi(t)dt = \infty$. (We shall use similar expressions, like $\int^b \phi(t)dt < \infty$, with the obvious meanings.)

Exercises 1.9

1 (Integration by substitution) Let $\Phi : [a, b] \to \mathbf{R}$ be a primitive on $[a, b]$ of a regulated function, let ψ be a function from an interval containing the range of Φ into a Banach space Y, and let $c = \Phi(a), d = \Phi(b)$. Prove that if either ψ is continuous, or ψ is regulated and Φ is monotone, then

$$\int_c^d \psi(u)du = \int_a^b \psi(\Phi(t))\Phi'(t)dt.$$

† If $b = \infty$, then $\Phi(b-)$ means $\lim_{c \to \infty} \Phi(c)$.

2 (Integration by parts) Let $\Phi:[a,b]\to\mathbf{R}$ be a primitive on $[a,b]$ of a regulated function, let ψ be a regulated function from $[a,b]$ into a Banach space Y, and let Ψ be a primitive of ψ on $[a,b]$. Prove that

$$\int_a^b \Phi(t)\psi(t)dt = \Phi(b)\Psi(b) - \Phi(a)\Psi(a) - \int_a^b \Phi'(t)\Psi(t)dt.$$

3 (Taylor's theorem with integral remainder) Let Y be a Banach space, and let $\phi:[a,b]\to Y$ be a function whose $(r-1)$th derivative $\phi^{(r-1)}$ is continuous on $[a,b]$ and whose rth derivative $\phi^{(r)}$ is regulated on $[a,b]$. Prove that

$$\phi(b) = \phi(a) + (b-a)\phi'(a) + \ldots + \frac{(b-a)^{r-1}}{(r-1)!}\phi^{(r-1)}(a)$$
$$+ \frac{1}{(r-1)!}\int_a^b (b-t)^{r-1}\phi^{(r)}(t)dt.$$

4 Let X be a normed space and Y a Banach space, and let $v:[a,b]\to\mathcal{L}(X,Y)$ be regulated. Prove that for all $h\in X$

$$\int_a^b v(t)h\,dt = \left(\int_a^b v(t)dt\right)h.$$

5 Let I be an interval in $\mathbf{R}$, and let $\phi:I\to Y$ be a continuous function such that the derivative $\phi'_+(t)$ exists and satisfies

$$\|\phi'_+(t)\| \le \mu(t)h(\|\phi(t)\|)$$

for nearly all $t\in I$, where $\mu:I\to[0,\infty[$ and $h:[0,\infty[\to[0,\infty[$ are continuous, $h(x)>0$ for all $x>0$, $\int_0 du/h(u)<\infty$, and $\int^\infty du/h(u)=\infty$. Prove that for all $t,t_0\in I$

$$\|\phi(t)\| \le H^{-1}\left(H(\|\phi(t_0)\|) + \left|\int_{t_0}^t \mu(s)ds\right|\right),$$

where $H(x) = \int_0^x du/h(u)$ $(x\ge 0)$, and that

$$\|\phi(t)\| \ge H^{-1}\left(H(\|\phi(t_0)\|) - \left|\int_{t_0}^t \mu(s)ds\right|\right)$$

whenever the right-hand side is defined.

[Hint. The function H is a continuous strictly increasing function from $[0,\infty[$ onto itself, and $H'(x)=1/h(x)$ for all $x>0$.]

6 Prove that if $\psi:[a,b]\to Y$ has a continuous derivative on $[a,b]$, then ψ is a rectifiable path and its length is equal to $\int_a^b\|\psi'(t)\|dt$.

7 Let J be a compact interval in $\mathbf{R}$, let $R(J,Y)$ be the linear space of regulated functions from J into Y, and for each $\phi\in R(J,Y)$ let

$$\|\phi\| = \sup_{t\in J}\|\phi(t)\|.$$

Prove that $\|\ \|$ is a norm on $R(J,Y)$, and that if Y is complete so is $R(J,Y)$.

1.10 Further monotonicity theorems and increment inequalities

We return now to the ideas of §1.4, and we prove two generalizations of (1.4.3) in which we allow larger exceptional sets where the condition $D^+\phi(t)\ge 0$ is not satisfied.

(1.10.1) *Let $\phi : [a,b] \to R$ be a function such that*
  (i) $\lim\sup_{s\to t-}\phi(s) \leq \phi(t)$ *for all* $t\in\,]a,b]$ *and* $\phi(t) \leq \lim\sup_{s\to t+}\phi(s)$ *for all* $t\in[a,b[$,
  (ii) $D^+\phi(t) \geq 0$ *for almost all* $t\in[a,b[$,
  (iii) $D^+\phi(t) > -\infty$ *for nearly all* $t\in[a,b[$.
Then ϕ is increasing.

Since a countable set has measure 0, this theorem clearly contains the result of (1.4.3) as a special case. The proof depends on the following lemma.

(1.10.2) **Lemma.** *Let E be a subset of $[a,b[$ of measure 0, and let $\varepsilon > 0$. Then there is a continuous increasing function χ such that $\chi(b) - \chi(a) \leq \varepsilon$ and that $D_+\chi(t) = \infty$ for each $t\in E$.*

Suppose first that G is a countable union of mutually disjoint open intervals contained in the interval $I = [a-1,b]$ of total length l, and for each $t\in I$ let $g(t)$ be the total length of intervals of G to the left of t. Obviously $0 \leq g(t') - g(t) \leq t' - t$ whenever $t,t'\in I$ and $t \leq t'$, so that g is continuous and increasing on I. Moreover, $g(a-1) = 0$, $g(b) = l$, and $D_+g(t) = 1$ for each $t\in G$.

Next, since E has measure 0, we can find a contracting sequence of sets G_n, each containing E and contained in I, such that for each n the set G_n is a countable union of mutually disjoint open intervals of total length $\varepsilon 2^{-n-1}$. For each n let g_n be the function associated as above with G_n, and let $\chi(t) = \sum_{n=1}^{\infty} g_n(t)$. Since $0 \leq g_n(t) \leq \varepsilon 2^{-n-1}$ for all $t\in I$, it follows that the series converges uniformly and χ is continuous on I, and clearly χ is increasing and $\chi(b) - \chi(a) \leq \chi(b) - \chi(a-1) = \varepsilon$. Further, if $t_0\in E$ and $t_0 < t \leq b$, then for each positive integer N

$$\chi(t) - \chi(t_0) \geq \sum_{n=1}^{N} (g_n(t) - g_n(t_0)),$$

and since $t_0 \in G_n$ for each n this gives

$$D_+\chi(t_0) \geq \sum_{n=1}^{N} D_+g_n(t_0) = N,$$

whence $D_+\chi(t_0) = \infty$.

The proof of (1.10.1) is now easy; let E be the subset of $[a,b[$ where $-\infty < D^+\phi(t) < 0$ (so that E has measure 0), let $\varepsilon > 0$, and let χ be constructed as in the lemma. Then $\phi + \chi$ obviously satisfies the conditions of (1.4.3), so that

$$\phi(b) + \chi(b) \geq \phi(a) + \chi(a).$$

Since $\chi(b) - \chi(a) \leq \varepsilon$, and ε is arbitrary, this gives the result.

Our second generalization of (1.4.3) concerns absolutely continuous functions. A function $\phi : [a, b] \to \mathbf{R}$ is said to be *absolutely continuous* if for each $\varepsilon > 0$ there exists $\delta > 0$ such that, if $]\alpha_1, \beta_1[, \ldots,]\alpha_N, \beta_N[$ is any finite family of mutually disjoint open subintervals of $[a, b]$ with total length $\sum_{n=1}^{N} (\beta_n - \alpha_n) \leq \delta$, then

$$\sum_{n=1}^{N} |\phi(\beta_n) - \phi(\alpha_n)| \leq \varepsilon.$$

Clearly an absolutely continuous function is continuous.

(1.10.3) Lemma. *If $\phi : [a, b] \to \mathbf{R}$ is absolutely continuous, and E is a subset of $[a, b]$ of measure 0, then $\phi(E)$ has measure 0.*

To prove this, let $\varepsilon > 0$ and let δ be chosen as above. Let also E be a subset of $]a, b[$ of measure 0, let (I_n) be a sequence of mutually disjoint open subintervals of $[a, b]$ such that $E \subseteq \bigcup I_n$ and that $\Sigma |I_n| < \delta$, where $|I_n|$ is the length of I_n. Since ϕ is continuous, $\phi(I_n)$ is an interval with length $\phi(\xi_n) - \phi(\eta_n)$, where ξ_n and η_n are points of $\bar{I}_n$ at which ϕ attains its supremum and infimum on $\bar{I}_n$. For each positive integer N we have

$$\sum_{n=1}^{N} |\xi_n - \eta_n| \leq \sum_{n=1}^{N} |I_n| \leq \delta,$$

and therefore $\sum_{n=1}^{N} |\phi(I_n)| \leq \varepsilon$. Since N is arbitrary, this implies that $\sum_{n=1}^{\infty} |\phi(I_n)| < \varepsilon$, and since the measure of $\phi(E)$ does not exceed this last sum, the result follows.

(1.10.4) *If $\phi : [a, b] \to \mathbf{R}$ is an absolutely continuous function such that $D^+\phi(t) \geq 0$ for almost all $t \in [a, b[$, then ϕ is increasing.*

Let $\varepsilon > 0$, and let $\psi(t) = \phi(t) + \varepsilon t$. Then ψ is absolutely continuous, and $D^+\psi(t) \geq \varepsilon > 0$ for almost all $t \in [a, b[$. The set E of t for which $D^+\psi(t) \leq 0$ therefore has measure 0, whence so also has $\psi(E)$, and the result follows from this and (1.4.1).

By using (1.10.1) instead of (1.4.3), we obtain generalizations of (1.4.4, 5) and of the results of §§1.5–8. For instance, the generalization of (1.6.2) is as follows.

(1.10.5) *Let p be a continuous sublinear functional on Y, let $\phi : [a, b] \to Y$ and $\psi : [a, b] \to \mathbf{R}$ be continuous functions whose right-hand derivatives*

$\phi'_+(t), \psi'_+(t)$ *exist for nearly all* $t \in [a, b[$ *and satisfy* $p(\phi'_+(t)) \le \psi'_+(t)$ *for almost all* $t \in [a, b[$. *Then*

$$p(\phi(b) - \phi(a)) \le \psi(b) - \psi(a). \tag{1}$$

Moreover, there is equality in (1) *if and only if*

$$p(\phi(t) - \phi(a)) = \psi(t) - \psi(a)$$

for all $t \in [a, b]$, *and then also* $p(\phi'_+(t)) \ge \psi'_+(t)$ *whenever* $\phi'_+(t), \psi'_+(t)$ *exist (so that* $p(\phi'_+(t)) = \psi'_+(t)$ *almost everywhere). Further, if* Y *is real and* p *is linear, and equality holds in* (1), *then* $p(\phi'_+(t)) = \psi'_+(t)$ *whenever* $\phi'_+(t)$, $\psi'_+(t)$ *exist.*

(Here the condition that $\phi'_+(t)$ exists for nearly all $t \in [a, b[$ implies that $D^+ p(\phi(t)) < \infty$ for nearly all $t \in [a, b]$.)

The corresponding generalization of (1.6.3) is perhaps of more interest.

(1.10.6) *Let* p *be a continuous sublinear functional on* Y, *let* $\phi : [a, b] \to Y$ *and* $\psi : [a, b] \to \mathbf{R}$ *be continuous functions whose right-hand derivatives* $\phi'_+(t), \psi'_+(t)$ *exist for nearly all* $t \in [a, b[$, *and let* $\psi(a) < \psi(b)$. *Then there is a set of* $\xi \in [a, b[$ *of positive measure for which*

$$p(\phi(b) - \phi(a))\psi'_+(\xi) \le (\psi(b) - \psi(a))p(\phi'_+(\xi)). \tag{2}$$

Moreover, either (2) *holds with strict inequality for a set of* ξ *of positive measure, or* (2) *holds whenever* $\phi'_+(\xi), \psi'_+(\xi)$ *both exist (and with equality almost everywhere).*

One point which should be noted here is that under the conditions of (1.10.6) the functions $t \mapsto \psi'_+(t)$ and $t \mapsto p(\phi'_+(t))$ are measurable, so that the set of points ξ satisfying (2) is measurable. For instance, to show that the second function is measurable we have only to note that

$$p(\phi'_+(t)) = \lim_{n \to \infty} p\left(\frac{\phi(t + 1/n) - \phi(t)}{1/n} \right),$$

so that $t \mapsto p(\phi'_+(t))$ is the limit of continuous functions.

We mention also the extension of (1.6.4) corresponding to (1.10.5).

(1.10.7) *Let* E *be a closed convex set in* Y, *and let* $\phi : [a, b] \to Y$ *be a continuous function such that* $\phi'_+(t)$ *exists for nearly all* $t \in [a, b[$ *and belongs to* E *for almost all* $t \in [a, b[$. *Then* $(\phi(b) - \phi(a))/(b - a) \in E$ *(so that also* $\phi'_+(t) \in E$ *whenever it exists).*

It follows in particular from this result that if $\phi : [a, b] \to Y$ is a continuous function such that $\phi'_+(t)$ exists for nearly all $t \in [a,b[$, and H is a subset of the domain of ϕ'_+ of measure 0, then the range D of ϕ'_+ and the set $A \backslash \phi'_+(H)$ have the same closed convex hulls.

We conclude this section with an application of (1.10.1) which gives a sufficient condition for a continuous function to be the Lebesgue integral of one of its Dini derivatives.

We recall that a measurable extended real-valued function f on $[a,b]$ is said to *have a definite integral on* $[a,b]$ (in the Lebesgue sense) if at least one of the functions f^+, f^- has a finite Lebesgue integral on $[a,b]$.†
The integral of f is then defined by the relation

$$\int_a^b f(t)dt = \int_a^b f^+(t)dt - \int_a^b f^-(t)dt,$$

the values $\infty, -\infty$ being permitted. If the integral is finite, then f is (Lebesgue) integrable on $[a,b]$.

In the proof of our theorem, in addition to the linearity, additivity, and positivity of the integral for integrable functions, we use the facts that

(i) if f possesses a definite integral on $[a,b]$ not equal to ∞ and $f_n = \max\{f, -n\}$, then f_n is integrable on $[a,b]$ and its integral on $[a,b]$ tends to that of f as $n \to \infty$,

(ii) if f is integrable on $[a,b]$, then the function $t \mapsto \int_a^t f(s)ds$ is continuous on $[a,b]$ and has a derivative equal to $f(t)$ for almost all t.

(1.10.8) *Let* $\phi : [a,b] \to \mathbf{R}$ *be a continuous function such that* $D_+\phi(t) < \infty$ *nearly everywhere in* $[a,b[$ *and that* $D_+\phi$ *has a definite integral on* $[a,b]$ *in the Lebesgue sense, infinite values being permitted. Then*

$$\phi(b) - \phi(a) \le \int_a^b D_+\phi(t)dt. \qquad (3)$$

We may suppose that the integral on the right is not ∞, otherwise there is nothing to prove. For each positive integer n let $f_n = \max\{D_+\phi, -n\}$. Then f_n is integrable on $[a,b]$ and $f_n(t) \ge D_+\phi(t)$ for all $t \in [a,b[$. Further, if $F_n(t)$ is the integral of f_n over $[a,t]$, where $a \le t \le b$, then $F_n(b) \to \int_a^b D_+\phi(t)dt$ as $n \to \infty$.

We observe now that for all $t \in [a,b[$

$$D^+(F_n(t) - \phi(t)) \ge D_+F_n(t) - D_+\phi(t)$$

† As usual f^+, f^- are the positive and negative parts of f, i.e.

$$f^+ = \max\{f, 0\}, \quad f^- = \max\{-f, 0\}.$$

and, by (ii) above, the expression on the right is almost everywhere equal to $f_n(t) - D_+\phi(t) \geq 0$. Further, if $a \leq t < t + h \leq b$, then

$$\frac{F_n(t+h) - F_n(t)}{h} = \frac{1}{h}\int_t^{t+h} f_n(s)ds \geq -n,$$

so that $D_+ F_n(t) \geq -n$ for all $t \in [a,b[$. Hence

$$D^+(F_n(t) - \phi(t)) \geq -n - D_+\phi(t) > -\infty$$

for nearly all t, so that, by (1.10.1), $F_n - \phi$ is increasing on $[a,b]$. Hence

$$F_n(b) - \phi(b) \geq F_n(a) - \phi(a) = -\phi(a)$$

and on making $n \to \infty$ we obtain the required result.

By applying this result to $-\phi$, we deduce that if $D^+\phi(t) > -\infty$ nearly everywhere and $D^+\phi$ has a definite integral on $[a,b]$, then

$$\int_a^b D^+\phi(t)dt \leq \phi(b) - \phi(a).$$

Further, since the condition that $D_+\phi$ has a definite integral on $[a,b]$, combined with the inequality (3), implies that $D^+\phi$ has a definite integral on $[a,b]$, we obtain the following corollaries.

(1.10.8. Corollary 1) *Let $\phi : [a,b] \to \mathbf{R}$ be a continuous function such that $D_+\phi(t) < \infty$ and $D^+\phi(t) > -\infty$ nearly everywhere in $[a,b[$ and that $D_+\phi$ has a definite integral on $[a,b]$. Then*

$$\int_a^b D_+\phi(t)dt = \int_a^b D^+\phi(t)dt = \phi(b) - \phi(a).$$

(1.10.8. Corollary 2) *Let $\phi : [a,b] \to \mathbf{R}$ be a continuous function such that $\phi'_+(t)$ exists nearly everywhere on $[a,b[$ and that ϕ'_+ has a definite integral on $[a,b]$. Then*

$$\int_a^b \phi'_+(t)dt = \phi(b) - \phi(a).$$

Exercises 1.10

1 Prove that if $\phi : [a,b] \to \mathbf{R}$ is absolutely continuous, ψ is an extended real-valued function Lebesgue integrable on $[a,b]$, and $\phi'(t) \leq \psi(t)$ for almost all $t \in [a,b]$, then

$$\phi(b) - \phi(a) \leq \int_a^b \psi(t)dt.$$

2 Let I be an interval in $\mathbf{R}$, and let $\phi : I \to Y$ be a continuous function such that the

derivative $\phi'_+(t)$ exists for nearly all $t \in I$ and satisfies

$$\| \phi'_+(t) \| \leq \mu(t)h(\| \phi(t) \|)$$

for almost all $t \in I$, where μ is a non-negative function Lebesgue integrable on each compact subinterval of I, $h : [0, \infty[\rightarrow [0, \infty[$ is continuous, $h(x) > 0$ for all $x > 0$, $\int_0 du/h(u) < \infty$, and $\int^\infty du/h(u) = \infty$. Prove that the result of Exercise 1.9.5 holds for ϕ.

3 Let $q : Y \rightarrow \mathbf{R}$ be a continuous quasi-convex function, and let $\phi : [a, b] \rightarrow Y$ be a continuous function possessing a right-hand derivative nearly everywhere in $[a, b]$. Prove that there exists a set of $\xi \in [a, b[$ of positive measure for which

$$q\left(\frac{\phi(b) - \phi(a)}{b - a} \right) \leq q(\phi'_+(\xi)).$$

1.11 Historical notes on the classical mean value theorems, monotonicity theorems and increment inequalities

The two classical mean value theorems (1.3.3) and (1.3.10. Corollary 1) which provide the starting point for much of the discussion in this chapter, are usually attributed to Lagrange and Cauchy.[†] However, the proofs of both theorems depend ultimately on the completeness of the set of real numbers, and, at the time when Lagrange and Cauchy obtained their results, no proof of the completeness property had been given, nor indeed was it recognized (except possibly by Bolzano; see Grattan-Guinness (1970, p. 76)) that any proof must depend on a study of the structure of the set of real numbers.

Lagrange's version of the mean value theorem appeared in his *Théorie des Fonctions Analytiques* (Lagrange, 1797) and in his *Leçons sur le Calcul des Fonctions* (Lagrange, 1801). In these two works Lagrange presented a treatment of the calculus which took as its basic premise that an arbitrary real-valued function ϕ can be developed in a power series about each point of its domain.[‡] The derivative of ϕ at a is introduced in terms of the coefficient of $t - a$ in the power series for $\phi(t)$ about the point a, and similarly for higher derivatives.

In his discussion of the mean value theorem, Lagrange was primarily

---

[†] Vacca (1901) has observed that the geometrical form of Cauchy's mean value theorem, which concerns tangents to a plane curve, was known to Cavalieri in 1635: 'Si curva linea quaecunque data tota sit in eodem plano, cui occurrat recta in duobus punctis ... poterimus aliam rectam lineam prefatae aequidistantem ducere, quae tangat portionem curvae lineae inter duos praedictos occursus continuatam.'

    The special case of the mean value theorem known as Rolle's theorem (see (1.3.2)) was proved by the French mathematician Michel Rolle for polynomials in 1691 (see Voss and Molk, 1909, p. 269).

[‡] This premise rests rather on the interpretation of the term 'function' than on any mathematical justification.

concerned to obtain explicit forms for the remainder after n terms in a Taylor series, and the mean value theorem appeared simply as the special case $n = 1$ of the general result. The key to Lagrange's treatment is a monotonicity theorem; if we allow for his unusual standpoint and insert all his hypotheses, this result is as follows (Lagrange, 1797, §48; 1801, Leçon 9).

(A) *If ϕ is developable in a power series about each point of $]\alpha, \beta[$ and $\phi'(t) > 0$ for all $t \in]\alpha, \beta[$, then ϕ is strictly increasing.*

Lagrange's attempted proof of this theorem is incorrect. By using the power series expansion of ϕ, he purports to show that if $\alpha < a < b < \beta$, then for each $\varepsilon > 0$ there exists a positive integer n such that, for $r = 1, \ldots, n$,

$$(\phi'(a_{r-1}) - \varepsilon)(a_r - a_{r-1}) < \phi(a_r) - \phi(a_{r-1}) < (\phi'(a_{r-1}) + \varepsilon)(a_r - a_{r-1}),$$

where $a_r = a + r(b - a)/n$ ($r = 0, 1, \ldots, n$). The left-hand inequalities give

$$\phi(b) - \phi(a) > \sum_{r=1}^{n} \phi'(a_{r-1})(a_r - a_{r-1}) - (b - a)\varepsilon, \tag{1}$$

and the proof is concluded with the remark that ε can be chosen so small that $(b - a)\varepsilon$ is less than the sum on the right.

It is in fact true that under Lagrange's hypotheses there exists an integer n with the specified property, but Lagrange completely fails to prove this. His last remark is, of course, nonsense, but since ε is arbitrary we can certainly conclude from (1) that $\phi(b) - \phi(a) \geq 0$, so that ϕ is increasing.

Given the result of (A), the rest of Lagrange's work on the remainder problem is substantially correct. The argument in his *Leçons* is particularly elegant,[†] and is briefly as follows (Lagrange, 1801, Leçon 9).

Let $[a, b]$ be an interval contained in the domain of ϕ, and for $n = 1$, $2, \ldots$ let m_n and M_n be the least and greatest values of $\phi^{(n)}$ on $[a, b]$. Since

$$\frac{d}{dt}(\phi(t) - \phi(a) - m_1(t - a)) = \phi'(t) - m_1 \geq 0,$$

the monotonicity theorem (with some goodwill over $>$ and $\geq$) shows that

$$\phi(b) - \phi(a) \geq m_1(b - a), \tag{2}$$

and similarly

$$\phi(b) - \phi(a) \leq M_1(b - a). \tag{3}$$

Next,

$$\frac{d}{dt}(\phi(t) - \phi(a) - \phi'(a)(t - a) - \tfrac{1}{2}m_2(t - a)^2) = \phi'(t) - \phi'(a) - m_2(t - a) \geq 0,$$

† The argument in his earlier book (1797, §§47, 49) is similar in principle.

by (2) applied to ϕ' on the interval $[a, t]$, and therefore, again by (A),

$$\phi(b) - \phi(a) - \phi'(a)(b - a) \geq \tfrac{1}{2} m_2 (b - a)^2.$$

A similar argument gives the reversed inequality with M_2 in place of m_2, and, by an obvious induction, we obtain the general result that

$$\frac{m_n}{n!}(b - a)^n \leq \phi(b) - \phi(a) - \phi'(a)(b - a) - \ldots - \frac{\phi^{n-1}(a)}{(n-1)!}(b - a)^{n-1}$$

$$\leq \frac{M_n}{n!}(b - a)^n.$$

Since $\phi^{(n)}$ takes in $[a, b]$ all values between m_n and M_n, it follows that there exists $\xi \in [a, b]$ for which

$$\phi(b) - \phi(a) - \phi'(a)(b - a) - \ldots - \frac{\phi^{(n-1)}(a)}{(n-1)!}(b - a)^{n-1} = \frac{\phi^{(n)}(\xi)}{n!}(b - a)^n,$$

and this is Lagrange's form of the remainder. In particular, the case $n = 1$ gives

$$\phi(b) - \phi(a) = \phi'(\xi)(b - a)$$

for some $\xi \in [a, b]$, and this is his form of the mean value theorem.

Lagrange seems to have regarded it as self-evident that the function $\phi^{(n)}$ has greatest and least values on $[a, b]$, and that $\phi^{(n)}$ attains all values between them. Lagrange was thus effectively appealing to the theorem that a continuous function on a bounded closed interval is bounded and attains its bounds, and to the intermediate value theorem, and in doing so he anticipated by many years the proper formulation and proof of these theorems.

The intermediate value theorem was the first of these two theorems to be obtained, by Bolzano (1817) and Cauchy (1821, Note III). Bolzano's proof is an obvious application of the following principle, which is essentially Dedekind's theorem. *Let $M(t)$ be a property, holding for some but not all real t, such that if $M(u)$ holds then $M(t)$ holds for all $t \leq u$. Then there exists a greatest U such that $M(t)$ holds for all $t < U$.*† To prove this principle, Bolzano chooses numbers u, D such that $M(t)$ holds for $t = u$ but not for $t = u + D$. If there is no $v > u$ such that $M(v)$ holds, then $U = u$; otherwise there exists $v > u$ such that $M(v)$ holds, and then there is a least positive integer m such that $M(u + D/2^m)$ holds (so that $M(t)$ does not hold for $t = u + D/2^{m-1}$). Similarly, either $U = u + D/2^m$, or there is a least positive integer n such that $M(u + D/2^m + D/2^{m+n})$ holds, and so on.

---

† This statement represents a fair reading of Bolzano's original *taken in conjunction with his proof*: his statement alone is a good deal less clear.

If this process does not terminate, then U is the sum of the series

$$u + D/2^m + D/2^{m+n} + \dots.$$

To show that this series converges, Bolzano uses the sufficiency of what is now known as the Cauchy convergence criterion.[†] He states this sufficiency and attempts to prove it, though without success since his treatment does not involve any study of the structure of the set of real numbers.

Cauchy's proof of the intermediate value theorem is more direct, and is astonishingly modern in tone; given a continuous $\phi : [a, b] \to \mathbf{R}$ with $\phi(a) > 0, \phi(b) < 0$, he constructs by repeated dissection a contracting sequence of subintervals $[a_n, b_n]$ with length tending to 0 such that $\phi(a_n) > 0, \phi(b_n) < 0$. To show that the sequences $(a_n), (b_n)$ converge, he appeals to the principle of monotone sequences, which he seems to have regarded as self-evident, and the proof is completed with the remark that, if c is the common limit of these sequences, then

$$f(c) = \lim_{n \to \infty} f(a_n) \geq 0 \quad \text{and} \quad f(c) = \lim_{n \to \infty} f(b_n) \leq 0,$$

whence $f(c) = 0$.[‡]

In contrast to the intermediate value theorem, the theorem on the bounds of a continuous function does not appear to have been explicitly formulated by the early nineteenth-century mathematicians, and according to Pringsheim (1916, p. 19) it was first obtained by Weierstrass, who gave it special emphasis in his lectures.[*]

Cauchy's treatment of the mean value theorem stemmed, not from Lagrange's work, but from a paper published in 1806 by Ampère, a pupil of Lagrange who later achieved fame for his researches in electricity and magnetism.[¶] In the first part of his paper Ampère attempted to justify the basic premise of Lagrange's theory by showing that every real-valued function has a derivative everywhere, an attempt which naturally met with no success. At the same time, Ampère attempted, again without success, to prove the following result, which implies Lagrange's inequalities (2) and (3).

---

† Cauchy stated the criterion in 1821.

‡ It has been suggested that Cauchy derived a number of his ideas from Bolzano's paper, but there seems to be little evidence for this suggestion in the two treatments of the intermediate value theorem. Cauchy's use of repeated dissection (he actually used a subdivision into m subintervals at each stage) has more in common with the work of Ampère (see below) than with that of Bolzano.

* Weierstrass began his lectures at Berlin in 1857.

¶ Ampère was a colleague of Cauchy's at l'École Royale Polytechnique, having been appointed to a chair there in 1809. Cauchy studied at l'École Royale Polytechnique and obtained his chair there in 1816.

(B) *Let $\phi:[a,b] \to \mathbf{R}$ be differentiable on $[a,b]$, and let $d = (\phi(b) - \phi(a))/$
$(b - a)$. Then there exist $\eta, \zeta \in [a, b]$ such that $\phi'(\eta) \leq d \leq \phi'(\zeta)$.*

Although Ampère's argument as a whole is nonsensical, the following result obtained in the course of his argument contains the germ of a successful proof, not only of (B) but also of the mean value theorem in the form (1.3.3).

(C) *Let $\phi:[a,b] \to \mathbf{R}$ be continuous, let $d = (\phi(b) - \phi(a))/(b - a)$, let m be a positive integer, and let $\delta = (b - a)/m$. Then there exists $\alpha \in [a, b - \delta]$ such that $(\phi(\alpha + \delta) - \phi(\alpha))/\delta = d$.*†

The special case of (C) where $\phi(a) = \phi(b)$ shows that if ϕ is a continuous function on an interval and its graph has a horizontal chord of length l, then the graph also has a horizontal chord of length l/m for each integer $m \geq 2$. This is the positive part of what is nowadays known as the *horizontal chord theorem*. The theorem also has a negative part which is rather striking, namely that the graph of ϕ does not necessarily have a horizontal chord of length l/k if k is not an integer. Both the positive and negative parts of the theorem were obtained by Lévy (1934), and he is usually credited with the entire result.‡

Ampère's proof of (C) is as follows (Ampère, 1806, pp. 151–4). It is easily verified that, if $a_r = a + r\delta$ $(r = 0, 1, \ldots, m)$, then

$$\min_{1 \leq r \leq m} (\phi(a_r) - \phi(a_{r-1}))/\delta \leq d \leq \max_{1 \leq r \leq m} (\phi(a_r) - \phi(a_{r-1}))/\delta.$$

Hence either the function $t \mapsto (\phi(t + \delta) - \phi(t))/\delta$ takes the value d at one of the points $a_0, \ldots, a_{m-1}$, or its values at these points include one greater than d and one less than d. The function therefore takes the value d at some point of $[a, b - \delta]$, and this is the required result.

In fact, more than this is true. If the function $t \mapsto (\phi(t + \delta) - \phi(t))/\delta$ takes a value greater than d at one of the points $a_0, \ldots, a_{m-1}$, it must also take a value less than d at one of these points, and similarly if it takes a value less than d it must take a value greater than d. Hence either it takes the value d at *each* of the points $a_0, \ldots, a_{m-1}$, or its values at these points include one greater than d and one less than d. In particular, this implies that if $m \geq 3$ the number α in (C) can be chosen to belong to the *open* interval $]a, b - \delta[$.

---

† Ampère does not formulate (C) as a separate proposition; moreover, he regards it as self-evident that if a (continuous) function on an interval takes a value greater than d and a value less than d then it takes the value d.

‡ See, for example, Boas (1960, pp. 77–80, 158). Lévy actually obtained a generalization of the positive part for a general plane curve.

Suppose now that ϕ, in addition to being continuous on $[a, b]$, has a derivative at each point of $]a, b[$. By repeated applications of this sharpened version of (C) with $m = 3$, we obtain a sequence of subintervals $[a_n, b_n]$ of $]a, b[$, each contained in the interior of its predecessor, such that $b_n - a_n = (b - a)/3^n$ and that

$$(\phi(b_n) - \phi(a_n))/(b_n - a_n) = d.$$

These subintervals shrink down to a point ξ, and clearly $a_n < \xi < b_n$ for all n. Applying now the result of Exercise 1.1.1, that

$$\phi'(\xi) = \lim_{\substack{s, t \to \xi \\ s < \xi < t}} \frac{\phi(t) - \phi(s)}{t - s}$$

(this result was first proved by Peano (1892a)), we deduce that $\phi'(\xi) = d$, and this is (1.3.3).

A proof of (1.3.3) on these lines, though rather more complicated and not employing the sharpened version of (C), was given by Kowalewski (1900).

Cauchy's first attack on the mean value theorem appears in his *Résumé des Leçons données à l'École Royale Polytechnique sur le Calcul Infinitésimal* (Cauchy, 1823). Cauchy's own extension of the mean value theorem was discovered after the *Résumé* was printed, and was inserted in an *Addition* to the *Résumé*, and in the main text (Leçon 7) he included only a version for one function, namely that if ϕ is continuously differentiable on $[a, b]$ then there exists $\xi \in [a, b]$ such that

$$\phi(b) - \phi(a) = (b - a)\phi'(\xi). \tag{4}$$

Like Lagrange, Cauchy deduced this result from a pair of increment inequalities by an application of the intermediate value theorem to ϕ'. The increment inequalities used by Cauchy are essentially the same as Lagrange's inequalities (2) and (3), and can be stated as follows.

(D) *If* $\phi : [a, b] \to \mathbf{R}$ *has a derivative at each point of* $[a, b]$ *and* $m \le \phi'(t) \le M$ *for all* $t \in [a, b]$, *then*

$$m(b - a) \le \phi(b) - \phi(a) \le M(b - a).\dagger \tag{5}$$

In a footnote to his discussion of (D), Cauchy remarks that 'On peut consulter sur ce sujet un Mémoire de M. Ampère', and his treatment follows that of Ampère in that he attempts to prove the result directly

---

† The hypotheses in (D) are those used by Cauchy in his arguments. In his formulation of the result he took m and M to be the least and greatest values of ϕ' on $[a, b]$ and tacitly assumed that they exist. Moreover, in applying the intermediate value theorem to ϕ' to deduce (4) from (5) he tacitly assumed that ϕ' is bounded and attains its bounds on $[a, b]$.

without using an analogue of Lagrange's monotonicity theorem (A) (such an analogue does in fact appear in the *Résumé* prior to the discussion of the mean value theorem–see (E) below).

Cauchy's proof of (D) is as follows. Given $\varepsilon > 0$ we can find $\delta > 0$ such that

$$\phi'(t) - \varepsilon < \frac{\phi(t + h) - \phi(t)}{h} < \phi'(t) + \varepsilon \tag{6}$$

whenever $0 < |h| < \delta$ and $t, t + h \in [a, b]$. Hence, if $a = a_0 < a_1 < \ldots < a_n = b$ and max $\{a_r - a_{r-1}\} < \delta$, we have

$$\phi'(a_{r-1}) - \varepsilon < \frac{\phi(a_r) - \phi(a_{r-1})}{a_r - a_{r-1}} < \phi'(a_{r-1}) + \varepsilon \qquad (r = 1, \ldots, n),$$

and therefore also

$$(m - \varepsilon)(a_r - a_{r-1}) < \phi(a_r) - \phi(a_{r-1}) < (M + \varepsilon)(a_r - a_{r-1}).$$

By addition we now have

$$(m - \varepsilon)(b - a) < \phi(b) - \phi(a) < (M + \varepsilon)(b - a),$$

and since ε is arbitrary this gives (5).

This proof is, unfortunately, invalid, since the first step in the argument requires the *uniform* convergence of the difference quotient $(\phi(t + h) - \phi(t))/h$ to $\phi'(t)$ on $[a, b]$, whereas the hypotheses imply only the pointwise convergence of the difference quotient. It is true that the uniformity of the convergence of the difference quotient is implied by the continuity of ϕ' (which Cauchy assumed for his deduction of (4)), but it is not possible to rescue the proof of (D) by inserting this additional hypothesis, since the proof that the continuity of ϕ' implies the uniformity requires at the least a result equivalent to (D).

Cauchy is not seriously to be blamed for his error, for at this period the notion of ordinary convergence had only just emerged with any clarity, and the definition of uniform convergence was still some twenty years in the future, awaiting the work of Weierstrass, Stokes and Seidel (see Grattan-Guinness, 1970, Ch. 6). Perhaps the most interesting point in Cauchy's argument is the clarity with which he expressed the uniformity in his assertion (6): 'Désignons par δ, ε deux nombres très-petits, le premier étant choisi de telle sorte que, pour des valeurs numériques de i inférieures à δ, et pour une valeur quelconque de t comprise entre les limites a, b, le rapport $(\phi(t + i) - \phi(t))/i$ reste toujours supérieur à $\phi'(t) - \varepsilon$ et inférieur à $\phi'(t) + \varepsilon$.'† Had Cauchy realized that this statement is not implied by the

† In this and the following quotation we have altered Cauchy's f and x to ϕ and t, and his x_0 and X to a and b.

pointwise convergence, he would have had here ready to hand the correct definition of uniform convergence.

We have already mentioned that Cauchy included in the *Résumé* (Leçon 6) a monotonicity theorem analogous to (A), viz.

(E) *If* $\phi:[a,b] \to \mathbf{R}$ *possesses a positive derivative at each point of* $[a,b]$, *then* ϕ *is strictly increasing.*

It is interesting to see how this fares in Cauchy's hands. His argument begins (correctly) with the observation that if $\phi'(t) > 0$ then $\phi(t_1) < \phi(t) < \phi(t_2)$ whenever $t_1 < t < t_2$ and t_1, t_2 are sufficiently close to t. From this he infers without further explanation that 'les différences infiniment petites $\Delta t, \Delta y$ étant de même signe, la fonction $y\ [=\phi(t)]$ croîtra ..., à partir de $t = a$, en même temps que la variable t'!

Cauchy's treatment of his generalized mean value theorem, given in the *Addition* to the *Résumé*, also depended on a pair of increment inequalities, which can be stated as:

(F) *If* ϕ, ψ *possess derivatives at each point of* $[a,b]$, *and* $\psi'(t) > 0$ *and* $m \le \phi'(t)/\psi'(t) \le M$ *for all* $t \in [a,b]$, *then*

$$m \le \frac{\phi(b) - \phi(a)}{\psi(b) - \psi(a)} \le M.$$

By applying the intermediate value theorem to ϕ'/ψ', he then deduced his generalized mean value relation

$$\frac{\phi(b) - \phi(a)}{\psi(b) - \psi(a)} = \frac{\phi'(\xi)}{\psi'(\xi)} \tag{7}$$

for some $\xi \in [a,b]$, subject to the continuity of ϕ' and ψ' and the non-vanishing of ψ' on $[a,b]$.

Cauchy remarked that (F) can be proved by arguments entirely similar to those in his (fallacious) proof of (D) given in the main text of the *Résumé*. He observed further that (F) can also be proved by the application of the monotonicity theorem (E) to the functions $\phi - m\psi$ and $M\psi - \phi$. This alternative proof of (F) is thus a straightforward generalization of Lagrange's proof of (2) and (3), and indeed both Cauchy's and Lagrange's arguments show the same disregard for the distinction between $>$ and $\ge$ in the application of their monotonicity theorems. It therefore seems likely that Cauchy owed his inspiration here to Lagrange's treatment of the remainder problem in the *Leçons sur le Calcul des Fonctions*.

In 1829 Cauchy published a revised version of the *Résumé* under the title *Leçons sur le Calcul Différentiel* (Cauchy, 1829), and in this the fallacious proof of (D) is suppressed, and only the generalization of Lagrange's argument is given (the monotonicity theorem (E) appears with the same incomplete proof as before). As in the *Résumé*, the mean value relations (4) and (7) require the continuity of the derivative(s) on $[a,b]$.

The standard modern proofs of the mean value relations given in §1.3, which require only the existence of the derivative(s) on the open interval $]a,b[$, first appear in print in 1868 in Serret's *Cours de Calcul Différentiel et Intégral* (Serret, 1868, vol. 1, pp. 17–23), and are attributed by Serret to Bonnet. They depend, of course, on Weierstrass's theorem on the bounds of a continuous function, and this was still comparatively recent when Bonnet obtained his proof.

Although these improved versions of the mean value theorems appeared in various books from 1868 onwards, they escaped the notice of several writers of this period who persisted with the treatment due to Lagrange and Cauchy. One of these writers was Jordan, who in the first edition of his *Cours d'Analyse*, published in 1882, based his treatment on Cauchy's theorem (D) (thus requiring the continuity of the derivative in the mean value theorem). More surprisingly, Jordan resurrected Cauchy's fallacious proof of (D), though the error was rather better concealed, in a passage (Jordan, 1882, p. 21) running as follows: '… donnons successivement à t une série de valeurs $a_1, \ldots, a_{n-1}$ intermédiaires entre a et $a + h$. Posons

$$\phi(a_r) - \phi(a_{r-1}) = (a_r - a_{r-1})[\phi'(a_{r-1}) + \varepsilon_r].$$

…. Supposons maintenant les valeurs intermédiaires $a_1, \ldots, a_{n-1}$ indéfiniment multipliées. Les quantités $\varepsilon_1, \varepsilon_2, \ldots$ tendront toutes vers zéro, car ε_1, par exemple, est le différence entre $(\phi(a_1) - \phi(a))/(a_1 - a)$ et sa limite $\phi'(a)$…'.

Jordan's error would not be worth mentioning had it not resulted in a curious theorem and some missed opportunities. The error was pointed out in a short note by Peano (1884a), which produced a rejoinder in defence of Jordan by Gilbert (1884). Gilbert observed that Jordan did not require $\varepsilon_1, \varepsilon_2, \ldots$ to tend to 0 for *all* choices of the a_r, but only for *some* choice. However, Gilbert gave no proof that such a choice can be made, and in a subsequent note Peano (1884b) commented 'le théorème résultera démontré lorsque M. Gilbert aura trouvé ce mode particulier de division, pour lequel la condition précédente est satisfaite. Et … ce mode existe, mais je laisserai le soin de la trouver à M. Gilbert …'. Peano then proceeded to state the following theorem, the proof of which he left as an exercise.

(G) *Let* $\phi:[a,b] \to \mathbf{R}$ *be differentiable on* $[a,b]$, *and let* $\varepsilon > 0$. *Then there exists a partition* $\{a_0, \ldots, a_n\}$ *of* $[a,b]$ *(cf. Exercise 1.7.5) such that, for* $r = 1, \ldots, n$,

$$\left| \frac{\phi(a_r) - \phi(a_{r-1})}{a_r - a_{r-1}} - \phi'(a_{r-1}) \right| \le \varepsilon.\dagger$$

At first sight, (G) enables us to rescue Jordan's (and therefore Cauchy's) proof of (D). However, the proof of (G) seems to require Bonnet's version of the mean value theorem, so that (G) provides a devious route to the goal of (D). In stating (G), Peano was, of course, attempting only to deal with Gilbert's observation concerning Jordan's proof, rather than to rescue the proof itself. But had Peano really wished to rescue the proof, then he might easily have been led to the following modification of (G), which also implies (D).

(H) *Let* $\phi:[a,b] \to \mathbf{R}$ *be differentiable on* $[a,b]$, *and let* $\varepsilon > 0$. *Then there exists a partition* $\{a_0, \ldots, a_n\}$ *of* $[a,b]$, *and numbers* ξ_r *equal to either* a_{r-1} *or* a_r $(r = 1, \ldots, n)$ *such that, for* $r = 1, \ldots, n$,

$$\left| \frac{\phi(a_r) - \phi(a_{r-1})}{a_r - a_{r-1}} - \phi'(\xi_r) \right| \le \varepsilon.$$

Although only slight alterations are required in the proof of (G) to give (H), there are significant differences in the two theorems, and in particular the proof of (H) does not involve Bonnet's version of the mean value theorem. Moreover, the argument obtained by modifying the proof of (G) can be recognized as being essentially the same as that used to prove the Heine–Borel theorem for a bounded closed interval in $\mathbf{R}$. It is an interesting thought that had Peano pursued the questions raised by his theorem (G) he might have anticipated Borel in the discovery of the covering theorem.‡

In the 1870s, a resurgence of interest in the subject of integration, following the work of Riemann and Darboux, led to the study of a new type of theorem in differentiation theory, which provided sufficient conditions for two continuous functions to differ by a constant. The

† Theorem (G) implies in particular that $\phi(b) - \phi(a)$ is the limit of a sequence of Riemann sums of ϕ' of the form $\sum_{r=1}^{n} \phi'(a_{r-1})(a_r - a_{r-1})$. For further information, see Exercise 1.11.1.

‡ The covering theorem was first stated by Borel in 1894 (see Zoretti, 1912, p. 128), though the idea is implicit in a paper of Heine of 1872 (see §2.14).

first result of this type to go beyond the immediate consequence of Bonnet's mean value theorem (cf. (1.3.5)) was the following result of Dini (1878, p. 72), which provides also the first application of countable exceptional sets in differentiation theory.

(I) *If* $\phi, \psi : [a, b] \to \mathbf{R}$ *are continuous, and* $\phi'(t), \psi'(t)$ *exist and are equal nearly everywhere, then* $\phi - \psi$ *is constant.*

Dini (1878, p. 190) also introduced the upper and lower derivatives which bear his name, and among other results he proved (1878, p. 193):

(J) *If* $\phi : [a, b] \to \mathbf{R}$ *is continuous and D denotes any one of* $D^{\pm}$ *and* $D_{\pm}$, *then*

$$(b - a) \inf D\phi \leq \phi(b) - \phi(a) \leq (b - a) \sup D\phi.$$

However, Dini failed to amalgamate his generalized derivatives with the idea of countable exceptional sets, and the first result combining the two ideas was obtained by Scheeffer (1884b, Satz IV), who proved:

(K) *If* $\phi, \psi : [a, b] \to \mathbf{R}$ *are continuous, and* $D\phi(t) = D\psi(t)$ *for nearly all* $t \in \,]a, b[$, *where D denotes a fixed Dini derivative, then* $\phi - \psi$ *is constant.*†

A new proof of Scheeffer's theorem was given by Lebesgue (1904, p. 78). Lebesgue's argument employs a covering theorem (proved by transfinite induction), which asserts that if with each $t \in [a, b[$ there is associated a subinterval $I_t = [t, t + h_t[$ of $[a, b[$, then there is a countable subset of mutually disjoint subintervals I_t whose union is $[a, b[$. Lebesgue also used this covering theorem to obtain an extension of Scheeffer's result in which the condition that $D\phi, D\psi$ are equal nearly everywhere is replaced by the condition that $D\phi, D\psi$ are bounded and are equal almost everywhere.

Lebesgue commented that his proof of Scheeffer's theorem is less artificial than that of Scheeffer, but Scheeffer's argument is both natural and brilliantly simple. What it essentially proves is the following result, first made explicit by de la Vallée Poussin (1909, vol. 1, pp. 79–80), which trivially implies the continuous case of the monotonicity theorem (1.4.3):

(L) *If* $\phi : [a, b] \to \mathbf{R}$ *is a continuous function such that* $\phi(a) > \phi(b)$, *then the set* $A = \{t \in [a, b[: D^{+}\phi(t) < 0\}$ *is uncountable.*

Scheeffer's argument, as presented by de la Vallée Poussin, is briefly as follows: we may obviously suppose that $\phi(a) > 0 > \phi(b)$. Let $I = \,]0, -\phi(b)/(b - a)\,[$, and for each $c \in I$ let t_c be the greatest element of $[a, b]$

---

† The notation $D^{\pm}, D_{\pm}$ for the Dini derivatives is also due to Scheeffer (1884a).

such that $\phi(t_c) + c(t_c - a) = 0$. Then $t_c \in]a, b[$ and $\phi(t) + c(t - a) < 0$ for $t_c < t \le b$, whence $D^+\phi(t_c) \le -c < 0$, so that $t_c \in A$. The function $c \mapsto t_c$ is therefore a one-to-one mapping of I into A, so that A is uncountable.

This argument appears in the text in a simplified form in the proof of (1.4.1), which is due to Zygmund, as is also the proof of (1.4.2) and the general case of (1.4.3) (see Saks, 1937, pp. 203–4).

Mean value inequalities and results for vector-valued functions of the type of (1.3.8) and (1.6.3–6) first appear in the guise of theorems for complex-valued functions. The prototype of (1.3.8), (1.6.3) and (1.6.6) is a theorem of Darboux (1876), namely:

(M) *If* $\phi : [a, b] \to \mathbf{C}$ *and* $\psi : [a, b] \to \mathbf{R}$ *possess continuous derivatives on* $[a, b]$ *and* $\psi'(t) \ne 0$ *for each* $t \in [a, b]$, *there exists* $\xi \in [a, b]$ *such that*

$$\left| \frac{\phi(b) - \phi(a)}{\psi(b) - \psi(a)} \right| \le \left| \frac{\phi'(\xi)}{\psi'(\xi)} \right|.$$

Darboux's proof is couched in geometrical–kinematical terms, but essentially consists of a reduction to the special case $\psi(t) = t$, as in the remark following (1.3.10. Corollary 1), this special case then being disposed of by means of the inequality

$$|\phi(b) - \phi(a)| = \left| \int_a^b \phi'(t) dt \right| \le (b - a) \max_{a \le \xi \le b} |\phi'(\xi)|.$$

The prototype of (1.6.4) and (1.6.5. Corollary) is due to Weierstrass, and was presumably obtained prior to the appearance of Darboux's theorem, though it was not published until some time between 1882 and 1891 in the *Cours* of Hermite.† Hermite's version of the theorem is as follows (Bieberbach (1934, p. 114) gives a version involving contour integrals).

(N) *If* $f : [a, b] \to \mathbf{C}$ *and* $g : [a, b] \to [0, \infty[$ *are continuous, and* E *is a closed convex set in* $\mathbf{C}$ *containing the range of* f, *then*

$$\int_a^b f(t) g(t) dt \bigg/ \int_a^b g(t) dt \in E.$$

In fact, the ratio of the two integrals is the limit of ratios of Riemann

---

† The theorem appears in the 3rd edition of Hermite's *Cours* (1891, p. 58), which was '*rédigé*' in 1882. It is not in the first edition of 1873.

sums of the form

$$\sum f(a_r)g(a_r)(a_r - a_{r-1})/\sum g(a_r)(a_r - a_{r-1}),$$

and obviously these ratios of Riemann sums belong to E.

From (N) we easily deduce the following result, which corresponds more closely to (1.6.4).

(O) *If $\phi:[a,b] \to C$ and $\psi:[a,b] \to R$ possess continuous derivatives on $[a,b]$, $\psi'(t) > 0$ for all $t \in [a,b]$, and E is a closed convex set in C containing the range of ϕ'/ψ', then $(\phi(b) - \phi(a))/(\psi(b) - \psi(a)) \in E$.*

Various attempts were made during the nineteenth century to create a differential calculus for vector-valued functions, for instance by Grassmann in his *Geometrische Analyse* and the second edition of his *Die Ausdehnungslehre* (Grassmann, 1847, 1862). However, most of these early attempts were in the spirit of the eighteenth-century analysts, and it is not until the work of Peano on ordinary differential equations (1887a, 1890) and on geometric calculus (1888, 1891a, 1896) that we find a theory which meets present-day standards of rigour. Peano's first publication in the field of geometric calculus, *Calcolo geometrico secondo l'Ausdehnungslehre di H. Grassmann* (Peano, 1888),[†] contains the following generalization of Weierstrass's result in which the assumption of the continuity of the derivative is removed.

(P) *If $\phi:[a,b] \to R^3$ (or R^2 or R) is a continuous function possessing a derivative at each point of $]a,b[$, and E is a closed convex set containing the range of ϕ', then $(\phi(b) - \phi(a))/(b - a) \in E$.*

As with many of the results given by Peano in his book, this theorem was stated without proof, but in a later paper Peano (1896, p. 186) gave a proof in which he reduced the theorem to the real case by considering the signed volume of the tetrahedron whose vertices are the point $\phi(t)$ and three arbitrary points P, Q, R (for R^2 a triangle is used). A much more powerful argument would have been available to Peano had he pursued the questions raised by his theorem (G) and obtained the result of (H), for (H), unlike (G), continues to hold when ϕ takes its values in any normed space Y and $|\;\;|$ is the norm in Y. For such a ϕ we can therefore express the difference quotient $d = (f(b) - f(a))/(b - a)$ as the limit (in the norm topology) of a sequence of sums of the form

$$\frac{1}{b - a} \sum_{r=1}^{n} \phi'(\xi_r)(a_r - a_{r-1}),$$

---

† In spite of its title, this was in reality a completely new work in both form and substance.

where ξ_r is either a_r or a_{r-1} $(r = 1, \ldots, n)$. Since each sum clearly belongs to any closed convex set E in Y containing the range of ϕ', we deduce that $d \in E$. Applying now a simple continuity argument to pass from the case where ϕ' exists on $[a, b]$ to the case where ϕ' exists only on $]a, b[$, we thus obtain the result of (P) with Y in place of $\mathbf{R}^3$.†

Peano's generalization (P) of Weierstrass's theorem (O) seems to have remained effectively unknown, and the further extension of (P) to either $\mathbf{R}^n$ for $n \geq 4$ or to a normed space Y was curiously long-delayed. As far as I know, the first such extension is a result of Ważewski (1949, 1951b), which contains (1.6.4) as a special case (see the notes on this chapter).

An extension of Darboux's theorem (M) to functions with values in $\mathbf{R}^n$ was obtained by Mie (1893) in a paper on differential equations, namely:

(Q) *If $\phi : [a, b] \to \mathbf{R}^n$ is differentiable on $[a, b]$, there exists $\xi \in [a, b]$ such that $\| \phi(b) - \phi(a) \| \leq (b - a) \| \phi'(\xi) \|$.*

Mie's proof, which uses a repeated bisection argument and which applies equally to functions with values in a general normed space, appears also to have remained unknown, and it was rediscovered by Halperin (1954). The further extension to the case where ϕ' exists only on $]a, b[$ was made by Aziz and Diaz (1963a), using an argument essentially similar to that used above to deduce (1.3.3) from Ampère's theorem (C) (for further references see the notes on this chapter).

Exercises 1.11

1 Let $\phi : [a, b] \to \mathbf{R}$ be a function possessing a derivative on $[a, b]$, and let $\varepsilon > 0$. Prove that there exist a positive integer n and points $a_0, \ldots, a_n$ satisfying $a = a_0 < a_1 < \ldots < a_n = b$ such that for $r = 1, \ldots, n$

$$\left| \frac{\phi(a_r) - \phi(a_{r-1})}{a_r - a_{r-1}} - \phi'(a_{r-1}) \right| \leq \varepsilon.$$

[This result implies in particular that $\phi(b) - \phi(a)$ is the limit of a sequence of Riemann sums of ϕ' of the form $\sum_{r=1}^n \phi'(a_{r-1})(a_r - a_{r-1})$. The result is false if the hypothesis that $\phi'(t)$ exists for all $t \in [a, b]$ is replaced by the hypothesis that ϕ is continuous, and $\phi'(t)$ exists for all $t \in [a, b[$ (take, for example, $a = 0$, $b = 1$, $\phi(t) = (1 - t)^{1/2}$). The result also fails for complex-valued or vector-valued functions (take, for example, $a = -1, b = 0, \phi(t) = t^2 e^{i/t}$ $(-1 \leq t < 0)$, $\phi(0) = 0$).]

2 Let Y be a normed space, let $\phi : [a, b] \to Y$ be a function possessing a derivative on $[a, b]$, and let $\varepsilon > 0$. Prove that there exist a partition $\{a_0, \ldots, a_n\}$ of $[a, b]$, and

† A further extension for topological linear spaces can also be obtained by this method.

numbers ξ_r equal to either a_{r-1} or a_r $(r = 1,\ldots,n)$, such that for $r = 1,\ldots,n$

$$\left|\frac{\phi(a_r) - \phi(a_{r-1})}{a_r - a_{r-1}} - \phi'(\xi_r)\right| \leq \varepsilon.$$

Deduce that if the range of ϕ' is contained in a closed convex set E in Y, then $(\phi(b) - \phi(a))/(b - a) \in E$.

Extend this last result to the case where $\phi : [a,b] \to Y$ is a continuous function such that $\phi'(t)$ exists and belongs to E for all $t \in \,]a,b[$.

[The last part of this exercise gives another form of the 'mean value' theorem for vector-valued functions; generalizations can be found in §2.6.

Hint. For the first part use the Heine–Borel theorem.]

2

Ordinary differential equations

Throughout this chapter Y denotes a Banach space, either real or complex. We consider functions of a real variable with values in Y, and we use I and J to denote intervals in $\mathbf{R}$; usually J will be a compact interval. When Y is over the complex field, the space $\mathbf{R} \times Y$ will be regarded as the product of the normed space $\mathbf{R}$ with the real Banach space underlying Y; when Y is real, then $\mathbf{R} \times Y$ has, of course, its usual meaning.

For any compact metric space X we use $C(X, Y)$ to denote the vector space of continuous functions from X into Y with the sup norm.

2.1 Definitions

The term 'ordinary differential equation of the nth order' is normally used to mean a formal relation of the type

$$F(t, y, y', \ldots, y^{(n)}) = 0 \tag{1}$$

involving an 'unknown' function of a real variable and its first n derivatives. If the 'unknown' function is supposed to take its values in the Banach space Y, then the domain of the function F must be a subset of $\mathbf{R} \times Y^{n+1}$, and in most cases F takes its values in Y. To make this notion precise, it is easier–and more useful–to define what is meant by 'a solution of the differential equation (1)' rather than to define the term 'differential equation' itself.

Thus let n be a positive integer, let $E \subseteq \mathbf{R} \times Y^{n+1}$, and let $F : E \to Y$ be a given function. We say that a function ϕ *is a solution of the differential equation* (1) *on an interval* I if ϕ is an n-times differentiable function from I into Y such that for all $t \in I$

$$(t, \phi(t), \phi'(t), \ldots, \phi^{(n)}(t)) \in E$$

and

$$F(t, \phi(t), \phi'(t), \ldots, \phi^{(n)}(t)) = 0.$$

(A solution of a differential equation is sometimes referred to as an 'exact' solution, in contrast to an approximate solution (see §2.3).)

Although we do not attempt to define the term 'differential equation' by itself, we nevertheless find it convenient to speak of the formal expression (1) as a 'differential equation of the nth order'. When $Y = \mathbf{R}$, the differential equation is said to be *scalar*.

We shall be concerned particularly with a differential equation of the first order of the form

$$y' = f(t, y), \qquad (2)$$

where f is a function from a set $E \subseteq \mathbf{R} \times Y$ into Y. As in §1.5, a function ϕ is a solution of (2) on I if ϕ is a continuous function from I into Y such that, for each $t \in I$, the point $(t, \phi(t)) \in E$ (i.e. the graph of ϕ lies in E) and the derivative $\phi'(t)$ exists and satisfies

$$\phi'(t) = f(t, \phi(t)).$$

We usually consider the solutions ϕ of the equation (2) that satisfy an initial condition of the form $\phi(t_0) = y_0$, where (t_0, y_0) is a given point of E. The most important case is that in which f is continuous, and in this case $t \mapsto f(t, \phi(t))$ is continuous on I, so that also ϕ' is continuous on I.

When Y is the product $Y_1 \times Y_2 \times \ldots \times Y_n$, where $Y_1, \ldots, Y_n$ are Banach spaces, we can express f in terms of its components $f_i : E \to Y_i$ $(i = 1, \ldots, n)$, i.e. for all $(t, y) \in E$

$$f(t, y) = (f_1(t, y), \ldots, f_n(t, y)).$$

The equation (2) is then equivalent to the system of n simultaneous equations

$$y'_i = f_i(t, y_1, \ldots, y_n) \qquad (i = 1, \ldots, n). \qquad (3)$$

More precisely, if ϕ is a solution of (2) on I, and ϕ has components $\phi_i : I \to Y_i$ $(i = 1, \ldots, n)$, i.e. for each $t \in I$

$$\phi(t) = (\phi_1(t), \ldots, \phi_n(t)),$$

then for all $t \in I$

$$(t, \phi_1(t), \ldots, \phi_n(t)) \in E \qquad (4)$$

and the derivative $\phi'_i(t)$ exists and satisfies

$$\phi'_i(t) = f_i(t, \phi_1(t), \ldots, \phi_n(t)) \qquad (i = 1, \ldots, n). \qquad (5)$$

Conversely, if (4) and (5) hold for all $t \in I$, then the function $\phi : I \to Y$ with components $\phi_1, \ldots, \phi_n$ is a solution of (2) on I. Further, in this case an initial condition of the form $\phi(t_0) = y_0$ is equivalent to the n initial conditions $\phi_i(t_0) = y_{i0}$ $(i = 1, \ldots, n)$, where $t_0 \in I$ and $y_0 = (y_{10}, \ldots, y_{n0}) \in Y$.

In particular, if $Y_i = \mathbf{R}$ for each i (so that $Y = \mathbf{R}^n$), then the f_i and ϕ_i are real-valued functions. Thus a system of n simultaneous differential equations in n real 'unknown' functions can be reduced to a single vector differential equation in one vector-valued 'unknown' function.

Another important case which is covered by the differential equation (2) is an equation of the nth order of the form

$$y^{(n)} = G(t, y, y', \dots, y^{(n-1)}), \tag{6}$$

where G is a function from a set $E \subseteq \mathbf{R} \times Y^n$ into Y. In fact, this equation is equivalent to the system of n equations

$$\left. \begin{aligned} z_1' &= z_2, \\ z_2' &= z_3, \\ &\cdots\cdots \\ z_{n-1}' &= z_n, \\ z_n' &= G(t, z_1, z_2, \dots, z_n). \end{aligned} \right\} \tag{7}$$

For, if ϕ is a solution of (6) on I, and

$$\psi_1 = \phi, \psi_2 = \phi', \dots, \psi_n = \phi^{(n-1)},$$

then $\psi_1, \dots, \psi_n$ are functions from I into Y that satisfy the system (7) on I. Conversely, if $\psi_1, \dots, \psi_n$ are functions from I into Y that satisfy (7) on I, then ψ_1 is n-times differentiable and is a solution of (6) on I, and $\psi_2, \dots, \psi_n$ are $\psi_1', \dots, \psi_1^{(n-1)}$, respectively. If we now set $z = (z_1, \dots, z_n)$ and define $g : E \to Y^n$ by

$$g(t, z) = g(t, z_1, \dots, z_n) = (z_2, z_3, \dots, z_n, G(t, z_1, \dots, z_n)),$$

then the system (7) can be written as

$$z' = g(t, z), \tag{8}$$

i.e. it is of the form (2) with Y^n in place of Y. Further, an initial condition of the form $\psi(t_0) = z_0$ for (8) is equivalent to the n initial conditions $\psi_i(t_0) = z_{i0}$ for (7), and therefore to the n initial conditions

$$\phi(t_0) = z_{10}, \phi'(t_0) = z_{20}, \dots, \phi^{(n-1)}(t_0) = z_{n0}$$

for (6). Moreover, g is continuous if and only if G is continuous, and then any solution ϕ of (6) is necessarily C^n (i.e. has continuous derivatives of all orders up to and including n).

Example (a)

Consider the system of differential equations

$$y_i' = \sum_{j=1}^n a_{ij}(t)y_j + b_i(t) \qquad (i = 1, \dots, n), \tag{9}$$

where a_{ij} and b_i $(i, j = 1, \dots, n)$ are real-valued functions continuous on an interval I. Let $b : I \to \mathbf{R}^n$ be the function with components $b_1, \dots, b_n$, and for each $t \in I$ let $A(t)$ be the linear function from $\mathbf{R}^n$ into itself whose value at the point y is $z = A(t)y$, where

$$z_i = \sum_{j=1}^{n} a_{ij}(t)y_j \qquad (i = 1, \ldots, n).$$

Then the system of equations (9), subject to the initial conditions $y_i(t_0) = y_{i0}$ $(i = 1, \ldots, n)$ is clearly equivalent to the equation

$$y' = A(t)y + b(t) \tag{10}$$

subject to the initial condition $y(t_0) = (y_{10}, \ldots, y_{n0})$.

In making these definitions, we have in effect produced a function A from the interval I into the space $\mathscr{L}(\mathbf{R}^n; \mathbf{R}^n)$, and this function A is continuous with respect to the usual norm topology in $\mathscr{L}(\mathbf{R}^n; \mathbf{R}^n)$, for if $t, t' \in I$ then

$$\| A(t) - A(t') \| \leq \left(\sum_{i,j} (a_{ij}(t) - a_{ij}(t'))^2 \right)^{1/2}.$$

We therefore obtain a natural generalization of the equation (10), in which the unknown function y now takes its values in a general normed space Y, by defining A to be a continuous function from the interval I into the space $\mathscr{L}(Y; Y)$ of continuous linear maps of Y into itself, and b to be a continuous function from I into Y. An equation

$$y' = A(t)y + b(t)$$

of this more general form is called a *linear equation of the first order*, and we return to such equations in §§2.7–9.

We remark that in the finite-dimensional case it is often helpful to regard the equation (10) as a matrix equation, where $A(t)$ denotes the $n \times n$ matrix $[a_{ij}(t)]$ and $y, y', b(t)$ denote column matrices whose entries are the components of $y(t), y'(t), b(t)$, and $A(t)y$ is the matrix product.

Example (b)

Consider the nth-order differential equation

$$y^{(n)} + a_1(t)y^{(n-1)} + a_2(t)y^{(n-2)} + \ldots + a_n(t)y = q(t), \tag{11}$$

where $a_1, \ldots, a_n, q$ are continuous real-valued functions on an interval I. From (6) and (7), we see that this equation is equivalent to the linear system of equations

$$\left. \begin{aligned} z_1' &= z_2, \\ z_2' &= z_3, \\ &\cdots\cdots \\ z_{n-1}' &= z_n, \\ z_n' &= -a_n(t)z_1 - a_{n-1}(t)z_2 - \ldots - a_1(t)z_n + q(t). \end{aligned} \right\} \tag{12}$$

More precisely, if y is a solution of (11) on I, and

$$z_1 = y, z_2 = y', \ldots, z_n = y^{(n)},$$

then $z_1, \ldots, z_n$ are solutions of (12) on I, and conversely, if $z_1, \ldots, z_n$ are solutions of (12) on I, then $y = z_1$ is a solution of (11) on I and $z_2, \ldots, z_n$ are $y', \ldots, y^{(n-1)}$ respectively.

Now let $A(t)$ be the linear map of $\mathbf{R}^n$ into itself whose matrix with respect to the natural basis in $\mathbf{R}^n$ is

$$\begin{bmatrix} 0 & 1 & 0 & \cdots & 0 & 0 \\ 0 & 0 & 1 & \cdots & 0 & 0 \\ \cdots & \cdots & \cdots & \cdots & \cdots & \cdots \\ 0 & 0 & 0 & \cdots & 0 & 1 \\ -a_n(t) & -a_{n-1}(t) & -a_{n-2}(t) & \cdots & -a_2(t) & -a_1(t) \end{bmatrix}$$

and let $b : I \to \mathbf{R}^n$ be the function with components $0, \ldots, 0, q$. Then the equation (11) is equivalent to the linear equation

$$z' = A(t)z + b(t), \tag{13}$$

in the sense that if y is a solution of (11) on I, then the function from I into $\mathbf{R}^n$ with components $y, y', \ldots, y^{(n-1)}$ is a solution of (13) on I, and if z is a solution of (13) on I, then the first component y of z is a solution of (11) on I and the remaining components are $y', \ldots, y^{(n-1)}$.

An equation of the form (11) is called a *linear equation of the nth order*. More generally, this name is applied to an equation of the form

$$y^{(n)} + A_1(t)y^{(n-1)} + \ldots + A_n(t)y = q(t)$$

where the unknown function y takes its values in a normed space Y, q is a continuous function from an interval I into Y, and $A_1, \ldots, A_n$ are continuous functions from I into $\mathscr{L}(Y; Y)$.

2.2 Preliminary results

We collect here two simple results that will be used frequently in later sections.

In the first lemma, which is fundamental to our treatment, the two relations expressed by the differential equation $y' = f(t, y)$ and the initial condition that $y = y_0$ when $t = t_0$ are embodied in a single integral equation. Moreover, the conditions of the lemma require only the continuity of the function ϕ, and not its differentiability.

(2.2.1) **Lemma.** *Let $E \subseteq \mathbf{R} \times Y$, let $f : E \to Y$ be continuous, let $(t_0, y_0) \in E$, and let I be an interval in $\mathbf{R}$ containing t_0. Then a function $\phi : I \to Y$ is a*

solution of the differential equation $y' = f(t, y)$ on I satisfying the initial
condition $\phi(t_0) = y_0$ if and only if
  (i) ϕ is continuous,
  (ii) $(t, \phi(t)) \in E$ for all $t \in I$,

(iii) $$\phi(t) = y_0 + \int_{t_0}^{t} f(u, \phi(u)) du \qquad (t \in I).$$

If ϕ is a solution of the differential equation on I such that $\phi(t_0) = y_0$,
then (i) and (ii) are obviously satisfied, and (iii) follows by integration of
the relation $\phi'(t) = f(t, \phi(t))$. Conversely, if (i) and (ii) hold, then
$t \mapsto f(t, \phi(t))$ is continuous on I, and therefore for all $t \in I$

$$\frac{d}{dt} \left(\int_{t_0}^{t} f(u, \phi(u)) du \right) = f(t, \phi(t)),$$

by (1.9.4). Hence, if also (iii) holds, then $\phi'(t) = f(t, \phi(t))$.

It is an immediate consequence of (2.2.1) that if f is continuous and
bounded on E, say, $\| f(t, y) \| \leq M$ for all $(t, y) \in E$, and ϕ is a solution of
the differential equation $y' = f(t, y)$ on an interval I satisfying the condition
$\phi(t_0) = y_0$, then for all $s, t \in I$

$$\| \phi(s) - \phi(t) \| \leq M | s - t | \qquad (1)$$

and

$$\| \phi(t) \| \leq \| y_0 \| + M | t - t_0 |. \qquad (2)$$

Alternatively, this follows directly from the increment inequality (1.6.2.
Corollary 1) (and then the condition that f is continuous can be omitted).
  In view of lemma (2.2.1) (iii), the next result is of obvious application.

(2.2.2) **Lemma.** *Let $E \subseteq \mathbf{R} \times Y$, let $f : E \to Y$ be continuous, and let (ϕ_n)
be a sequence of continuous functions from a compact interval J in $\mathbf{R}$ into Y
converging uniformly on J to a (continuous) function ϕ, such that the graphs
of ϕ_n $(n = 1, 2, \ldots)$ and of ϕ lie in E. Then $f(t, \phi_n(t)) \to f(t, \phi(t))$ as $n \to \infty$,
uniformly for t in J.*

If this is false, we can find a positive number ε, a strictly increasing
sequence of positive integers (n_r), and a sequence (t_r) of points of J, such
that for all r

$$\| f(t_r, \phi_{n_r}(t_r)) - f(t_r, \phi(t_r)) \| > \varepsilon. \qquad (3)$$

Since J is compact, we can find a subsequence of (t_r) converging to a point
$t \in J$, and we now replace (t_r) by this subsequence. Since (ϕ_{n_r}) converges
uniformly to ϕ, and

$$\| \phi_{n_r}(t_r) - \phi(t) \| \leq \| \phi_{n_r}(t_r) - \phi(t_r) \| + \| \phi(t_r) - \phi(t) \|,$$

we see that $\phi_{n_r}(t_r) \to \phi(t)$ as $r \to \infty$, and since f is continuous at $(t, \phi(t))$, it follows that $f(t_r, \phi_{n_r}(t_r))$ and $f(t_r, \phi(t_r))$ both tend to $f(t, \phi(t))$ as $r \to \infty$. This contradicts (3) and proves the statement.

If f is continuous on a set $E \subseteq \mathbf{R} \times Y$, and J is a compact interval in $\mathbf{R}$, we can regard the set of all solutions on J of the differential equation $y' = f(t, y)$ as a subset Δ of $C(J, Y)$ (possibly empty). Since f is continuous, any such solution on J is necessarily continuously differentiable, and we can therefore regard Δ also as a subset of $C^1(J, Y)$. Lemma (2.2.2) shows that if $\phi_n, \phi \in \Delta$ and $\phi_n \to \phi$ in $C(J, Y)$, then also $\phi'_n \to \phi'$ uniformly on J, so that $\phi_n \to \phi$ in $C^1(J, Y)$. In other words, the identity map $\phi \mapsto \phi$ from Δ equipped with the metric of $C(J, Y)$ onto Δ equipped with the metric of $C^1(J, Y)$ is continuous. Since the inverse of this map is trivially continuous, it follows that the map is a homeomorphism. Certain results concerning Δ as a subspace of $C^1(J, Y)$ can therefore be inferred directly from the corresponding results concerning Δ as a subspace of $C(J, Y)$, and in the sequel we work consistently in $C(J, Y)$ and take such extensions to $C^1(J, Y)$ for granted.

2.3 Approximate solutions

Let $E \subseteq \mathbf{R} \times Y$, let $f : E \to Y$ be continuous, let $\varepsilon \geq 0$, and let I be an interval in $\mathbf{R}$. We say that a function $\psi : I \to Y$ is an *ε-approximate solution* of the equation

$$y' = f(t, y) \tag{1}$$

on I if

(i) ψ is continuous on I,

(ii) $(t, \psi(t)) \in E$ for all $t \in I$,

(iii)   there exists a finite subset H of I such that, for all $t \in I \backslash H, \psi'(t)$ exists and satisfies

$$\| \psi'(t) - f(t, \psi(t)) \| \leq \varepsilon.$$

Thus an ε-approximate solution of the differential equation is a function that approximately satisfies the equation. If $\varepsilon = 0, \psi$ is an *exact solution*.

If $g : E \to Y$ is a function such that $\| g(t, y) - f(t, y) \| \leq \varepsilon$ for all $(t, y) \in E$, and ψ is an exact solution of the equation $y' = g(t, y)$ on I, then ψ is an ε-approximate solution of the equation (1) on I. In fact

$$\| \psi'(t) - f(t, \psi(t)) \| = \| g(t, \psi(t)) - f(t, \psi(t)) \| \leq \varepsilon$$

for all $t \in I$.

We observe that, if ψ is an ε-approximate solution of the equation (1) on I, and $t_0, t \in I$, then, by the increment inequality (1.6.2. Corollary 1),

$$\left\| \psi(t) - \psi(t_0) - \int_{t_0}^{t} f(u, \psi(u)) du \right\| \leq \varepsilon |t - t_0|. \tag{2}$$

We prove now the local existence of approximate solutions of the equation (1). For simplicity, we state and prove the result only for solutions to the right of the initial point t_0.

(2.3.1) *Let $(t_0, y_0) \in \mathbf{R} \times Y$, let B be the closed ball in Y with centre y_0 and radius $\rho > 0$, and let $C = [t_0, t_0 + \alpha] \times B$ where $\alpha > 0$. Let also $f : C \to Y$ be continuous and bounded, let $M = \sup\limits_C \| f(t, y) \|$, and let $J = [t_0, t_0 + \eta]$, where $\eta = \min \{ \alpha, \rho / M \}$. Then for each $\varepsilon > 0$ the equation*

$$y' = f(t, y) \tag{3}$$

has an ε-approximate solution ψ on J such that $\psi(t_0) = y_0$. Moreover, ψ can be chosen so that for all $s, t \in J$

$$\| \psi(s) - \psi(t) \| \leq M |s - t|. \tag{4}$$

The function ψ that we construct has a graph consisting of a finite number of segments, with successive vertices at the points $t_0, t_1, \ldots,$ $t_N = t_0 + \eta$, and, for $t \in [t_n, t_{n+1}]$, ψ is given by

$$\psi(t) = \psi(t_n) + (t - t_n) f(t_n, \psi(t_n)). \tag{5}$$

Thus the slope of ψ over this interval $[t_n, t_{n+1}]$ is equal to the slope at t_n of any solution of (3) having the same value at t_n as ψ. The points t_n are defined inductively to achieve the desired degree of approximation. It is obvious that a function ψ of this form necessarily satisfies (4); it is also obvious that if a function ψ satisfies (4) on J, then $\psi(t) \in B$ for all $t \in J$.

There is a simple proof when Y is finite-dimensional, for in this case the cylinder C is compact, and therefore f is uniformly continuous on C. Thus, given $\varepsilon > 0$ we can find $\delta > 0$ such that

$$\| f(t, y) - f(t', y') \| \leq \varepsilon \tag{6}$$

whenever $(t, y), (t', y')$ are points of E such that $|t - t'| \leq \delta, \| y - y' \| \leq \delta$. Divide J into N subintervals of equal length by points $t_0 < t_1 < \ldots < t_N = t_0 + \eta$, where N is chosen so that $t_{n+1} - t_n \leq \min \{ \delta, \delta / M \}$ for each n. On the interval $[t_0, t_1]$, ψ is defined by the formula

$$\psi(t) = y_0 + (t - t_0) f(t_0, y_0)$$

(in particular, this implies that $\psi(t_1) \in B$). We now proceed inductively; given ψ defined on $[t_0, t_n]$ with $\psi(t_n)$ in B, we define ψ on $]t_n, t_{n+1}]$ by (5). Then for all $t \in]t_n, t_{n+1}[$ we have both $0 < t - t_n < \delta$ and $\| \psi(t) - \psi(t_n) \| \leq M(t - t_n) < \delta$, and therefore, by (6),

$$\| \psi'(t) - f(t, \psi(t)) \| = \| f(t_n, \psi(t_n)) - f(t, \psi(t)) \| \leq \varepsilon,$$

as required.

To prove the theorem in the general case, where Y may be infinite-dimensional, we have to select the points t_n with more care.

Again let $\varepsilon > 0$, and consider first the function $\zeta_0 : J \to \mathbf{R}$ given by

$$\zeta_0(t) = \| f(t_0, y_0) - f(t, y_0 + (t - t_0)f(t_0, y_0)) \|.$$

This function ζ_0 is continuous on J and $\zeta_0(t_0) = 0$, whence there is a greatest $t_1 \in J$ such that $\zeta_0(t) \le \varepsilon$ for all $t \in [t_0, t_1]$. With this t_1 we define ψ on $[t_0, t_1]$ by

$$\psi(t) = y_0 + (t - t_0)f(t_0, y_0),$$

so that $\psi(t_1) \in B$, and, for $t_0 < t < t_1$

$$\| \psi'(t) - f(t, \psi(t)) \| \le \varepsilon. \tag{7}$$

If $t_1 = t_0 + \eta$, then ψ is the required function. If $t_1 < t_0 + \eta$, we proceed inductively to define $t_2, \dots$. Given ψ defined on $[t_0, t_n]$ with $\psi(t_n) \in B$, and satisfying (7) for all $t \in [t_0, t_n]$ except $t_0, \dots, t_n$, we choose t_{n+1} to be the greatest element of $[t_n, t_0 + \eta]$ such that

$$\| f(t_n, y_n) - f(t, y_n + (t - t_n)f(t_n, y_n)) \| \le \varepsilon$$

for all $t \in [t_n, t_{n+1}]$, where $y_n = \psi(t_n)$; we then define ψ on $]t_n, t_{n+1}]$ by

$$\psi(t) = y_n + (t - t_n)f(t_n, y_n).$$

It is immediate that if $t_0 \le t \le s \le t_n$ $(n = 1, 2, \dots)$, then

$$\| \psi(s) - \psi(t) \| \le M(s - t). \tag{8}$$

We have now to show that there exists N for which $t_N = t_0 + \eta$. Suppose on the contrary that the process above does not terminate. Then the numbers t_n form a strictly increasing sequence, which converges to a point $\tau \in J$. Clearly (8) holds for $t_0 \le t \le s < \tau$, and this implies that $\psi(s) \in B$ for $t_0 \le s < \tau$ and that the points $y_n = \psi(t_n)$ form a Cauchy sequence in B. Hence this sequence (y_n) converges to a point $c \in B$ for which

$$\| y_n - c \| \le M(\tau - t_n).$$

Since f is continuous at (τ, c), we can find $\delta > 0$ such that

$$\| f(t, y) - f(\tau, c) \| \le \tfrac{1}{2}\varepsilon \tag{9}$$

whenever $(t, y) \in C$ and $|t - \tau| \le \delta$, $\| y - c \| \le \delta$. If now n is chosen so that $\tau - t_n \le \tfrac{1}{2} \min \{\delta, \delta/M\}$, then $\| y_n - c \| < \delta$ and, for $t_n \le t < \tau$,

$$\| y_n - (t - t_n)f(t_n, y_n) - c \| \le \| y_n - c \| + M(t - t_n) \le 2M(\tau - t_n) \le \delta.$$

It therefore follows from (9) that for $t_n \le t < \tau$

$$\| f(t_n, y_n) - f(t, y_n + (t - t_n)f(t_n, y_n)) \| \le \varepsilon.$$

But this implies that $t_{n+1} \ge \tau$, and gives the required contradiction.

(2.3.2) *Let* $E \subseteq \mathbf{R} \times Y$, *let* J *be a compact interval in* $\mathbf{R}$, *and suppose that*
(i) $f : E \to Y$ *is continuous.*

(ii) (ϕ_n) *is a sequence of continuous functions from J into Y converging uniformly on J to a (continuous) function* ϕ,

(iii) *the graphs of* ϕ_n $(n = 1, 2, ...)$ *and of* ϕ *lie in E*,

(iv) ϕ_n *is an* ε_n-*approximate solution of the equation*

$$y' = f(t, y) \tag{10}$$

on J, where $\varepsilon_n \to 0$ *as* $n \to \infty$.

Then ϕ *is an exact solution of* (10) *on J.*

By (2), for all n and all $t, t_0 \in J$ we have

$$\left\| \phi_n(t) - \phi_n(t_0) - \int_{t_0}^t f(u, \phi_n(u)) du \right\| \leq \varepsilon_n |t - t_0|. \tag{11}$$

By (2.2.2), $f(u, \phi_n(u)) \to f(u, \phi(u))$ as $n \to \infty$, uniformly for u in J, and hence (11) gives

$$\phi(t) - \phi(t_0) - \int_{t_0}^t f(u, \phi(u)) du = 0.$$

The result therefore follows from (2.2.1).

(2.3.2. Corollary) *Let* $E \subseteq \mathbf{R} \times Y$, *and let* (f_n) *be a sequence of continuous functions from E into Y converging uniformly to a (continuous) function f on E. Let also J be a compact interval in* $\mathbf{R}$, *and suppose that, for each n,* ϕ_n *is a solution of the equation* $y' = f_n(t, y)$ *on J, and that the sequence* (ϕ_n) *converges uniformly on J to a function* ϕ *whose graph lies in E. Then* ϕ *is a solution of the equation* $y' = f(t, y)$ *on J.*

In fact, ϕ_n is an ε_n-approximate solution of $y' = f(t, y)$ on J, where

$$\varepsilon_n = \sup_E \| f_n(t, y) - f(t, y) \|,$$

and $\varepsilon_n \to 0$ as $n \to \infty$, since $f_n \to f$ uniformly on E.

Exercise 2.3

1 Let $f : J \times Y \to Y$ be continuous and bounded, where Y is finite-dimensional and J is a compact interval in $\mathbf{R}$, let $(t_0, y_0) \in J \times Y$, and suppose that the equation $y' = f(t, y)$ has a unique solution ϕ on J satisfying the condition $\phi(t_0) = y_0$. Suppose further that ϕ_n $(n = 1, 2, ...)$ is an ε_n-approximate solution of $y' = f(t, y)$ on J satisfying the condition that $\phi_n(t_n) = y_n$, where $\varepsilon_n \to 0$ and $(t_n, y_n) \to (t_0, y_0)$ as $n \to \infty$. Prove that $\phi_n \to \phi$ uniformly on J.

2.4 Existence theorems for $y' = f(t, y)$ when Y is finite-dimensional

It has been shown in §1.5 that, if f is a bounded continuous function from the strip $[t_0, t_0 + \alpha] \times \mathbf{R}$, then for each $y_0 \in \mathbf{R}$ the scalar equation $y' = f(t, y)$ has at least one solution taking the value y_0 at t_0. Moreover, there may be an infinity of such solutions.

In this section we consider the more general case where f is a continuous function from a subset of $\mathbf{R} \times Y$ into Y, the dimension of Y being finite. The arguments of §1.5 do not extend to this more general situation, and instead we use the Ascoli–Arzelà theorem (A.1.2) together with the results of §2.3 on approximate solutions.

The key theorem is the following, which gives the local existence of a solution.

(2.4.1) *Let Y be finite-dimensional, let C be the cylinder $\{(t,y) : t_0 \leq t \leq t_0 + \alpha, \|y - y_0\| \leq \rho\}$ in $\mathbf{R} \times Y$, where $\rho > 0$, and let $f : C \to Y$ be a continuous function such that $\|f(t,y)\| \leq M$ for all $(t,y) \in C$. Then, if $\eta = \min\{\alpha, \rho/M\}$, the equation $y' = f(t,y)$ has at least one solution ϕ on $[t_0, t_0 + \eta]$ satisfying the initial condition that $\phi(t_0) = y_0$.*

A similar result holds for solutions to the left of t_0.

By (2.3.1), for each positive integer n we can find a $1/n$-approximate solution ψ_n of the equation on $[t_0, t_0 + \eta]$ such that $\psi_n(t_0) = y_0$ and that $\|\psi_n(s) - \psi_n(t)\| \leq M|s - t|$ for all $s, t \in [t_0, t_0 + \eta]$. The sequence (ψ_n) is obviously equicontinuous and uniformly bounded, and hence, by the Ascoli–Arzelà theorem, we can select a subsequence of (ψ_n) converging uniformly on $[t_0, t_0 + \eta]$ to a function ϕ. By (2.3.2), ϕ is a solution of the equation on $[t_0, t_0 + \eta]$, and this completes the proof.

From (2.4.1) we obtain the following generalization of the 'existence' part of (1.5.1), where now the domain of f is a slab in $\mathbf{R} \times Y$.

(2.4.2) *Let Y be finite-dimensional, let J be a compact interval in $\mathbf{R}$, let $f : J \times Y \to Y$ be bounded and continuous, and let $(t_0, y_0) \in J \times Y$. Then the equation $y' = f(t,y)$ has at least one solution ϕ on J satisfying the initial condition that $\phi(t_0) = y_0$. Further, the set Σ of all such solutions is compact in $C(J, Y)$.*†

If t_0 is not an endpoint of J, then, by (2.4.1) with ρ sufficiently large, we can find a solution ϕ_1 of the equation on the part of J to the right of t_0 satisfying $\phi_1(t_0) = y_0$, and a solution ϕ_2 on the part of J to the left of t_0 satisfying $\phi_2(t_0) = y_0$. The solution ϕ is obtained by piecing together ϕ_1 and ϕ_2, i.e. it is the function on J that agrees with ϕ_1 and ϕ_2 on their respective domains.‡ If t_0 is an endpoint of J, we simply take ϕ to be ϕ_1 or ϕ_2.

---

† We show in (2.10.5) that Σ is also connected in $C(J, Y)$, so that it is a continuum.

‡ Note that the solutions ϕ_1 and ϕ_2 match at t_0, since

$$\phi_1'(t_0) = f(t_0, \phi_1(t_0)) = f(t_0, y_0) = f(t_0, \phi_2(t_0)) = \phi_2'(t_0).$$

This piecing together of solutions occurs frequently in the sequel, and in future we take the details for granted.

Next, any sequence of solutions belonging to Σ is obviously equicontinuous and uniformly bounded, and therefore, by the Ascoli–Arzelà theorem (A.1.2), has a subsequence converging uniformly on J, i.e. converging in $C(J, Y)$. By (2.3.2), the limit of this subsequence belongs to Σ, so that Σ is compact in $C(J, Y)$.

(2.4.2. Corollary) *Let* Y *be finite-dimensional, let* I *be an interval in* $\mathbf{R}$, *let* $f : I \times Y \to Y$ *be a continuous function which is bounded on* $J \times Y$ *for each compact subinterval* J *of* I, *and let* $(t_0, y_0) \in I \times Y$. *Then the equation* $y' = f(t, y)$ *has at least one solution* ϕ *on* I *satisfying the initial condition that* $\phi(t_0) = y_0$.

We deduce this directly from the main theorem by piecing together solutions in the obvious way.

If the condition of boundedness in (2.4.2. Corollary) is omitted, there may be no solution of the equation with domain I. For example, let f be the function with domain $\mathbf{R}^2$ given by $f(t, y) = e^{-y}$, so that the equation is

$$y' = e^{-y}.$$

For any given $(t_0, y_0) \in \mathbf{R}^2$, the only solutions of this equation taking the value y_0 at t_0 are restrictions of the solutions given by

$$y = \log(t - t_0 + e^{y_0}),$$

with domain $]t_0 - e^{y_0}, \infty[$.

We turn now to the case where the domain of f is an arbitrary open set in $\mathbf{R} \times Y$, and for simplicity we confine our attention to solutions to the right of the initial point t_0.

Let E be an open set in $\mathbf{R} \times Y$, let $f : E \to Y$ be continuous (or, more generally, let f be defined and continuous on the intersection of E with the half-space $[t_0, \infty[\times Y)$, and let ϕ be a solution of the equation $y' = f(t, y)$ on an interval $[t_0, b[$, where $t_0 < b \le \infty$. We say that ϕ *reaches to the boundary of* E *on the right* if for each compact subset C of E there exists $\tau \in [t_0, b[$ such that $(t, \phi(t)) \in E \backslash C$ for all $t \in [\tau, b[$.

(2.4.3) *Let* Y *be finite-dimensional, let* E *and* f *be as specified above, and let* ϕ *be a solution of the equation* $y' = f(t, y)$ *on an interval* $[t_0, b[$, *where* $t_0 < b \le \infty$. *Then the following statements are equivalent*:

(i) ϕ *reaches to the boundary of* E *on the right*;

(ii) *the graph of* ϕ *is not contained in any compact subset of* E;

(iii) *either* $b = \infty$ *or* ϕ *has no extension to* $[t_0, b]$ *that is a solution of the equation on this interval*;

(iv) *there is no sequence of points* $t_n \in [t_0, b[$ *tending to* b *such that the sequence* $(t_n, \phi(t_n))$ *converges to a point of* E.

If in addition E is bounded, then each of (i)–(iv) *is also equivalent to*
(v) *the distance of the point* $(t, \phi(t))$ *from the complement of E tends to* 0
as $t \to b-$.

It is obvious that (i) implies (ii). Also (ii) implies (iii), for if $b < \infty$ and ϕ
has an extension of the type specified in (iii), then the graph of this extension
is a compact subset of E containing the graph of ϕ. Next, (iii) implies (iv),
for if (iv) is false, we can find a sequence of points $t_n \in [t_0, b[$ tending to b
such that the sequence $(t_n, \phi(t_n))$ converges to a point $(b, c) \in E$ (so that
also $b < \infty$). Since f is continuous at (b, c), f is bounded on a neighbourhood
of this point, and since $\phi'(t) = f(t, \phi(t))$, (1.7.3) shows that $\phi(t) \to c$ as
$t \to b-$. Then also $\phi'(t) = f(t, \phi(t)) \to f(b, c)$ as $t \to b-$, and hence, by
(1.7.4), the extension of ϕ to $[t_0, b]$ taking the value c at b is a solution of
the equation on $[t_0, b]$, contrary to (iii). Also (iv) implies (i), for if (i) is
false, there exists a compact set $C \subseteq E$ and a sequence of points $s_n \in [t_0, b[$
tending to b such that $(s_n, \phi(s_n)) \in C$ for all n, whence there exists a subsequence (s_{nr}) for which $(s_{nr}, \phi(s_{nr}))$ converges to a point of C, contrary
to (iv).

Suppose next that E is bounded. Then (i) implies (v), for if $\varepsilon > 0$ and C_ε is
the set of points of $\mathbf{R} \times Y$ whose distance from the complement of E is
greater than or equal to ε, then C_ε is a closed subset of E, and is therefore
compact. If now (i) holds, there exists $\tau \in [t_0, b[$ such that $(t, \phi(t)) \in E \backslash C_\varepsilon$
for all $t \in [\tau, b[$, so that (v) holds. Conversely, (v) obviously implies (i), and
this completes the proof.

This theorem implies in particular that if f is continuous on the slab
$S = [t_0, a[\times Y, t_0 < b \le a$, and ϕ is a solution of the equation $y' = f(t, y)$
on $[t_0, b[$ reaching to the boundary of S on the right, then either $b = a$,
or $\|\phi(t)\| \to \infty$ as $t \to b-$ (if $b < a$, take C in (iv) to be $\{(t, y) : t_0 \le t \le b,$
$\|y\| \le \rho\}$).

Similar definitions and results hold for solutions to the left of t_0.

The next theorem is the main result of this section.

(2.4.4) *Let* Y *be finite-dimensional, let* E *be an open set in* $\mathbf{R} \times Y$, *let*
$(t_0, y_0) \in E$, *and let* $f : E \to Y$ *be continuous (or, more generally, let* f *be
defined and continuous on the intersection of* E *with the half-space*
$H = [t_0, \infty[\times Y)$. *Then there is at least one solution* ϕ *of the equation*

$$y' = f(t, y) \tag{1}$$

satisfying the initial condition that $\phi(t_0) = y_0$ *and reaching to the boundary
of* E *on the right. Further, if* ψ *is any solution of* (1), *then* ψ *has an extension*
$\bar\psi$ *that reaches to the boundary of* E *on the right.*

A similar result holds to the left of t_0.

We note that $\bar\psi$ may be ψ itself, and that neither ϕ nor $\bar\psi$ need be unique.

To prove the existence of ϕ, let (G_n) be a sequence of bounded open subsets of E, containing (t_0, y_0) and with union E, such that $\bar{G}_n \subseteq G_{n+1}$ for all n. By Tietze's extension theorem (A.1.1), for each n we can find a bounded continuous function $f_n : \mathbf{R} \times Y \to Y$ that agrees with f on $\bar{G}_n \cap H$ and takes the value 0 everywhere on the complement of G_{n+1}. Let $[t_0, t_0 + \alpha_1] \times Y$ be a slab containing $\bar{G}_1 \cap H$. By (2.4.2) applied to f_1 on this slab, we can find a solution ω_1 of the equation $y' = f_1(t, y)$ on $[t_0, t_0 + \alpha_1]$ satisfying $\omega_1(t_0) = y_0$. Let $t_1 (> t_0)$ be the first point of $[t_0, t_0 + \alpha_1]$ where the graph of ω_1 meets $E \backslash G_1$, and let ϕ_1 be the restriction of ω_1 to $[t_0, t_1]$. Clearly ϕ_1 is a solution of (1) on $[t_0, t_1]$ satisfying $\phi_1(t_0) = y_0$ and $(t_1, \phi_1(t_1)) \in E \backslash G_1$.

We now repeat the preceding construction with G_2, f_2, but starting from the point (t_1, y_1), where $y_1 = \phi_1(t_1)$. We thus obtain a solution ϕ_2 of (1) on an interval $[t_1, t_2]$ such that $\phi_2(t_1) = y_1$ and that $(t_2, \phi_2(t)) \in E \backslash G_2$. Proceeding in this manner, we obtain a sequence of functions $\phi_n : [t_{n-1}, t_n] \to Y$, each a solution of (1) on its domain, such that $\phi_{n+1}(t_n) = \phi_n(t_n)$ and that $(t_n, \phi_n(t_n)) \in E \backslash G_n$ for all n. The solution ϕ is now obtained by piecing together these functions ϕ_n in the obvious manner (the domain of ϕ is the interval $[t_0, b[$, where b is the limit of the (increasing) sequence of points t_n). Since any compact subset of E is contained in some G_N, and $(t_N, \phi(t_N)) = (t_N, \phi_N(t_N)) \in E \backslash G_N$, it follows from (2.4.3) (iv) that ϕ reaches to the boundary of E on the right, as required.

Now let ψ be any solution of (1). If ψ does not reach to the boundary of E on the right, then, by (2.4.3) (iii), ψ has an extension ψ_1 to a closed interval $[t_0, b]$ that is a solution of (1) on this interval. By the first part of the proof we can find a solution ψ_2 of (1) on an interval $[b, b'[$ satisfying $\psi_2(b) = \psi_1(b)$ and reaching to the boundary of E on the right, and to obtain $\bar{\psi}$ we have only to piece together ψ_1 and ψ_2.

We consider next the special case of a scalar equation, and we prove the existence of a maximal solution reaching to the boundary of E. The statement of the result has to be worded rather carefully, since the interval of existence of this maximal solution may be smaller than that of some other solutions satisfying the same initial condition. At the same time, we extend the differential inequality of (1.5.1).

(2.4.5) *Let E be an open set in $\mathbf{R}^2$, let $(t_0, y_0) \in E$, and let $f : E \to \mathbf{R}$ be continuous (or, more generally, let f be defined and continuous on the intersection of E with the half-plane $[t_0, \infty[\times \mathbf{R})$. Let also Σ be the class of all solutions ϕ of the differential equation $y' = f(t, y)$ satisfying the initial condition that $\phi(t_0) = y_0$ whose domain is a closed or left-half-closed interval I_ϕ with t_0 as* *left-hand endpoint. Then there exists an element* Φ *of* Σ, *reaching to the boundary of E on the right, that is maximal in the sense that* $\phi(t) \le \Phi(t)$ *whenever* $\phi \in \Sigma$ *and* $t \in I_\phi \cap I_\Phi$. *Further, if* ψ *is a continuous function on an interval* I_ψ *containing* t_0 *such that the graph of* ψ *lies in E, that* $\psi(t_0) \le y_0$, *and that* $D\psi(t) \le f(t, \psi(t))$ *for nearly all* $t \in I_\psi \cap I_\Phi$, *where D is some fixed Dini derivative, then* $\psi(t) \le \Phi(t)$ *for all* $t \in I_\psi \cap I_\Phi$.

The proof follows that of (2.4.4), but at each stage in the continuation process we select the maximal solution of the appropriate equation (the existence of this maximal solution follows from (1.5.1)). Thus with the notation used in the proof of (2.4.4), at the first stage we choose ω_1 to be the maximal solution of $y' = f_1(t, y)$ on $[t_0, t_0 + \alpha_1]$ satisfying $\omega_1(t_0) = y_0$, and we define t_1 and ϕ_1 as before. Then ϕ_1 obviously belongs to Σ, and it has the additional property that $\phi(t) \le \phi_1(t)$ whenever $\phi \in \Sigma$ and $t \in I_\phi \cap [t_0, t_1]$. To prove this last property, suppose on the contrary that there exist $\phi \in \Sigma$ and $\tau \in [t_0, t_1]$ for which $\phi(\tau) > \phi_1(\tau)$ (note that ϕ may not satisfy $y' = f_1(t, y)$ on $[t_0, \tau]$, since its graph on this interval may lie partly in $E \backslash \bar{G}_1$; see Fig. 2.1). If σ is the greatest element of $[t_0, \tau]$ for which $\phi(\sigma) = \phi_1(\sigma)$, then since $(\sigma, \phi(\sigma)) = (\sigma, \phi_1(\sigma)) \in G_1$ and G_1 is open, we can find $\eta \in]\sigma, \tau]$ such that $(t, \phi(t)) \in G_1$ for all $t \in [\sigma, \eta]$. It now follows from (1.5.1) that the continuous function on $[t_0, t_0 + \alpha_1]$ that agrees with ϕ_1 on $[t_0, \sigma]$ and with ϕ on $[\sigma, \eta]$, and whose graph on $[\eta, t_0 + \alpha_1]$ is a segment of slope $\inf f_1$, does not exceed ϕ_1, and this gives a contradiction.

We now repeat the construction with G_2, f_2 as before, and ϕ_2 here has the additional property that $\phi(t) \le \phi_2(t)$ whenever $\phi \in \Sigma$ and $t \in I_\phi \cap [t_1, t_2]$ (since $\phi(t_1) \le y_1 = \phi_2(t_1)$, the preceding argument applies). Proceeding in this manner, we obtain a sequence of solutions, and as in (2.4.4) we piece these together to obtain Φ.

Next, let ψ have the properties specified in the theorem, and let

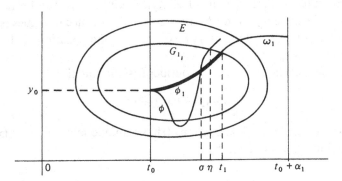

Figure 2.1

$t \in I_\psi \cap I_\Phi$. Then we can find an integer n such that the points $(s, \psi(s))$ and $(s, \Phi(s))$ belong to G_n for all $s \in [t_0, t]$. Applying (1.5.1) to f_n on $[t_0, t] \times \mathbf{R}$, we deduce that $\psi(t) \le \Phi(t)$, and this completes the proof.

Corresponding results hold for the minimal solution in Σ, and in the sequel, when we refer to the maximal or minimal solution of $y' = f(t, y)$ to the right of the initial point, these solutions are assumed to reach to the boundary of E on the right.

The last part of (2.4.5) can be sharpened into an 'epidermis' property (cf. (1.5.2)). Corresponding results hold for the minimal solution in Σ. There is also a result for solutions to the left of t_0, the form of which may be inferred from Exercise 1.5.1.

We conclude this section with an example which shows that when Y is infinite-dimensional, the continuity of f does not imply the existence of either global or local solutions of the differential equation $y' = f(t, y)$.

Let c_0 denote the Banach space of sequences $y = (y_n)_{n \ge 1}$ of real numbers that converge to the limit 0, with the sup norm $\|y\| = \sup_n |y_n|$, and for each $t \in \mathbf{R}$ and each sequence $y = (y_n)_{n \ge 1} \in c_0$ let $f(t, y)$ be the sequence $(3y_n^{2/3} + 3/n^2)$. Then f maps $\mathbf{R} \times c_0$ into c_0, and is clearly continuous.

Consider now the equation

$$y' = f(t, y), \tag{2}$$

subject to the condition that $y = 0$ when $t = 0$. If ϕ is a solution of the equation (2) on some interval I containing 0, and, for each $t \in I$, $\phi(t)$ is the sequence $(\phi_n(t))$ in c_0, then (cf. §1.1, Example (e), p. 5) for each n the function ϕ_n is a solution on I of the scalar equation

$$x' = 3x^{2/3} + 3/n^2 \tag{3}$$

satisfying $\phi_n(0) = 0$. However, although there exists a unique sequence of functions (ϕ_n) such that ϕ_n $(n = 1, 2, \ldots)$ is a solution of (3) on I satisfying $\phi_n(0) = 0$, for these functions ϕ_n the associated sequence of values $(\phi_n(t))$ does not belong to c_0 for any $t \in I \setminus \{0\}$, so that the equation (2) has no solution $\phi : I \to c_0$.

To prove these last statements, consider first the equation

$$\frac{dt}{dx} = (3x^{2/3} + 3/n^2)^{-1}. \tag{4}$$

It is easily verified that this equation (4) has a unique solution $t = \psi_n(x)$ on $\mathbf{R}$ satisfying $\psi_n(0) = 0$, given by

$$\psi_n(x) = x^{1/3} - n^{-1} \tan^{-1}(nx^{1/3}), \tag{5}$$

and clearly ψ_n is strictly increasing and has range $\mathbf{R}$. Further, if J is an interval containing 0, then any solution of (4) on J taking the value 0 at 0 is the restriction of ψ_n to J. It therefore follows that the equation (3) has a unique solution ϕ_n on I satisfying $\phi_n(0) = 0$, namely the restriction of ψ_n^{-1} to I. Moreover, by (5), for all $t \in I$ we have

$$t = (\phi_n(t))^{1/3} - n^{-1}\tan^{-1}(n(\phi_n(t))^{1/3}),$$

and this clearly implies that $\phi_n(t)$ does not tend to 0 unless $t = 0$.

Exercises 2.4

1 Construct a proof of (1.5.1) on the following lines.
(i) Let $J = [t_0, t_0 + \alpha]$, and let $f, g : J \times \mathbf{R} \to \mathbf{R}$ be continuous functions such that $f(t, y) > g(t, y)$ for all $(t, y) \in J \times \mathbf{R}$. Suppose also that $\chi, \omega : J \to \mathbf{R}$ are continuous functions such that $\chi(t_0) > \omega(t_0)$ and that, for some fixed Dini derivative,
$$D\chi(t) \geq f(t, \chi(t)) \quad \text{and} \quad D\omega(t) \leq g(t, \omega(t))$$
for nearly all $t \in J$. Prove that $\chi(t) > \omega(t)$ for all $t \in [t_0, t_0 + \alpha[$.
(ii) Let $f : J \times \mathbf{R} \to \mathbf{R}$ be continuous and bounded, let (ε_n) be a decreasing sequence of positive numbers converging to 0, and for each n let ψ_n be a solution on J of the equation $y' = f(t, y) + \varepsilon_n$ satisfying $\psi_n(t_0) = y_0 + \varepsilon_n$. Prove that the sequence (ψ_n) has a subsequence converging uniformly on J to a function Φ that is a solution of $y' = f(t, y)$ on J satisfying $\Phi(t_0) = y_0$.
(iii) Let $\phi : J \to \mathbf{R}$ be a continuous function such that $\phi(t_0) \leq y_0$ and that $D\phi(t) \leq f(t, \phi(t))$ for nearly all $t \in J$, where D is a fixed Dini derivative. Prove that $\phi(t) \leq \Phi(t)$ for all $t \in J$, and deduce that Φ is the maximal solution of the equation $y' = f(t, y)$ on J taking the value y_0 at t_0.
[This proof, unlike that of §1.5, uses the existence theorem (2.4.2).]
2 Suppose that the conditions of (2.4.5) are satisfied, that the maximal and minimal solutions Φ, Ψ of $y' = f(t, y)$ are defined on $[t_0, t_1]$, and that $E \supseteq \{(t, y) \in \mathbf{R}^2 : t_0 \leq t \leq t_1, \Psi(t) \leq y \leq \Phi(t)\}$. Prove that if $\Psi(t_1) < y_1 < \Phi(t_1)$, there exists an element ϕ_1 of Σ defined on $[t_0, t_1]$ such that $\phi_1(t_1) = y_1$.
[For a generalization of this result, see (2.10.5).]
3 Let $E \subseteq \mathbf{R}^2$, and let $f : E \to \mathbf{R}$ be a continuous function such that $x \mapsto f(t, x)$ is decreasing for each t. Prove that if $(t_0, x_0) \in E$, then the equation $x' = f(t, x)$ has exactly one solution with initial point (t_0, x_0) and reaching to the boundary of E on the right.
4 Prove that if $\lambda > 0$ then the equation
$$x' = (1 + 2\lambda t - x^2)^{1/2}$$
has a unique solution ϕ on $[0, \infty[$ satisfying $\phi(0) = 0$, and that
$$\sin t \leq \phi(t) \leq (\sin t)(1 + 2\lambda t)^{1/2} \quad \text{if } 0 \leq t \leq \tfrac{1}{2}\pi,$$
$$1 \leq \phi(t) \leq (1 + 2\lambda t)^{1/2} \qquad \text{if } t > \tfrac{1}{2}\pi.$$

2.5 Some global existence theorems and other comparison theorems

Let Y be finite-dimensional, and let f be continuous on a slab $S = [t_0, b[\times Y$, where $t_0 < b \leq \infty$. The example following (2.4.2. Corollary) shows that it is possible for the domain of each solution ϕ of the equation $y' = f(t, y)$ taking a given value at t_0 and reaching to the boundary of S on the right to be a proper subinterval of $[t_0, b[$ (and in this case each ϕ is unbounded, by the remark following the proof of (2.4.3)). There are many theorems in the literature that give sufficient conditions for the existence of one or more solutions of the equation $y' = f(t, y)$ on the entire interval $[t_0, b[$, and we present here three such results.

In each of these results we suppose that f satisfies for certain (t, y) a condition of the form

$$\| f(t, y) \| \le g(t, \| y \|),$$

where the scalar equation $x' = g(t, x)$ is known to have a non-negative solution ψ on the interval $[t_0, b[$. Results of this type, where information about solutions of the equation $y' = f(t, y)$ is derived from information about the solutions of a related scalar equation $x' = g(t, x)$, are known as 'comparison theorems', and such theorems have many applications to problems of existence, uniqueness and stability. In addition to the three results on global existence mentioned already, we give two comparison theorems which deal with the behaviour of the solutions of $y' = f(t, y)$ as $t \to b -$. Other examples can be found in the exercises at the end of the section.

As in §2.4 we consider only solutions to the right of the initial point; similar results hold for solutions to the left.

The first two results are elementary, and employ only the existence theorem (2.4.2. Corollary).

(2.5.1) *Let Y be finite-dimensional, let I be the interval $[t_0, b[$, where $t_0 < b \le \infty$, and let $f : I \times Y \to Y$ be continuous. Suppose further that the scalar equation $x' = g(t, x)$ has a non-negative solution ψ on I, and that $\| f(t, y) \| \le g(t, \| y \|)$ whenever $t \in I$ and $\| y \| = \psi(t)$. Then for each $y_0 \in Y$ such that $\| y_0 \| \le \psi(t_0)$ the equation $y' = f(t, y)$ has at least one solution ϕ on I satisfying the conditions that $\phi(t_0) = y_0$ and that $\| \phi(t) \| \le \psi(t)$ for all $t \in I$.*

The hypotheses imply that $g(t, \psi(t)) \ge 0$ for all $t \in I$, so that ψ is increasing. Let F be defined on $I \times Y$ by

$$F(t, y) = f(t, y) \quad (\| y \| \le \psi(t)), \quad F(t, y) = f(t, \psi(t)y / \| y \|) \quad (\| y \| > \psi(t)).$$

It is easily verified that F is continuous. Further, since F is bounded on $J \times Y$ for each compact subinterval J of I, (2.4.2. Corollary) shows that for each $y_0 \in Y$ the equation $y' = F(t, y)$ has a solution ϕ on I satisfying $\phi(t_0) = y_0$.

Now let $\| y_0 \| \le \psi(t_0)$, and let ϕ be such a solution of $y' = F(t, y)$.

Suppose further that $\|\phi(\tau)\| > \psi(\tau)$ for some $\tau \in I$, and let σ be the greatest element of $[t_0, \tau[$ for which $\|\phi(\sigma)\| = \psi(\sigma)$. Then for all $t \in]\sigma, \tau[$ we have $\|\phi(t)\| > \psi(t)$, and therefore, by (1.6.1),

$$D^+ \|\phi(t)\| \leq \|\phi'(t)\| = \|F(t, \phi(t))\|$$
$$= \|f(t, \psi(t)\phi(t)/\|\phi(t)\|)\| \leq g(t, \psi(t)) = \psi'(t).$$

From (1.4.4) we deduce that $\|\phi\| - \psi$ is decreasing on $[\sigma, \tau]$, and, since $\|\phi(\sigma)\| = \psi(\sigma)$, this contradicts the hypothesis that $\|\phi(\tau)\| > \psi(\tau)$. Hence $\|\phi(t)\| \leq \psi(t)$ for all $t \in I$, whence also $F(t, \phi(t)) = f(t, \phi(t))$, and therefore ϕ is a solution of $y' = f(t, y)$ on I.

The next result shows that we can impose a further condition on the solution ϕ of (2.5.1) provided that the equation $x' = -g(t, x)$ has a positive solution on I.

(2.5.2) *Let Y be finite-dimensional, let I be the interval $[t_0, b[$, where $t_0 < b \leq \infty$, and let $f : I \times Y \to Y$ be continuous. Suppose further that the scalar equations $x' = g(t, x)$ and $x' = -g(t, x)$ have positive solutions ψ, χ on I such that $\chi(t_0) \leq \psi(t_0)$ and that $\|f(t, y)\| \leq g(t, \|y\|)$ whenever $t \in I$ and $\|y\| = \psi(t), \chi(t)$. Then for each $y_0 \in Y$ such that $\chi(t_0) \leq \|y_0\| \leq \psi(t_0)$ the equation $y' = f(t, y)$ has at least one solution ϕ on I satisfying the conditions that $\phi(t_0) = y_0$ and that $\chi(t) \leq \|\phi(t)\| \leq \psi(t)$ for all $t \in I$.*

The hypotheses imply that $g(t, \psi(t)) \geq 0$ and $g(t, \chi(t)) \geq 0$ for all $t \in I$, and therefore ψ is increasing and χ is decreasing, so that $\chi(t) \leq \psi(t)$ for all $t \in I$. Let $\bar{f}$ be defined on $I \times Y$ by

$$\bar{f}(t, y) = \begin{cases} f(t, y) & \text{if } \|y\| \geq \chi(t), \\ f(t, \chi(t)y/\|y\|) & \text{if } \tfrac{1}{2}\chi(t) \leq \|y\| < \chi(t), \\ f(t, 2y) & \text{if } \|y\| < \tfrac{1}{2}\chi(t). \end{cases}$$

Then $\bar{f}$ is continuous and satisfies the hypotheses of (2.5.1), whence for each $y_0 \in Y$ satisfying $\|y_0\| \leq \psi(t_0)$ we can find a solution ϕ of $y' = \bar{f}(t, y)$ on I such that $\phi(t_0) = y_0$ and that $\|\phi(t)\| \leq \psi(t)$ for all $t \in I$.

Now let $\|y_0\| \geq \chi(t_0)$. Then the solution ϕ must satisfy $\|\phi(t)\| \geq \chi(t)$ for all $t \in I$, for otherwise we can find a subinterval $[\sigma, \tau]$ of I such that $\|\phi(\sigma)\| = \chi(\sigma)$ and that $\tfrac{1}{2}\chi(t) \leq \|\phi(t)\| < \chi(t)$ for all $t \in]\sigma, \tau]$. Then for all $t \in]\sigma, \tau[$ we have

$$D^+ \|\phi(t)\| \geq -\|\phi'(t)\| = -\|\bar{f}(t, \phi(t))\|$$
$$= -\|f(t, \chi(t)\phi(t)/\|\phi(t)\|)\| \geq -g(t, \chi(t)) = \chi'(t),$$

and this gives a contradiction, as in the proof of (2.5.1). Hence $\chi(t) \leq \|\phi(t)\| \leq \psi(t)$ for all $t \in I$, whence also $\bar{f}(t, \phi(t)) = f(t, \phi(t))$, so that ϕ is a solution of $y' = f(t, y)$ on I.

In the next result we assume that the given solution ψ of $x' = g(t,x)$ is the *maximal* solution, and we deduce correspondingly more about the solutions of $y' = f(t,y)$. In contrast with (2.5.1,2) we now require that g is continuous and that the inequality

$$\| f(t,y) \| \le g(t, \| y \|) \tag{1}$$

holds for all $(t,y) \in I \times Y$.

(2.5.3) *Let Y be finite-dimensional, let I be the interval $[t_0, b[$, where $t_0 < b \le \infty$, and let $f : I \times Y \to Y$ and $g : I \times [0, \infty[\to [0, \infty[$ be continuous functions satisfying (1) for all $(t,y) \in I \times Y$. Let also $x_0 \ge 0$, and suppose that the maximal solution ψ of the scalar equation $x' = g(t,x)$ satisfying $\psi(t_0) = x_0$ has domain I. Then every solution ϕ of $y' = f(t,y)$ satisfying $\| \phi(t_0) \| \le x_0$ and reaching to the boundary of $I \times Y$ on the right has domain I and satisfies $\| \phi(t) \| \le \psi(t)$ for all $t \in I$. Further, if $\phi(t_0) \ne 0$ and χ is the minimal solution of $x' = -g(t,x)$ such that $\chi(t_0) = \| \phi(t_0) \|$, then $\| \phi(t) \| \ge \chi(t)$ for all t in the domain of χ.*

Let ϕ be a solution of $y' = f(t,y)$ satisfying $\| \phi(t_0) \| \le x_0$ and reaching to the boundary of $I \times Y$ on the right, and let the domain of ϕ be $[t_0, a[$. Since $-\| \phi'(t) \| \le D^+ \| \phi(t) \| \le \| \phi'(t) \|$ and $\phi'(t) = f(t, \phi(t))$ for all $t \in [t_0, a[$, we have

$$-g(t, \| \phi(t) \|) \le D^+ \| \phi(t) \| \le g(t, \| \phi(t) \|).$$

Hence, by (1.5.1), $\| \phi(t) \| \le \psi(t)$ for all $t \in [t_0, a[$, and $\| \phi(t) \| \ge \chi(t)$ for all such t in the domain of χ. By the remark following the proof of (2.4.3), we deduce that $a = b$, and this completes the proof.

The most important case of (2.5.3) arises when g is of the form $g(t,x) = \mu(t)h(x)$, where we have the following result.

(2.5.4) *Let Y be finite-dimensional, let $I = [a,b[$, where $a < b \le \infty$, and let $f : I \times Y \to Y$ be a continuous function such that for all $(t,y) \in I \times Y$*

$$\| f(t,y) \| \le \mu(t)h(\| y \|), \tag{2}$$

where $\mu : I \to [0, \infty[$ and $h : [0, \infty[\to [0, \infty[$ are continuous, $h(x) > 0$ for all $x > 0$, $\int_0 du/h(u) < \infty$,† and $\int^\infty du/h(u) = \infty$. Then every solution ϕ of $y' = f(t,y)$ reaching to the boundary of $I \times Y$ on the right and left has domain I. Moreover, for all $t, t_0 \in I$

$$\| \phi(t) \| \le H^{-1} \left(H(\| \phi(t_0) \|) + \left| \int_{t_0}^t \mu(s)ds \right| \right), \tag{3}$$

---

† The condition $\int_0 du/h(u) < \infty$ is relevant only in so far as obtaining the best estimate in (3) is concerned, for in the existence part of the theorem, h can be replaced by $1 + h$.

where $H(x) = \displaystyle\int_0^x du/h(u) \ (x \geq 0)$, *and*

$$\|\phi(t)\| \geq H^{-1}\left(H(\|\phi(t_0)\|) - \left|\int_{t_0}^t \mu(s)ds\right|\right)$$

for all $t, t_0 \in I$ *for which* $\left|\int_{t_0}^t \mu(s)ds\right| \leq H(\|\phi(t_0)\|)$.

The 'comparison' equations to be considered here are $x' = \mu(t)h(x)$ and $x' = -\mu(t)h(x)$, and these have solutions which are easily obtained by separation of variables (see Exercises 2.5.1,2). The results of the theorem then follow from (2.5.3) and its analogue for solutions to the left of the initial point. Alternatively, the inequality (2) implies that each solution ϕ of $y' = f(t, y)$ satisfies

$$\|\phi'(t)\| \leq \mu(t)h(\|\phi(t)\|)$$

for all t in its domain, and we can therefore employ Exercise 1.9.5 (and indeed we can weaken the hypotheses on μ as in Exercise 1.10.2).

(2.5.4. Corollary) *Suppose that the hypotheses of the main theorem are satisfied, and that in addition* $\int_a^b \mu(t)dt < \infty$. *Then*

(i) *for each* $y_0 \in Y$ *there exists* $K > 0$ *such that if* ϕ *is a solution of* $y' = f(t, y)$ *on* I *taking the value* y_0 *somewhere on* I, *then* $\|\phi(t)\| \leq K$ *for all* $t \in I$,

(ii) *each solution* ϕ *of* $y' = f(t, y)$ *on* I *tends to a limit in* Y *as* $t \to b-$,

(iii) *for each* $y_0 \in Y$ *there exists a solution* ϕ *of* $y' = f(t, y)$ *on* I *such that* $\phi(t) \to y_0$ *as* $t \to b-$.

Here (i) is an immediate consequence of (3), and implies that each solution ϕ of $y' = f(t, y)$ on I is bounded. Hence $t \mapsto h(\|\phi(t)\|)$ is bounded on I, and therefore also

$$\|\phi'(t)\| = \|f(t, \phi(t))\| \leq \mu(t)h(\|\phi(t)\|) \leq M\mu(t),$$

say. It follows that

$$\|\phi(t) - \phi(s)\| = \left\|\int_s^t \phi'(v)dv\right\| \leq \int_s^t \|\phi'(v)\| dv \leq M\int_s^t \mu(v)dv \qquad (4)$$

whenever $a \leq s < t < b$, and since the last integral is arbitrarily small when s is sufficiently near b, $\phi(t)$ tends to a limit in Y as $t \to b-$.

It remains to prove (iii). Let $y_0 \in Y$, let (t_n) be a sequence of points of I increasing to the limit b, and for each n choose a solution ϕ_n of the equation on I such that $\phi_n(t_n) = y_0$. By (i), the inequalities (4) hold for each ϕ_n with an M that is independent of n, and clearly we can include $t = b$ provided that we define $\phi_n(b)$ to be the limit of $\phi_n(t)$ as $t \to b-$. Suppose first that

$t \mapsto \int_a^t \mu(v)dv$ is strictly increasing on I, and let

$$\rho(s,t) = \left| \int_s^t \mu(v)dv \right|.$$

Then ρ is a metric on the set $E = [a, b]$ (if $b = \infty$, E is to be regarded as a subset of the extended real line), and clearly with this metric E is compact. The inequalities (4) for the ϕ_n assert that the sequence (ϕ_n) is equicontinuous on E with respect to the metric ρ, and since the ϕ_n are uniformly bounded on E (by (4) or (i)), we can select a subsequence (ϕ_{n_r}) converging uniformly on E to a function ϕ (A.1.2). Clearly ϕ satisfies (4) on E, and therefore $\phi(t)$ tends to a limit as $t \to b-$, and this limit is obviously y_0. Moreover, by (2.3.2), ϕ is a solution of $y' = f(t, y)$ on I, so that ϕ is the required function.

This completes the proof where $t \mapsto \int_a^t \mu(v)dv$ is strictly increasing, and for the case where this function is increasing non-strictly, we have only to replace ρ by

$$\rho(s,t) = \left| \int_s^t \mu(v)dv \right| + \sigma(s,t),$$

where σ is the usual metric for the extended real line $\bar{\mathbf{R}}$.

It is natural to seek a generalization of this corollary in which the condition $\| f(t,y) \| \leq \mu(t)h(\| y \|)$ is replaced by a condition of the more general form $\| f(t,y) \| \leq g(t, \| y \|)$ considered in (2.5.1–3). The following result is in this direction, but it does not include (2.5.4. Corollary), since we have to impose the condition that $x \mapsto g(t, x)$ is increasing, and we obtain a slightly weaker result.

(2.5.5) *Let Y be finite-dimensional, let $I = [a, b[$, where $a < b \leq \infty$, let $f : I \times Y \to Y$ be continuous, let $g : I \times [0, \infty[\to [0, \infty[$ be a function such that $x \mapsto g(t, x)$ is increasing for each $t \in I$, and let*

$$\| f(t, y) \| \leq g(t, \| y \|)$$

for all $(t, y) \in I \times Y$. Suppose further that for each $x_0 > 0$ the equation $x' = g(t, x)$ has a bounded solution on I taking the value x_0 at a. Then
(i) *for each $t_0 \in I$ every solution of $y' = f(t, y)$ defined at t_0 and reaching to the boundary of $I \times Y$ on the right is defined on $[t_0, b[$ and tends to a limit as $t \to b-$,*
(ii) *for each $y_0 \in Y$ there is a solution ϕ of $y' = f(t, y)$ such that $\phi(t) \to y_0$ as $t \to b-$.*

In contrast to (2.5.4. Corollary), the solution ϕ in (ii) may exist only on a proper subinterval of I (see Exercise 2.5.8).

Let $t_0 \in I$, let ϕ be a solution of $y' = f(t, y)$ defined at t_0 and reaching to the boundary of $I \times Y$ on the right and left, and let ψ be a bounded solution of $x' = g(t, x)$ on I such that $\psi(a) = \| \phi(t_0) \| + 1$. Since g is non-negative, ψ is increasing, and therefore $\psi(t)$ tends to a finite limit M as $t \to b-$.

Let J be the greatest subinterval of I containing t_0 such that $\| \phi(t) \| < \psi(t)$ for all $t \in J$. If $t \in J$, then

$$\| \phi'(t) \| = \| f(t, \phi(t)) \| \le g(t, \| \phi(t) \|) \le g(t, \psi(t)) = \psi'(t),$$

and therefore, by (1.6.2),

$$\| \phi(t) - \phi(s) \| \le \psi(t) - \psi(s) \tag{5}$$

whenever $s, t \in \bar{J}$ and $s < t < b$. If the right-hand endpoint d of J is less than b, then (5) holds for $t = d$, whence

$$\| \phi(d) \| \le \psi(d) - \psi(t_0) + \| \phi(t_0) \| < \psi(d).$$

But then $\| \phi(t) \| < \psi(t)$ for all t in some neighbourhood of d, contrary to the definition of J, and hence $d = b$, so that $J \supseteq [t_0, b[$. Moreover, since $\psi(t)$ tends to a limit as $t \to b-$, (5) implies that $\phi(t)$ also tends to a limit.

We observe next that, if the left-hand endpoint c of J is greater than a, then $\| \phi(c) \| = \psi(c)$, and hence, by (5) with $s = c, t = t_0$,

$$\psi(c) = \| \phi(c) \| \le \psi(t_0) - \psi(c) + \| \phi(t_0) \|,$$

so that

$$\psi(c) \le \tfrac{1}{2}(\| \phi(t_0) \| + \psi(t_0)) \le \tfrac{1}{2}(\| \phi(t_0) \| + M).$$

It therefore follows that $J \supseteq [t_1, b[$, whence t_1 is the greatest element of I for which $\psi(t_1) \le \tfrac{1}{2}(\| \phi(t_0) \| + M)$ (note that t_1 depends only on the value of ϕ at t_0 and not on t_0 itself).

Now let $y_0 \in Y$, choose ψ so that $\psi(a) = \| y_0 \| + 1$, let t_1 be defined as above, and let $(t_n)_{n \ge 2}$ be a sequence of points of $[t_1, b[$ increasing to the limit b. For each positive integer n choose ϕ_n to be a solution of $y' = f(t, y)$ on $[t_1, b[$ such that $\phi_n(t_n) = y_0$. The inequality (5) holds for each ϕ_n whenever $t_1 \le s < t < b$, and we can obviously include also $t = b$ provided we define $\phi_n(b)$ and $\psi(b)$ to be the limits of $\phi_n(t)$ and $\psi(t)$ as $t \to b-$. If now ψ is strictly increasing, we argue as in the proof of (2.5.4. Corollary), with I replaced by $[t_1, b[$ and with

$$\rho(s, t) = | \psi(s) - \psi(t) |. \tag{6}$$

Similarly, when ψ is increasing non-strictly we add $\sigma(s, t)$ to the right-hand side in (6), and the proof is completed as before.

It follows without difficulty from the analogue of (2.5.1) for solutions to the left of the initial point that, if in (2.5.5) we impose the additional hypothesis that for each $t_1 \in I$ and $x_1 \geq 0$ the equation $x' = -g(t,x)$ has a solution ξ on $[a, t_1]$ such that $\xi(t_1) = x_1$, then the solution in (ii) can be chosen to have domain I.

Exercises 2.5

1 Let $I = [t_0, b[$, where $t_0 < b \leq \infty$, let $x_0 \geq 0$, and let $\mu : I \to [0, \infty[$ and $h : [0, \infty[\to [0, \infty[$ be continuous. Suppose also that $h(x) > 0$ for all $x > 0$, that $\int_0 du/h(u) < \infty$, and that for all $t \in I$

$$\int_{t_0}^t \mu(s)dt < \int_{x_0}^\infty du/h(u) \leq \infty.$$

Prove that
(i) if $x_0 > 0$ the scalar equation $x' = \mu(t)h(x)$ has a unique solution ψ on I satisfying $\psi(t_0) = x_0$, given by

$$\psi(t) = H^{-1}\left(H(x_0) + \int_{t_0}^t \mu(s)ds \right), \qquad (*)$$

where $H(x) = \int_0^x du/h(u)$ $(x \geq 0)$,
(ii) if $x_0 = 0$ the function ψ defined by $(*)$ has domain I and is the maximal solution of $x' = \mu(t)h(x)$ taking the value 0 at t_0,
(iii) $\psi(t)$ (as defined by $(*)$) tends to a finite limit as $t \to b-$ if and only if

$$\int_{t_0}^b \mu(s)ds < \int_{x_0}^\infty du/h(u) \leq \infty.$$

[The solution $x = \psi(t)$ is given formally by separation of variables, i.e.

$$\int_{x_0}^x du/h(u) = \int_{t_0}^t \mu(s)ds.$$

The exercise gives a precise formulation of this process and gives information about the domain of ψ which is important in connection with (2.5.4).
    The example $x' = 3x^{2/3}$ shows that the solutions of $x' = \mu(t)h(x)$ may not be unique if $x_0 = 0$.]

2 Let $I = [t_0, b[$, where $t_0 < b \leq \infty$, let $x_0 > 0$, and let $\mu : I \to [0, \infty[$ and $h : [0, \infty[\to [0, \infty[$ be continuous. Suppose also that $h(x) > 0$ for all $x > 0$ and that $\int_0 du/h(u) < \infty$. Prove that the equation $x' = -\mu(t)h(x)$ has a unique solution χ satisfying $\chi(t_0) = x_0$, given by

$$\chi(t) = H^{-1}\left(H(x_0) - \int_{t_0}^t \mu(s)ds \right),$$

where $H(x) = \int_0^x du/h(u)$ $(x \geq 0)$, the domain of χ being the set of $t \in I$ for which $\int_{t_0}^t \mu(s)ds \leq H(x_0)$.

3 Let $I = [t_0, b[$, where $t_0 < b \leq \infty$, let $x_0 > 0$, let $\mu : I \to [0, \infty[$ and $h :]0, \infty[\to]0, \infty[$ be continuous, and suppose that $\int_0 du/h(u) = \infty$. Prove that the scalar equation $x' = -\mu(t)h(x)$ has a unique solution χ on I satisfying $\chi(t_0) = x_0$, and that $\chi(t) \to 0$ as $t \to b-$ if and only if $\int_{t_0}^b \mu(s)ds = \infty$.

4 Let Y be a finite-dimensional inner product space, let $I = [t_0, b[$, where $t_0 < b \leq \infty$, and let $f : I \times Y \to Y$ be continuous. Suppose further that the scalar equation

$x' = g(t, x)$ has a positive solution ψ on I and that

$$\text{Re} \langle f(t, y), y \rangle \le \| y \| g(t, \| y \|)$$

whenever $t \in I$ and $\| y \| = \psi(t)$, where $\langle , \rangle$ is the inner product on Y. Prove that for each $y_0 \in Y$ such that $\| y_0 \| \le \psi(t_0)$ the equation $y' = f(t, y)$ has at least one solution ϕ on I satisfying the conditions that $\phi(t_0) = y_0$ and that $\| \phi(t) \| \le \psi(t)$ for all $t \in I$.

5 Let Y be a finite-dimensional inner product space, let $I = [t_0, b[$, where $t_0 < b \le \infty$, and let $f : I \times Y \to Y$ and $g : I \times [0, \infty[\to \mathbf{R}$ be continuous functions satisfying

$$\text{Re} \langle f(t, y), y \rangle \le \| y \| g(t, \| y \|)$$

for all $(t, y) \in I \times Y$. Let also $x_0 \ge 0$, and suppose that the maximal solution ψ of the equation $x' = g(t, x)$ satisfying $\psi(t_0) = x_0$ has domain I and is non-negative. Prove that every solution ϕ of $y' = f(t, y)$ satisfying $\| \phi(t_0) \| \le x_0$ and reaching to the boundary of $I \times Y$ on the right has domain I and satisfies $\| \phi(t) \| \le \psi(t)$ for all $t \in I$.

6 Let Y be a finite-dimensional inner product space, let $I = [t_0, b[$, where $t_0 < b \le \infty$, let $f : I \times Y \to Y$ be continuous, and for all $(t, y) \in I \times Y$ let

$$\text{Re} \langle f(t, y), y \rangle \le - \mu(t) \| y \| h(\| y \|),$$

where the functions $\mu : I \to [0, \infty[$ and $h : [0, \infty[\to [0, \infty[$ are continuous, $h(0) = 0$, $h(x) > 0$ for all $x > 0$, and

$$\int_0 du/h(u) = \infty, \qquad \int_{t_0}^b \mu(t)dt = \infty.$$

Prove that for each $y_0 \in Y$ every solution ϕ of the equation $y' = f(t, y)$ satisfying $\phi(t_0) = y_0$ and reaching to the boundary of $I \times Y$ on the right has domain I and tends to 0 as $t \to b -$.

7 Let Y be a finite-dimensional inner product space, let $I = [t_0, b[$, where $t_0 < b \le \infty$, and let $f : I \times Y \to Y$ be a continuous function such that for all $(t, y) \in I \times Y$

$$\text{Re} \langle f(t, y), y \rangle \le \mu(t) \| y \| h(\| y \|),$$

where $\mu : I \to \mathbf{R}$ and $h : [0, \infty[\to [0, \infty[$ are continuous, $h(x) > 0$ for all $x > 0$, $\int_0 du/h(u) < \infty$, and $\int^\infty du/h(u) = \infty$. Prove that every solution ϕ of $y' = f(t, y)$ defined at t_0 and reaching to the boundary of $I \times Y$ on the right has domain I. Prove also that if in addition the integral $\int_{t_0}^b \mu(s)ds$ is convergent, then $\| \phi(t) \|$ tends to a finite limit as $t \to b -$.

[Hint. Let H be defined as in Exercises 2.5.1,2. To prove the existence of ϕ on I use the fact that $dH(\| \phi(t) \|)/dt \le \mu(t)$ at each point t for which $\| \phi(t) \| \ne 0$. For the last part, observe that the convergence of the integral $\int_{t_0}^b \mu(s)ds$ implies that ϕ is bounded on I (consider separately the cases where (i) ϕ has no zeros near b, (ii) ϕ has zeros arbitrarily near b). Then $\| \phi(t) \| h(\| \phi(t) \|) \le M$ for all $t \in I$, and therefore also

$$\frac{d}{dt} \| \phi(t) \|^2 \le 2M\mu(t).$$

Hence

$$\| \phi(w) \|^2 - \| \phi(v) \|^2 \le 2M \int_v^w \mu(s)ds$$

whenever $t_0 \le v \le w < b$, and the proof is completed by using a 'one-sided' form of Cauchy's convergence criterion.]

8 Let $I = [0, \infty[$, and let $g : I \times I \to I$ and $f : I \times \mathbf{R} \to \mathbf{R}$ be given by

$$g(t,x) = \begin{cases} (x+1)^2 & \text{if either } t = 0, x \geq 0, \text{ or } 0 < t \leq 1/2, 0 \leq x \leq 1/t - 1, \\ 1/t^2 & \text{otherwise,} \end{cases}$$

$$f(t,x) = g(t,|x|).$$

Prove that f, g satisfy the hypotheses of (2.5.5) with $Y = \mathbf{R}$, $a = 0$, $b = \infty$, and that for each $C \leq 1$ the only solution ϕ of $y' = f(t, y)$ such that $\phi(t) \to C$ as $t \to \infty$ and reaching to the boundary of $I \times \mathbf{R}$ on the left has domain $]0, \infty[$.

2.6 Peano's linear differential inequality, and the integral inequalities of Gronwall and Bellman

In the next section, where we discuss equations satisfying a Lipschitz condition, our arguments depend on the special case of the differential inequality $\phi'(t) \leq f(t, \phi(t))$ used in (2.4.5) in which the equation $y' = f(t, y)$ is linear. Since the direct proof of this special case is simple, we give it separately here, so that our treatment of the Lipschitz case is independent of the results of §2.4. We obtain also some associated integral inequalities which are of frequent use in the theory of stability of solutions of differential equations. Particular cases of these results were obtained by Peano, Gronwall and Bellman (see §2.14).

(2.6.1) *Let* $J = [t_0, t_0 + \alpha]$, *let* $p, q : J \to \mathbf{R}$ *be continuous, and let* $\phi : J \to \mathbf{R}$ *be a continuous function such that*

$$D_+\phi(t) \leq p(t)\phi(t) + q(t)$$

nearly everywhere in J. *Then* ϕ *does not exceed the solution of the scalar linear equation* $x' = p(t)x + q(t)$ *taking the value* $\phi(t_0)$ *at* t_0, *i.e. for all* $t \in J$

$$\phi(t) \leq \phi(t_0) \exp\left(\int_{t_0}^t p(v)dv\right) + \int_{t_0}^t q(u)\exp\left(\int_u^t p(v)dv\right)du. \quad (1)$$

The proof follows the method of solution of the equation $x' = p(t)x + q(t)$ by means of an integrating factor.

Let $r(t) = \exp(-\int_{t_0}^t p(v)dv)$ (so that $r(t) > 0$). Then for nearly all $t \in J$

$$D_+(\phi(t)r(t)) = \liminf_{s \to t+}\left(\frac{\phi(s)r(s) - \phi(t)r(t)}{s - t}\right)$$

$$= r(t)\liminf_{s \to t+}\left(\frac{\phi(s) - \phi(t)}{s - t}\right) + \lim_{s \to t+}\left(\frac{\phi(s)(r(s) - r(t))}{s - t}\right)$$

$$= r(t)D_+\phi(t) + \phi(t)r'(t)$$

$$= r(t)D_+\phi(t) - \phi(t)r(t)p(t)$$

$$\leq q(t)r(t).$$

Hence, by (1.4.4),

$$\phi(t)r(t) - \phi(t_0) \leq \int_{t_0}^{t} q(u)r(u)du,$$

and this gives (1).

A similar argument gives the following analogous result for intervals to the left of the initial point.

(2.6.1)′ *Let* $J = [t_0 - \alpha, t_0]$, *let* $p, q : J \to \mathbf{R}$ *be continuous, and let* $\phi : J \to \mathbf{R}$ *be a continuous function such that*

$$D^+\phi(t) \geq - p(t)\phi(t) - q(t)$$

nearly everywhere in J. *Then for all* $t \in J$

$$\phi(t) \leq \phi(t_0) \exp\left(\int_{t}^{t_0} p(v)dv\right) + \int_{t}^{t_0} q(u)\exp\left(\int_{t}^{u} p(v)dv\right)du.$$

(2.6.2) (*Gronwall's inequality*) *Let* I *be an interval in* $\mathbf{R}$, *let* $h : I \to [0, \infty[$ *and* $g, \phi : I \to \mathbf{R}$ *be continuous, and let* $t_0 \in I$. *If for all* $t \in I$ *with* $t \geq t_0$

$$\phi(t) \leq g(t) + \int_{t_0}^{t} h(u)\phi(u)du,$$

then for all such t

$$\phi(t) \leq g(t) + \int_{t_0}^{t} g(u)h(u)\exp\left(\int_{u}^{t} h(v)dv\right)du.$$

Similarly, if for all $t \in I$ *with* $t \leq t_0$

$$\phi(t) \leq g(t) + \int_{t}^{t_0} h(u)\phi(u)du,$$

then for all such t

$$\phi(t) \leq g(t) + \int_{t}^{t_0} g(u)h(u)\exp\left(\int_{t}^{u} h(v)dv\right)du.$$

To prove the result for $t \geq t_0$, let $\psi(t) = \int_{t_0}^{t} h(u)\phi(u)du$. Then $\psi'(t) = h(t)\phi(t) \leq h(t)g(t) + h(t)\psi(t)$, and $\psi(t_0) = 0$, and we now have only to apply (2.6.1). The result for $t \leq t_0$ follows similarly from (2.6.1)′.

We note explicitly the case where g is constant.

(2.6.2. Corollary) (*Bellman's inequality*) *Let* I *be an interval in* $\mathbf{R}$, *let* $h : I \to [0, \infty[$ *and* $\phi : I \to \mathbf{R}$ *be continuous, let* $c \in \mathbf{R}$, *and let* $t_0 \in I$. *If for* *all $t \in I$ with $t \geq t_0$*

$$\phi(t) \leq c + \int_{t_0}^{t} h(u)\phi(u)du,$$

then for all such t

$$\phi(t) \leq c \exp\left(\int_{t_0}^{t} h(u)du\right).$$

Similarly, if for all $t \in I$ with $t \leq t_0$

$$\phi(t) \leq c + \int_{t}^{t_0} h(u)\phi(u)du,$$

then for all such t

$$\phi(t) \leq c \exp\left(\int_{t}^{t_0} h(u)du\right).$$

Applications of this corollary are given in Exercises 2.8.

The argument employed in the proof of (2.6.2) extends to give the following non-linear generalization of Bellman's inequality.

(2.6.3) *Let* $J = [t_0, t_0 + \alpha]$, *let* $f: J \times \mathbf{R} \to \mathbf{R}$ *be a continuous function such that* $x \mapsto f(t, x)$ *is increasing on* $\mathbf{R}$ *for each* $t \in J$, *let* $x_0 \in \mathbf{R}$, *and suppose that the maximal solution* Φ *of* $x' = f(t, x)$ *taking the value* x_0 *at* t_0 *exists on* J. *Let also* $\phi: J \to \mathbf{R}$ *be a continuous function such that*

$$\phi(t) \leq x_0 + \int_{t_0}^{t} f(u, \phi(u))du$$

for all $t \in J$. *Then* $\phi(t) \leq \Phi(t)$ *for all* $t \in J$.

Let $\psi(t) = x_0 + \int_{t_0}^{t} f(u, \phi(u))du$. Then $\psi'(t) = f(t, \phi(t)) \leq f(t, \psi(t))$ for all $t \in J$, and $\psi(t_0) = x_0$, and we now have only to apply (2.4.5).

Exercises 2.6

1 Let $J = [t_0, t_0 + \alpha]$, let p, q be extended real-valued functions Lebesgue integrable on J, and let $\phi: J \to \mathbf{R}$ be an absolutely continuous function such that

$$\phi'(t) \leq p(t)\phi(t) + q(t)$$

for almost all $t \in J$. Prove that ϕ satisfies the inequality (1) of (2.6.1).
   [Hint. Follow the argument of (2.6.1), using Exercise 1.10.1 in place of (1.4.4); the integrating factor r is absolutely continuous, whence so is ϕr.]
2 Let $J = [t_0, t_0 + \alpha]$, let p be a non-negative and q an extended real-valued function, each Lebesgue integrable on J, and let $\phi: J \to [0, \infty[$ be a continuous function

such that $D_+\phi(t) < \infty$ nearly everywhere in J and that

$$D_+\phi(t) \le p(t)\phi(t) + q(t)$$

for almost all $t\in J$. Prove that ϕ satisfies the inequality (1) of (2.6.1).

[Hint. Follow the argument of (2.6.1), using (1.10.8) in place of (1.4.4); note that the conditions that p and ϕ are non-negative imply that $D_+(\phi(t)r(t)) < \infty$ nearly everywhere.]

3 Prove the following generalization of Gronwall's inequality (2.6.2). Let $I = [t_0, b[$, where $t_0 < b \le \infty$, and let $h : I \to [0, \infty[$ and $g, \phi : I \to \mathbf{R}$ be functions such that h, gh, and $h\phi$ are Lebesgue integrable on each compact subinterval of I and that

$$\phi(t) \le g(t) + \int_{t_0}^t h(u)\phi(u)du$$

for almost all $t\in I$. Then

$$\phi(t) \le g(t) + \int_{t_0}^t g(u)h(u) \exp\left(\int_u^t h(v)dv \right)du$$

for almost all $t\in I$.

Obtain a similar generalization of Bellman's inequality (2.6.2. Corollary).

[Hint. Use Exercise 1.]

4 Let $J = [t_0, t_0 + \alpha]$, let $f : J \times \mathbf{R} \to \mathbf{R}$ be continuous, let $x_0\in\mathbf{R}$, and suppose that the maximal solution Φ of $x' = f(t, x)$ taking the value x_0 at t_0 exists on J. Prove that if $\phi : J \to \mathbf{R}$ is a continuous function such that

$$\phi(t) \le \phi(s) + \int_s^t f(u, \phi(u))du$$

whenever $s, t\in J$ and $s \le t$, then $\phi(t) \le \Phi(t)$ for all $t\in J$.

2.7 Lipschitz conditions

Let f be a function from a set $E \subseteq \mathbf{R} \times Y$ into Y, and let $K \ge 0$. Then $f(t, y)$ is said to be K-*Lipschitzian in y in E* if for all $(t, y), (t, z)\in E$

$$\| f(t, y) - f(t, z) \| \le K \| y - z \|.$$

We say also that $f(t, y)$ is *Lipschitzian in y in E* if it is K-Lipschitzian for some $K \ge 0$.

In this section we prove that the Lipschitz condition together with the continuity of f implies the existence and uniqueness of solutions of the equation

$$y' = f(t, y). \tag{1}$$

When the dimension of Y is finite, the existence of solutions is, of course, a consequence of the continuity of f alone. However, the addition of the Lipschitz condition ensures the existence of solutions in the infinite-dimensional case also. Although the Lipschitz condition appears fairly restrictive, it covers many important cases, and we cannot depart too far from it without losing at least the uniqueness of the solutions.

The principal results of this section are (2.7.4) and (2.7.6). The other results are essentially preliminaries to the proofs of the main results, though all have some interest in their own right.

Our first result proves the uniqueness of the solutions of the equation (1) and provides the key to the proof of existence.

(2.7.1) *Let $E \subseteq \mathbf{R} \times Y$, and let $f : E \to Y$ be a continuous function such that $f(t, y)$ is K-Lipschitzian in y in E, where $K > 0$. Let also $\varepsilon_1, \varepsilon_2 \geq 0$, and let ψ_1 and ψ_2 be respectively ε_1- and ε_2-approximate solutions of the equation* (1) *on an interval I. Then for all $t, t_0 \in I$*

$$\| \psi_1(t) - \psi_2(t) \| \leq \| \psi_1(t_0) - \psi_2(t_0) \| \exp(K|t - t_0|)$$
$$+ (\varepsilon_1 + \varepsilon_2) K^{-1} (\exp(K|t - t_0|) - 1).$$

In particular, if ψ_1, ψ_2 are exact solutions of (1) *on I such that $\psi_1(t_0) = \psi_2(t_0)$ for some $t_0 \in I$, then $\psi_1 = \psi_2$.*

Let $\phi(t) = \| \psi_1(t) - \psi_2(t) \|$ $(t \in I)$. Then for all except a finite number of points $t \in I$ we have

$$- \| \psi_1'(t) - \psi_2'(t) \| \leq D^+ \phi(t) \leq \| \psi_1'(t) - \psi_2'(t) \|.$$

(cf. (1.6.1)). The second of these inequalities gives

$$D^+ \phi(t) \leq \| f(t, \psi_1(t)) - f(t, \psi_2(t)) \| + \| \psi_1'(t) - f(t, \psi_1(t)) \|$$
$$+ \| \psi_2'(t) - f(t, \psi_2(t)) \| \leq K\phi(t) + \varepsilon_1 + \varepsilon_2,$$

and by applying (2.6.1) with $p(t) = K, q(t) = \varepsilon_1 + \varepsilon_2$, we obtain the required result for all $t \in I$ such that $t \geq t_0$. Similarly, by using the left-hand inequality for $D^+ \phi(t)$ and (2.6.1)$'$, we obtain the result for $t \leq t_0$.

(2.7.2) *Let C be the cylinder $\{ (t, y) : t_0 \leq t \leq t_0 + \alpha, \| y - y_0 \| \leq \rho \}$ in $\mathbf{R} \times Y$, where $\rho > 0$, and let $f : C \to Y$ be a continuous function with the properties that $\| f(t, y) \| \leq M$ for all $(t, y) \in C$ and that $f(t, y)$ is Lipschitzian in y in C. Then, if $\eta = \min \{ \alpha, \rho / M \}$, the equation* (1) *has one and only one solution ϕ on $[t_0, t_0 + \eta]$ satisfying the condition $\phi(t_0) = y_0$.*

A similar result holds for solutions to the left of t_0.

By (2.3.1), for each positive integer n we can find a $1/n$-approximate solution ψ_n of the equation (1) on $[t_0, t_0 + \eta]$ satisfying $\psi_n(t_0) = y_0$. Further, by the inequality of (2.7.1) applied to ψ_m, ψ_n, the sequence (ψ_n) converges uniformly on $[t_0, t_0 + \eta]$. If now $\phi = \lim \psi_n$, then $\phi(t_0) = y_0$, and the graph of ϕ lies in C. It therefore follows from (2.3.2) that ϕ is an exact solution of (1) on $[t_0, t_0 + \eta]$. Moreover, ϕ is unique, by (2.7.1), and this completes the proof.

(2.7.3) *Let $J = [a, b]$ be a compact interval in $\mathbf{R}$, let $f : J \times Y \to Y$ be* *a continuous function such that* $f(t, y)$ *is* K-*Lipschitzian in* y *in* $J \times Y$, *where* $K > 0$, *and let* $(t_0, y_0) \in J \times Y$. *Then the equation* $y' = f(t, y)$ *has one and only one solution on* J *taking the value* y_0 *at* t_0.

Suppose first that $a \leq t_0 < b$, and consider the solution on $[t_0, b]$. We observe that if $(\tau, \xi) \in [t_0, b[\times Y, \ 0 < \mu < 1/K$, and $\delta = \min \{b - \tau, \mu\}$, then the equation has a solution on the interval $[\tau, \tau + \delta]$ taking the value ξ at τ. To prove this, let $\sigma = \sup_J \| f(t, \xi) \|$. If $\sigma = 0$, the constant function $t \mapsto \xi$ is the required solution, so that we may suppose $\sigma > 0$. Then, if $t \in J$ and $\| y - \xi \| \leq \rho$ we have

$$\| f(t, y) \| \leq \| f(t, y) - f(t, \xi) \| + \| f(t, \xi) \| \leq K \| y - \xi \| + \sigma \leq K\rho + \sigma.$$

It therefore follows from (2.7.2) that, if $\eta = \min \{b - \tau, \rho/(K\rho + \sigma)\}$, then the equation has a solution on the interval $[\tau, \tau + \eta]$ taking the value ξ at τ, and since $0 < \delta < 1/K$ we can choose ρ so that $\rho/(K\rho + \sigma) = \delta$.

We now divide the interval $[t_0, b]$ into m subintervals $[t_0, t_1], \ldots,$ $[t_{m-1}, t_m]$ of equal length not exceeding $1/K$, where $t_m = b$. By the preceding remark, we can define inductively functions $\phi_i : [t_i, t_{i+1}] \to Y$ $(i = 0, \ldots, m - 1)$ such that each ϕ_i is a solution of the equation on its domain and that $\phi_0(t_0) = y_0$ and $\phi_i(t_i) = \phi_{i-1}(t_i)$ $(i = 1, \ldots, m - 1)$. By piecing together these solutions, we thus obtain a solution on $[t_0, b]$ taking the value y_0 at t_0. A similar proof applies to the interval $[a, t_0]$ when $a < t_0 \leq b$, and hence we obtain the required solution on J. Moreover, by (2.7.1), there is at most one such solution, and this completes the proof.

(2.7.4) *Let* I *be an interval in* **R**, *let* $f : I \times Y \to Y$ *be a continuous function such that, for each compact subinterval* J *of* I, *there exists a positive number* K *(possibly depending on* J) *such that* $f(t, y)$ *is* K-*Lipschitzian in* y *in* $J \times Y$, *and let* $(t_0, y_0) \in I \times Y$. *Then the equation* $y' = f(t, y)$ *has one and only one solution on* I *taking the value* y_0 *at* t_0.

Let b be the right-hand endpoint of I. If $b \in I$, then, by (2.7.3), there is a solution of the equation on $[t_0, b]$ taking the value y_0 at t_0. If $b \notin I$, let $(t_n)_{n \geq 1}$ be a strictly increasing sequence of points of $[t_0, b[$ with the limit b. By (2.7.3), we can define inductively a sequence of functions $\phi_n : [t_n, t_{n+1}] \to Y$, each a solution of the equation on its domain, such that $\phi_0(t_0) = y_0$ and that $\phi_n(t_n) = \phi_{n-1}(t_n)$ for $n = 1, 2, \ldots$. By piecing together these solutions, we obtain a solution of the equation on $[t_0, b[$ taking the value y_0 at t_0. A similar argument applies on the left of t_0, and we piece together the two solutions thus obtained to provide the required solution on I. By (2.7.1), the restriction of this solution to any compact subinterval of I containing t_0 is uniquely determined, and hence the solution itself is unique.

Example

Theorem (2.7.4) provides the principal existence and uniqueness theorem in the theory of linear differential equations (cf. §2.1, Example (*a*), p. 71). Let *I* be an interval in **R**, let *A* be a continuous function from *I* into the space $\mathcal{L}(Y; Y)$ of continuous linear maps of *Y* into itself, and let *b* be a continuous function from *I* into *Y*. Then *for each point* $(t_0, y_0) \in I \times Y$ *the linear equation*

$$y' = A(t)y + b(t)$$

has exactly one solution ϕ *on I such that* $\phi(t_0) = y_0$. To prove this, let $f : I \times Y \to Y$ be given by

$$f(t, y) = A(t)y + b(t).$$

Since *A* and *b* are continuous, so is *f*. Moreover, if *J* is a compact sub-interval of *I*, then $t \mapsto \| A(t) \|$ is continuous and therefore bounded on *J*, say $\| A(t) \| \leq K$, and hence for all $t \in J$ and $y, z \in Y$ we have

$$\| f(t, y) - f(t, z) \| = \| A(t)y - A(t)z \| = \| A(t)(y - z) \|$$
$$\leq \| A(t) \| \, \| y - z \| \leq K \| y - z \|.$$

The stated result therefore follows directly from (2.7.4).

Let $E \subseteq \mathbf{R} \times Y$, and let $f : E \to Y$. We say that $f(t, y)$ is *locally Lipschitzian in y in E* if for each point $(s, z) \in E$ there exists a neighbourhood *U* of (s, z) in *E* and a number $K \geq 0$ such that $f(t, y)$ is *K*-Lipschitzian in *y* in *U*.

The next theorem is the global uniqueness theorem for the local Lipschitz case.

(2.7.5) *Let* $E \subseteq \mathbf{R} \times Y$, *and let* $f : E \to Y$ *be a continuous function such that* $f(t, y)$ *is locally Lipschitzian in y in E. If* ϕ_1, ϕ_2 *are solutions of the equation* $y' = f(t, y)$ *on some interval I taking the same value at some point of I, then* $\phi_1 = \phi_2$.

Let *A* be the set of points $t \in I$ such that $\phi_1(t) = \phi_2(t)$. Since *I* is connected and *A* is non-empty, it is enough to prove that *A* is both open and closed in *I*, for then $A = I$. Since ϕ_1, ϕ_2 are continuous, *A* is evidently closed in *I*. To prove that *A* is open in *I*, let $t_0 \in A$ and let $y_0 = \phi_1(t_0) = \phi_2(t_0)$. Since $f(t, y)$ is locally Lipschitzian in *y* in *E*, we can find positive δ, K such that $f(t, y)$ is *K*-Lipschitzian in *y* in the set $\{(t, y) \in E : |t - t_0| < \delta,$ $\| y - y_0 \| < \delta\}$. It then follows from the last part of (2.7.1) that $\phi_1(t) = \phi_2(t)$ for all $t \in]s - \delta, s + \delta[\cap I$, so that *A* is open in *I*.

(2.7.6) *Let* $E \subseteq \mathbf{R} \times Y$, *let* $f : E \to Y$ *be a continuous function such that* $f(t, y)$ *is locally Lipschitzian in y in E, and let* (t_0, y_0) *be an interior point* *of E. Then there is a greatest interval I containing t_0 on which the equation $y' = f(t, y)$ has a solution taking the value y_0 at t_0, and, moreover, this solution is unique. Further, t_0 is an interior point of I, and if E is open so is I.*

The hypotheses imply that there exists a cylinder $C = \{(t, y) : |t - t_0| \leq \delta, \|y - y_0\| \leq \delta\}$ contained in E such that f is bounded on C and that $f(t, y)$ is Lipschitzian in y in C. By (2.7.2), we can therefore find an interval I_1 with t_0 as an interior point such that the equation has one and only one solution on I_1 taking the value y_0 at t_0.

Let Ω be the set of all pairs (I_ω, ϕ_ω) such that I_ω is an interval containing t_0 and that ϕ_ω is a solution of the equation on I_ω satisfying $\phi_\omega(t_0) = y_0$. By the preceding remark, Ω is non-empty. Moreover, by (2.7.5), two solutions ϕ_ω, ϕ_μ agree on $I_\omega \cap I_\mu$. Let $I = \bigcup_\Omega I_\omega$, and define $\phi : I \to Y$ by setting $\phi(t) = \phi_\omega(t)$ whenever $t \in I_\omega$. Then ϕ is a well-defined function that is a solution of the equation on I satisfying $\phi(t_0) = y_0$, and, by (2.7.5), it is the only such solution. Since there is clearly no larger interval containing t_0 on which the equation has a solution taking the value y_0 at t_0, this proves the first part of the theorem.

Next, since $I_1 \subseteq I$, t_0 is an interior point of I. Further, if E is open, then I is open, for if, say, the right-hand endpoint b of I belongs to I, then $(b, \phi(b))$ is an interior point of E. By the first part of the proof, there is therefore a solution ψ of the equation on a non-degenerate interval $[b, c]$ such that $\psi(b) = \phi(b)$. The function on $I \cup [b, c]$ that agrees with ϕ on I and with ψ on $[b, c]$ is then a solution of the equation on $I \cup [b, c]$ taking the value y_0 at t_0, and this contradicts the definition of I.

The interval I of (2.7.6) is called the *maximal interval of existence* of the solution of the equation $y' = f(t, y)$ satisfying the given initial condition.

When E is open and the dimension of Y is finite, the solution $\phi : I \to Y$ reaches to the boundary of E in both directions, and the results of (2.4.3) apply. When Y is infinite-dimensional, the notion of reaching to the boundary of E in the sense defined in §2.4 is not useful, since an open set E in $\mathbf{R} \times Y$ is not now the union of a sequence of compact sets. In comparison with (2.4.3), all that we can assert here is that the following statements are equivalent:

(i) $[t_0, b[$ is the right-hand part of the maximal interval of existence of ϕ;

(ii) the graph of ϕ is not contained in any bounded closed subset of E on which f is bounded;

(iii) either $b = \infty$, or ϕ has no extension to $[t_0, b]$ that is a solution of the equation on this interval;

(iv) there is no sequence of points $t_n \in [t_0, b[$ converging to b such that the sequence $(t_n, \phi(t_n))$ converges to a point of E.

The proof is similar to that of (2.4.3), using (1.7.1–3).

As an example of an equation to which (2.7.6) can be applied, consider the scalar equation

$$y' = 3y^{2/3} \tag{2}$$

in the upper half-plane $\Pi = \{(t, y) \in \mathbf{R}^2 : y > 0\}$. Here the function f is given by $f(t, y) = 3y^{2/3}$, and $f(t, y)$ is locally Lipschitzian in y in Π. In fact, if $y, z > 0$, then, by the mean value theorem,

$$|3y^{2/3} - 3z^{2/3}| = 2\xi^{-1/3}|y - z|$$

for some ξ between y and z, and therefore

$$|3y^{2/3} - 3z^{2/3}| \le 2\delta^{-1/3}|y - z|$$

whenever $y, z \ge \delta > 0$. Hence $f(t, y)$ is $2\delta^{-1/3}$-Lipschitzian in y in $\{(t, y) : y \ge \delta\}$ for each $\delta > 0$, and is therefore locally Lipschitzian in Π.

It follows from (2.7.6) that if $(t_0, y_0) \in \Pi$, the equation (2) has one and only one solution taking the value y_0 at t_0, with graph in Π and with maximal interval of existence. In fact this solution is given by

$$y = (t - t_0 + y_0^{1/3})^3, \tag{3}$$

and its maximal interval of existence is $]t_0 - y_0^{1/3}, \infty[$.

It is easy to see that the function $(t, y) \mapsto 3y^{2/3}$ is not locally Lipschitzian in any open subset of $\mathbf{R}^2$ containing a point of the t-axis, and indeed we have already seen in §1.5 that for each $t_0 \in \mathbf{R}$ the equation (2) has an infinity of solutions on $\mathbf{R}$ taking the value 0 at t_0. The discussion of §1.5 shows also that the solution (3) has an infinity of extensions to the whole real line $\mathbf{R}$, each satisfying (2) on $\mathbf{R}$ but with graph lying partly outside Π. Thus as soon as the graph of the solution passes outside Π we lose uniqueness.

The function $(t, y) \mapsto 3y^{2/3}$ is also locally Lipschitzian in the lower half-plane $\{(t, y) : y < 0\}$, and similar remarks apply there.

Note the difference between the hypotheses of (2.7.4) and the hypothesis that $f : I \times Y \to Y$ is a continuous function such that $f(t, y)$ is locally Lipschitzian in y in $I \times Y$. Under the latter hypothesis the maximal interval of existence of a solution may be a proper subset of I. This is easily seen by considering the scalar equation $y' = e^{-y}$ in the plane $\mathbf{R}^2$. An argument similar to that for (2) shows that e^{-y} is locally Lipschitzian in $\mathbf{R}^2$ and the solution taking the value y_0 at t_0 has already been shown in

§2.4 (p. 80) to have maximal interval of existence $]t_0 - e^{y_0}, \infty [$. On the other hand, in (2.7.4) every solution has maximal interval of existence I.

Exercises 2.7

1   For each of the functions $f : E \to \mathbf{R}$ defined below, show that for each $(t_0, x_0) \in E$ the equation $x' = f(t, x)$ has exactly one solution ϕ satisfying $\phi(t_0) = x_0$ and reaching to the boundary of E in both directions, and find the domain of ϕ. In each case sketch the family of solutions.

        (a) $E = \mathbf{R}^2$, $f(t, x) = x^2$;

        (b) $E = \mathbf{R} \times]0, \infty[$, $f(t, x) = t/x$;

        (c) $E = \mathbf{R}^2$, $f(t, x) = 1 + x^2$;

        (d) $E = \mathbf{R}^2$, $f(t, x) = \sin x$;

        (e) $E = \mathbf{R}^2$, $f(t, x) = 2tx/(1 + t^2 + x^2)$;

        (f) $E = \mathbf{R}^2$, $f(t, x) = (1 + t^2)/(1 + x^2)$.

2   Let $J = [t_0, t_0 + \alpha]$, and let $f : I \times Y \to Y$ be a continuous function such that

$$\| f(t, y) - f(t, z) \| \le \mu(t) \| y - z \|$$

for almost all $t \in J$ and all $y, z \in Y$, where μ is a non-negative function Lebesgue integrable on J. Let also

$$M(t) = \exp \left(\int_{t_0}^{t} \mu(s) ds \right) \qquad (t \in J),$$

let X be the linear space of continuous $\psi : J \to Y$ with the norm

$$\| \psi \|^* = \sup_{J} \| \psi(t) \| / M(t),$$

let $y_0 \in Y$, and let $T : X \to X$ be given by

$$T\psi(t) = y_0 + \int_{t_0}^{t} f(u, \psi(u)) du \qquad (\psi \in X, t \in J).$$

Prove that

  (i) X is complete, and T is a contraction of X into itself,

  (ii)   the equation $y' = f(t, y)$ has exactly one solution ϕ on J such that $\phi(t_0) = y_0$,

  (iii) if $\psi \in X$ and (ψ_n) is the sequence in X defined by $\psi_0 = \psi$, $\psi_n = T\psi_{n-1}$ $(n = 1, 2, \ldots)$, then $\psi_n \to \phi$ in X,

  (iv)   if $\psi \in X$, then for all $t \in J$

$$\| \psi(t) - \phi(t) \| \le (M(t))^2 \sup_{u \in J} \left(\left\| \psi(u) - y_0 - \int_{t_0}^{u} f(s, \psi(s)) ds \right\| / M(u) \right).$$

[The special case of (ii) obtained by taking $\mu(t) = K$ provides an alternative proof of (2.7.3), and (iv) provides a substitute for the inequality of (2.7.1). A local result of the form of (2.7.2) can be deduced from (ii) by using the extension lemma (2.10.2) below.

The functions ψ_n in (iii) are known as the *successive approximations* to ϕ, and we return to them in §2.12.]

3   By using Exercise 2.6.2, show that the arguments of (2.7.1, 2) can be extended to cover the generalized Lipschitz condition considered in Exercise 2.

2.8 Linear equations

We recall (§2.1, Example (*a*), p. 71) that a *linear equation of the first order*
is an equation of the form

$$y' = A(t)y + b(t), \tag{1}$$

where *b* is a continuous function from an interval *I* in **R** into a normed
space *Y* and *A* is a continuous function from *I* into the space $\mathscr{L}(Y; Y)$ of
continuous linear maps of *Y* into itself (the value of *A*(*t*) at the point
y of *Y* being denoted by *A*(*t*)*y*). We have also shown (§2.7, Example,
p. 100) that the solutions of this equation are unique; more precisely:

(2.8.1) *Let A, b satisfy the above conditions. Then for each* $(t_0, y_0) \in I \times Y$
the equation (1) *has exactly one solution* ϕ *on I such that* $\phi(t_0) = y_0$. *Moreover, every solution of* (1) *has maximal interval of existence I.*

The linear equation

$$y' = A(t)y \tag{2}$$

is called *homogeneous* (or sometimes the *homogeneous* (linear) *equation*
associated with the equation (1)). The zero function from *I* into *Y* is
obviously a solution of (2), and, by the uniqueness part of (2.8.1), this
function is the only solution of (2) on *I* that takes the value 0 at any point
of *I*.

The linearity, for each $t \in I$, of the map $A(t) : Y \to Y$ has many important
consequences. For instance, it is obvious that if η is a solution of (1) on *I*
and ϕ is a solution of (2) on *I*, then $\phi + \eta$ is a solution of (1) on *I*, and
that if η_1, η_2 are solutions of (1), then $\eta_1 - \eta_2$ is a solution of (2).

For the homogeneous equation (2) we have:

(2.8.2) *The set S of solutions of* (2) *on I forms a vector space over the underlying field* (**R** *or* **C**) *of Y, and, if Y has finite dimension n, so does S. Further, if*
$\phi_1, \ldots, \phi_p \in S$, *then the following statements are equivalent:*
  (i) $\phi_1, \ldots, \phi_p$ *are linearly independent in S;*
  (ii) $\phi_1(t_0), \ldots, \phi_p(t_0)$ *are linearly independent in Y for some* $t_0 \in I$;
  (iii) $\phi_1(t_0), \ldots, \phi_p(t_0)$ *are linearly independent in Y for each* $t_0 \in I$.

The linearity of *A*(*t*) trivially implies that, if $\phi_1, \phi_2 \in S$ and α_1, α_2 are
scalars, then $\alpha_1 \phi_1 + \alpha_2 \phi_2 \in S$, so that *S* is a vector space. Further, if
$t_0 \in I$ and $F : S \to Y$ is given by $F(\phi) = \phi(t_0)$ ($\phi \in S$), then *F* is a linear
isomorphism of *S* onto *Y*, for it is clearly linear, and, by (2.8.1), it is one-to-
one and onto *Y*. Since an isomorphism preserves both dimension and
linear independence, the remaining results now follow.

This theorem shows that if Y has finite dimension n and $\phi_1, \ldots, \phi_n$ are n linearly independent solutions of (2) on I, then every solution of (2) on I is of the form $\sum \alpha_j \phi_j$ for some (unique) scalars $\alpha_1, \ldots, \alpha_n$. In view of this, the function $\sum \alpha_j \phi_j$, where $\alpha_1, \ldots, \alpha_n$ are arbitrary scalars, is sometimes called the *general solution* of (2). We deduce also that if η is a given solution of (1) on I, then every solution ψ of (1) on I is of the form $\sum \alpha_j \phi_j + \eta$ for some (unique) scalars $\alpha_1, \ldots, \alpha_n$ (for $\psi - \eta \in S$).

The study of the equations (1) and (2) is facilitated by the introduction of the further differential equation

$$X' = A(t) \circ X, \tag{3}$$

where the unknown function X takes its values in $\mathscr{L}(Y; Y)$, and $\circ$ denotes as usual the operation of composition. This equation (3) is also a linear equation, for if for each $t \in I$ we define the mapping $\mathscr{A}(t)$ of $\mathscr{L}(Y; Y)$ into itself by

$$\mathscr{A}(t)X = A(t) \circ X \qquad (X \in \mathscr{L}(Y; Y)),$$

then $\mathscr{A}(t)$ is clearly linear, and it is continuous on $\mathscr{L}(Y; Y)$, for if $X, X' \in \mathscr{L}(Y; Y)$ then

$$\| \mathscr{A}(t)X - \mathscr{A}(t)X' \| = \| A(t) \circ (X - X') \| \le \| A(t) \| \, \| X - X' \|.$$

Moreover, $t \mapsto \mathscr{A}(t)$ is continuous on I, for if $t, t' \in I$ and $X \in \mathscr{L}(Y; Y)$ then

$$\| (\mathscr{A}(t) - \mathscr{A}(t'))X \| = \| (A(t) - A(t')) \circ X \| \le \| A(t) - A(t') \| \, \| X \|,$$

so that $\| \mathscr{A}(t) - \mathscr{A}(t') \| \le \| A(t) - A(t') \|$.

It therefore follows from (2.8.1) that for each $t_0 \in I$ and each $X_0 \in \mathscr{L}(Y; Y)$ the equation (3) has exactly one solution on I whose value at t_0 is X_0. Moreover, every solution of (3) has maximal interval of existence I.

The reasons for the usefulness of the equation (3) are to be found in the following lemma.

(2.8.3) **Lemma.** *Let $A : I \to \mathscr{L}(Y; Y)$ be continuous, and let $X = U(t)$ be a solution of (3) on I. Then*
(i) *for each $y_0 \in Y$ the function $t \mapsto U(t)y_0$ is a solution of (2) on I,*
(ii) *for each $X_0 \in \mathscr{L}(Y; Y)$ the function $t \mapsto U(t) \circ X_0$ is a solution of (3) on I.*
These results follow immediately from the easily verified relations

$$\frac{d}{dt}(U(t)y_0) = U'(t)y_0 \quad \text{and} \quad \frac{d}{dt}(U(t) \circ X_0) = U'(t) \circ X_0. \dagger \tag{4}$$

Now let $t_0 \in I$, and let $t \mapsto R(t; t_0)$ be the solution of (3) on I whose value at t_0 is the identity map 1_Y of Y onto itself. Then, by the second

† It should be noted that in general the function $t \mapsto X_0 \circ U(t)$ is not a solution of (3).

part of the lemma (2.8.3) and the uniqueness part of (2.8.1), the solution of (3) on I taking a given value X_0 at t_0 is $t \mapsto R(t;t_0) \circ X_0$. In particular, if $t_1 \in I$, then for all $t \in I$ we have

$$R(t;t_1) = R(t;t_0) \circ R(t_0;t_1).$$

Taking $t_1 = t$, we thus obtain that

$$1_Y = R(t;t_0) \circ R(t_0;t),$$

whence $R(t;t_0)$ maps Y onto itself. On interchanging t_0 and t we obtain also that

$$1_Y = R(t_0;t) \circ R(t;t_0),$$

and this implies that $R(t;t_0)$ is one-to-one, i.e. for each $t \in I$ the map $R(t;t_0)$ is a linear homeomorphism of Y onto itself. Since any solution U of (3) on I necessarily satisfies the relation

$$U(t) = R(t;t_0) \circ U(t_0)$$

(by the remark above), we therefore obtain the following result.

(2.8.4) *Let* $A:I \to \mathcal{L}(Y;Y)$ *be continuous, and let* U *be a solution of* (3) *on* I. *If for some* $t_0 \in I$ *the continuous linear map* $U(t_0):Y \to Y$ *is a homeomorphism, then the map* $U(t):Y \to Y$ *is a homeomorphism for each* $t \in I$.

We say that a solution U of the equation (3) on I is a *fundamental kernel* for the equations (1) and (2) if for all $t \in I$ the map $U(t):Y \to Y$ is a homeomorphism. Any fundamental kernel U for (1) and (2) determines the function A uniquely, since $A(t) = U'(t) \circ U(t)^{-1}$.

Some information about the function $t \mapsto U(t)^{-1}$ can be derived from yet another differential equation related to (1) and (2), namely

$$X' = -X \circ A(t), \tag{5}$$

where again the unknown function X takes its values in $\mathcal{L}(Y;Y)$. An argument similar to that above shows that this equation too is linear, so that the result of (2.8.1) applies. We prove further:

(2.8.5) *Let* $A:I \to \mathcal{L}(Y;Y)$ *be continuous, and let* U *be a fundamental kernel for the equations* (1) *and* (2). *Then the function* $t \mapsto U(t)^{-1}$ *is a solution of the equation* (5) *on* I *(so that it is continuously differentiable there), and its derivative at the point* t *is* $-U(t)^{-1} \circ U'(t) \circ U(t)^{-1}$.

Let $t_0 \in I$. Then, by (2.8.1), the equation (5) has a unique solution V on I such that $V(t_0) = U(t_0)^{-1}$, and

$$\frac{d}{dt}(V(t) \circ U(t)) = V(t) \circ U'(t) + V'(t) \circ U(t)$$
$$= V(t) \circ A(t) \circ U(t) - V(t) \circ A(t) \circ U(t) = 0.$$

Hence $t \mapsto V(t) \circ U(t)$ is constant, and since it has the value 1_Y at t_0, $V(t) \circ U(t) = 1_Y$ for all $t \in I$, so that $V(t) = U(t)^{-1}$. Hence also $V'(t) = -V(t) \circ A(t) = -V(t) \circ U'(t) \circ U(t)^{-1}$, and this gives the result.

The next result, which is known as the 'variation of parameters formula', provides one of the main applications of fundamental kernels (see also Exercise 2.8.12).

(2.8.6) *Let* $A : I \to \mathscr{L}(Y; Y)$ *and* $b : I \to Y$ *be continuous, let* U *be a fundamental kernel for the equation* (1) *on* I, *and let* $(t_0, y_0) \in I \times Y$. *Then the solution of* (1) *on* I *taking the value* y_0 *at* t_0 *is given by*

$$y = U(t)U(t_0)^{-1}y_0 + U(t) \int_{t_0}^{t} U(u)^{-1}b(u)du. \dagger \qquad (6)$$

This result can be verified by direct differentiation of the formula (6), but we prefer to give here an alternative proof which explains the phrase 'variation of parameters'.

By the lemma (2.8.3) and the uniqueness part of (2.8.1), the solution of the equation $y' = A(t)y$ taking the value y_0 at t_0 is $t \mapsto U(t)U(t_0)^{-1}y_0$. We wish here to find the corresponding solution of the equation $y' = A(t)y + b(t)$, and to do this we consider the function $t \mapsto U(t)U(t_0)^{-1}w(t)$, where we have replaced the fixed vector y_0 above by a function w. Thus let y be the solution of $y' = A(t)y + b(t)$ on I taking the value y_0 at t_0, and let $w : I \to Y$ be defined by

$$y = U(t)U(t_0)^{-1}w(t),$$

so that $w(t_0) = y_0$. Then

$$\frac{dy}{dt} = U(t)U(t_0)^{-1}w'(t) + U'(t)U(t_0)^{-1}w(t)$$

$$= U(t)U(t_0)^{-1}w'(t) + A(t)U(t)U(t_0)^{-1}w(t)$$

$$= U(t)U(t_0)^{-1}w'(t) + A(t)y.$$

Since $y' = A(t)y + b(t)$, we obtain that

$$U(t)U(t_0)^{-1}w'(t) = b(t),$$

and since $U(t)$ and $U(t_0)$ are homeomorphisms, this gives

$$w'(t) = U(t_0)U(t)^{-1}b(t).$$

---

† Following our convention that $A(t)y$ denotes the value of $A(t)$ at y, here $U(t)U(t_0)^{-1}y_0$ denotes the element

$$U(t)(U(t_0)^{-1}(y_0)) = (U(t) \circ U(t_0)^{-1})(y_0)$$

of Y.

Since $t \mapsto U(t)^{-1}$ is continuous (by (2.8.5)), we therefore obtain

$$w(t) = y_0 + U(t_0) \int_{t_0}^{t} U(u)^{-1} b(u) du,$$

so that y satisfies (6).

When Y has finite dimension n, there is a simple relationship between the fundamental kernels for the equation

$$y' = A(t)y \tag{7}$$

and linearly independent sets of n solutions of this equation. Thus *let U be a fundamental kernel for* (7), *let* $\mathbf{U}(t)\,(t \in I)$ *be the matrix of $U(t)$ with respect to a given basis* $B = \{e_1, \ldots, e_n\}$ *for Y, and for $j = 1, \ldots, n$ let* $\phi_j : I \to Y$ *be the function such that the coordinates of $\phi_j(t)$ with respect to B form the jth column of* $\mathbf{U}(t)$. *Then $\phi_1, \ldots, \phi_n$ are linearly independent solutions of* (7) *on I (and therefore form a basis for the space S of* (2.8.2)). In fact, $\phi_j(t) = U(t)e_j$, so that ϕ_j is a solution of (7) on I, by (2.8.3)(i). Moreover, since $U(t)$ is an isomorphism of Y onto itself, the vectors $\phi_1(t), \ldots, \phi_n(t)$ are linearly independent in Y, whence $\phi_1, \ldots, \phi_n$ are linearly independent in S.

Conversely, *if $\phi_1, \ldots, \phi_n$ are linearly independent solutions of* (7) *on I, $\mathbf{U}(t)$ is the matrix whose columns are the coordinates of $\phi_1(t), \ldots, \phi_n(t)$, and $U(t)$ has matrix $\mathbf{U}(t)$, then U is a fundamental kernel for* (7). Here $U(t)$ is clearly an isomorphism of Y onto itself. Moreover, $U'(t) = A(t) \circ U(t)$, for this definition of $U(t)$ is equivalent to the definition

$$U(t) \sum \alpha_j e_j = \sum \alpha_j \phi_j(t)$$

for all scalars α_j, and hence for all $t \in I$ we have (cf. the first relation in (4))

$$U'(t) \sum \alpha_j e_j = \sum \alpha_j \phi'_j(t) = \sum \alpha_j A(t) \phi_j(t) = A(t) \sum \alpha_j \phi_j(t)$$

$$= A(t) \left(U(t) \sum \alpha_j e_j \right) = A(t) U(t) \sum \alpha_j e_j.$$

Another result in the same circle of ideas is the *Abel–Liouville identity*. Let $\phi_1, \ldots, \phi_n$ be solutions of $y' = A(t)y$ on I (not necessarily linearly independent), and for each $t \in I$ let $\mathbf{V}(t)$ be the matrix whose jth column $(j = 1, \ldots, n)$ consists of the coordinates $\phi_{1j}(t), \ldots, \phi_{nj}(t)$ of $\phi_j(t)$ with respect to the basis $\{e_1, \ldots, e_n\}$. Then for all $t, t_0 \in I$ we have

$$\det \mathbf{V}(t) = \det \mathbf{V}(t_0) \exp \left(\int_{t_0}^{t} \operatorname{tr} A(u) du \right). \tag{8}$$

(This identity provides an alternative proof of the equivalence of (2.8.2)(ii) and (iii).)

Let $W(t) = \det \mathbf{V}(t)$. Then to prove (8) it is enough to show that for all $t \in I$

$$W'(t) = W(t) \operatorname{tr} A(t),$$

or, equivalently, that if the matrix of $A(t)$ with respect to $\{e_1, \ldots, e_n\}$ is $[a_{ij}(t)]$, then

$$W'(t) = W(t) \sum_{i=1}^{n} a_{ii}(t). \tag{9}$$

In matrix form, the equation $y' = A(t)y$ is

$$y' = [a_{ij}(t)]y,$$

and since each ϕ_j is a solution we have

$$\phi'_{ij}(t) = \sum_{k=1}^{n} a_{ik}(t)\phi_{kj}(t) \qquad (i,j = 1, \ldots, n). \tag{10}$$

We observe now that $W'(t)$ is the sum of the n determinants obtained from $W(t)$ by differentiating one row at a time, the other rows remaining unchanged. Thus the ith such determinant $W_i(t)$ differs from $W(t)$ only in its ith row, which is

$$\phi'_{i1}(t), \ldots, \phi'_{in}(t),$$

and, by (10), this can be written as

$$\sum_{k=1}^{n} a_{ik}(t)\phi_{k1}(t), \ldots, \sum_{k=1}^{n} a_{ik}(t)\phi_{kn}(t).$$

Hence this row is the sum of $a_{ii}(t)$ times the ith row of $W(t)$ and a linear combination of the other rows of $W_i(t)$, whence $W_i(t) = a_{ii}(t)W(t)$. On summing from $i = 1$ to $i = n$ we thus obtain (9).

A number of inequalities for solutions of the linear equations

$$y' = A(t)y + b(t)$$

and

$$y' = A(t)y$$

(where Y is now an arbitrary real or complex normed space) can be derived from the differential inequalities of (2.6.1) and (2.6.1)'. For example, if ϕ is a solution of the non-homogeneous equation above, then, since

$$-\|\phi'(t)\| \le D^+ \|\phi(t)\| \le \|\phi'(t)\|$$

(cf. (1.6.1)), we have

$$-\|A(t)\| \, \|\phi(t)\| - \|b(t)\| \le D^+ \|\phi(t)\| \le \|A(t)\| \, \|\phi(t)\| + \|b(t)\|$$

and therefore for all $t, t_0 \in I$ we have

$$\| \phi(t) \| \leq \| \phi(t_0) \| \exp \left(\left| \int_{t_0}^{t} \| A(v) \| \, dv \right| \right)$$

$$+ \left| \int_{t_0}^{t} \| b(u) \| \exp \left(\left| \int_{u}^{t} \| A(v) \| \, dv \right| \right) du \right|. \quad (11)$$

A sharper result is obtained by using the *logarithmic norm* μ on $\mathscr{L}(Y; Y)$, defined by the relation

$$\mu(T) = \lim_{h \to 0+} \frac{\| 1_Y + hT \| - 1}{h} \quad (T \in \mathscr{L}(Y; Y)),$$

where 1_Y is the identity map of Y onto itself. The limit on the right exists, for it is the right-hand derivative at 0 of the convex function $h \mapsto \| 1_Y + hT \|$ from $\mathbf{R}$ into itself. Moreover, we obviously have

$$|\mu(T)| \leq \| T \|,$$
$$\mu(\alpha T) = \alpha \mu(T) \text{ for all } \alpha \geq 0,$$
$$\mu(T_1 + T_2) \leq \mu(T_1) + \mu(T_2)$$

(to prove the last, write $(\| 1_Y + hT_1 + hT_2 \| - 1)/h$ as $(\| 21_Y + 2hT_1 + 2hT_2 \| - 2)/(2h)$).†

We prove now:

(2.8.7) *Let* $A: I \to \mathscr{L}(Y; Y)$ *and* $b: I \to Y$ *be continuous, and let* ϕ *be a solution of the non-homogeneous equation* $y' = A(t)y + b(t)$ *on* I. *Then, for all* $t, t_0 \in I$ *with* $t \geq t_0$,

$$\| \phi(t) \| \leq \| \phi(t_0) \| \exp \left(\int_{t_0}^{t} \mu(A(v)) dv \right)$$

$$+ \int_{t_0}^{t} \| b(u) \| \exp \left(\int_{u}^{t} \mu(A(v)) dv \right) du. \quad (12)$$

If $t, t + h \in I$ and $h > 0$, then

$$\| \phi(t+h) \| - \| \phi(t) \| = \| \phi(t) + h\phi'(t) + o(h) \| - \| \phi(t) \|$$
$$= \| \phi(t) + hA(t)\phi(t) + hb(t) + o(h) \| - \| \phi(t) \|$$
$$\leq \| \phi(t) + hA(t)\phi(t) \| - \| \phi(t) \| + h \| b(t) \| + o(h)$$
$$\leq (\| 1_Y + hA(t) \| - 1) \| \phi(t) \| + h \| b(t) \| + o(h),$$

so that

$$D^+ \| \phi(t) \| \leq \mu(A(t)) \| \phi(t) \| + \| b(t) \|.$$

† The function μ is not a norm in the strict sense, since it may take negative values.

The function $t \mapsto \mu(A(t))$ is continuous (since

$$|\mu(A(t)) - \mu(A(t'))| \leq \|A(t) - A(t')\|,$$

by the properties of μ above), and hence we may apply (2.6.1) as before.

For a solution ϕ of the homogeneous equation $y' = A(t)y$, (12) gives

$$\|\phi(t)\| \leq \|\phi(t_0)\| \exp\left(\int_{t_0}^t \mu(A(u))du \right) \qquad (t, t_0 \in I, t \geq t_0), \qquad (13)$$

and this in turn implies that if U is a fundamental kernel for the equation then

$$\|U(t)\| \leq \|U(t_0)\| \exp\left(\int_{t_0}^t \mu(A(u))du \right) \qquad (t, t_0 \in I, t \geq t_0).$$

In fact, (13) shows that for all $y_0 \in Y$

$$\|U(t)y_0\| \leq \|U(t_0)y_0\| \exp\left(\int_{t_0}^t \mu(A(u))du \right)$$

$$\leq \|U(t_0)\| \|y_0\| \exp\left(\int_{t_0}^t \mu(A(u))du \right),$$

and this gives the required result.

We note the following corollary.

(2.8.7. Corollary) *Let I be the interval $[a, b[$, where $a < b \leq \infty$, and let $A : I \to \mathscr{L}(Y ; Y)$ be continuous. If*

$$\limsup_{t \to \infty} \int_a^t \mu(A(u))du < \infty,$$

then every solution ϕ of $y' = A(t)y$ on I is bounded. If in addition the integral $\int_a^\infty \mu(A(u))du$ is convergent (not necessarily absolutely), or diverges to $-\infty$, then for every ϕ the norm $\|\phi(t)\|$ tends to a finite limit as $t \to \infty$.

The inequality (13) trivially implies the boundedness of ϕ, and shows also that, if $\int_a^\infty \mu(A(u))du = -\infty$, then $\|\phi(t)\| \to 0$ as $t \to \infty$. To prove the remaining result, we may obviously suppose that ϕ is not the zero solution, so that ϕ nowhere vanishes on I. Then (13) gives

$$\log \|\phi(t)\| - \log \|\phi(t_0)\| \leq \int_{t_0}^t \mu(A(u))du,$$

and the one-sided form of Cauchy's convergence criterion (cf. Exercise 2.5.7) now shows that $\log \|\phi(t)\|$ tends to a finite limit as $t \to b-$, whence so also does $\|\phi(t)\|$.

It is possible to obtain some estimates for the logarithmic norm $\mu(T)$. For instance, $\mu(T) \geq \operatorname{Re}(\lambda)$ for every eigenvalue of T, for if y is an eigenvector of T of norm 1 corresponding to the eigenvalue λ, then

$$\| 1_Y + hT \| \geq \| (1_Y + hT)y \| = \| (1 + h\lambda)y \| = |1 + h\lambda|$$
$$= 1 + h\operatorname{Re}(\lambda) + O(\lambda^2)$$

as $h \to 0+$.

If Y is a Hilbert space, then $\mu(T)$ is equal to the greatest element M of the spectrum of the self-adjoint transformation $W = \frac{1}{2}(T + T^*)$, where T^* is the adjoint of T. In fact, it is easily verified that for all $y \in Y$ and $h > 0$

$$\| (1_Y + hT)y \|^2 = \| y \|^2 + 2h\langle Wy, y \rangle + h^2 \| Ty \|^2.$$

Since $M = \sup_{\|y\| = 1} \langle Wy, y \rangle$ (by (A.2.7)), we obviously have

$$\| 1_Y + hT \|^2 \leq 1 + 2hM + h^2 \| T \|^2,$$

whence $\mu(T) \leq M$. On the other hand, if $\varepsilon > 0$ we can find $y \in Y$ with $\| y \| = 1$ such that $\langle Wy, y \rangle \geq M - \varepsilon$, and then also

$$\| 1_Y + hT \|^2 \geq \| (1_Y + hT)y \|^2 \geq 1 + 2h(M - \varepsilon) - h^2 \| T \|^2,$$

so that $\mu(T) \geq M$.

We conclude this section by considering the application of the preceding results to a linear equation of the nth order, and for simplicity we confine ourselves here to the scalar case, i.e. to the case of an equation

$$y^{(n)} + a_1(t)y^{(n-1)} + a_2(t)y^{(n-2)} + \ldots + a_n(t)y = q(t), \tag{14}$$

where $a_1, \ldots, a_n, q$ are continuous real-valued functions on the interval I.

We have already seen in §2.1, Example (b), (p. 72), that if $A(t): \mathbf{R}^n \to \mathbf{R}^n$ is the linear map whose matrix with respect to the natural basis in $\mathbf{R}^n$ is

$$\begin{bmatrix} 0 & 1 & 0 & \ldots & 0 & 0 \\ 0 & 0 & 1 & \ldots & 0 & 0 \\ \cdot\cdot & \cdot\cdot & \cdot\cdot & \ldots & \cdot\cdot & \cdot\cdot \\ 0 & 0 & 0 & \ldots & 0 & 1 \\ -a_n(t) & -a_{n-1}(t) & -a_{n-2}(t) & \ldots & -a_2(t) & -a_1(t) \end{bmatrix},$$

and $b: I \to \mathbf{R}^n$ is the function with components $0, \ldots, 0, q$, then the equation (14) is equivalent to the first-order linear equation

$$z' = A(t)z + b(t). \tag{15}$$

More precisely, if $\psi: I \to \mathbf{R}$ is a solution of (14) on I, then the function from I into $\mathbf{R}^n$ with components $\psi, \psi', \ldots, \psi^{(n-1)}$ is a solution of (15) on I, and conversely, if ϕ is a solution of (15) on I, then the first component

ψ of ϕ is a solution of (14) on I and the remaining components of ϕ are $\psi', \ldots, \psi^{(n-1)}$.† Moreover, ψ satisfies the conditions

$$\psi(t_0) = y_{10}, \psi'(t_0) = y_{20}, \ldots, \psi^{(n-1)}(t_0) = y_{n0} \qquad (16)$$

if and only if $\phi(t_0) = y_0 = (y_{10}, y_{20}, \ldots, y_{n0})$.

From (2.8.1), we deduce immediately:

(2.8.8) *Let* $a_1, \ldots, a_n, q : I \to \mathbf{R}$ *be continuous. Then for each* $t_0 \in \mathbf{R}$ *and each set of real numbers* $y_{10}, \ldots, y_{n0}$ *the equation* (14) *has exactly one solution* ψ *on* I *satisfying the conditions* (16). *Moreover, every solution of* (14) *has maximal interval of existence* I.

As with the first-order equation, an equation

$$y^{(n)} + a_1(t)y^{(n-1)} + \ldots + a_n(t)y = 0, \qquad (17)$$

in which the right-hand side is zero, is called a *homogeneous* (linear) *equation of the nth order.* Exactly as above, this is equivalent to the first-order homogeneous equation

$$z' = A(t)z. \qquad (18)$$

It is obvious that the set of solutions of (17) on I forms a real vector space $\mathscr{S}$. It is also obvious (cf. the footnote preceding (2.8.8)) that if $\psi_1, \ldots, \psi_p$ are solutions of (17) on I, and, for each j, ϕ_j is the function from I into $\mathbf{R}^n$ with components $\psi_j, \psi_j', \ldots, \psi_j^{(n-1)}$ (so that ϕ_j is a solution of (18)), then $\psi_1, \ldots, \psi_p$ are linearly independent in $\mathscr{S}$ if and only if $\phi_1, \ldots, \phi_p$ are linearly independent in the space S of solutions of (18) on I.

On combining the 'if' in this remark with the first part of (2.8.2), we obtain immediately the following result.

(2.8.9) *Let* $a_1, \ldots, a_n : I \to \mathbf{R}$ *be continuous. Then the set* $\mathscr{S}$ *of solutions of the homogeneous equation* (17) *on* I *form a real vector space of dimension n.*

The remark above shows also that if $\psi_1, \ldots, \psi_n$ are linearly independent solutions of (17) on I, and $U(t)$ is the linear map of $\mathbf{R}^n$ into itself whose matrix with respect to the natural basis in $\mathbf{R}^n$ is

$$\begin{bmatrix} \psi_1(t) & \psi_2(t) & \cdots & \psi_n(t) \\ \psi_1'(t) & \psi_2'(t) & \cdots & \psi_n'(t) \\ .. & .. & \cdots & .. \\ \psi_1^{(n-1)}(t) & \psi_2^{(n-1)}(t) & \cdots & \psi_n^{(n-1)}(t) \end{bmatrix}, \qquad (19)$$

† The maps $\psi \mapsto \phi$ and $\phi \mapsto \psi$ are in fact inverses of each other, and are clearly linear.

then U is a fundamental kernel for the equation (18). Conversely, if U is a fundamental kernel for (18), then its matrix is of the form (19), where $\psi_1,\ldots,\psi_n$ are linearly independent solutions of (17) on I.

The determinant $W(t)$ of this matrix (19) is known as the *Wronskian* of $\psi_1,\ldots,\psi_n$. Since the vectors in $\mathbf{R}^n$ whose coordinates form the columns of this matrix are linearly independent in $\mathbf{R}^n$ if and only if $W(t) \neq 0$, the second part of (2.8.2) gives:

(2.8.10) *Let* $\psi_1,\ldots\psi_n$ *be solutions of* (17) *on* I, *and let* $W(t)$ *be their Wronskian. Then the following statements are equivalent:*
(i) $\psi_1,\ldots,\psi_n$ *are linearly independent in* $\mathscr{S}$;
(ii) $W(t_0) \neq 0$ *for some* $t_0 \in I$;
(iii) $W(t_0) \neq 0$ *for each* $t_0 \in I$.

The equivalence of (ii) and (iii) can also be deduced from the Abel–Liouville identity, which here takes the form

$$W(t) = W(t_0)\exp\left(-\int_{t_0}^{t} a_1(u)du \right). \tag{20}$$

Exactly similar results hold in the complex case; the theory can also be extended without difficulty to the more general nth order linear equation mentioned in §2.1, Example (*b*) (p. 72).

Exercises 2.8

1 Let Y be finite-dimensional, let $I = [a,\infty[$, let $A:I \to \mathscr{L}(Y;Y)$ be continuous, and suppose that every non-zero solution of the equation $y' = A(t)y$ on I has a non-zero limit as $t \to \infty$. Prove that for each $y_0 \in Y$ there is exactly one solution that has the limit y_0 as $t \to \infty$.
2 Let I be an interval in $\mathbf{R}$, let $A:I \to \mathscr{L}(Y;Y)$ be continuous, and let U be a fundamental kernel for the equation $y' = A(t)y$ on I. Prove that for all $t, t_0 \in I$

$$\| U(t) \| \le \| U(t_0) \| \exp\left(\left| \int_{t_0}^{t} \| A(u) \| du \right| \right)$$

and

$$\| U(t)^{-1} \| \le \| U(t_0)^{-1} \| \exp\left(\left| \int_{t_0}^{t} \| A(u) \| du \right| \right).$$

[Hint. Use the equations $X' = A(t) \circ X$ and $X' = -X \circ A(t)$ in combination with (2.6.1) and (2.6.1)′.]
3 Let I be the interval $[a,\infty[$, let $A:I \to \mathscr{L}(Y;Y)$ be a continuous function such that $K = \int_a^\infty \| A(t) \| dt < \infty$, and let U be a fundamental kernel for $y' = A(t)y$ on I. Prove that
(i) there exists $M \ge 0$ such that $\| U(t) \| \le M$ and $\| U(t)^{-1} \| \le M$ for all $t \in I$,

(ii) there exists a linear homeomorphism Z of Y onto itself such that $U(t) \to Z$ and $U(t)^{-1} \to Z^{-1}$ in $\mathscr{L}(Y; \underline{Y})$ as $t \to \infty$,

(iii) if $W(t) = U(t) \circ Z^{-1}$ $(t \geq a)$, then W is a fundamental kernel for $y' = A(t)y$ on I and $W(t)$ and $W(t)^{-1}$ tend to 1_Y in $\mathscr{L}(Y; Y)$ as $t \to \infty$,

(iv) every solution ϕ of $y' = A(t)y$ on I tends to a limit $\phi(\infty)$ as $t \to \infty$, and for any given $t_0 \in I$ the map $\phi(t_0) \mapsto \phi(\infty)$ is a linear homeomorphism of Y onto itself.

[Hint. For (i), use Exercise 2. For (ii) note that since U and $t \mapsto U(t)^{-1}$ are solutions of $X' = A(t) \circ X$ and $X' = -X \circ A(t)$ we have

$$\left\| \frac{d}{dt} U(t) \right\| \leq M \| A(t) \| \quad \text{and} \quad \left\| \frac{d}{dt} U(t)^{-1} \right\| \leq M \| A(t) \|,$$

and these imply that $Z = \lim U(t)$ and $\bar{Z} = \lim U(t)^{-1}$ exist. Then show that $Z \circ \bar{Z} = 1_Y = \bar{Z} \circ Z$.]

4 Let I be an interval in $\mathbf{R}$, let $A, B : I \to \mathscr{L}(Y; Y)$ be continuous, let U be a fundamental kernel for $y' = A(t)y$ on I, and let

$$C(t) = U(t)^{-1} \circ (B(t) - A(t)) \circ U(t).$$

Prove that V is a fundamental kernel for $y' = B(t)y$ on I if and only if $V(t) = U(t) \circ W(t)$ $(t \in I)$, where W is a fundamental kernel for $y' = C(t)y$ on I.

[Hint. For the 'only if', use (2.8.5).]

5 Let $A : I \to \mathscr{L}(Y; Y)$ be continuous, and let U be a fundamental kernel for the equation $y' = A(t)y$ on I. Prove that

(i) if there exists $M > 0$ such that $\| U(t)^{-1} \| \geq M$ for all $t \in I$, then every solution ϕ of the equation on I satisfies $\| \phi(t) \| \geq M^{-1} \| U(t_0) \|^{-1} \| \phi(t_0) \|$ for all $t, t_0 \in I$,

(ii) if for each $t_0 \in I$ there exists $L > 0$ such that every solution ϕ of the equation on I satisfies $\| \phi(t) \| \geq L \| \phi(t_0) \|$ for all $t \in I$, then $\| U(t)^{-1} \| \leq L \| U(t_0)^{-1} \|$.

6 Prove that if ϕ is a solution of $y' = A(t)y$ on I, then for all $t \in I$

$$D^- \| \phi(t) \| \geq -\mu(-A(t)) \| \phi(t) \|,$$

where μ is the logarithmic norm defined in the text (p. 110). Deduce that if $t_0, t \in I$ and $t \geq t_0$ then

$$\| \phi(t) \| \geq \| \phi(t_0) \| \exp \left(-\int_{t_0}^{t} \mu(-A(u)) du \right)$$

and

$$\| U(t)^{-1} \| \leq \| U(t_0)^{-1} \| \exp \left(-\int_{t_0}^{t} \mu(-A(u)) du \right).$$

7 Prove that if Y is finite-dimensional, t_0 is a given point of I, and for all $t \in I$

$$\| U(t) \| \leq M \quad \text{and} \quad \mathrm{Re} \int_{t_0}^{t} \mathrm{tr}\, A(u) du \geq -L,$$

then there exists $K \geq 0$ such that $\| U(t)^{-1} \| \leq K$ for all $t \in I$.

[Hint. Recall that if $\mathbf{C} = [c_{ij}]$ is a non-singular $n \times n$ matrix, then the i,jth entry in $\mathbf{C}^{-1}$ is $\delta_{ij}/\det \mathbf{C}$, where δ_{ij} is the cofactor of c_{ji} in $\mathbf{C}$. Combine this with the Abel–Liouville identity.]

Exercises on stability of solutions of linear equations and perturbations of linear equations

The theory of linear equations, and the differential and integral inequalities considered in §2.6 together lead to important results concerning stability of solutions of differential equations, and we include a number of these results as exercises, both here and in §2.9. We consider four of the simplest types of stability, and since there are difficulties in the infinite-dimensional case arising from the possible non-existence of solutions, we begin with the case where Y is finite-dimensional. Thus let $I = [a, \infty[$, let E be a subset of $I \times Y$ open in $I \times Y$, let $f : E \to Y$ be continuous, and suppose first that the zero function from I into Y is a solution on I of the equation

$$y' = f(t, y) \qquad\qquad (*)$$

(so that $f(t, 0) = 0$ for all $t \in I$). Then the zero solution of $(*)$ is
(S) *stable* if for each $\varepsilon > 0$ and each $t_0 \in I$ there exists $\delta > 0$ such that, if ϕ is a solution of $(*)$ satisfying $\| \phi(t_0) \| < \delta$ and reaching to the boundary of E on the right, then ϕ is defined on $[t_0, \infty[$ and $\| \phi(t) \| < \varepsilon$ for all $t \geq t_0$,
(AS) *asymptotically stable* if it is stable and in addition the δ in (S) can be chosen so that $\phi(t) \to 0$ as $t \to \infty$,
(US) *uniformly stable* if it is stable and the δ in (S) can be chosen to be independent of t_0, i.e. if for each $\varepsilon > 0$ there exists $\delta > 0$ such that, if $t_0 \in I$ and ϕ is a solution of $(*)$ satisfying $\| \phi(t_0) \| < \delta$ and reaching to the boundary of E on the right, then ϕ is defined on $[t_0, \infty[$ and $\| \phi(t) \| < \varepsilon$ for all $t \geq t_0$,
(SS) *strictly stable* if for each $\varepsilon > 0$ there exists $\delta > 0$ such that, if $t_0 \in I$ and ϕ is a solution of $(*)$ satisfying $\| \phi(t_0) \| < \delta$ and reaching to the boundary of E in both directions, then ϕ is defined on I and $\| \phi(t) \| < \varepsilon$ for all $t \in I$.

The corresponding definitions for a general solution ψ of $(*)$ (where now $f(t, 0)$ may be non-zero) are similar. For example, ψ is said to be *stable* if for each $\varepsilon > 0$ and each $t_0 \in I$ there exists $\delta > 0$ such that, if ϕ is a solution of $(*)$ satisfying $\| \phi(t_0) - \psi(t_0) \| < \delta$ and reaching to the boundary of E on the right, then ϕ is defined on $[t_0, \infty[$ and $\| \phi(t) - \psi(t) \| < \varepsilon$ for all $t \geq t_0$. Questions concerning the various forms of stability of ψ can obviously be reduced to questions concerning those of a zero solution by the substitution $y = \psi + z$.

When Y is infinite-dimensional, both the existence and the extension of solutions present difficulties, and we therefore restrict ourselves to situations where for each $(t_0, y_0) \in E$ the equation $(*)$ has exactly one solution ϕ on I satisfying $\phi(t_0) = y_0$ and every other solution $\bar{\phi}$ such that $\bar{\phi}(t_0) = y_0$ is a restriction of ϕ. In this case the definitions above can be reworded as follows: the zero solution of $(*)$ is
(S)' *stable* if for each $\varepsilon > 0$ and each $t_0 \in I$ there exists $\delta > 0$ such that, if ϕ is a solution of $(*)$ on I satisfying $\| \phi(t_0) \| < \delta$, then $\| \phi(t) \| < \varepsilon$ for all $t \geq t_0$,
(AS)' *asymptotically stable* if it is stable and in addition the δ in (S)' can be chosen so that $\phi(t) \to 0$ as $t \to \infty$,
(US)' *uniformly stable* if for each $\varepsilon > 0$ there exists $\delta > 0$ such that, if $t_0 \in I$ and ϕ is a solution of $(*)$ on I satisfying $\| \phi(t_0) \| < \delta$, then $\| \phi(t) \| < \varepsilon$ for all $t \geq t_0$,
(SS)' *strictly stable* if for each $\varepsilon > 0$ there exists $\delta > 0$ such that, if $t_0 \in I$ and ϕ is a solution of $(*)$ on I satisfying $\| \phi(t_0) \| < \delta$, then $\| \phi(t) \| < \varepsilon$ for all $t \in I$.

These definitions are, for instance, appropriate to a homogeneous linear equation $y' = A(t)y$, where $A : I \to \mathscr{L}(Y; Y)$ is continuous and Y may be either finite- or infinite-dimensional. It is easy to verify that if any solution of a homogeneous linear equation is stable, then so are all solutions, and in this case we say that the equation itself is *stable* (to prove that the equation is stable, it is, of course, enough to verify that the zero solution is stable). Similar remarks apply to the other stability properties. In particular we note that the equation $y' = A(t)y$ is asymptotically stable if and only if it is stable and every solution tends to 0 as $t \to \infty$.

In the following exercises, we assume without further reference that I is an interval in **R**, arbitrary unless otherwise specified, that $A : I \to \mathscr{L}(Y; Y)$ is continuous, and that U is a fundamental kernel for the equation $y' = A(t)y$ on I.

8 Prove that the following statements are equivalent:
   (i)   every solution of $y' = A(t)y$ on I is bounded;
   (ii)   for each $t_0 \in I$ there exists $L > 0$ such that every solution ϕ of $y' = A(t)y$ on I satisfies $\|\phi(t)\| \leq L \|\phi(t_0)\|$ for all $t \in I$;
   (iii)   there exists $M > 0$ such that $\|U(t)\| \leq M$ for all $t \in I$.
Prove also that if I is the interval $[a, \infty[$ then each of (i)–(iii) is equivalent to
   (iv)   the equation $y' = A(t)y$ is stable (thus for homogeneous linear equations boundedness and stability are equivalent).
   [Hint. To prove that (i) implies (iii) use the uniform boundedness principle.]
9 Prove that if I is the interval $[a, \infty[$ then the following statements are equivalent:
   (i)   there exists $M > 0$ such that $\|U(t) \circ U(t_0)^{-1}\| \leq M$ whenever $a \leq t_0 \leq t$;
   (ii)   the equation $y' = A(t)y$ is uniformly stable.
   [This result implies in particular that for a homogeneous linear equation with constant coefficients (see §2.9) the properties of boundedness, stability, and uniform stability are equivalent.]
10 Prove that the following statements are equivalent:
   (i)   for each $y_0 \in Y$ there exists $K > 0$ such that each solution ϕ of $y' = A(t)y$ on I taking the value y_0 somewhere on I satisfies $\|\phi(t)\| \leq K$ for all $t \in I$;
   (ii)   there exists $M > 0$ such that $\|U(t)\| \leq M$ and $\|U(t)^{-1}\| \leq M$ for all $t \in I$.
Prove also that if I is the interval $[a, \infty[$ then each of (i) and (ii) is equivalent to
   (iii)   the equation $y' = A(t)y$ is strictly stable.
   [Hint. To prove that (i) implies that $\|U(t)^{-1}\| \leq M$ use the uniform boundedness principle.]
11 Let $I = [a, \infty[$, let $A, B : I \to \mathscr{L}(Y; Y)$ be continuous, let U be a fundamental kernel for $y' = A(t)y$ on I, and suppose that

$$\int_a^\infty \|U(t)^{-1} \circ (B(t) - A(t)) \circ U(t)\| \, dt < \infty.$$

Prove that there is a fundamental kernel V for $y' = B(t)y$ on I such that
   (i) $\|V(t)\| \leq M \|U(t)\|$ and $\|V(t)^{-1}\| \leq M \|U(t)^{-1}\|$ for all $t \in I$,
   (ii) $\|V(t) - U(t)\| = o(\|U(t)\|)$ as $t \to \infty$,
   (iii)   the integral condition above holds with V in place of U.
   Deduce that the equation $y' = A(t)y$ is (a) stable, (b) asymptotically stable, (c) strictly stable, if and only if the equation $y' = B(t)y$ has the same property.

   Prove further that, if either equation is strictly stable, then for each solution ψ of $y' = B(t)y$ on I there exists a unique solution ϕ of $y' = A(t)y$ on I such that $\psi(t) - \phi(t) \to 0$ as $t \to \infty$, and that for each $t_0 \in I$ the map $\psi(t_0) \mapsto \phi(t_0)$ is a linear homeomorphism of Y onto itself.
   [It is easy to see that the results concerning strict stability continue to hold if the

integral condition is replaced by

$$\int_a^\infty \| B(t) - A(t) \| \, dt < \infty.$$

Hint. For (i)–(iii) use Exercises 3 and 4. For the stability results use Exercises 8, 10, and 5.]

12 Let $E \subseteq I \times Y$, let $F : E \to Y$ be continuous, and let $(t_0, y_0) \in E$. Prove that ϕ is a solution of the equation

$$y' = A(t)y + F(t, y)$$

on I satisfying the condition $\phi(t_0) = y_0$ if and only if
   (i) ϕ is continuous,
   (ii) $(t, \phi(t)) \in E$ for all $t \in I$,
   (iii) for all $t \in I$

$$\phi(t) = U(t)U(t_0)^{-1}y_0 + U(t)\int_{t_0}^t U(u)^{-1}F(u, \phi(u))du.$$

[The term $F(t, y)$ can be regarded as a perturbation of the linear equation $y' = A(t)y$. There is a considerable body of results which assert that, if the perturbation is in some sense small, then the solutions of the perturbed equation behave in a similar manner to those of the linear equation. In such results the formula (iii) is of frequent application, and examples are given in Exercises 13 and 14 below.

Hint. The 'if' is proved as in (2.2.1). For the 'only if' follow the argument of (2.8.6).]

13 Let $F : I \times Y \to Y$ be a continuous function such that for all $(t, y) \in I \times Y$

$$\| F(t, y) \| \le \lambda(t) \| y \|,$$

where $\lambda : I \to \,]0, \infty[$ is a continuous function satisfying the condition that

$$K = \int_I \| U(t)^{-1} \| \lambda(t) \| U(t) \| \, dt < \infty.$$

Prove that each solution ϕ of the equation

$$y' = A(t)y + F(t, y)$$

on I satisfies

$$\| \phi(t) \| \le e^K \| U(t) \| \, \| U(t_0) \|^{-1} \| \phi(t_0) \|$$

for all $t, t_0 \in I$, and that if $I = [a, \infty[$ there exists $y_1 \in Y$ such that

$$\phi(t) = U(t)y_1 + o(\| U(t) \|)$$

as $t \to \infty$.

Deduce that, if Y is finite-dimensional, $I = [a, \infty[$, and the equation $y' = A(t)y$ is (a) stable, (b) asymptotically stable, (c) strictly stable, then the zero solution of the perturbed equation $y' = A(t)y + F(t, y)$ has the same property.

[It is easy to see that the results concerning stability and asymptotic stability continue to hold if the condition $K < \infty$ is replaced by

$$\int_a^\infty \| U(t)^{-1} \| \lambda(t)dt < \infty,$$

and that for strict stability it can be replaced by $\int_a^\infty \lambda(t)dt < \infty$.

Hint. For the inequality for $\| \phi(t) \|$, use Exercise 12, set $\psi(t) = \| \phi(t) \| / \| U(t) \|$, and apply Bellman's inequality (2.6.2. Corollary); for the asymptotic relation observe that the integral $\int_a^\infty U(u)^{-1} F(u, \phi(u))du$ is convergent.

For the stability results use Exercises 8 and 10.]

14 Let $F : I \times Y \to Y$ be a continuous function such that for all $(t, y) \in I \times Y$

$$\| F(t, y) \| \leq \lambda(t) \| y \|,$$

where $\lambda : I \to]0, \infty[$ is continuous and $L = \int_I \lambda(t) dt < \infty$. Prove that if

$$\| U(t) \| \leq M \quad \text{and} \quad \| U(t) \circ U(u) \|^{-1} \leq M$$

whenever $u, t \in I$ and $u \leq t$, then each solution ϕ of the equation

$$y' = A(t)y + F(t, y)$$

on I satisfies

$$\| \phi(t) \| \leq M e^{LM} \| \phi(t_0) \|$$

whenever $t, t_0 \in I$ and $t \geq t_0$.

Deduce that if Y is finite-dimensional and I is the interval $[a, \infty[$, and $y' = A(t)y$ is uniformly stable, then so is the zero solution of $y' = A(t)y + F(t, y)$.

[Hint. For the first part use Exercise 12 and Bellman's inequality. For the second part use Exercise 9.]

15 Let $I = [a, \infty[$ and let $B : I \to \mathscr{L}(Y; Y)$ be a continuous function such that $\int_a^\infty \| B(t) - A(t) \| dt < \infty$. Prove that if $y' = A(t)y$ is uniformly stable, so is $y' = B(t)y$.

[Here Y may be infinite-dimensional; however, the argument of Exercise 14 still applies.]

2.9 Linear equations with constant coefficients

We consider now the special case of a linear equation of the form

$$y' = Ay, \tag{1}$$

where A is a given continuous linear map from Y into itself (independent of t). If $Y = \mathbf{R}^n$ or $\mathbf{C}^n$, and the matrix of A with respect to the natural basis in Y is $[a_{ij}]$, then (1) is equivalent to the linear system

$$y_i' = \sum_{j=1}^{n} a_{ij} y_j \qquad (i = 1, \dots, n)$$

in which the coefficients a_{ij} are constants. For this reason, the equation (1) is usually known as a *linear equation with constant coefficients*.

It is obvious that each solution of (1) has maximal interval of existence $\mathbf{R}$.

The fundamental kernels for an equation of this type have a particularly simple form. Let $A \in \mathscr{L}(Y; Y)$, and for $t \in \mathbf{R}$ let $S_n(t)$ be the element of $\mathscr{L}(Y; Y)$ given by

$$S_n(t) = 1_Y + tA + \frac{t^2}{2!} A^2 + \dots + \frac{t^n}{n!} A^n,$$

where A^r denotes the r-fold composite $A \circ A \circ \dots \circ A$. The sequence $(S_n(t))$ is Cauchy, for if $n > m$ then

$$\| S_n(t) - S_m(t) \| = \left\| \sum_{r=m+1}^{n} \frac{t^r}{r!} A^r \right\| \leq \sum_{r=m+1}^{n} \frac{|t|^r \| A \|^r}{r!}, \tag{2}$$

and here the last expression is arbitrarily small for all sufficiently large m. Hence $S_n(t)$ converges to a limit in $\mathscr{L}(Y; Y)$, and we denote this limit by $\exp(tA)$. In particular, $\exp(0A) = 1_Y$. We observe further that $S'_n(t) = A \circ S_n(t)$, and from (2) we see easily that the sequence $(S'_n(t))$ converges, uniformly on each compact interval in $\mathbf{R}$. Applying (1.7.1), we deduce that the function $t \mapsto \exp(tA)$ thus defined is differentiable and that for all $t \in \mathbf{R}$

$$\frac{d}{dt}(\exp(tA)) = \lim_{n \to \infty} S'_n(t) = \lim_{n \to \infty} A \circ S_n(t) = A \circ (\lim_{n \to \infty} S_n(t)) = A \circ \exp(tA).$$

Since $\exp(0A) = 1_Y$ is a homeomorphism of Y onto itself, the function $t \mapsto \exp(tA)$ is therefore the fundamental kernel for the equation $y' = Ay$ taking the value 1_Y at 0.

It follows easily from the preceding argument that if Y has finite dimension n, and the matrix of the r-fold composite A^r with respect to a given basis for Y is $[a_{ij}^{(r)}]$ $(r = 1, 2, \ldots)$, then for each i,j and each $t \in \mathbf{R}$ the series $1 + \sum_{r=1}^{\infty} a_{ij}^{(r)} t^r / r!$ is convergent, to $c_{ij}(t)$, say, and the matrix of $\exp(tA)$ with respect to B is $[c_{ij}(t)]$. In fact, this provides a definition of the matrix $\exp(tA)$ for any $n \times n$ matrix $\mathbf{A} = [a_{ij}]$ with real or complex entries. In general, this formula for $\exp(tA)$ is of little use for computational purposes, but we will meet one important exception to this in the proof of (2.9.1) below.

The generalized exponential function defined above satisfies the addition theorem

$$\exp(tA) \circ \exp(uA) = \exp((t + u)A) \qquad (t, u \in \mathbf{R}), \tag{3}$$

for the functions $t \mapsto \exp(tA) \circ \exp(uA)$ and $t \mapsto \exp((t + u)A)$ are both solutions of the equation $X' = A \circ X$ taking the value $\exp(uA)$ at $t = 0$ (cf. (2.8.3)(ii)), and are therefore identical. In particular, (3) gives

$$(\exp(tA))^{-1} = \exp(-tA) \qquad (t \in \mathbf{R}). \tag{4}$$

From (4), (3), and (2.8.6), we deduce that if $b : \mathbf{R} \to Y$ is continuous then the solution of the non-homogeneous equation

$$y' = Ay + b(t)$$

taking the value y_0 at t_0 is

$$y = \exp((t - t_0)A)y_0 + \int_{t_0}^{t} \exp((t - u)A)b(u)du.$$

We note also that if $A, B \in \mathscr{L}(Y; Y)$ and $A \circ B = B \circ A$, then

$$\exp(A + B) = \exp A \circ \exp B = \exp B \circ \exp A \tag{5}$$

for the commutativity of A and B enables us to carry over to this case the standard proof of the addition theorem for the exponential function of a complex variable. If A and B do not commute, then (5) need not hold.

We observe next that if A has an eigenvalue λ with corresponding eigenvector z, then $\exp(tA)z = \exp(\lambda t)z$. Hence, if Y has finite dimension n and A has distinct eigenvalues $\lambda_1, \ldots, \lambda_n$ with linearly independent eigenvectors $z_1, \ldots, z_n$, and we take these eigenvectors as a basis for Y, then the matrix $\exp(tA)$ with respect to this basis is the diagonal matrix

$$\begin{bmatrix} e^{\lambda_1 t} & 0 & \cdots & 0 \\ 0 & e^{\lambda_2 t} & \cdots & 0 \\ \cdot\cdot & \cdot\cdot & \cdots & \cdot\cdot \\ 0 & 0 & \cdots & e^{\lambda_n t} \end{bmatrix}.$$

Moreover, the functions $t \mapsto \exp(\lambda_j t)z_j$ form a linearly independent set of solutions of the equation $y' = Ay$ on $\mathbf{R}$ (cf. (2.8.3)(i) and (2.8.2)(i) and (ii)).

We can obtain a corresponding result for a general A (again in the finite-dimensional case) by using the Jordan decomposition of a linear map. Since there are difficulties in the real case arising from the non-existence of eigenvalues, we begin by considering the complex case, and we suppose for simplicity that $Y = \mathbf{C}^n$.

Thus let A be a linear map of $\mathbf{C}^n$ into itself, let $\lambda_1, \ldots, \lambda_p$ be the distinct eigenvalues of A, and let m_k $(k = 1, \ldots, p)$ be the multiplicity of λ_k as a root of the characteristic equation of A. Then, by the Jordan decomposition of A, there is a basis $\{z_1, \ldots, z_n\}$ of $\mathbf{C}^n$ such that the matrix of A with respect to this basis is

$$\begin{bmatrix} \mathbf{A}_1 & 0 & \cdots & 0 \\ 0 & \mathbf{A}_2 & \cdots & 0 \\ \cdot\cdot & \cdot\cdot & \cdots & \cdot\cdot \\ 0 & 0 & \cdots & \mathbf{A}_p \end{bmatrix}, \tag{6}$$

where each $\mathbf{A}_k$ is a matrix with m_k rows and columns, of the form

$$\mathbf{A}_k = \begin{bmatrix} \mathbf{A}_{k1} & 0 & \cdots & 0 \\ 0 & \mathbf{A}_{k2} & \cdots & 0 \\ \cdot\cdot & \cdot\cdot & \cdots & \cdot\cdot \\ 0 & 0 & \cdots & \mathbf{A}_{kq_k} \end{bmatrix} \tag{7}$$

and each $\mathbf{A}_{kl}$ is a square matrix with all the leading diagonal entries equal to λ_k, with 1s immediately below the leading diagonal, and with 0s elsewhere. The sizes of the $\mathbf{A}_{kl}$ decrease in the wide sense as l increases, and the possibility $\mathbf{A}_{kl} = \lambda_k$ is allowed.

For $v = 2, 3, \ldots$, the matrix of the map A^v with respect to the basis $\{z_1, \ldots, z_n\}$ is the vth power of the matrix (6), and this is easily seen to be

$$
\begin{bmatrix}
\mathbf{A}_1^\nu & 0 & \cdots & 0 \\
0 & \mathbf{A}_2^\nu & \cdots & 0 \\
\cdot\cdot & \cdot\cdot & \cdots & \cdot\cdot \\
0 & 0 & \cdots & \mathbf{A}_n^\nu
\end{bmatrix}.
$$

The νth power of each $\mathbf{A}_k$ is obtained similarly from (7), and hence the matrix of $\exp(tA)$ with respect to $\{z_1,\ldots,z_n\}$ is

$$
\begin{bmatrix}
\exp(t\mathbf{A}_1) & 0 & \cdots & 0 \\
0 & \exp(t\mathbf{A}_2) & \cdots & 0 \\
\cdot\cdot & \cdot\cdot & \cdots & \cdot\cdot \\
0 & 0 & \cdots & \exp(t\mathbf{A}_n)
\end{bmatrix}, \tag{8}
$$

where

$$
\exp(t\mathbf{A}_k) =
\begin{bmatrix}
\exp(t\mathbf{A}_{k1}) & 0 & \cdots & 0 \\
0 & \exp(t\mathbf{A}_{k2}) & \cdots & 0 \\
\cdot\cdot & \cdot\cdot & \cdots & \cdot\cdot \\
0 & 0 & \cdots & \exp(t\mathbf{A}_{kq_k})
\end{bmatrix}. \tag{9}
$$

If $\mathbf{A}_{kl} = \lambda_k$, then obviously $\exp(t\mathbf{A}_{kl}) = \exp(\lambda_k t)$. On the other hand, if $\mathbf{A}_{kl}$ has r rows and columns, where $r \geq 2$, we can write

$$
\mathbf{A}_{kl} = \lambda_k \mathbf{I}_r + \mathbf{Z}_r,
$$

where $\mathbf{I}_r$ is the $r \times r$ identity matrix and $\mathbf{Z}_r$ is the $r \times r$ matrix with 1s in the subdiagonal immediately below the leading diagonal and 0s elsewhere. It is easily verified that $\mathbf{Z}_r^2$ is obtained from $\mathbf{Z}_r$ by moving the 1s down to the next lower subdiagonal, that $\mathbf{Z}_r^3$ is obtained from $\mathbf{Z}_r^2$ by moving the 1s down a further place, and so on, so that $\mathbf{Z}_r^{r-1}$ has 1 in the bottom left-hand corner and 0s elsewhere, and $\mathbf{Z}_r^r = 0$. Since $\mathbf{I}_r$ and $\mathbf{Z}_r$ commute, we therefore have

$$
\exp(t\mathbf{A}_{kl}) = \exp(\lambda_k t\mathbf{I}_r + t\mathbf{Z}_r) = \exp(\lambda_k t\mathbf{I}_r)\exp(t\mathbf{Z}_r)
$$

$$
= \exp(\lambda_k t)(\mathbf{I}_r + t\mathbf{Z}_r + \frac{t^2}{2!}\mathbf{Z}_r^2 + \ldots + \frac{t^{r-1}}{(r-1)!}\mathbf{Z}_r^{r-1})
$$

$$
= \exp(\lambda_k t)
\begin{bmatrix}
1 & 0 & 0 & \cdots & 0 & 0 \\
t & 1 & 0 & \cdots & 0 & 0 \\
\frac{1}{2}t^2 & t & 1 & \cdots & 0 & 0 \\
\cdot\cdot & \cdot\cdot & \cdot\cdot & \cdots & \cdot\cdot & \cdot\cdot \\
\dfrac{t^{r-1}}{(r-1)!} & \dfrac{t^{r-2}}{(r-2)!} & \dfrac{t^{r-3}}{(r-3)!} & \cdots & t & 1
\end{bmatrix}. \tag{10}
$$

Combining (8), (9), (10), we have thus determined the matrix of the map

$\exp(tA): \mathbf{C}^n \to \mathbf{C}^n$ with respect to the basis $\{z_1, \dots, z_n\}$.

Next, by (2.8.3)(i) and the equivalence of (2.8.2)(i) and (ii), the functions $t \mapsto \exp(tA)z_j$ $(j = 1, \dots, n)$ form a linearly independent set of solutions of the equation $y' = Ay$ on $\mathbf{R}$, and the coordinates of $\exp(tA)z_j$ with respect to the basis $\{z_1, \dots, z_n\}$ form the jth column of the matrix (8). It therefore follows that $\exp(tA)z_j$ is of the form $\exp(\lambda_k t)P(t)$, where k is the index of the matrix $\exp(tA_k)$ containing entries in the jth column of the matrix (8), and P is a polynomial of degree at most $m_k - 1$ with values in $\mathbf{C}^n$. Hence we have proved:

(2.9.1) *Let A be a linear map of $\mathbf{C}^n$ into itself, let $\lambda_1, \dots, \lambda_p$ be the distinct eigenvalues of A, and let $m_1, \dots, m_p$ be the multiplicities of $\lambda_1, \dots, \lambda_p$. Then the equation $y' = Ay$ has a linearly independent set of solutions of the form*

$$t \mapsto e^{\lambda_k t} P_{ks}(t) \qquad (k = 1, \dots, p; s = 1, \dots, m_k),$$

where P_{ks} is a polynomial of degree at most $m_k - 1$ with values in $\mathbf{C}^n$, and these functions form a basis for the space of solutions.

To deal with the real case, it is simplest to revert to the matrix point of view, and as a first step we reinterpret (2.9.1) in terms of a linear system.

Thus let $a_{ij}(i, j = 1, \dots, n)$ be complex numbers, and consider the linear system

$$y_i' = \sum_{j=1}^{n} a_{ij} y_j \qquad (i = 1, \dots, n). \tag{11}$$

The problem of determining complex-valued solutions $y_1, \dots, y_n$ of (11) is equivalent to the problem of finding solutions of the equation $y' = Ay$ with values in $\mathbf{C}^n$, where A is the linear map of $\mathbf{C}^n$ into itself whose matrix with respect to the natural basis in $\mathbf{C}^n$ is $[a_{ij}]$. Since the eigenvalues of A are those of the matrix $[a_{ij}]$, it therefore follows from (2.9.1) that if $\lambda_1, \dots, \lambda_p$ are the distinct eigenvalues of the matrix $[a_{ij}]$, with multiplicities $m_1, \dots, m_p$, then the linear system (11) has a linearly independent set of solutions

$$y_1 = e^{\lambda_k t} P_{ks1}(t), \dots, y_n = e^{\lambda_k t} P_{ksn}(t) \qquad (k = 1, \dots, p; s = 1, \dots, m_k),$$

where $P_{ks1}, \dots, P_{ksn}$ are complex-valued polynomials of degree at most m_{k-1}.

Now suppose that the a_{ij} are *real*. The foregoing analysis still applies, and the system (11) has a linearly independent set of solutions of the form (12), but these solutions are in general complex-valued, and we should like to obtain real-valued solutions. Fortunately, the fact that the a_{ij} are real enables us to assert a little more about the Jordan decomposition of the map $A : \mathbb{C}^n \to \mathbb{C}^n$ defined by the matrix $[a_{ij}]$. First, if an eigenvalue λ_k of A (i.e. of $[a_{ij}]$) is real, and the columns of the matrix (6) containing A_k are the κth, ..., $(\kappa + m_k)$th (note that these are also the columns of (8) containing $\exp(tA_k)$), we can choose the basis vectors $z_\kappa, \dots, z_{\kappa + m_k}$ in the decomposition to have real coordinates with respect to the natural basis in $\mathbb{C}^n$. This implies that for $j = \kappa, \dots, \kappa + m_k$ the coordinates of $\exp(tA)z_j$ with respect to the *natural* basis in $\mathbb{C}^n$ are real, and hence in (12) the m_k solutions $y_1, \dots, y_n$ corresponding to this k are real-valued.

Next, complex eigenvalues of A occur in conjugate pairs. If $\lambda_c, \lambda_d (= \bar{\lambda}_c)$ are such a pair, and the columns of (6) containing A_c and A_d are respectively the γth, ..., $(\gamma + m_c)$th and the δth, ..., $(\delta + m_d)$th (note that $m_d = m_c$), we can choose the basis vectors $z_\delta, \dots, z_{\delta + m_c}$ in the decomposition so that their coordinates with respect to the natural basis in $\mathbb{C}^n$ are the complex conjugates of those of $z_\gamma, \dots, z_{\gamma + m_c}$, and so that A_d is the complex conjugate of A_c. It follows that the solutions $v_1, \dots, v_n$ of (11) of the form (12) with the factor $\exp(\lambda_d t)$ are the complex conjugates of the solutions $u_1, \dots, u_n$ with the factor $\exp(\lambda_c t)$. We may obviously replace these solutions $u_1, \dots, u_n$ and $v_1, \dots, v_n$ by

$$\tfrac{1}{2}(u_1 + v_1), \dots, \tfrac{1}{2}(u_n + v_n) \quad \text{and} \quad \frac{1}{2i}(u_1 - v_1), \dots, \frac{1}{2i}(u_n - v_n),$$

without affecting linear independence, and these new solutions are real-valued. We have thus shown that *when the a_{ij} are real, the system* (11) *has a linearly independent set of* **real-valued** *solutions of the following forms*: (i) *corresponding to a real eigenvalue λ of the matrix $[a_{ij}]$ with multiplicity m, there are m solutions of the form*

$$y_i = e^{\lambda t} P_i(t) \qquad (i = 1, \dots, n),$$

where the P_i are real-valued polynomials of degree at most $m - 1$; (ii) *corresponding to complex eigenvalues $\mu + iv, \mu - iv$ of $[a_{ij}]$ with multiplicity M, there are M solutions of the form*

$$y_i = e^{\mu t}(Q_i(t) \cos vt + R_i(t) \sin vt) \qquad (i = 1, \dots, n),$$

and M solutions of the form

$$y_i = e^{\mu t}(Q_i(t) \cos vt - R_i(t) \sin vt) \qquad (i = 1, \dots, n),$$

where the Q_i and R_i are real-valued polynomials of degree at most $M - 1$.

Exercises 2.9

1 Let $I = [a, \infty[$, and let $A \in \mathscr{L}(Y; Y)$. Prove that the following statements are equivalent:
   (i) all solutions of $y' = Ay$ on I are bounded;

(ii) $y' = Ay$ is stable;

(iii) $y' = Ay$ is uniformly stable.

Prove also that (i)–(iii) imply

(iv) if $b : I \to Y$ is a continuous function such that $\int_a^\infty \| b(t) \| \, dt < \infty$, then all solutions of $y' = Ay + b(t)$ on I are bounded,

and

(v) if $C : I \to \mathscr{L}(Y ; Y)$ is a continuous function such that $\int_a^\infty \| C(t) \| \, dt < \infty$, then $y' = (A + C(t))y$ is uniformly stable.

[Hint. For the equivalence of (i)–(iii) use Exercises 2.8.8,9. For (v) use Exercise 2.8.15.]

2 Prove that if Y is finite-dimensional, $A \in \mathscr{L}(Y ; Y)$, and every solution of the equation $y' = Ay$ on $\mathbf{R}$ tends to 0 as $t \to \infty$, then there exists $\mu > 0$ such that $\| \exp(tA) \| \leq \exp(-\mu t)$ for all $t \in \mathbf{R}$.

[Hint. Use (2.9.1).]

3 Let Y be finite-dimensional, let $I = [a, \infty[$, let $C = \{(t, y) : t \in I, \| y \| < \rho\}$, where $\rho > 0$, and let $F : C \to Y$ be a continuous function with the properties that

(i) there exists $K \geq 0$ such that $\| F(t, y) \| \leq K \| y \|$ for all $(t, y) \in C$,

(ii) for each $\varepsilon > 0$ there exist $\eta > 0$ and $T > a$ such that $\| F(t, y) \| \leq \varepsilon \| y \|$ whenever $\| y \| \leq \eta$ and $t \geq T$.

Prove that if $A \in \mathscr{L}(Y ; Y)$ and $y' = Ay$ is asymptotically stable (i.e. every solution tends to 0 as $t \to \infty$), then the zero solution of the equation

$$y' = Ay + F(t, y)$$

on I is asymptotically stable.

[An interesting particular case of this result is that in which $F(t, y) = B(t)y$, where $B : I \to \mathscr{L}(Y ; Y)$ is continuous and $B(t) \to 0$ in $\mathscr{L}(Y ; Y)$ as $t \to \infty$.

Hint. Let $\mu(> 0)$ be chosen as in Exercise 2, let $t_0 \in I$, and let ϕ be a solution of $y' = Ay + F(t, y)$ reaching to the boundary of C on the right, with domain $J = [t_0, b[$. Use Exercise 2.8.12, and Bellman's inequality (2.6.2. Corollary) applied to $e^{\mu t} \| \phi(t) \|$, to show that for all $t \in J$

$$\| \phi(t) \| \leq e^{(K - \mu)(t - t_0)} \| \phi(t_0) \|. \qquad (*)$$

If $K < \mu$, this gives the result. If $K \geq \mu$, let $0 < \varepsilon < \mu$, and choose $\eta(< \rho)$ and T as in (ii). Then if $\| \phi(t_0) \| < \eta \exp(-(K - \mu)(T - t_0))$, we have $\| \phi(t) \| < \eta$ whenever $t \in J$ and $t_0 \leq t \leq T$, whence $[t_0, T] \subseteq J$. Now let c be the greatest point of J such that $\| \phi(t) \| < \eta$ for all $t \in [t_0, c[$ (so that $c > T$). If $c < b$, then $\| \phi(c) \| = \eta$. Now apply $(*)$ on $[T, c[$ with t_0 and K replaced by T and ε to show that $c = b = \infty$.]

2.10 Dependence on initial conditions and parameters

In this section we study the behaviour of the solutions of the equation $y' = f(t, y)$ when the initial point (t_0, y_0) varies. We also allow the differential equation itself to depend on a parameter λ belonging to a topological or metric space Λ, that is to say we consider the equation $y' = f(t, y, \lambda)$, and we study the behaviour of the solutions when both the initial point (t_0, y_0) and the parameter λ vary.

We begin with the case where $f(t, y, \lambda)$ is locally Lipschitzian in y.

Let Λ be a topological space, let f be a continuous function from an open set Γ in the topological space $\Omega = \mathbf{R} \times Y \times \Lambda$ into Y, and suppose that $f(t, y, \lambda)$ is locally Lipschitzian in y in Γ, i.e. for each point of Γ there exist a neighbourhood N of the point in Γ and a positive number K such that

$$\| f(t, y, \lambda) - f(t, z, \lambda) \| \leq K \| y - z \| \tag{1}$$

for all (t, y, λ), $(t, z, \lambda) \in N$. For each $\lambda \in \Lambda$ let $E(\lambda)$ be the set of points $(t, y) \in \mathbf{R} \times Y$ for which $(t, y, \lambda) \in \Gamma$. Then $E(\lambda)$ is open in $\mathbf{R} \times Y$, since it is the inverse image of Γ by the continuous map $(t, y) \mapsto (t, y, \lambda)$ from $\mathbf{R} \times Y$ into Ω. It therefore follows from (2.7.6) that for each point $\gamma = (t_0, y_0, \lambda) \in \Gamma$ the equation $y' = f(t, y, \lambda)$ has a unique solution $t \mapsto \phi(t, \gamma)$ having the value y_0 at t_0 and with maximal interval of existence $I(\gamma)$, this interval being open. We prove here:

(2.10.1) *Suppose that the above conditions are satisfied, and let* $\bar{\gamma} = (\bar{t}_0, \bar{y}_0, \bar{\lambda}) \in \Gamma$. *Then for each compact subinterval J of $I(\bar{\gamma})$ containing $\bar{t}_0$ there exists a neighbourhood V of $\bar{\gamma}$ in Γ such that $J \subseteq I(\gamma)$ for all $\gamma \in V$. Moreover,* $\phi(t, \gamma) \to \phi(t, \bar{\gamma})$ *as* $\gamma \to \bar{\gamma}$, *uniformly for t in J.*

We note immediately the following corollary of (2.10.1).

(2.10.1. Corollary) *Under the conditions of the main theorem, the domain A of the function ϕ, i.e. the set of (t, γ) such that $\gamma \in \Gamma$ and $t \in I(\gamma)$, is open in $\mathbf{R} \times \Omega$.*

To prove this, let $(\bar{t}, \bar{\gamma}) \in A$, so that $\bar{t} \in I(\bar{\gamma})$. Since $I(\bar{\gamma})$ is open, we can find a compact subinterval J of $I(\bar{\gamma})$ containing $\bar{t}$ and $\bar{t}_0$ as interior points, and then by the main theorem we can find a neighbourhood V of $\bar{\gamma}$ in Γ such that $J \subseteq I(\gamma)$ for all $\gamma \in V$. Hence $J \times V \subseteq A$, and since $J \times V$ is a neighbourhood of $(\bar{t}, \bar{\gamma})$ in $\mathbf{R} \times \Omega$, this proves the result.

Consider now the proof of (2.10.1). We shall see that it is enough to prove the result when $\bar{y}_0 = 0$ and $\phi(t, \bar{\gamma}) = 0$ for all $t \in I(\bar{\gamma})$ (so that also $f(t, 0, \bar{\lambda}) = 0$ for all $t \in I(\bar{\gamma})$). Assuming this for the present, let J be a compact subinterval of $I(\bar{\gamma})$ containing $\bar{t}_0$ as an interior point. Then a simple covering argument shows that there exist a closed ball B in Y with centre 0 and radius $\rho > 0$, a neighbourhood W of $\bar{\lambda}$ in Λ, and a positive number K, such that $f(t, y, \lambda)$ is K-Lipschitzian in y in $J \times B \times W$ (i.e. f satisfies (1) for all $(t, y, \lambda), (t, z, \lambda) \in J \times B \times W$).

Next, let g be the function from $J \times Y \times W$ into Y given by

$$g(t, y, \lambda) = \begin{cases} f(t, y, \lambda) & \text{if } y \in B, \\ f(t, \rho y / \| y \|, \lambda) & \text{if } y \in Y \backslash B. \end{cases} \tag{2}$$

Clearly g is continuous and agrees with f on $J \times B \times W$. Moreover,
$g(t, y, \lambda)$ is $2K$-*Lipschitzian* in y in $J \times Y \times W$. Assuming this too for the
moment, we obtain from (2.7.3) that for each point $\gamma = (t_0, y_0, \lambda) \in J \times Y \times W$ the equation $y' = g(t, y, \lambda)$ has a unique solution $t \mapsto \psi(t, \gamma)$ on J
taking the value y_0 at t_0. Further, $\psi(t, \gamma) \to 0$ as $\gamma \to \bar{\gamma}$, uniformly for
t in J, for if $t \in J$ then

$$\left\| \frac{d}{dt} \psi(t, \gamma) \right\| = \| g(t, \psi(t, \gamma), \lambda) \|$$
$$\le \| g(t, \psi(t, \gamma), \lambda) - g(t, 0, \lambda) \| + \| g(t, 0, \lambda) \|$$
$$\le 2K \| \psi(t, \gamma) \| + \sigma(\lambda),$$

where $\sigma(\lambda) = \sup_J \| g(t, 0, \lambda) \|$, and by (1.6.1), (2.6.1), and (2.6.1)', this gives

$$\| \psi(t, \gamma) \| \le \| y_0 \| \exp(2K|t - t_0|) + \tfrac{1}{2}\sigma(\lambda)K^{-1}(\exp(2K|t - t_0|) - 1).$$

By (A.1.3) with $X = J$ and $Z = W$, $g(t, 0, \lambda) \to g(t, 0, \bar{\lambda}) = 0$ as $\lambda \to \bar{\lambda}$, uniformly for t in J, whence $\sigma(\lambda) \to 0$ as $\lambda \to \bar{\lambda}$, and therefore $\psi(t, \gamma)$ tends
uniformly to 0 as $\gamma \to \bar{\gamma}$.

It follows now that we can find a neighbourhood V of $\bar{\gamma}$ in Γ such that
$\psi(t, \gamma) \in B$ whenever $\gamma \in V$ and $t \in J$. This implies that $t \mapsto \psi(t, \gamma)$ is a solution
of the equation $y' = f(t, y, \lambda)$ on J, whence it is the restriction of $t \mapsto \phi(t, \gamma)$
to J. It follows that $J \subseteq I(\gamma)$ whenever $\gamma \in V$, and that $\phi(t, \gamma) \to 0$ as $\gamma \to \bar{\gamma}$,
uniformly for t in J, and this is the required result.

To justify the reduction to the special case considered in the proof,
let $\bar{\gamma} \in \Gamma$, and let $t \mapsto \phi(t, \bar{\gamma})$ and $I(\bar{\gamma})$ be the corresponding solution and
its interval of existence, where now $\phi(t, \bar{\gamma})$ is not necessarily 0. We may
obviously suppose that $\Gamma \subseteq I(\bar{\gamma}) \times Y \times \Lambda$ (else we replace Γ by its intersection with this set). Define $f^* : \Gamma^* \to Y$ by

$$f^*(t, y, \lambda) = f(t, y + \phi(t, \bar{\gamma}), \lambda) - \frac{d}{dt}\phi(t, \bar{\gamma}),$$

where Γ^* is the inverse image of Γ in $I(\bar{\gamma}) \times Y \times \Lambda$ by the mapping $(t, y, \lambda) \mapsto$
$(t, y + \phi(t, \bar{\gamma}), \lambda)$. Clearly Γ^* is open in $\mathbf{R} \times Y \times \Lambda$, f^* is continuous, and
$f^*(t, y, \lambda)$ is locally Lipschitzian in y in Γ^*. Moreover, the equation

$$y' = f(t, y, \lambda) \tag{3}$$

has the solution $t \mapsto \phi(t, \gamma)$ on $I(\gamma) \cap I(\bar{\gamma})$ taking the value y_0 at t_0 if and
only if the equation

$$y' = f^*(t, y, \lambda) \tag{4}$$

has the solution $t \mapsto \phi(t, \gamma) - \phi(t, \bar{\gamma})$ on $I(\gamma) \cap I(\bar{\gamma})$ taking the value $y_0 -$
$\phi(t_0, \bar{\gamma})$ at t_0. Since $\gamma \to \bar{\gamma}$ if and only if $(t_0, y_0 - \phi(t_0, \bar{\gamma}), \lambda) \to (\bar{t}_0, 0, \bar{\lambda})$, we

may replace (3) by (4), and this is equivalent to taking $\bar{y}_0 = 0$ and $\phi(t, \bar{y}) = 0$ for all $t \in I(\bar{y})$.

It remains to prove the italicized statement (p. 127) concerning the function g defined by (2), and this is a consequence of the following lemma.

(2.10.2) *Let Y be a normed space, let B be the closed ball in Y with centre 0 and radius $\rho > 0$, and let h be the radial retraction of Y onto B given by $h(y) = y$ if $y \in B$ and $h(y) = \rho y / \|y\|$ if $y \in Y \backslash B$. Then h is 2-Lipschitzian on Y.†*

To prove this, we observe first that if $\|x\| \geq \|x'\|$ and $\lambda \geq \mu \geq 0$, then

$$\lambda \|x - x'\| = \|\lambda x - \mu x' - (\lambda - \mu)x'\| \leq \|\lambda x - \mu x'\| + (\lambda - \mu)\|x'\|$$

$$\leq \|\lambda x - \mu x'\| + \lambda \|x\| - \mu \|x'\| \leq 2\|\lambda x - \mu x'\|.$$

If $y, z \in Y \backslash B$ and $\|y\| \geq \|z\|$, we now take $x = h(y)$, $x' = h(z)$, $\lambda = \|y\|/\rho$, $\mu = \|z\|/\rho$, while if $y \in Y \backslash B$ and $z \in B$ we take x, x', λ as before and $\mu = 1$ (so that in either case $\lambda \geq \mu \geq 1$).

If now $\bar{f}$ is the restriction of f to $J \times B \times W$ and g is defined by (2), then

$$g(t, y, \lambda) = \bar{f}(t, h(y), \lambda),$$

and this trivially implies the required Lipschitzian property of g.

We note in passing that if Y is an inner product space then the radial retraction h is 1-Lipschitzian on Y, for here

$$\|\lambda x - \mu x'\|^2 = \lambda \mu \|x - x'\|^2 + (\lambda - \mu)(\lambda \|x\|^2 - \mu \|x'\|^2)$$

whenever $\lambda, \mu \geq 0$, and therefore $\|\lambda x - \mu x'\| \geq \|x - x'\|$ whenever $\|x\| \geq \|x'\|$ and $\lambda \geq \mu \geq 1$. Hence in this case the function g defined by (2) has the same Lipschitz constant as $\bar{f}$.

We consider next the case where f is continuous, and since our arguments employ sequences we require that Λ is a metric space instead of a topological space. We also require Y to be finite-dimensional.

In the first theorem below we avoid any difficulties arising from varying domains of existence of solutions by working in a slab $S = J \times Y \times \Lambda$, where J is a compact interval in **R**, and by assuming that $f : S \to Y$ is bounded and continuous. In this case (2.4.2) shows that for each point

---

† The constant 2 cannot be improved; consider, for instance, **R**² with the norm $\|(\xi, \eta)\| = |\xi| + |\eta|$.

$\gamma = (t_0, y_0, \lambda) \in S$ the equation $y' = f(t, y, \lambda)$ has at least one solution on J taking the value y_0 at t_0, and that the set $\Sigma(\gamma)$ of all such solutions is compact in $C(J, Y)$. We prove that this set varies continuously with γ.

(2.10.3) *Let Y be finite-dimensional, let Λ be a metric space, let J be a compact interval in $\mathbf{R}$, and let $S = J \times Y \times \Lambda$. Let also $f : S \to Y$ be bounded and continuous, and for each point $\gamma = (t_0, y_0, \lambda) \in S$ let $\Sigma(\gamma)$ be the subset of $C(J, Y)$ consisting of all solutions of the equation $y' = f(t, y, \lambda)$ on J taking the value y_0 at t_0. Then for each point $\bar\gamma = (\bar{t}_0, \bar{y}_0, \bar\lambda) \in S$ and each open subset G of $C(J, Y)$ containing $\Sigma(\bar\gamma)$, there exists a neighbourhood V of $\bar\gamma$ in S such that $\Sigma(\gamma) \subseteq G$ whenever $\gamma \in V$.*

If the result is false we can find $\bar\gamma \in S$, an open set G in $C(J, Y)$ containing $\Sigma(\bar\gamma)$, and a sequence (γ_n) of points of S converging to $\bar\gamma$, such that, for each n, $\Sigma(\gamma_n)$ is not contained in G. For $n = 1, 2, \ldots$, choose $\phi_n \in \Sigma(\gamma_n) \backslash G$. Further, for each $\chi \in C(J, Y)$ let $d(\chi)$ be the distance of χ from $\Sigma(\bar\gamma)$, and let δ be the distance of $\Sigma(\bar\gamma)$ from the complement of G. Then $d(\phi_n) \geq \delta$ for all n, and $\delta > 0$ since $\Sigma(\bar\gamma)$ is compact.

Since f is bounded, inequalities (1) and (2) of §2.2 imply that the sequence (ϕ_n) is equicontinuous and uniformly bounded. It therefore follows from the Ascoli–Arzelà theorem (A.1.2) that (ϕ_n) has a subsequence converging in $C(J, Y)$ to a function ϕ, and obviously $d(\phi) \geq \delta > 0$. Let B be a closed ball in Y such that $\phi_n(t) \in B$ for all n and all $t \in J$, so that also $\phi(t) \in B$ for all $t \in J$. By (A.1.3) with $X = J \times B, Z = \Lambda, f(t, y, \lambda_n) \to f(t, y, \bar\lambda)$ as $n \to \infty$, uniformly for (t, y) in $J \times B$, and therefore $\phi \in \Sigma(\bar\gamma)$, by (2.3.2. Corollary). Hence $d(\phi) = 0$, and this gives the required contradiction.

As an immediate consequence of this result we have:

(2.10.3. Corollary) *Let S, f be as in the main theorem, and suppose that for each point $\gamma = (t_0, y_0, \lambda) \in S$ the equation $y' = f(t, y, \lambda)$ has exactly one solution ϕ_γ on J taking the value y_0 at t_0. Then the function $\gamma \mapsto \phi_\gamma$ from S into $C(J, Y)$ is continuous.*

We consider finally the case where the domain of f is an arbitrary open set in Ω, and here we confine ourselves to a result corresponding to (2.10.3. Corollary).

(2.10.4) *Let Y be finite-dimensional, let Λ be a metric space, let J be a compact interval in $\mathbf{R}$, and let f be a continuous function from an open set Γ in the metric space $\Omega = \mathbf{R} \times Y \times \Lambda$ into Y. For each $\lambda \in \Lambda$ let $E(\lambda)$ be the* (*open*) *set of points* $(t, y) \in \mathbf{R} \times Y$ *for which* $(t, y, \lambda) \in \Gamma$, *and suppose that for each point* $\gamma = (t_0, y_0, \lambda) \in \Gamma$ *the equation* $y' = f(t, y, \lambda)$ *has exactly one solution taking the value* y_0 *at* t_0 *and reaching to the boundary of* $E(\lambda)$ *in both directions. Let this solution be* $t \mapsto \phi(t, \gamma)$, *and let its domain be the* (*open*) *interval* $I(\gamma)$.† *Then for each* $\bar{\gamma} = (\bar{t}_0, \bar{y}_0, \bar{\lambda}) \in \Gamma$ *and each compact subinterval* J *of* $I(\bar{\gamma})$ *containing* $\bar{t}_0$ *as an interior point there exists a neighbourhood* V *of* $\bar{\gamma}$ *in* Γ *such that* $J \subseteq I(\gamma)$ *for all* $\gamma \in V$. *Moreover,* $\phi(t, \gamma) \to \phi(t, \bar{\gamma})$ *as* $\gamma \to \bar{\gamma}$, *uniformly for* t *in* J.

Exactly as in the proof of (2.10.1), we may assume that $\bar{y}_0 = 0$ and that $\phi(t, \bar{\gamma}) = 0$ for all $t \in I(\bar{\gamma})$ (so that also $f(t, 0, \bar{\lambda}) = 0$ for all $t \in I(\bar{\gamma})$). Let J be a compact subinterval of $I(\bar{\gamma})$ containing $\bar{t}_0$ as an interior point. Then there exist a closed ball B in Y with centre 0 and radius $\rho > 0$ and a neighbourhood W of $\bar{\lambda}$ in Λ such that f is bounded on $J \times B \times W$. In fact, since the set $\{(t, 0, \bar{\lambda}) : t \in J\}$ in Ω is compact, we can find a neighbourhood U of $(0, \bar{\lambda})$ in $Y \times \Lambda$ such that the cylinder $J \times U$ is contained in Γ. By (A.1.3) with $X = J, Z = U$, $f(t, y, \lambda) \to f(t, 0, \bar{\lambda}) = 0$ as $(y, \lambda) \to (0, \bar{\lambda})$, uniformly for t in J, and this implies the statement.

Next, let $g : J \times Y \times W \to Y$ be defined as in (2). Clearly g is bounded and continuous, and agrees with f on $J \times B \times W$. We now apply (2.10.3) to g; thus we can find a neighbourhood V of $\bar{\gamma}$ in Γ such that, if $\gamma \in V$, then every solution of $y' = g(t, y, \lambda)$ on J taking the value y_0 at t_0 maps J into B. Hence such a solution is also a solution of $y' = f(t, y, \lambda)$ on J, and is therefore the restriction of $t \mapsto \phi(t, \gamma)$ to J. Hence $J \subseteq I(\gamma)$ for all $\gamma \in V$, and, again by (2.10.3), $\phi(t, \gamma) \to \phi(t, \bar{\gamma})$ as $\gamma \to \bar{\gamma}$, uniformly for t in J.

We remark here that if Λ is a subset of a normed space, we can regard the equation $y' = f(t, y, \lambda_0)$, subject to the initial condition that $y = y_0$ when $t = t_0$, as equivalent to the pair of equations $y' = f(t, y, \lambda)$, $\lambda' = 0$, subject to the initial conditions that $y = y_0$ and $\lambda = \lambda_0$ when $t = t_0$. Thus in this case the discussion of the equation $y' = f(t, y, \lambda)$, where both λ and the initial point vary, can be reduced to the discussion of a new equation $z' = g(t, z)$, where $z = (y, \lambda)$, and where now only the initial point varies. In this more special situation, some of the difficulties encountered above can be avoided.

We conclude this section with an application of the preceding results to the proof of the following supplement to (2.4.2).

† The uniqueness of the solution implies that, for each subinterval J of $I(\gamma)$ containing t_0, the restriction of $t \mapsto \phi(t, \gamma)$ to J is the only solution of $y' = f(t, y, \lambda)$ on J taking the value y_0 at t_0 (for, by (2.4.4), any solution on J has an extension reaching to the boundary of $E(\lambda)$ in both directions).

(2.10.5) *Let Y be finite-dimensional, let J be a compact interval in* **R**, *let* $f : J \times Y \to Y$ *be bounded and continuous, let* $(t_0, y_0) \in J \times Y$, *and let* Σ *be the subset of* $C(J, Y)$ *consisting of all solutions of the equation* $y' = f(t, y)$ *on J taking the value* y_0 *at* t_0. *Then* Σ *is connected.*

We require a lemma.

(2.10.6) Lemma. *Let Y be finite-dimensional, let B be a closed ball in Y, and let J be compact. Let also* $f : J \times B \to Y$ *be continuous, let* $M = \sup_{J \times B} \| f(t, y) \|$, *and let* ϕ *be a solution of the equation* $y' = f(t, y)$ *on J. Then for each* $\varepsilon > 0$ *we can find a continuous function* $g : J \times Y \to Y$ *such that*

(i) $g(t, y)$ *is Lipschitzian in y in* $J \times Y$,

(ii) $\| g(t, y) \| \leq M + \frac{1}{2}\varepsilon$ *for all* $(t, y) \in J \times Y$,

(iii) $\| f(t, y) - g(t, y) \| \leq \varepsilon$ *for all* $(t, y) \in J \times B$,

(iv) ϕ *is a solution of the equation* $y' = g(t, y)$ *on J.*

By (A.1.5), we can find a continuous function $g_1 : J \times B \to Y$ such that $\| g_1(t, y) \| \leq M$ and $\| f(t, y) - g_1(t, y) \| \leq \frac{1}{2}\varepsilon$ for all $(t, y) \in J \times B$ and that $g_1(t, y)$ is Lipschitzian in y in $J \times B$. The function $g_2 : J \times B \to Y$ given by $g_2(t, y) = g_1(t, y) + f(t, \phi(t)) - g_1(t, \phi(t))$ then clearly satisfies (i) and (ii) on $J \times B$ and has the properties (iii) and (iv). If now we take $g(t, y) = g_2(t, h(y))$, where h is the radial retraction of Y into B (see (2.10.2)), we obtain the required result.

Consider now the proof that Σ is connected. If the result is false, then Σ is the union of two non-empty disjoint sets Σ_1, Σ_2, each closed in Σ and therefore in $C(J, Y)$. For all $\chi \in C(J, Y)$ let $d_1(\chi), d_2(\chi)$ be the distances of χ from Σ_1, Σ_2, and let $d(\chi) = d_1(\chi) - d_2(\chi)$ (so that d is continuous on $C(J, Y)$). We show that there exists $\chi \in \Sigma$ for which $d(\chi) = 0$. This implies that either $d_1(\chi) = d_2(\chi) > 0$, in which case $\chi \notin \Sigma_1 \cup \Sigma_2$, or that $d_1(\chi) = d_2(\chi) = 0$, i.e. $\chi \in \Sigma_1 \cap \Sigma_2$, and in either case we obtain a contradiction.

Let $\phi \in \Sigma_1, \psi \in \Sigma_2$, and let B be the closed ball in Y with centre y_0 and radius $(M + 1)l$, where l is the length of J. By (2.10.6), for each positive integer n we can find a function g_n having the properties (i)–(iv) with $\varepsilon = 1/n$ and with B as specified, and similarly we can find h_n related to ψ in the same way. Consider now the equation

$$y' = (1 - \sigma)g_n(t, y) + \sigma h_n(t, y),$$

where $0 \leq \sigma \leq 1$ and n is fixed. It follows from (2.7.3) that for each $\sigma \in [0, 1]$ this equation has a unique solution $\chi_{n,\sigma}$ on J taking the value y_0 at t_0, and, by (ii) and (iii) of (2.10.6) and our choice of the radius of B, $\chi_{n,\sigma}$ is a $1/n$-approximate solution of $y' = f(t, y)$ on J. By (2.10.1) (or (2.10.3. Corollary)), the function $\sigma \mapsto \chi_{n,\sigma}$ from $[0, 1]$ into $C(J, Y)$ is continuous, whence so also is $\sigma \mapsto d(\chi_{n,\sigma})$. But $d(\chi_{n,0}) = d(\phi) = -d_2(\phi) < 0$, and similarly $d(\chi_{n,1}) = d(\psi) > 0$. Hence there exists $\tau_n \in [0,1]$ for which $d(\chi_{n,\tau_n}) = 0$. By (ii) of (2.10.6), the sequence (χ_{n,τ_n}) is equicontinuous and uniformly bounded, and hence, by $(A.1.2)$ and $(2.3.2)$, it has a subsequence converging in $C(J, Y)$ to a function $\chi \in \Sigma$. Since d is continuous, $d(\chi) = 0$, and this is the required result.

(2.10.5. Corollary) *If the conditions of the main theorem are satisfied, then for each $\tau \in J$ the set $E(\tau) = \{\phi(\tau) : \phi \in \Sigma\}$ is compact and connected in Y.*

This follows since $E(\tau)$ is the image of Σ by the continuous function $\phi \mapsto \phi(\tau)$ from $C(J, Y)$ into Y.

Exercise 2.10

1 Let J be a compact interval in $\mathbf{R}$, let $f : J \times \mathbf{R} \to \mathbf{R}$ be continuous and bounded, and for each point $P_0 = (t_0, y_0) \in J \times \mathbf{R}$ let Φ_{P_0} be the maximal solution of $y' = f(t, y)$ with domain J and taking the value y_0 at t_0. Let also $\bar{P}_0 = (\bar{t}_0, \bar{y}_0)$ be a given point of $J \times \mathbf{R}$, let $\bar{\Phi}$ be the corresponding maximal solution, and let $U = \{(t, y) \in J \times \mathbf{R} : y \geq \bar{\Phi}(t)\}$. Prove that $\Phi_{P_0} \to \bar{\Phi}$ uniformly as $P_0 \to \bar{P}_0$ in U.

Use this result to obtain a corresponding analogue of (2.10.4).

[Hint. Observe that $\Phi_{P_0} \geq \bar{\Phi}$ whenever $P_0 \in U$, and use (2.10.3).]

2.11 Further existence and uniqueness theorems

The theorems of this section are generalizations of the theorems of §2.7 involving Lipschitz conditions, and are of two types. In the first type, the Lipschitz condition
$$\| f(t, y) - f(t, z) \| \leq K \| y - z \|$$
is replaced by a condition of the form
$$\| f(t, y) - f(t, z) \| \leq g(t, \| y - z \|), \tag{1}$$
while in the second type, which applies only when Y is a Hilbert space, the Lipschitz condition is replaced by a 'monotonicity' condition
$$\mathrm{Re} \langle f(t, y) - f(t, z), y - z \rangle \leq \| y - z \| g(t, \| y - z \|).$$
The conditions on g are the same in both types of theorem, and require that the zero function is a solution of the scalar equation $x' = g(t, x)$ and that this solution is unique in a certain specified sense. More precisely, we suppose that g satisfies the following condition, introduced by Iyanaga and Kamke.

(I–K) *g is a continuous function from the rectangle $]t_0, t_0 + \alpha] \times [0, \beta]$ in*

$\mathbf{R}^2$ into $[0, \infty[$, *satisfying* $g(t,0) = 0$ *for all* $t \in]t_0, t_0 + \alpha]$, *with the property that, for each* $t_1 \in]t_0, t_0 + \alpha]$, $\chi = 0$ *is the only solution of* $x' = g(t, x)$ *on* $]t_0, t_1]$ *such that*

$$\chi(t_0 +) = 0 \quad \text{and that} \quad \lim_{t \to t_0+} \frac{\chi(t)}{t - t_0} = 0. \tag{2}$$

Our results depend on the following differential inequality.

(2.11.1) Lemma. *Let* g *satisfy the condition* (I–K) *and let* $\omega : [t_0, t_0 + \alpha] \to [0, \beta]$ *be a continuous function such that* $\omega(t_0) = \omega'(t_0) = 0$ *and that*

$$D_+ \omega(t) \leq g(t, \omega(t))$$

for nearly all $t \in]t_0, t_0 + \alpha[$. *Then* $\omega = 0$.

Suppose on the contrary that $\omega(t_1) = b > 0$ for some $t_1 \in]t_0, t_0 + \alpha]$. Let $\bar{g}$ be the extension of g to $]t_0, t_0 + \alpha] \times \mathbf{R}$ given by

$$\bar{g}(t, x) = g(t, \beta) \quad (x > \beta), \qquad \bar{g}(t, x) = g(t, 0) = 0 \quad (x < 0),$$

and let χ be the minimal solution of the equation $x' = \bar{g}(t, x)$ to the left of t_1 satisfying the condition $\chi(t_1) = b$ and reaching to the boundary of $]t_0, t_1] \times \mathbf{R}$ on the left. Since g is non-negative, χ is increasing. Further, $\chi(t) \geq 0$ for all t, for if $\chi(\tau) < 0$ for some $\tau \in]t_0, t_1]$, there is a least σ between τ and t_1 for which $\chi(\sigma) = 0$; then $\chi(u) < 0$ for $\tau \leq u < \sigma$, whence $\chi'(u) = \bar{g}(u, \chi(u)) = 0$, and therefore $\chi(\tau) = 0$, giving a contradiction. It follows that χ is defined on the whole interval $]t_0, t_1]$, and its range is contained in $[0, b]$.

Next, since $D_+ \omega(t) \leq g(t, \omega(t)) = \bar{g}(t, \omega(t))$ for nearly all $t \in]t_0, t_1]$, it follows from Exercise 1.5.1 that $\chi(t) \leq \omega(t)$ for all $t \in]t_0, t_1]$. Since χ is non-negative and $\omega(t_0) = \omega'(t_0) = 0$, it follows that $\chi(t_0 +) = 0$ and that

$$0 \leq \lim_{t \to t_0+} \chi(t)/(t - t_0) \leq \omega'(t_0) = 0.$$

Hence χ satisfies (2), and since χ is obviously a solution of $x' = g(t, x)$, the condition (I–K) implies that $\chi = 0$, contrary to our supposition that $\chi(t_1) > 0$.

Before passing to the applications of (2.11.1), we note some cases where the condition is satisfied.

(a) The condition (I–K) is obviously implied by the following condition of Perron.

(P) g *is a continuous function from the rectangle* $[t_0, t_0 + \alpha] \times [0, \beta]$

in $\mathbf{R}^2$ into $[0, \infty[$, satisfying $g(t, 0) = 0$ for all $t \in [t_0, t_0 + \alpha]$, with the property that, for each $t_1 \in]t_0, t_0 + \alpha]$, the zero function is the only solution of $x' = g(t, x)$ on $[t_0, t_1]$ taking the value 0 at t_0.

In particular, the function g given by $g(t, x) = Kx$ satisfies (P) (put $x = e^{Kt}u$) and therefore also (I–K). Hence the condition (1) with g satisfying (I–K) includes the Lipschitz condition.

(b) *If* $g(t, x) = \mu(t - t_0)x$, *where* μ *is continuous and non-negative on* $]0, \alpha]$, *then* g *satisfies* (I–K) *if and only if*

$$\liminf_{t \to 0+} \left(\log t + \int_t^\alpha \mu(u) du \right) < \infty. \tag{3}$$

For instance, $g(t, x) = x/(t - t_0)$ satisfies (I–K) while $g(t, x) = Cx/(t - t_0)$ does not satisfy (I–K) for any $C > 1$.

In proving this, we may obviously suppose $t_0 = 0$. Then, if $0 < \gamma < \alpha$ and χ is a solution of $x' = \mu(t)x$ on $]0, \gamma]$, we have

$$\chi(t) = \chi(\gamma) \exp\left(-\int_t^\gamma \mu(u) du \right)$$

for all $t \in]0, \gamma]$, and therefore also

$$\chi(t)/t = \chi(\gamma) \exp\left(-\log t - \int_t^\gamma \mu(u) du \right).$$

Hence if (3) holds and $\chi(t)/t \to 0$ as $t \to 0+$, then $\chi(\gamma) = 0$, so that $\chi = 0$. On the other hand, if (3) does not hold, then $\log t + \int_t^\gamma \mu(u) du \to \infty$ as $t \to 0+$, so that also $\int_t^\gamma \mu(u) du \to \infty$ as $t \to 0+$, and hence there exists a non-zero χ satisfying (2).

(c) *Let* $g(t, x) = \mu(t - t_0)h(x)$, *where* μ *is a continuous non-negative function on* $]0, \alpha]$ *such that* $\int_0^\alpha \mu(s) ds < \infty$, *and* h *is a continuous function on* $[0, \beta]$ *such that* $h(0) = 0, h(x) > 0$ *for* $0 < x \leq \beta$, *and* $\int_0^\beta (1/h(u)) du = \infty$. *Then* g *satisfies* (I–K).

In proving this, we may again suppose that $t_0 = 0$. Let $0 < \gamma < \alpha$, and let χ be a solution of $x' = \mu(t)h(x)$ on $]0, \gamma]$ satisfying $\chi(0+) = 0$. Then χ is increasing, whence there exists a least $\delta \in [0, \gamma]$ such that $\chi(t) > 0$ for $\delta < t \leq \gamma$. If $\delta < \gamma$, then for $\delta < t \leq \gamma$

$$\int_{\chi(t)}^{\chi(\gamma)} \frac{du}{h(u)} = \int_t^\gamma \mu(s) ds$$

and this is clearly impossible, since $\chi(t) \to 0$ as $t \to \delta+$. Hence $\delta = \gamma$, so that $\chi = 0$.

(2.11.2) *Let* E *be a subset of* $[t_0, t_0 + \alpha] \times Y$ *containing* (t_0, y_0), *and let* $f : E \to Y$ *be a function with the property that, for nearly all* $t\in\,]t_0, t_0 + \alpha]$,

$$\| f(t, y) - f(t, z) \| \le g(t, \| y - z \|) \qquad (4)$$

whenever $(t, y), (t, z)\in E$ *and* $\| y - z \| \le \beta$, *where g satisfies condition* (I − K). *Then the equation* $y' = f(t, y)$ *has at most one solution on* $[t_0, t_0 + \alpha]$ *taking the value* y_0 *at* t_0.

If Y is a complex Hilbert space, the condition (4) *can be replaced by*

$$\mathrm{Re} \langle f(t, y) - f(t, z), y - z \rangle \le \| y - z \| g(t, \| y - z \|). \qquad (5)$$

Let ϕ_1, ϕ_2 be solutions of $y' = f(t, y)$ on $[t_0, t_0 + \alpha]$ such that $\phi_1(t_0) = \phi_2(t_0) = y_0$, and let $\omega(t) = \| \phi_1(t) - \phi_2(t) \|$. Then ω is continuous, $\omega(t_0) = 0$, and

$$\omega'(t_0) = \lim_{t\to t_0 +} \frac{\| \phi_1(t) - \phi_1(t_0) - \phi_2(t) + \phi_2(t_0) \|}{t - t_0} = \| \phi_1'(t_0) - \phi_2'(t_0) \| = 0.$$

Further, we can obviously find a positive number $\alpha_1 \le \alpha$ such that $\omega(t) < \beta$ for all $t\in[t_0, t_0 + \alpha_1[$ and that $\omega(\alpha_1) = \beta$ if $\alpha_1 < \alpha$.

Suppose now that f satisfies (4). Then for nearly all $t\in\,]t_0, t_0 + \alpha_1[$,

$$D^+\omega(t) \le \| \phi_1'(t) - \phi_2'(t) \| = \| f(t, \phi_1(t)) - f(t, \phi_2(t)) \| \le g(t, \omega(t)).$$

Hence ω satisfies the conditions of (2.11.1), and therefore $\omega(t) = 0$ for all $t\in[t_0, t_0 + \alpha_1]$. Hence $\alpha_1 = \alpha$, so that $\phi_1 = \phi_2$.

When Y is a Hilbert space and f satisfies (5), we argue as follows. If $\phi_1(t) = \phi_2(t)$, then $\phi_1'(t) = \phi_2'(t)$, and therefore

$$D^+\omega(t) \le \| \phi_1'(t) - \phi_2'(t) \| = 0 = g(t, 0) = g(t, \omega(t)).$$

On the other hand, if $\phi_1(t) \ne \phi_2(t)$ and (5) holds for t, then

$$\omega'(t) = \frac{d}{dt}(\| \phi_1(t) - \phi_2(t) \|^2)^{1/2}$$
$$= \frac{\mathrm{Re} \langle \phi_1'(t) - \phi_2'(t), \phi_1(t) - \phi_2(t) \rangle}{\| \phi_1(t) - \phi_2(t) \|}$$
$$= \frac{\mathrm{Re} \langle f(t, \phi_1(t)) - f(t, \phi_2(t)), \phi_1(t) - \phi_2(t) \rangle}{\| \phi_1(t) - \phi_2(t) \|}$$
$$\le g(t, \omega(t)).$$

Hence $D^+\omega(t) \le g(t, \omega(t))$ for nearly all $t\in\,]t_0, t_0 + \alpha_1[$, and the argument is completed as before.

In the existence theorem corresponding to (2.11.2), we require that f is continuous, and therefore the result is of interest only when Y is infinite-dimensional, since the finite-dimensional case is contained in (2.4.1).

(2.11.3) *Let* $y_0\in Y$, *let B be the closed ball in Y with centre* y_0 *and radius*

$\rho > 0$, and let $f : [t_0, t_0 + \alpha] \times B \to Y$ be a continuous function such that $\|f(t, y)\| \le M$ for all $(t, y) \in [t_0, t_0 + \alpha] \times B$ and that

$$\|f(t, y) - f(t, z)\| \le g(t, \|y - z\|) \qquad (6)$$

whenever $(t, y), (t, z) \in \,]t_0, t_0 + \alpha] \times B$, where g satisfies condition (I–K) with $\beta = 2\rho$. Then, if $\eta = \min \{\alpha, \rho/M\}$, the equation $y' = f(t, y)$ has exactly one solution on $[t_0, t_0 + \eta]$ taking the value y_0 at t_0.

If Y is a complex Hilbert space, then (6) can be replaced by

$$\mathrm{Re} \, \langle f(t, y) - f(t, z), y - z \rangle \le \|y - z\| g(t, \|y - z\|). \qquad (7)$$

It should be noted that, although the inequalities (6) and (7) have only to be satisfied on $]t_0, t_0 + \alpha] \times B$, we require that f is continuous on the closed cylinder $[t_0, t_0 + \alpha] \times B$. The uniqueness of the solution is, of course, a consequence of (2.11.2), so that we have only to prove existence.

We require two lemmas.

(2.11.4) Lemma. *Let* (ϕ_n) *be a sequence of continuous functions from* J *into* $\mathbf{R}$ *converging uniformly on* J *to a function* ϕ. *Let also* E *be a set in* $\mathbf{R}^2$ *containing the graphs of* ϕ_n $(n = 1, 2, \dots)$ *and of* ϕ, *let* $f : E \to \mathbf{R}$ *be continuous, and suppose that for each* n

$$D_+\phi_n(t) \le f(t, \phi_n(t))$$

nearly everywhere in J. *Then for all* $t \in J^\circ$

$$D_+\phi(t) \le f(t, \phi(t)).$$

Let $\varepsilon > 0$, and let $t \in J^\circ$. Since f is continuous at $(t, \phi(t))$, we can find $\delta > 0$ such that $|f(s, y) - f(t, \phi(t))| \le \varepsilon$ whenever $(s, y) \in E$ and $|s - t| \le \delta$, $|y - \phi(t)| \le \delta$. Also, since ϕ is continuous and $\phi_n \to \phi$ uniformly on J, we can find η satisfying $0 < \eta \le \delta$ and a positive integer q such that $s \in J$ and $|\phi_n(s) - \phi(t)| \le \delta$ whenever $|s - t| \le \eta$ and $n \ge q$, and then also $|f(s, \phi_n(s)) - f(t, \phi(t))| \le \varepsilon$. Hence for each $n \ge q$

$$D_+\phi_n(s) \le f(t, \phi(t)) + \varepsilon$$

nearly everywhere in $[t, t + \eta]$, and therefore, by (1.4.4),

$$\phi_n(s) - \phi_n(t) \le (f(t, \phi(t)) + \varepsilon)(s - t)$$

whenever $s \in [t, t + \eta]$. Hence also

$$\phi(s) - \phi(t) \le (f(t, \phi(t)) + \varepsilon)(s - t),$$

so that $D_+\phi(t) \le f(t, \phi(t)).\dagger$

(2.11.5) Lemma. *Let* $M \ge 0$, *let* Δ *be a class of uniformly bounded con-*

$\dagger$ Indeed $D^+\phi(t) \le f(t, \phi(t))$ (cf. (1.4.5)).

tinuous functions $\psi : J \to \mathbf{R}$ *with the property that for all* $s, t \in J$

$$|\psi(s) - \psi(t)| \leq M|s - t|, \tag{8}$$

and let $\Psi = \sup \psi$. *Let also E be a set in* $\mathbf{R}^2$ *containing the graphs of each* $\psi \in \Delta$ *and of* $\overset{\Delta}{\Psi}$, *let* $f : E \to \mathbf{R}$ *be continuous, and suppose that for each* $\psi \in \Delta$

$$D_+ \psi(t) \leq f(t, \psi(t)) \tag{9}$$

nearly everywhere in J. Then for all $s, t \in J$

$$|\Psi(s) - \Psi(t)| \leq M|s - t| \tag{10}$$

(so that Ψ *is continuous), and for all* $t \in J^\circ$

$$D_+ \Psi(t) \leq f(t, \Psi(t)).$$

We remark first that if $\psi_1, \ldots, \psi_k \in \Delta$ and $\psi = \max \{\psi_1, \ldots, \psi_k\}$, then ψ satisfies (9) nearly everywhere in J. Here the case $k = 2$ is almost immediate (cf. the proof of (1.5.1); note that, by (1.4.5), ψ_1 and ψ_2 satisfy (9) with D^+ in place of D_+), and the general case follows from this by induction.

Next, from (8) we obtain (10), and this, together with (8), implies that for all $\psi \in \Delta$ and all $s, t \in J$

$$0 \leq \Psi(t) - \psi(t) \leq \Psi(s) - \psi(s) + 2M|s - t|.$$

From this it follows easily that for each positive integer n we can find a positive integer k, a partition of J into k subintervals of equal length, and k functions $\psi_1, \ldots \psi_k \in \Delta$ such that in the jth subinterval

$$0 \leq \Psi(t) - \psi_j(t) \leq 1/n.$$

Let $\psi^{(n)} = \max \{\psi_1, \ldots, \psi_k\}$. Then $0 \leq \Psi(t) - \psi^{(n)}(t) \leq 1/n$ for all $t \in J$, so that the sequence $(\psi^{(n)})$ converges uniformly to Ψ on J. Also $D_+ \psi^{(n)}(t) \leq f(t, \psi^{(n)}(t))$ for nearly all $t \in J$ (by the remark above), and the required result therefore follows from (2.11.4).

Consider now the proof of (2.11.3), and suppose first that f satisfies (6). Let $J = [t_0, t_0 + \eta]$, and let (ε_n) be a decreasing sequence of positive numbers with the limit 0. By (2.3.1), for each positive integer n we can find an ε_n-approximate solution ψ_n of the equation $y' = f(t, y)$ on J, satisfying $\psi_n(t_0) = y_0$, with the property that for all $s, t \in J$

$$\|\psi_n(s) - \psi_n(t)\| \leq M|s - t|. \tag{11}$$

Let $\sigma_{m,n}(t) = \|\psi_m(t) - \psi_n(t)\|$, where $t \in J$ and $m > n \geq 1$. Obviously $\sigma_{m,n}(t_0) = 0$, and for all $s, t \in J$

$$|\sigma_{m,n}(s) - \sigma_{m,n}(t)| \leq 2M|s - t|. \tag{12}$$

Further, for all except a finite number of points of J

$$D_+ \sigma_{m,n}(t) \leq \| \psi'_m(t) - \psi'_n(t) \|$$
$$\leq \| f(t, \psi_m(t)) - f(t, \psi_n(t)) \| + \varepsilon_m + \varepsilon_n$$
$$\leq g(t, \sigma_{m,n}(t)) + 2\varepsilon_n. \qquad (13)$$

For each positive integer n, let $\omega_n = \sup_{m>n} \sigma_{m,n}$. Then $\omega_n(t_0) = 0$, and, by (12), (13), and (2.11.5) (applied to each compact subinterval of $]t_0, t_0 + \eta]$),

$$|\omega_n(s) - \omega_n(t)| \leq 2M |s - t|$$

for all $s, t \in J$ and

$$D_+ \omega_n(t) \leq g(t, \omega_n(t)) + 2\varepsilon_n \qquad (14)$$

for all $t \in J^\circ$. The sequence (ω_n) is therefore equicontinuous and uniformly bounded, and hence (A.1.2) it has a subsequence (ω_{n_r}) converging uniformly on J to a function ω, and clearly $\omega(t_0) = 0$. By (14) and (2.11.4),

$$D_+ \omega(t) \leq g(t, \omega(t)) + 2\varepsilon_n$$

for all $t \in J^\circ$, and therefore also

$$D_+ \omega(t) \leq g(t, \omega(t)).$$

We show next that $\omega'(t_0) = 0$. Since f is continuous at (t_0, y_0), given $\varepsilon > 0$ we can find $\delta > 0$ such that $\| f(t, y) - f(t_0, y_0) \| \leq \varepsilon$ whenever $t_0 \leq t \leq t_0 + \delta$ and $\| y - y_0 \| \leq \delta$. Let $\lambda = \min \{ \delta, \delta/M \}$. By (11), $\| \psi_n(t) - y_0 \| \leq \delta$ for all n and all $t \in [t_0, t_0 + \lambda]$, and therefore

$$\| f(t, \psi_m(t)) - f(t, \psi_n(t)) \| \leq 2\varepsilon$$

whenever $m > n \geq 1$ and $t \in [t_0, t_0 + \lambda]$. By the penultimate inequality in (13),

$$D_+ \sigma_{m,n}(t) \leq 2\varepsilon + 2\varepsilon_n$$

for all but a finite number of points $t \in [t_0, t_0 + \lambda]$, and hence, by (1.4.4),

$$0 \leq \sigma_{m,n}(t) = \sigma_{m,n}(t) - \sigma_{m,n}(t_0) \leq (2\varepsilon + 2\varepsilon_n)(t - t_0)$$

whenever $t \in [t_0, t_0 + \lambda]$. This implies in turn that

$$0 \leq \omega_n(t) \leq (2\varepsilon + 2\varepsilon_n)(t - t_0),$$

and so also

$$0 \leq \omega(t) \leq (2\varepsilon + 2\varepsilon_n)(t - t_0),$$

whence $\omega'(t_0) = 0$.

From (2.11.1) we deduce now that $\omega = 0$, and this implies that the sequence (ψ_n) is uniformly convergent on J. The limit of this sequence is then the required solution.

To prove the result when f satisfies (7), we choose the sequence (ψ_n)

and define $\sigma_{m,n}$ as before, so that $\sigma_{m,n}(t_0) = 0$, and (12) holds for all $s, t \in J$. If t is a point such that $\psi'_m(t), \psi'_n(t)$ exist, and $\psi_m(t) \neq \psi_n(t)$, then, as in the proof of (2.11.2),

$$\sigma'_{m,n}(t) = \frac{d}{dt}(\|\psi_m(t) - \psi_n(t)\|^2)^{1/2}$$

$$= \frac{\mathrm{Re}\,\langle \psi'_m(t) - \psi'_n(t), \psi_m(t) - \psi_n(t)\rangle}{\|\psi_m(t) - \psi_n(t)\|}$$

$$= \frac{\mathrm{Re}\,\langle f(t, \psi_m(t)) - f(t, \psi_n(t)), \psi_m(t) - \psi_n(t)\rangle}{\|\psi_m(t) - \psi_n(t)\|}$$

$$+ \frac{\mathrm{Re}\,\langle \psi'_m(t) - f(t, \psi_m(t)) - \psi'_n(t) + f(t, \psi_n(t)), \psi_m(t) - \psi_n(t)\rangle}{\|\psi_m(t) - \psi_n(t)\|}$$

$$\leq g(t, \sigma_{m,n}(t)) + \varepsilon_m + \varepsilon_n$$

$$\leq g(t, \sigma_{m,n}(t)) + 2\varepsilon_n, \tag{15}$$

while if $\psi_m(t) = \psi_n(t)$ then

$$D_+\sigma_{m,n}(t) = \|\psi'_m(t) - \psi'_n(t)\|$$

$$= \|\psi'_m(t) - f(t, \psi_m(t)) - \psi'_n(t) + f(t, \psi_n(t))\|$$

$$\leq \varepsilon_m + \varepsilon_n \leq 2\varepsilon_n, \tag{16}$$

and therefore the final inequality in (13) holds for all but a finite number of $t \in J$.

If now $\omega_n = \sup_{m>n} \sigma_{m,n}$, then exactly as before we see that there exists a subsequence (ω_{n_r}) of (ω_n) converging uniformly on J to a function ω such that $\omega(t_0) = 0$ and that $D_+\omega(t) \leq g(t, \omega(t))$ for all $t \in J^\circ$. The third identity in (15) and the final inequality in (16) show (again as before) that $\omega'(t_0) = 0$, and therefore $\omega = 0$, as required.

Exercises 2.11

1 Let g satisfy condition (I–K), and let $\omega : [t_0, t_0 + \alpha] \to \mathbf{R}$ be a continuous function such that $\omega(t_0) = \omega'(t_0) = 0$ and that

$$D_+\omega(t) \leq g(t, \omega(t))$$

for nearly all $t \in [t_0, t_0 + \alpha[$ for which $0 < \omega(t) \leq \beta$. Prove that $\omega(t) \leq 0$ for all $t \in [t_0, t_0 + \alpha]$.

2 Let p be a continuous sublinear functional on Y such that $p(y) = p(-y) = 0$ if and only if $y = 0$, let g satisfy condition (I–K), let E be a subset of $[t_0, t_0 + \alpha] \times Y$ containing (t_0, y_0), and let $f : E \to Y$ be a function with the property that, for nearly all $t \in]t_0, t_0 + \alpha]$,

$$p(f(t, y) - f(t, z)) \leq g(t, p(y - z))$$

whenever $(t, y), (t, z) \in E$ and $0 < p(y - z) \leq \beta$. Prove that the equation $y' = f(t, y)$ has at most one solution on $[t_0, t_0 + \alpha]$ taking the value y_0 at t_0.

3 Let μ be an extended-real-valued function Lebesgue integrable on each compact subinterval of $]0,\alpha]$, let $h:]0,\beta] \to]0,\infty[$ be continuous, and suppose that for all a,b,c satisfying $0 \le a < b \le \alpha, 0 < c \le \beta$, we can find $\varepsilon > 0$ such that

$$\limsup_{v \to 0+} \left(\int_{\varepsilon v}^c \frac{du}{h(u)} - \int_{a+v}^b \mu(s)ds \right) > 0.$$

Let also $\omega:[0,\alpha] \to \mathbf{R}$ be a continuous function with the properties that $\omega(0) = 0$, that $\omega'_+(t) = 0$ whenever $\omega(t) = 0$, and that $\omega'_+(t)$ exists for nearly all $t \in]0,\alpha[$ for which $0 < \omega(t) \le \beta$ and satisfies $\omega'_+(t) \le \mu(t)h(\omega(t))$ for almost all such t. Prove that $\omega(t) \le 0$ for all $t \in [0,\alpha]$.

[Hint. Use (1.10.8. Corollary 2).]

4 Prove that the result of Exercise 2 holds with $t_0 = 0$ and $g(t,x) = \mu(t)h(x)$, where μ, h satisfy the conditions of Exercise 3.

5 Prove that if

(i) μ is an extended-real-valued function Lebesgue integrable on each compact subinterval of $]0,\alpha]$,

(ii) there exists $K > 0$ such that $\int_t^\alpha \mu(s)ds \le K$ for all $t \in]0,\alpha]$,

(iii) $h:]0,\beta] \to]0,\infty[$ is continuous and $\int_0^\beta du/h(u) = \infty$, then μ, h satisfy the conditions of Exercise 3.

[In particular, this exercise shows that Exercise 4 contains the uniqueness theorem for the generalized Lipschitz condition (see Exercise 2.7.2).]

6 Prove that if μ is a non-negative function Lebesgue integrable on each compact subinterval of $]0,\alpha]$, and $h(x) = x$, then μ and h satisfy the conditions of Exercise 3 if and only if

$$\liminf_{t \to 0+} \left(\log t + \int_t^\infty \mu(s)ds \right) < \infty.$$

[Hence Exercise 4 generalizes the case of Exercise 2 corresponding to the special case (b) of condition (I–K) (p. 134).]

7 Let q be a non-negative continuous quasi-convex function on Y such that $q(y) = 0$ if and only if $y = 0$, let E be a subset of $[t_0, t_0 + \alpha] \times Y$ containing (t_0, y_0), and let $f:E \to Y$ be a function with the property that, for almost all $t \in]t_0, t_0 + \alpha]$,

$$q(f(t,y) - f(t,z)) \le q\left(\frac{y - z}{t - t_0} \right)$$

whenever $(t,y), (t,z) \in E$. Prove that the equation $y' = f(t,y)$ has at most one solution on $[t_0, t_0 + \alpha]$ taking the value y_0 at t_0.

Prove also that the same conclusion holds if q is a continuous convex function on Y such that $q(y) = q(-y) = 0$ if and only if $y = 0$.

[Hint. For the case where q is quasi-convex, let ϕ_1, ϕ_2 be solutions of $y' = f(t,y)$ on $[t_0, t_0 + \alpha]$ such that $\phi_1(t_0) = \phi_2(t_0) = y_0$, let $\psi = \phi_1 - \phi_2$, and let $\chi(t) = \psi(t)/(t - t_0)$ $(t_0 < t \le t_0 + \alpha), \chi(t_0) = \psi'(t_0) = 0$. Use Exercise 1.10.3 to show that t_0 is the least point where $q \circ \chi$ attains its maximum on $[t_0, t_0 + \alpha]$, so that $q(\chi(t)) = 0$. For the convex case show similarly that $q(\chi(t)) \le 0$ and $q(-\chi(t)) \le 0$.]

8 Let $c > 0$, let $E = [0, \infty[\times \mathbf{R}$, and let $f_c:E \to \mathbf{R}$ be given by

$$f_c(t,x) = \frac{4xt^3}{c(x^2 + t^4)} \qquad ((t,x) \ne (0,0)), \qquad f_c(0,0) = 0.$$

(i) Prove that f_c is continuous on E and that $f_c(t,x)$ is locally Lipschitzian in x in $E \setminus \{(0,0)\}$.

(ii) Show that if $c \geq 2$ the zero function is the only solution of the equation $x' = f_c(t,x)$ taking the value 0 at 0 and reaching to the boundary of E on the right. Show also that if $c < 2$ the equation has non-unique solutions on $[0, \infty[$ taking the value 0 at 0, and that the maximal and minimal such solutions are $t \mapsto \lambda t^2$ and $t \mapsto -\lambda t^2$, where $\lambda = (2/c - 1)^{1/2}$. Show further that if $c = 1$ the family of all such solutions other than the zero function is given by

$$\phi(t) = \pm((A^2 + t^4)^{1/2} - A) \qquad (A \geq 0).$$

(iii) Prove that f_c satisfies the criterion

$$f_c(t,x) - f_c(t,y) \leq \frac{x-y}{t} \qquad (t > 0, x > y),$$

given by the particular case of Exercise 2 where $Y = \mathbf{R}$, $p(y) = y$, $g(t,x) = x/t$, if and only if $c \geq 4$.

9 Let $y_0 \in Y$, let B be the closed ball in Y with centre y_0 and radius $\rho > 0$, let $J = [t_0, t_0 + \alpha]$, and let $f : J \times B \to Y$ be a function with the properties that
(i) for all $(t,y),(t,z) \in J \times B$

$$\| f(t,y) - f(t,z) \| \leq g(t, \| y - z \|),$$

where g satisfies the condition (P) on p. 133,
(ii) there exists $\delta > 0$ such that for each y_1 satisfying $\| y_1 - y_0 \| < \delta$ the equation $y' = f(t,y)$ has a solution ϕ_1 on J such that $\phi_1(t_0) = y_1$ (the solution is necessarily unique, by (2.11.2)).
Prove that the solutions ϕ_1 are continuous in y_1 at y_0, i.e. $\phi_1 \to \phi$ in $C(J, Y)$.

[Hint. Let ϕ_0 be the solution of $y' = f(t,y)$ on J such that $\phi(t_0) = y_0$, consider $\psi = \| \phi_1 - \phi \|$, and note that (2.10.1) applies to the solutions of $x' = \bar{g}(t,x)$, where $\bar{g}$ is the extension of g to $J \times \mathbf{R}$ similar to that in (2.11.1).]

2.12 Successive approximations

Let I be an interval in $\mathbf{R}$, let $f : I \times Y \to Y$ be continuous, let $(t_0, y_0) \in I \times Y$, and let $\psi : I \to Y$ be a given continuous function. The functions $\psi_0, \psi_1, \ldots$ defined inductively by the formulae

$$\psi_0 = \psi, \qquad \psi_n(t) = y_0 + \int_{t_0}^{t} f(u, \psi_{n-1}(u)) du \qquad (n = 1, 2, \ldots), \qquad (1)$$

are called the *successive approximations* determined by ψ for the equation $y' = f(t,y)$ with the initial point (t_0, y_0). From (2.2.2), we see easily that if the sequence (ψ_n) converges to a function ϕ on I, uniformly on each compact subinterval of I, then for all $t \in I$

$$\phi(t) = y_0 + \int_{t_0}^{t} f(u, \phi(u)) du,$$

and therefore ϕ is a solution of the equation $y' = f(t,y)$ on I satisfying the initial condition $\phi(t_0) = y_0$. Hence to prove the existence of such a solution ϕ, it is enough to prove the uniform convergence of the sequence (ψ_n) on each compact subinterval of I.

Successive approximations can also be defined when the domain of f is an arbitrary subset E of $\mathbf{R} \times Y$, provided that the graphs of the successive approximations lie in E.

Exercise 2.7.2 shows that if $f : I \times Y \to Y$ is continuous and satisfies the generalized Lipschitz condition

$$\| f(t,y) - f(t,z) \| \le \mu(t) \| y - z \|,$$

for almost all $t \in I$ and all $y, z \in Y$, where μ is a non-negative function Lebesgue integrable on each compact subinterval of I, then for every continuous $\psi : I \to Y$ the successive approximations determined by ψ converge to the solution ϕ of $y' = f(t,y)$ satisfying $\phi(t_0) = y_0$, uniformly on each compact subinterval of I.

When f is merely continuous, the successive approximations ψ_n may fail to converge, even when the equation $y' = f(t,y)$ has a unique solution taking the value y_0 at t_0 (see Exercise 2.12.1). However, we can prove that under conditions closely similar to those of Perron mentioned in §2.11(a) (p. 133) all the successive approximations converge to the solution of the equation. More precisely, we have:

(2.12.1) *Let* $(t_0, y_0) \in \mathbf{R} \times Y$, *let* $I = [t_0, t_0 + \alpha]$, *let* B *be the closed ball in* Y *with centre* y_0 *and radius* $\rho > 0$, *and let* $f : I \times B \to Y$ *be a continuous function such that* $\| f(t,y) \| \le M$ *for all* $(t,y) \in I \times B$. *Suppose further that for all* $(t,y), (t,z) \in I \times B$

$$\| f(t,y) - f(t,z) \| \le h(t, \| y - z \|), \tag{2}$$

where h *is a continuous function from the rectangle* $I \times [0, 2\rho]$ *in* $\mathbf{R}^2$ *into* $\mathbf{R}$ *with the properties*
 (i) $h(t,0) = 0$ *for all* $t \in I$,
 (ii) $x \mapsto h(t,x)$ *is increasing on* $[0, 2\rho]$ *for all* $t \in I$,
 (iii) *for each* $t_1 \in]t_0, t_0 + \alpha]$, *the zero function is the only solution of* $x' = h(t,x)$ *on* $[t_0, t_1]$ *taking the value* 0 *at* t_0.
Then if $J = [t_0, t_0 + \eta]$, *where* $\eta = \min\{\alpha, \rho/M\}$, *and* $\psi : J \to Y$ *is continuous, the successive approximations* ψ_n *defined by* (1) *converge uniformly on* J *to the solution* ϕ *of* $y' = f(t,y)$ *such that* $\phi(t_0) = y_0$.

The solution ϕ exists and is unique, by (2.11.2,3).
Let $\bar{h} : I \times [0, \infty[\to \mathbf{R}$ be the extension of h given by
$$\bar{h}(t,x) = h(t,x) \quad (0 \le x \le 2\rho), \quad \bar{h}(t,x) = h(t,2\rho) \quad (x > 2\rho),$$
let $K = \sup |\bar{h}| = \sup |h|$, and define the functions $\chi_n : J \to \mathbf{R}$ by
$$\chi_0(t) = \max \{ \| \psi_1(t) - \psi_0(t) \|, K(t - t_0) \},$$

$$\chi_n(t) = \int_{t_0}^t \bar{h}(u, \chi_{n-1}(u))\,du \qquad (n = 1, 2, \ldots).$$

A simple induction argument shows that the functions χ_n are defined and non-negative on J, and that the sequence (χ_n) is decreasing (remember that $x \mapsto \bar{h}(t, x)$ is increasing for each $t \in J$). Further, the sequence (χ_n) is obviously equicontinuous, and therefore has a uniformly convergent subsequence (A.1.2). Since the sequence (χ_n) is decreasing, it follows that this sequence is itself uniformly convergent, to χ say, and therefore also

$$\chi(t) = \int_{t_0}^t \bar{h}(u, \chi(u))\,du.$$

Hence χ is a solution of the equation $x' = \bar{h}(t, x)$ on J such that $\chi(t_0) = 0$, and therefore $\chi = 0$, by the condition (iii). In particular, we deduce that $\bar{h}(t, \chi_n(t)) \to \bar{h}(t, \chi(t)) = \bar{h}(t, 0) = 0$ as $n \to \infty$, uniformly for t in J.

Consider now the sequence (ψ_n). Another easy induction argument shows that each function ψ_n is defined on J and takes its values in B, and is continuous. Further, for $n = 1, 2, \ldots$ and all $t \in J$

$$\|\psi_n(t) - \psi_{n-1}(t)\| \le \chi_{n-1}(t). \tag{3}$$

This is true by definition if $n = 1$. Also, if n is a positive integer for which (3) holds, then

$$\|\psi_{n+1}(t) - \psi_n(t)\| \le \int_{t_0}^t \|f(u, \psi_n(u)) - f(u, \psi_{n-1}(u))\|\,du$$

$$\le \int_{t_0}^t h(u, \|\psi_n(u) - \psi_{n-1}(u)\|)\,du$$

$$\le \int_{t_0}^t \bar{h}(u, \chi_{n-1}(u))\,du = \chi_n(t),$$

so that (3) holds for $n + 1$, and therefore for all n.

We observe now that for all $t \in J$

$$\|\psi_n'(t) - f(t, \psi_n(t))\| = \|f(t, \psi_{n-1}(t)) - f(t, \psi_n(t))\|$$

$$\le h(t, \|\psi_{n-1}(t) - \psi_n(t)\|) \le \bar{h}(t, \chi_{n-1}(t)).$$

Hence ψ_n is an ε_n-approximate solution of $y' = f(t, y)$ on J satisfying $\psi_n(t_0) = y_0$, where $\varepsilon_n = \sup_J \bar{h}(t, \chi_{n-1}(t)) \to 0$ as $n \to \infty$. But the proof of (2.11.3) shows that under these conditions the sequence (ψ_n) converges uniformly on J to the solution ϕ of $y' = f(t, y)$ such that $\phi(t_0) = y_0$, and this is the required result.

When Y is finite-dimensional, we can weaken the hypotheses of (2.12.1) as follows.

(2.12.2) *Let Y be finite-dimensional, and suppose that f satisfies the hypotheses of* (2.12.1), *except that in place of* (2) *we have*

$$\| f(t,y) - f(t,z) \| \le g(t, \| y - z \|)$$

for all (t,y), $(t,z) \in]t_0, t_0 + \alpha] \times B$, *where g satisfies condition* (I–K) *with* $\beta = 2\rho$. *Then the result of* (2.12.1) *holds.*

To prove this, let $h : I \times [0, 2\rho] \to [0, \infty[$ be given by

$$h(t,x) = \sup_{\| y - z \| \le x} \| f(t,y) - f(t,z) \|,$$

so that $0 \le h(t,x) \le g(t,x)$ for $t_0 < t \le t_0 + \alpha$, $0 \le x \le 2\rho$. Since f is uniformly continuous on $I \times B$, h is continuous, and obviously h possesses the properties (i) and (ii) of (2.12.1). It is therefore enough to prove that h possesses the property (iii) of (2.12.1).

Let $t_1 \in]t_0, t_0 + \alpha]$, and let ξ be a solution of $x' = h(t,x)$ on $[t_0, t_1]$ such that $\xi(t_0) = 0$; we have to show that $\xi = 0$. Suppose on the contrary that $\xi(\tau) > 0$ for some τ, and let ω be the minimal solution of $x' = g(t,x)$ with initial point $(\tau, \xi(\tau))$ and reaching to the boundary of $]t_0, \tau] \times [0, \infty[$ on the left. Then, as in (2.11.3), ω is defined on $]t_0, \tau]$. If $\omega(\sigma) > \xi(\sigma)$ for some $\sigma \in]t_0, \tau]$, let η be the least point of $[\sigma, \tau]$ for which $\omega(\eta) = \xi(\eta)$. Then $\omega(t) > \xi(t)$ for all $t \in [\sigma, \eta[$, and therefore also

$$\omega'(t) = g(t, \omega(t)) \ge h(t, \omega(t)) \ge h(t, \xi(t)) = \xi'(t),$$

whence $\omega(\sigma) \le \xi(\sigma)$, contrary to our choice of σ. Hence $\omega(t) \le \xi(t)$ for all $t \in]t_0, \tau]$, and a simple argument now shows that $\omega(t_0 +) = 0$ and that $\omega(t)/(t - t_0) \to 0$ as $t \to t_0 +$. Since g satisfies the condition (I–K) (p. 132) it follows that $\omega = 0$, and this contradicts the fact that $\omega(\tau) = \xi(\tau) > 0$. Hence $\xi = 0$, as required.

Exercise 2.12

1 Let $f : [0, 1] \times \mathbf{R} \to \mathbf{R}$ be given by

$$f(t,x) = 2t \quad (0 \le t \le 1, x \le 0), \quad f(t,x) = -2t \quad (0 \le t \le 1, x \ge t^2),$$
$$f(t,x) = 2t - 4x/t \quad (0 < t \le 1, 0 < x < t^2)$$

(so that f is continuous and bounded). Prove that
(i) the zero function is the unique solution of $x' = f(t,x)$ on $[0,1]$ taking the value 0 at 0,
(ii) the successive approximations $\psi_n : [0,1] \to \mathbf{R}$ defined by

$$\psi_0 = 0, \qquad \psi_n(t) = \int_0^t f(u, \psi_{n-1}(u)) du \qquad (n = 1, 2, \ldots),$$

do not converge on $[0,1]$.
[Hint. For (i), use Exercise 2.4.3.]

2.13 An existence theorem for a discontinuous function

We return to the general equation

$$y' = f(t, y) \tag{1}$$

but we consider a much wider class of functions f, which necessitates a wider definition of 'solution'. As before, we suppose that $E \subseteq \mathbf{R} \times Y$ and that $f : E \to Y$. We say that a continuous function ϕ is a *solution of* (1) *in the sense of Carathéodory on an interval I* if $(t, \phi(t)) \in E$ for $t \in I$, $\phi'(t)$ exists for almost all $t \in I$ and

$$\phi'(t) = f(t, \phi(t))$$

almost everywhere in I.

In what follows, Y will be finite dimensional and E will be of the form $I \times Y$ where $I \subseteq \mathbf{R}$ is an interval. The function f will be supposed measurable in t for fixed $y \in Y$ and continuous in y for fixed $t \in I$. Under these conditions, if $\phi : I \to Y$ is continuous, the mapping $t \mapsto f(t, \phi(t))$ is measurable ((A.6.2. Corollary 2) shows it is measurable on each compact subinterval of I, which is enough). If this function is, in addition, integrable over compact subintervals of I and ϕ is a solution of (1) satisfying $\phi(t_0) = y_0$ for some $(t_0, y_0) \in I \times Y$ we can integrate to obtain

$$\phi(t) = y_0 + \int_{t_0}^{t} f(u, \phi(u)) du \qquad (t \in I). \tag{2}$$

Conversely, if $t \mapsto f(t, \phi(t))$ is integrable over compact subintervals of I, and ϕ satisfies (2), then $\phi'(t)$ exists for almost all t in I, and ϕ is a solution of (1) in the Carathéodory sense. We conclude that, as in the case in which f is continuous, we need only consider the integral equation.

A condition on f which will ensure the integrability of $t \mapsto f(t, \phi(t))$ is that, for all $(t, y) \in I \times Y$,

$$\| f(t, y) \| \le \mu(t) h(\| y \|)$$

where μ is a real-valued function (measurable and) integrable on compact subintervals of I and $h : [0, \infty[\to [0, \infty[$ is continuous, for in this case $t \mapsto h(\| \phi(t) \|)$ is continuous and thus bounded on compact intervals, and so

$$\left| \int_{t_0}^{t} \| f(u, \phi(u)) \| du \right| \le \sup_{t_0 \le u \le t} h(\| \phi(u) \|) \left| \int_{t_0}^{t} \mu(u) du \right| < \infty.$$

We are now in a position to state the main result of this section. We have chosen as our model theorem (2.5.4); other results could have been taken instead, and indeed, they may be obtained from this one (see Exercises 2.13).

(2.13.1) *Let Y be finite-dimensional, let $I = [a, b[$ where $a < b \leq \infty$, let $f : I \times Y \to Y$ be a function with the properties that, for each $t \in I$, $y \mapsto f(t, y)$ is continuous on Y and that for each $y \in Y$, $t \mapsto f(t, y)$ is Lebesgue measurable on I. Suppose also that there exist a measurable function $\mu : I \to [0, \infty[$ which is integrable over compact subintervals of I and a continuous function $h : [0, \infty[\to [0, \infty[$ such that $h(x) > 0$ for $x > 0$, $\int_0 du/h(u) < \infty$ and $\int^\infty du/h(u) = \infty$, such that*

$$\| f(t, y) \| \leq \mu(t) h(\| y \|)$$

for all $(t, y) \in I \times Y$. Let $t_0 \in I$, $y_0 \in Y$. Then the equation

$$y' = f(t, y)$$

has a solution ϕ in the sense of Carathéodory on I satisfying $\phi(t_0) = y_0$. Moreover for all $t \in I$,

$$\| \phi(t) \| \leq H^{-1} \left(H(\| \phi(t_0) \|) + \left| \int_{t_0}^t \mu(u) du \right| \right) \qquad (3)$$

where $H(x) = \int_0^x du/h(u)$ for $x \geq 0$.

We begin the proof by making two simple reductions. First, it is enough to prove the result for a compact subinterval J of I; for once this is done, solutions on compact subintervals may be pieced together to produce a solution on the whole of I. Secondly, we may assume that μ is lower semicontinuous on J; for the integral of a measurable function is the infimum of the integrals of the lower semicontinuous functions which dominate it. The point of the latter step is that we can now find an increasing sequence (μ_n) of continuous functions on J with $\mu_n \to \mu$.

From (A.6.2 Corollary 1) we obtain a sequence (f_n) of continuous functions mapping $J \times Y$ into Y such that

$$\| f_n(t, y) \| \leq \mu_n(t) h(\| y \|) \qquad ((t, y) \in Y)$$

and that, for almost all $t \in I$,

$$\sup_{\| y \| \leq \rho} \| f_n(t, y) - f(t, y) \| \to 0$$

where ρ is chosen so that

$$\rho = \sup_{t \in J} H^{-1} \left(H(\| \phi(t_0) \|) + \left| \int_{t_0}^t \mu(u) du \right| \right)$$

(a supremum which is, of course, attained at one of the endpoints of J).

Theorems (2.4.4) and (2.5.4) may now be applied to each of the equations

$$y' = f_n(t, y);$$

for each n, a solution ϕ_n on J is obtained which satisfies

$$\| \phi_n(t) \| \leq H^{-1} \left(H(\| \phi(t_0) \|) + \left| \int_{t_0}^t \mu_n(u) du \right| \right)$$

$$\leq \rho$$

for $t \in J$. Writing $M = \sup_{\|y\| \le \rho} h(\|y\|)$, we find

$$\|f_n(t, \phi_n(t))\| \le \mu_n(t) h(\|\phi_n(t)\|) \le M\mu(t)$$

for $t \in J$. Therefore, for t, t' in J,

$$\|\phi_n(t) - \phi_n(t')\| \le \left| \int_{t'}^{t} \|f_n(u, \phi_n(u))\| \, du \right| \le M \left| \int_{t'}^{t} \mu(u) du \right|.$$

The sequence (ϕ_n) is therefore equicontinuous and bounded. By the Ascoli–Arzelà theorem (A.1.2), it has a uniformly convergent subsequence, say $\phi_{n_k} \to \phi$.

We saw above that the sequence $(f_{n_k}(t, \phi_{n_k}(t)))$ is dominated by $M\mu(t)$. The limit of the sequence is $f(t, \phi(t))$ since

$$\|f_{n_k}(t, \phi_{n_k}(t)) - f(t, \phi(t))\|$$
$$\le \sup_{\|y\| \le \rho} \|f_{n_k}(t, y) - f(t, y)\| + \|f(t, \phi_{n_k}(t)) - f(t, \phi(t))\| \to 0.$$

The function $M\mu$ is integrable. Therefore the dominated convergence theorem may be applied to obtain from the relation

$$\phi_n(t) = y_0 + \int_{t_0}^{t} f_n(u, \phi_n(u)) du \qquad (t \in J)$$

(which holds for each n and so for each n_k), the relation

$$\phi(t) = y_0 + \int_{t_0}^{t} f(u, \phi(u)) du \qquad (t \in J)$$

which is precisely what is required.

The inequality (3) for ϕ follows immediately from the corresponding formula for each ϕ_n; the latter holds by (2.5.4).

This theorem applies in particular to the linear equation

$$y' = A(t)y + b(t)$$

where Y is finite-dimensional, $t \mapsto A(t)$ is a measurable mapping of an interval I into $\mathcal{L}(Y, Y)$, and $t \mapsto \|A(t)\|, t \mapsto b(t)$ are integrable over compact subintervals of I.

Exercises 2.13

1 Let Y be finite-dimensional, let $E \subseteq \mathbf{R} \times Y$ be an open set, and let $(t_0, y_0) \in E$. Let $f : E \to Y$ be measurable in t and continuous in y for $(t, y) \in E$. Suppose there exist a function $\mu : \mathbf{R} \to [0, \infty[$ integrable over compact intervals and a continuous function $h : [0, \infty[\to [0, \infty[$ such that $\|f(t, y)\| \le \mu(t)h(\|y\|)$ for $(t, y) \in E$. Show that
   (i) there is a compact interval J with t_0 in its interior such that the equation $y' = f(t, y)$ has a solution ϕ in the sense of Carathéodory on J which satisfies $\phi(t_0) = y_0$,

(ii) $y' = f(t, y)$ has a solution ϕ in the sense of Carathéodory which satisfies $\phi(t_0) = y_0$ and which reaches to the boundary of E on the right,

(iii) any solution in the sense of Carathéodory of $y' = f(t, y)$ has an extension which reaches to the boundary of E on the right.

[Here, the local existence theorem (i) is a special case of (ii). Parts (ii) and (iii) generalize (2.4.4).

Hints. For (i), take a compact interval I with t_0 in its interior and a closed ball B with centre y_0 and radius ρ such that $I \times B \subseteq E$. Let r be the radial retraction onto B (r is the identity on B and $r(y) = (y - y_0)/\|y - y_0\|$ if $\|y - y_0\| > \rho$) and consider $\bar{f} : I \times Y \to Y$ defined by $\bar{f}(t, y) = f(t, r(y))$. Modify h in a suitable way, and apply (2.13.1). For (ii) and (iii), notice that (2.4.3) holds in our more general context (rely on the integral equation rather than the differential equation where necessary) and obtain the results by piecing together local solutions.]

2 Let $I \subseteq \mathbf{R}$ be an interval, let Y be finite-dimensional, let $f : I \times Y \to Y$ be measurable in the first coordinate and continuous in the second. Suppose that f satisfies the generalized Lipschitz condition

$$\|f(t, y_1) - f(t, y_2)\| \le \mu(t) \|y_1 - y_2\|$$

for $t \in I, y_1, y_2 \in Y$, where μ is integrable over compact subsets of I. Show that for each $t_0 \in I, y_0 \in Y$, the equation $y' = f(t, y)$ has a unique solution in the Carathéodory sense on I such that $\phi(t_0) = y_0$.

[This result can be generalized further on the lines of Exercises 2.11.2–6.

Hint. For existence, use (2.13.1). For uniqueness, let ϕ, ψ be two solutions and write $\omega(t) = \|\phi(t) - \psi(t)\|$. Both ϕ and ψ are absolutely continuous, and since $\| \|y_1\| - \|y_2\| \| \le \|y_1 - y_2\|$, ω is as well. Thus, using (1.6.1), $\omega'(t) \le \mu(t)\omega(t)$ almost everywhere, and Exercise 2.6.1 may be applied.]

2.14 Historical notes on existence and uniqueness theorems for differential equations and on differential and integral inequalities

The problem of determining the solution of a differential equation was formulated by Newton, and the method of solution by power series was suggested by him as early as 1692 (see Müller, 1928b). Although many techniques for the solution of particular equations were developed in the eighteenth century, the first general existence theorems for differential equations seem to have appeared only in the early years of the nineteenth century. The first such theorem was given by Cauchy in his lectures at the École Polytechnique between 1820 and 1830, and used a method of proof which subsequently became known as the Cauchy–Lipschitz method. Cauchy himself published only a brief outline of his method, in the introduction to a paper lithographed at Prague in 1835 and re-published in 1840 (Cauchy, 1840), and our knowledge of his treatment is derived from a paper published in 1837 by Coriolis (who was later to become known for his work on the ocean currents and winds caused by the rotation of the earth), and from the second volume of the Abbé

Moigno's *Leçons de Calcul Différentiel et de Calcul Intégral* published in 1844.

The account of the method given by Coriolis (1837) reads like a student's attempt to reproduce from memory a proof that he has not understood, and it is impossible to reconstruct Cauchy's argument from it. Fortunately, Moigno is a more accurate source, for he was a pupil and friend of Cauchy, and in the two substantial volumes of his *Leçons* he endeavoured to give a unified account of Cauchy's teachings on calculus, drawing heavily on Cauchy's own writings on the subject.† Moigno remarks (1844, p. 35) that Cauchy's treatment of the existence theorem had been printed, although the sheets containing it had not been made public, and since Moigno almost certainly had access to these sheets it is reasonable to assume that his account follows closely Cauchy's original treatment.

Cauchy's existence theorem, as presented by Moigno, is best understood when viewed as a generalization of the theorem, given in 1823 by Cauchy in his *Résumé*, that every continuous function f possesses a primitive, and indeed this latter theorem can be regarded as the existence theorem for the differential equation $y' = f(t)$. Before giving Cauchy's treatment of the general existence theorem we therefore set out his proof of the theorem on primitives, using modern language as an aid to brevity.‡

Let $f : [t_0, t_0 + \alpha] \to \mathbf{R}$ be continuous, and let $t_0 < t \le t_0 + \alpha$. Let also $P = \{t_0, \ldots, t_p\}$ be a partition of $[t_0, t]$,* let $\Delta t_i = t_{i+1} - t_i$ $(i = 0, \ldots, p - 1)$, let $m(P) = \max \Delta t_i$ be the *mesh* of P, and let

$$S_P(t) = \sum_{i=0}^{p-1} f(t_i)\Delta t_i.$$

Clearly

$$(t - t_0)\min_i f(t_i) \le S_P(t) \le (t - t_0)\max_i f(t_i),$$

and hence, by the intermediate value theorem, we can find $\theta \in [0, 1]$ such that

$$S_P(t) = (t - t_0)f(t_0 + \theta(t - t_0)). \tag{1}$$

---

† Much of the work on differential calculus in Moigno's book is taken from Cauchy's *Leçons sur le Calcul Différentiel* of 1829, while the treatment of the integral calculus follows the relevant sections of Cauchy's *Résumé des Leçons ... sur le Calcul Infinitésimal* of 1823.
‡ In the *Résumé* (Cauchy, 1823) the theorem on primitives is not given in a connected form; the proof of the existence of the limit of the sums $S_P(t)$ defined below occupies the whole of Leçon 21, the relations (3) and (4) are given in Leçon 22, equation (19), and Leçon 23, equation (7), respectively, and the deduction of (2) from (3) and (4) is given at the beginning of Leçon 26.
* So that $t_0 < t_1 < \ldots < t_p = t$ (see Exercise 1.7.5). Cauchy did not use the word partition, nor did he have symbols for intervals, absolute value, max and min.

We show that these sums $S_P(t)$ tend to a limit as $m(P) \to 0$, and for this we use Cauchy's convergence criterion.

First, let R be a partition of $[t_0, t]$ containing the points of P. By (1) applied to f on $[t_0, t_1]$, the change in the term in $S_P(t)$ for $i = 0$ caused by adding to P those points of R that lie in the subinterval $]t_0, t_1[$ is of the form

$$\{f(t_0 + \theta_0 \Delta t_0) - f(t_0)\}\Delta t_0$$

for some $\theta_0 \in [0, 1]$. On repeating this with $i = 1, \ldots, p - 1$, we therefore obtain that

$$S_R(t) - S_P(t) = \sum_{i=0}^{p-1} \varepsilon_i \Delta t_i,$$

where each ε_i is of the form $\varepsilon_i = f(t_i + \theta_i \Delta t_i) - f(t_i)$ for some $\theta_i \in [0, 1]$. Since f is continuous, $\varepsilon_0, \ldots, \varepsilon_{p-1}$ are arbitrarily small for all partitions P with sufficiently small mesh,[†] and hence the same is true of $S_R(t) - S_P(t)$.

Next, let P, Q be two partitions of $[t_0, t_1]$, and let $R = P \cup Q$. Since

$$S_P(t) - S_Q(t) = (S_R(t) - S_Q(t)) - (S_R(t) - S_P(t))$$

and R contains both P and Q, we deduce that $S_P(t) - S_Q(t)$ is arbitrarily small whenever the meshes of P, Q are sufficiently small, and, by Cauchy's convergence criterion, this implies that $S_P(t)$ tends to a limit as $m(P) \to 0$.

To complete the proof, we have to show that if $\int_{t_0}^{t} f(s)\,ds$ denotes the limit of the sums $S_P(t)$, then

$$\frac{d}{dt} \int_{t_0}^{t} f(s)\,ds = f(t). \qquad (2)$$

First, on taking the limit in (1) as $m(P) \to 0$, we obtain that

$$\int_{t_0}^{t} f(s)\,ds = (t - t_0)f(t_0 + \theta(t - t_0)) \qquad (3)$$

for some $\theta \in [0, 1]$. Further, since the limit of a sum is the sum of the limits, we have

$$\int_{t}^{t+h} f(s)\,ds = \int_{t_0}^{t} f(s)\,ds + \int_{t}^{t+h} f(s)\,ds \qquad (4)$$

whenever $t, t + h \in [t_0, t_0 + \alpha]$ (the value of $\int_{c}^{c} f(s)\,ds$ for any c being taken to be 0). From these two relations (3) and (4) we therefore deduce that

$$\int_{t_0}^{t+h} f(s)\,ds - \int_{t_0}^{t} f(s)\,ds = \int_{t}^{t+h} f(s)\,ds = hf(t + \theta h)$$

---

† The original wording, with *t*s for Cauchy's *x*s, is: 'si les éléments $t_1 - t_0, t_2 - t_1, \ldots, t_p - t_{p-1}$ ont des valeurs numériques très-petites, chacune des quantités $\varepsilon_0, \varepsilon_1, \ldots, \varepsilon_{p-1}$ différera très peu de zéro'.

for some $\theta \in [0, 1]$, and this implies (2), by virtue of the continuity of f.

This proof is deficient in two respects. Firstly, it is obvious that to prove that $S_R(t) - S_P(t)$ is arbitrarily small we have to use, not merely the continuity of f, but its *uniform* continuity. In fact Cauchy failed to perceive the difference between continuity and uniform continuity, and indeed it was not until 1870 that Heine first clearly distinguished the two properties and remarked that it was as yet unproved that continuity implies uniform continuity. The first proof of this proposition, for the case of a function on a bounded closed interval in **R**, was supplied by Heine himself in 1872, using an argument which contains implicitly the idea of Borel's covering theorem; an alternative argument was given by Lüroth in the following year for the case of a function on a bounded closed set in $\mathbf{R}^2$ (see Heine, 1870, 1872; and Lüroth, 1873).

The second deficiency in the proof above lies in the assertion that we can obtain (3) by a passage to the limit in (1), and here Cauchy's reasoning is obviously inadequate. The result of (3) is, of course, true, but it is not clear that (3) can be established at this stage in the proof without appealing to a theorem that was not available to Cauchy (for instance, Weierstrass's theorem on the bounds of a continuous function).

We turn now to Cauchy's proof of the general existence theorem for differential equations, and we shall see that again Cauchy implicitly relies on uniform continuity, and that his proof fails at precisely the point corresponding to (3) above.

Cauchy's theorem, as presented by Moigno, is as follows.

(A) *Let* $t_0, y_{01}, \ldots, y_{0n} \in \mathbf{R}$, *let* C *be the set of points* $(t, y_1, \ldots, y_n) \in \mathbf{R}^{n+1}$ *such that*

$$t_0 \leq t \leq t_0 + \alpha, |y_1 - y_{01}| \leq \rho, \ldots, |y_n - y_{0n}| \leq \rho,$$

and let $f_1, \ldots, f_n$ *be continuous real-valued functions (bounded) on* C *possessing continuous (bounded) first partial derivatives with respect to* $y_1, \ldots, y_n$ *on* C. *Then the system of differential equations*

$$y_i' = f_i(t, y_1, \ldots, y_n) \qquad (i = 1, \ldots, n)$$

has a set of solutions $y_1 = \phi_1(t), \ldots, y_n = \phi_n(t)$ *on* $[t_0, t_0 + \eta[$ *satisfying the conditions* $\phi_1(t_0) = y_{01}, \ldots, \phi_n(t_0) = y_{0n}$, *where* $\eta = \min\{\alpha, \rho/M\}$ *and* $M = \max_i |f_i|.$†

For simplicity, we confine ourselves to the case $n = 1$. Let $C =$

---

† Moigno treats first the case $n = 1$ (1844, pp. 385–96), and later (pp. 513–24) returns to a system of equations. That an nth order scalar equation can be reduced to a system of n first-order equations was known as early as 1750 (see Painlevé, 1910, p. 3).

$[t_0, t_0 + \alpha] \times [y_0 - \rho, y_0 + \rho]$, and let $f: C \to \mathbf{R}$ be a continuous function possessing a continuous first partial derivative with respect to y on C and such that

$$|f(t,y)| \le M \quad \text{and} \quad \left|\frac{\partial f(t,y)}{\partial y}\right| \le K$$

for all $(t,y) \in C$. By the mean value theorem applied to the function $y \mapsto f(t,y)$, the boundedness of the partial derivative implies that $f(t,y)$ is K-Lipschitzian in y in C, and this Lipschitzian property is all that is actually required by Cauchy.

Let $t_0 < t < t_0 + \eta$, where $\eta = \min\{\alpha, \rho/M\}$, let $P = \{t_0, \dots, t_p\}$ be a partition of $[t_0, t]$, let $\Delta t_i = t_{i+1} - t_i$ $(i = 0, \dots, p-1)$, and define $y_1, \dots, y_p$ by

$$y_1 = y_0 + \Delta t_0 f(t_0, y_0), \dots, y_p = y_{p-1} + \Delta t_{p-1} f(t_{p-1}, y_{p-1}). \quad (5)$$

The point y_p is clearly well-defined and depends only on t and P (note that $|y_i - y_0| \le M(t_i - t_0)$ for each i, so that each $(t_i, y_i) \in C$). Further,

$$(t - t_0) \min_i f(t_i, y_i) \le y_p - y_0 \le (t - t_0) \max_i f(t_i, y_i),$$

and hence, by the intermediate value theorem, we can find $\theta \in [0,1]$ and $\zeta \in [-1,1]$ such that

$$y_p - y_0 = (t - t_0)f(t_0 + \theta(t - t_0), y_0 + \zeta M(t - t_0)). \quad (6)$$

Similarly, if $\tau = t_m, \eta = y_m$, there exist $\theta' \in [0,1]$ and $\zeta' \in [-1,1]$ for which

$$y_p - \eta = (t - \tau)f(\tau + \theta'(t - \tau), \eta + \zeta' M(t - \tau)). \quad (7)$$

If for a given i we replace y_i in the formulae (5) by $y_i + \mu$, where μ is sufficiently small, the absolute value of the resulting change in y_{i+1} is

$$|y_i + \mu + \Delta t_i f(t_i, y_i + \mu) - y_i - \Delta t_i f(t_i, y_i)|,$$

and since $f(t,y)$ is K-Lipschitzian in y this expression does not exceed

$$|\mu|(1 + K\Delta t_i) \le |\mu| e^{K\Delta t_i}.$$

By repeating this argument, we infer that if y'_p is the new value of y_p arising from this change in y_i, then

$$|y'_p - y_p| \le |\mu| e^{K(t_p - t_i)} \le |\mu| e^{K\alpha}. \quad (8)$$

It is convenient now to change the notation and to denote y_p by $\phi_P(t)$. We wish to show that $\phi_P(t)$ tends to a limit as $m(P) \to 0$, and again we use Cauchy's convergence criterion.

First, let R be a partition of $[t_0, t]$ containing the points of P, and consider the effect on $\phi_P(t)$ of adding to P those points R that lie in the subinterval $]t_0, t_1[$. By (6), the change in y_1 is Δt_0 times an expression of the form

$$\varepsilon_0 = f(t_0 + \theta_0\Delta t_0, y_0 + \zeta_0 M\Delta t_0) - f(t_0, y_0),$$

where $\theta_0 \in [0, 1]$ and $\zeta_0 \in [-1, 1]$, and hence, by (8), the absolute value of the change in $y_p = \phi_p(t)$ arising from this change $\varepsilon_0\Delta t_0$ in y_1 does not exceed $|\varepsilon_0|\Delta t_0 e^{K\alpha}$. On repeating this argument with each subinterval $]t_i, t_{i+1}[$ in turn we deduce that

$$|\phi_R(t) - \phi_P(t)| \le e^{K\alpha} \sum_{i=0}^{p-1} |\varepsilon_i|\Delta t_i,$$

where each ε_i is of the form

$$\varepsilon_i = f(t_i + \theta_i\Delta t_i, y_i' + \zeta_i M\Delta t_i) - f(t_i, y_i')$$

for some $\theta_i \in [0, 1], \zeta_i \in [-1, 1]$, and $y_i' \in [y_0 - \rho, y_0 + \rho]$. Since f is continuous, we deduce, exactly as in the theorem on primitives, that $|\phi_R(t) - \phi_P(t)|$ is arbitrarily small for all P with sufficiently small $m(P)$, and therefore $\phi_P(t)$ tends to a limit $\phi(t)$ as $m(P) \to 0$.

This process defines a function ϕ on the interval $]t_0, t_0 + \eta[$, and we complete the definition of ϕ by setting $\phi(t_0) = y_0$. If $t_0 \le s < t < t_0 + \eta$, then from (7) with $\tau = s$ we obtain that

$$\phi(t) - \phi(s) = (t - s)f(s + \theta(t - s), \phi(s) + \zeta M(t - s)) \qquad (9)$$

for some $\theta \in [0, 1]$ and $\zeta \in [-1, 1]$. This clearly implies that ϕ is continuous, and that it is the desired solution of the equation $y' = f(t, y)$.

If we accept Cauchy's use of uniform continuity,† the only deficiency in this proof lies in the assertion that (7) implies (9), for it does not seem possible to justify (9) at this stage in the argument. However, the following argument enables us to avoid the use of (9) and therefore to complete Cauchy's proof.

We first make explicit the application of uniform continuity in the treatment of the expression $|\phi_R(t) - \phi_P(t)|$ above. Given $\varepsilon > 0$, we can find $\delta > 0$ such that $|f(t, y) - f(u, z)| \le \varepsilon$ whenever $(t, y), (u, z) \in C$ and $|t - u| \le \delta, |y - z| \le M\delta$. Hence if P, R are partitions of $[t_0, t]$ such that $R \supseteq P$ and $m(P) \le \delta$, then in the preceding argument we have $|\varepsilon_i| \le \varepsilon$ for each i, whence

$$|\phi_R(t) - \phi_P(t)| \le e^{K\alpha}(t - t_0)\varepsilon.$$

Cauchy's argument now shows that $\phi_P(t)$ tends to a limit $\phi(t)$ as $m(P) \to 0$, and hence we also have

$$|\phi(t) - \phi_P(t)| \le e^{K\alpha}(t - t_0)\varepsilon \qquad (10)$$

† It is obvious that Cauchy could have avoided an appeal to the uniform continuity of f by assuming the existence and boundedness of $\partial f/\partial t$.

whenever $m(P) \leq \delta$.

Now let $t_0 \leq s < t < t_0 + \eta$, let $\psi_W(t)$ be the point obtained by the application of the process described above to the interval $[s,t]$, using $(s, \phi(s))$ and a partition W of $[s,t]$ in place of (t_0, y_0) and P, and let $\psi(t)$ be the limit of $\psi_W(t)$ as $m(W) \to 0$. We show that $\psi(t) = \phi(t)$. In fact, if P is partition of $[t_0, t]$ containing s and with $m(P) \leq \delta$, and V and W are the intersections of P with $[t_0, s]$ and $[s, t]$, then, by (10),

$$|\phi(s) - \phi_V(s)| \leq e^{K\alpha}(s - t_0)\varepsilon.$$

Hence, by (8),

$$|\psi_W(t) - \phi_P(t)| \leq e^{2K\alpha}(s - t_0)\varepsilon,$$

and since $\psi_W(t) \to \psi(t)$ and $\phi_P(t) \to \phi(t)$ as $m(P) \to 0$, this proves the statement.

If now $(0 <)t - s \leq \delta$, then (10) applied to the interval $[s, t]$ with $W = \{s, t\}$ gives

$$|\phi(t) - \phi(s) - (t - s)f(s, \phi(s))| = |\psi(t) - \psi_W(t)| \leq e^{K\alpha}(t - s)\varepsilon.$$

This implies first that ϕ is continuous and that $\phi'_+(s) = f(s, \phi(s))$, and then that $\phi'_-(t) = f(t, \phi(t))$, so that ϕ satisfies the differential equation.

We note in passing that if y_0 is replaced by $y_0 + \mu$, and ϕ_μ is the corresponding solution, then, by applying (8) to the associated approximating functions and passing to the limit, we have

$$|\phi_\mu(t) - \phi(t)| \leq |\mu|e^{K\alpha} \qquad (t_0 \leq t < t_0 + \eta),$$

i.e. the solution ϕ varies continuously with y_0. This addition to the existence theorem is given by Moigno (1844, p. 396); it does, of course, assume the uniqueness of the solutions, and neither Cauchy nor Moigno mentions uniqueness.

We have already remarked that Cauchy used the boundedness of the partial derivative of f only to prove that $f(t, y)$ is Lipschitzian in y, and that the Lipschitzian property suffices for his arguments. The Lipschitz condition was first introduced explicitly by Lipschitz in a paper published in 1868, in which the main theorem is the case of (2.7.2) where $Y = \mathbf{R}^n$. Lipschitz was not aware of any previous existence theorems other than those involving the use of power series,† and his arguments closely parallel those of Cauchy. Like Cauchy, Lipschitz made implicit use of the uniform continuity of f, without perceiving the distinction between this property and that of continuity. Moreover, his proof that the approxi-

---

† Existence theorems involving power series were obtained by Cauchy in a series of papers published between 1831 and 1846 (see Painlevé, 1910, p. 16). Since such theorems are not considered in the text we exclude them from our discussion here.

mations $\phi_p(t)$ tend to a limit is less satisfactory than that of Cauchy, and his verification that the limit function ϕ satisfies the differential equation is certainly no more convincing than Cauchy's. On the other hand, Lipschitz did recognize the necessity of proving the uniqueness of the solution, although his arguments in this direction are worthless.† A corrected version of Lipschitz's theorem is given by Picard (1893, pp. 291–301).‡

An alternative proof of Lipschitz's theorem, using the successive approximations defined in §2.12, was given by Picard in 1890. The method of successive approximations seems to have been known to Cauchy, though the first account of his work in this direction occurs in Moigno's book (1844, pp. 702–3), where the method is applied to the equation $y'' = \lambda(t)y$. Liouville demonstrated the convergence of the successive approximations in another special case in 1838, and between 1864 and 1887 Caqué, Fuchs and Peano applied the method to linear equations (see Painlevé, 1910, p. 13). However, Picard seems to have been the first writer to apply the method to the general first-order system. Picard's treatment (1890, 1891), which contains the germ of the contraction mapping principle, yields a smaller domain of existence for the solution than do those of Cauchy and Lipschitz, and this defect was remedied by Bendixson (1893) and Lindelöf (1894).

The question whether the continuity of f alone is sufficient to ensure the local existence of solutions of the equation $y' = f(t, y)$ was first raised by Volterra (1882), and an affirmative answer was given in two papers published by Peano in 1886 and 1890. In the first of these two papers, Peano discussed only the scalar case, and his main theorem can be stated as follows.

(B) *Let $J = [t_0, t_0 + \alpha]$, let $f : J \times \mathbf{R} \to \mathbf{R}$ be continuous and bounded, let $y_0 \in \mathbf{R}$, and let Ω be the class of all differentiable functions $\omega : J \to \mathbf{R}$ such that $\omega(t_0) = y_0$ and that $\omega'(t) > f(t, \omega(t))$ for all $t \in J$. Then the function $\Phi = \inf_\Omega \omega$ is well-defined on J and is a solution of the equation $y' = f(t, y)$*

† In 1876 Lipschitz published in the *Bulletin des Sciences Mathématiques* a translation of his paper from the original Italian into French. This French version (which omits a short passage on the isoperimetric problem but is otherwise a direct translation) makes no mention of the earlier version, and is sometimes quoted as the original source for Lipschitz's theorem. Since the distinction between continuity and uniform continuity was clarified in the period between 1868 and 1876, a reader knowing only the French paper is likely to credit Lipschitz with more than he achieved.

‡ Picard's proof is still not quite complete, since he omits to verify that (with our notation) $\psi(t) = \phi(t)$.

on J taking the value y_0 at t_0; moreover, Φ is the maximal such solution.

The principal tool in Peano's account is a simple differential inequality, namely:

(C) *If $\omega\in\Omega$ and $\phi:J\to\mathbf{R}$ is a differentiable function such that $\phi(t_0)\le y_0$ and that $\phi'(t)\le f(t,\phi(t))$ for all $t\in J$, then $\phi(t)\le\omega(t)$ for all $t\in J$.*

It is obvious that given the result of (B) this differential inequality (C) implies:

(D) *If ϕ satisfies the hypotheses of (C), then $\phi(t)\le\Phi(t)$ for all $t\in J$.*

Thus the result of (1.5.1)(i) for differentiable ϕ is implicit in Peano's account, and Peano must be regarded as the originator of differential inequalities of this type.

Peano's argument is, for him, surprisingly unrigorous, but it is not difficult to make the necessary corrections and additions, and the argument still provides one of the most elegant and simple proofs of the scalar existence theorem. The most important correction that has to be made to the argument is the redefinition of the class Ω in both (B) and (C) as the class of all continuous $\omega:J\to\mathbf{R}$ with the properties that $\omega(t_0)=y_0$, that the right-hand derivative $\omega'_+(t)$ exists and satisfies $\omega'_+(t)>f(t,\omega(t))$ for all $t\in[t_0,t_0+\alpha[$, and that the left-hand derivative $\omega'_-(t)$ exists and satisfies $\omega'_-(t)>f(t,\omega(t))$ for all $t\in]t_0,t_0+\alpha]$. This new class Ω is non-empty, for if $K=\sup|f|$ then the function $t\mapsto y_0+(K+1)(t-t_0)$ belongs to it.

With this amendment, and taking for granted the result of (C), Peano's argument is as follows. First, the function $\Phi=\inf_\Omega\omega$ is well-defined and finite on J, for if $\phi(t)=y_0-K(t-t_0)(t\in J)$, then $\phi(t_0)=y_0$ and $\phi'(t)=-K\le f(t,\phi(t))$ for all $t\in J$, and therefore, by (C), $\phi(t)\le\omega(t)$ for all $\omega\in\Omega$ and all $t\in J$. It is also clear that $\Phi(t_0)=y_0$.

Now let $\tau\in[t_0,t_0+\alpha[$, let $f(\tau,\Phi(\tau))=m$, let $\varepsilon>0$, and let

$$\chi(t)=\Phi(\tau)+(m-\varepsilon)(t-\tau)\qquad(t\in J).$$

Since $f(\tau,\chi(\tau))=f(\tau,\Phi(\tau))=m$ and $t\mapsto f(t,\chi(t))$ is continuous, we can find $\delta>0$ such that $f(t,\chi(t))>m-\varepsilon$ whenever $\tau\le t\le\tau+\delta$, and then also $\chi'(t)=m-\varepsilon<f(t,\chi(t))$ for all such t. Let $\omega\in\Omega$. Then $\omega(\tau)\ge\Phi(\tau)=\chi(\tau)$, and hence, by (C), $\psi(t)\ge\chi(t)$ for all $t\in[\tau,\tau+\delta]$. Hence also $\Phi(t)\ge\chi(t)$ for all such t, and since $\Phi(\tau)=\chi(\tau)$ we infer that $D_+\Phi(\tau)\ge\chi'(\tau)=m-\varepsilon$, and therefore $D_+\Phi(\tau)\ge m$.

Next, $D^+\Phi(\tau)\le m$. To prove this, again let $\varepsilon>0$, and let

$$H(t,z)=m+\varepsilon-f(t,z+(m+\varepsilon)(t-\tau))\qquad(t\in J,z\in\mathbf{R}).$$

Clearly H is continuous on $J \times \mathbf{R}$ and $H(\tau, \Phi(\tau)) = \varepsilon > 0$, whence we can find $\eta > 0$ such that $H(t, z) > 0$ whenever $\tau \le t \le \tau + \eta$ and $|z - \Phi(\tau)| \le \eta$. Choose $\omega \in \Omega$ such that $\Phi(\tau) \le \omega(\tau) \le \Phi(\tau) + \eta$, and let θ be the continuous function on J that agrees with ω on $[t_0, \tau]$, whose graph on $[\tau, \tau + \eta]$ consists of a segment of slope $m + \varepsilon$, and whose graph on the rest of J consists of a segment of slope $K + 1$. On $[\tau, \tau + \eta]$,

$$0 < H(t, \omega(\tau)) = m + \varepsilon - f(t, \omega(\tau) + (m + \varepsilon)(t - \tau)) = \theta'(t) - f(t, \theta(t)),$$

so that $\theta \in \Omega$. Hence $\theta(t) \ge \Phi(t)$ for all $t \in J$, and in particular

$$\theta(t) = \omega(\tau) + (m + \varepsilon)(t - \tau) \ge \Phi(t)$$

whenever $\tau \le t \le \tau + \eta$. Hence also

$$\Phi(\tau) + (m + \varepsilon)(t - \tau) \ge \Phi(t),$$

so that $D^+ \Phi(\tau) \le m + \varepsilon$, i.e. $D^+ \Phi(\tau) \le m$, and therefore $\Phi'_+(t)$ exists and is equal to m.

To complete the proof, it is obviously necessary to prove the corresponding result for $\Phi'_-(t)$, but Peano does not touch on this point, and gives no indication to the reader that anything still remains to be done. Perhaps the easiest way to complete the proof is to show as in the proof of (1.5.1) that Φ is continuous and then to apply the second half of (1.4.5).

Another proof of this scalar case of Peano's existence theorem, in which the maximal solution Φ is again obtained as the infimum of the (redefined) class Ω above, was given by Perron (1915). The proof (which is correct) is essentially distinct from that of Peano, although it is similar in principle to his; it contains implicitly the differential inequality of (1.5.1)(i) for functions ϕ possessing right and left derivatives satisfying $\phi'_+(t) \le f(t, \phi(t))$ and $\phi'_-(t) \le f(t, \phi(t))$ at each point.

The differential inequalities of (1.5.1) were first made explicit by Montel (1926) and Kamke (1930b, pp. 82–3, 89–91) in the case of differentiable functions.[†] Dini derivatives seem first to have been introduced into such differential inequalities by Hukuhara (1940b, 1941), who gives a proof of the existence theorem rather similar to that of Perron. The introduction of countable exceptional sets seems to be due to Ważewski (1951a). In the proof of (1.5.1) given in the text we have borrowed Perron's proof of the continuity of Φ; the remainder of the proof, which is new, is due to J. S. Pym and the author. Yet another proof, using the existence theorem

† Their proof is outlined in Exercise 2.4.1. Parts (i) and (ii) of the exercise are due to Montel (1926). The exercise includes a proof of the existence of maximal and minimal solutions, but Montel and Kamke used the argument only to obtain the differential inequality of (1.5.1)(i), and employed different arguments (see Kamke, 1930b, pp. 78–81) to prove the existence of the maximal and minimal solutions.

for the Lipschitzian case combined with Weierstrass's theorem on the approximation of continuous functions by polynomials, is given by Corduneanu (1964).

The proofs of the scalar case of Peano's existence theorem by Perron and Hukuhara mentioned above, and also that given in §1.5, are 'elementary' in the sense that they make no use of the Ascoli–Arzelà theorem. Other such 'elementary' proofs have been given by Osgood (1898), Fukuhara (1928a), Grunsky (1961), W. Walter (1971), and J. Walter (1973). Osgood, Grunsky and W. Walter use a sequence of polygonal approximations to construct the maximal solution; Fukuhara obtains the maximal solution as the infimum of a family of polygonal approximations, and J. Walter uses polygonal approximations to prove the existence of *some* solution.

Peano's second paper (1890) on the existence problem for continuous f deals with the general case of the vector equation $y' = f(t, y)$ (using vector notation), and contains the result of (2.4.1). Peano's proof is both long and arduous, since what is essentially a proof of the Ascoli–Arzelà theorem is intricately embedded in it. His argument is also rendered more arduous than might be expected by being couched in the symbolic language of the logical calculus, though he does include a six-page '*résumé*' of the proof written in everyday mathematical language. A new proof of the scalar case is included in his treatment.

In the course of the proof of the main theorem, Peano introduces the class of ε-approximate solutions, and he essentially proves the finite-dimensional case of (2.3.1). He also gives a differential inequality very similar to (C) above. In addition, Peano gives also a new treatment of the Lipschitz case employing a special case of (2.6.1), proved via an integrating factor as in §2.6, namely:

(E) *If* $\phi(0) = 0$ *and* $\phi'(t) < a\phi(t) + b$ *for all* $t > 0$, *then* $\phi(t) < b(e^{at} - 1)/a$ *for* $t > 0$.†

The paper concludes with the example $y' = 3y^{2/3}$ (see §1.5) and the case $c = 1$ of the example in Exercise 2.11.8.

An account of Peano's proof of (2.4.1), freed from logical symbolism, was given by Mie (1893). Mie made a number of simplifications in Peano's arguments, and in particular he dispensed with the differential inequality (C) and replaced it with his generalization of Darboux's mean value inequality mentioned in §1.11 (p. 67). Mie also gave a new proof of the finite-dimensional case of (2.3.1) which is essentially that set out in §2.3.

---

† The case of (E) where $b = 0$ had already been used in Peano (1886).

However, Mie's account of Peano's Ascoli-type argument is a good deal less clear than Peano's original version.

Mie also used his theorem on the existence of ε-approximate solutions in combination with Peano's differential inequality (E) to give a simplified proof of the existence of the solution in the Lipschitz case which is essentially that given in §2.7. Peano (1892b) had already simplified his proof of uniqueness in the Lipschitz case, again using (E) as in §2.7. Thus the treatment of the Lipschitz case in (2.7.1,2) for finite-dimensional Y is essentially that of Peano and Mie. The existence of a maximal interval of existence of the solution when f is locally Lipschitzian seems first to have been discussed by Bliss (1905).

Simplifications of Peano's proof of his general existence theorem were made by la Vallée Poussin (1893) and Arzelà (1895, 1896). Arzelà made explicit use of the Ascoli–Arzelà theorem, while la Vallé Poussin's treatment includes a proof of a special case of this theorem. The method of proof of (2.4.1) used in the text, where the limit-function ϕ of the ε-approximate solutions is shown to satisfy the integral equation rather than the differential equation, is essentially that employed by la Vallée Poussin in his deduction of the existence theorem from the special case of the Ascoli–Arzelà theorem.

Differential equations in which the function f is not necessarily continuous seem first to have been studied by la Vallée Poussin (1893). His principal result gives necessary and sufficient conditions in terms of 'upper and lower sums' for the existence of a unique (continuous) ϕ such that

$$\phi(t) = y_0 + \int_{t_0}^{t} f(s, \phi(s))\,ds, \tag{11}$$

where f is bounded on some neighbourhood of the point (t_0, y_0) in $\mathbf{R} \times \mathbf{R}^n$ and the integral on the right is in the Riemann sense. The upper and lower sums employed by la Vallée Poussin are related to the sums employed in the Cauchy–Lipschitz existence theorem in the same manner that Darboux upper and lower sums are related to the sums used by Cauchy in his proof of the existence of a primitive. La Vallée Poussin proved in particular that if $f(t, y)$ is Lipschitzian in y, and $t \mapsto f(t, y)$ is Riemann integrable for each y, then there exists a unique ϕ satisfying (11). He also stated without proof that if f is Riemann integrable with respect to t and also with respect to each coordinate of y, then there is at least one ϕ satisfying (11). This last statement is erroneous, and the condition of integrability with respect to each coordinate of y should be replaced by the condition that $y \mapsto f(t, y)$ is continuous for each t.

The discovery of the Lebesgue integral made la Vallée Poussin's results of historical interest only, but the last result mentioned provided the clue to the generalization of Peano's theorem in (2.13.1), which is due to Carathéodory (1918, pp. 665–72). Carathéodory's original proof employed a new sequence of approximations; the treatment in the text is that of Alexiewicz and Orlicz (1955).

The extension of the local solutions obtained in Peano's existence theorem seems first to have been considered in detail by Kamke (1928; 1929; 1930b, pp. 75–7, 135–6), though the idea that solutions *could* be extended was obviously familiar (see, for example, the work of Perron (1928b) on stability, and also Mie (1893), who gave the result of Exercise 2.4.2).

Kamke proved that, if f is a continuous function from an open set E in $\mathbf{R}^{n+1}$ into $\mathbf{R}^n$ and $(t_0, y_0) \in E$, then there exists a solution ϕ of $y' = f(t, y)$ on an interval $[t_0, b[$ satisfying $\phi(t_0) = y_0$ and coming arbitrarily near to the boundary ∂E of E, in the sense that the graph of ϕ is not contained in any compact subset of E (i.e. either $b = \infty$, or $b < \infty$ and $\limsup_{t \to b-} \|\phi(t)\| = \infty$, or $b < \infty$ and there exists a sequence (t_n) increasing to the limit b such that the distance of $\phi(t_n)$ from ∂E tends to 0). Kamke (1930b, p. 81) also obtained the corresponding result for the maximal and minimal solutions in the scalar case. The definition of 'reaching to the boundary' adopted in §2.4 is that of Hartman (1964, pp. 12–4), and the equivalence of this definition with the apparently weaker property obtained by Kamke (see (2.4.3)) was proved by Wintner (1946a). The use of the Tietze extension theorem in the context of extension of solutions (see (2.4.4, 5)) may be new. Corresponding results for extension of solutions under the Carathéodory conditions are indicated by Coddington and Levinson (1955, p. 47).

Following Peano's discovery that the continuity of f is sufficient to ensure the local existence of (possibly non-unique) solutions of the equation $y' = f(t, y)$, it became of interest to obtain conditions less restrictive than that of Lipschitz which are sufficient to ensure the uniqueness of the solutions. In the following discussion we consider solutions on an interval $[t_0, T]$ taking a given value at t_0, and for brevity we write

$$D = f(t, y) - f(t, z) \quad \text{and} \quad d = y - z.$$

The function f is supposed continuous, though this condition can usually be relaxed.

The first advance was made by Osgood (1898), who proved that for a scalar equation the condition $|D| \le K|d|$ of Lipschitz can be replaced by the condition $|D| \leq h(|d|)$, where $h : [0, \infty[\rightarrow [0, \infty[$ is any continuous function such that $h(0) = 0, h(u) > 0$ for $u > 0$, and $\int_0 du/h(u) = \infty$. An alternative generalization of Lipschitz's condition, retaining the $|d|$ on the right but introducing a t, was found by Rosenblatt (1909), who proved, again for a scalar equation, that a sufficient condition for uniqueness is that $|D| \leq K|d|/(t - t_0)^c$, where either $0 < c < 1$ and $K \geq 0$, or $c = 1$ and $0 \leq K < 1$. Subsequently Carathéodory (1918, pp. 673–5) generalized the first of Rosenblatt's results by showing that for $Y = \mathbf{R}^n$ the condition $\| D \| \leq \lambda(t) \| d \|$ is sufficient provided that the function λ is non-negative and Lebesgue integrable on $[t_0, T]$.

The next group of uniqueness criteria were foreshadowed by Bendixson (1897) and Montel (1907), who proved the result of Exercise 2.4.3, i.e. that for a scalar equation the condition that $y \mapsto f(t, y)$ is decreasing ensures uniqueness. This condition can be written in the form that $D \leq 0$ whenever $d \geq 0$, and is thus one-sided, in contrast to the two-sided conditions of Osgood, Rosenblatt and Carathéodory. This result of Bendixson and Montel was extended by Tonelli (1925), who gave the one-sided condition (for a scalar equation) that $D \leq \lambda(t)h(d)$ whenever $d > 0$, where

(T_1) $h :]0, \infty[\rightarrow]0, \infty[$ is a continuous function such that $\int_0 du/h(u) = \infty$, and

(T_2) λ is an extended-real-valued function Lebesgue integrable on $[t, T]$ for each $t \in]t_0, T]$ such that the integral $\int_t^T \lambda(s)ds$ tends to a finite limit as $t \rightarrow t_0 +$.

This result was in turn extended to a system of n equations by Müller (1927), his condition being that $p(D) \leq \lambda(t)h(p(d))$ whenever $p(d) \geq 0$, where h and λ satisfy (T_1) and (T_2) and $p(y)$ is either max y_i or max $|y_i|$. Müller similarly extended Rosenblatt's second result by giving the condition $p(D) \leq Kp(d)/(t - t_0)$ whenever $p(d) \geq 0$, where $0 \leq K < 1$.

These results of Tonelli and Müller essentially contain all that was known previously, except for an improvement on Rosenblatt's second result for scalar equations obtained by Nagumo (1926), in which Rosenblatt's condition was replaced by the one-sided condition that $D < d/(t - t_0)$ whenever $d > 0$. Fukuhara (1928a) further extended this result by replacing the right-hand side by $\lambda(t)d$, where λ is continuous and bounded below on $[t, T]$ for each $t \in]t_0, T]$ and

$$\lim_{t \to t_0 +} \sup \left(\log(t - t_0) + \int_t^T \lambda(s)ds \right) < \infty.$$

Perron (1928a) replaced the strict inequality in Nagumo's condition by '$\leq$', but his condition, which was obtained for $Y = \mathbf{R}^n$ with the Euclidean norm, is two-sided, viz. $\|D\| \le \|d\|/(t - t_0)$.†

The next group of uniqueness criteria involve a scalar comparison equation $x' = g(t,x)$. The first such criterion, for a scalar equation, was obtained by Bompiani (1925), his condition being that $|D| \le g(t,|d|)$, where g satisfies the condition (P) of §2.11 (p. 133) and $x \mapsto g(t,x)$ is increasing. Perron (1926) extended this result to the case $Y = \mathbf{R}^n$ using the Euclidean norm, and at the same time removed the monotonicity hypothesis. The condition (I–K) of §2.11 (p. 132) was introduced by Iyanaga (1928), who gave the one-sided condition for a scalar equation that $D < g(t,d)$ whenever $d > 0$, where g satisfies (I–K). This result was partly improved by Kamke (1930a; 1930b, pp. 139–41), who gave the two-sided criterion $\|D\| \le g(t,\|d\|)$ for a system, where again g satisfies the condition (I–K) of §2.11, and this is the result now usually known as 'Kamke's criterion'. However, Kamke (1930c) subsequently extended his improvement of Iyanaga's result by giving the one-sided condition for a system that $p(D) \le g(t,p(d))$ whenever $p(d) \ge 0$, where g satisfies condition (I–K) and $p : Y \to \mathbf{R}$ has the property that for each differentiable $u : \mathbf{R} \to Y$ and for each $t \in \mathbf{R}$ the function $p \circ u$ has right and left derivatives at t not exceeding $p(u'(t))$. Finally Hukuhara (1940a) showed that a function p has this property if and only if it is sublinear (cf. §1.6), and thus obtained the criterion of Exercise 2.11.2.

When Y is an inner product space, alternative one-sided conditions can be obtained by using the inner product $\langle D,d \rangle$. The first such results were obtained by McShane (1939) and Giuliano (1940) for $Y = \mathbf{R}^n$; Giuliano's condition is that $\langle D,d \rangle \le \lambda(t)h(\|d\|^2)$, where λ,h satisfy (T_1) and (T_2); McShane treated the special case where λ is non-negative and $h(u) = u$.

Some mention should be made here of integral inequalities. The earliest that I have been able to find occurs in a proof by Lindelöf (1894, p. 119) of the uniqueness of the solutions in the Lipschitz case.‡ Lindelöf does not formulate the inequality explicitly, but what his argument gives is the following result.

(F) *If* $J = [t_0, t_0 + \alpha]$, $k \ge 0$, $\psi : J \to [0,\infty[$ *is continuous, and*

$$\psi(t) \le k \int_{t_0}^{t} \psi(s)ds \tag{12}$$

---

† This result of Perron's is now often known as 'Nagumo's criterion'. It should also be mentioned that the two-sided condition $\|D\| \le \lambda(t)h(\|d\|)$, where h has the properties specified in Osgood's condition (or, more generally, satisfies (T_1)) and λ is a continuous non-negative function on $]t_0, T]$ such that $\int_{t_0}^{T} \lambda(s)ds < \infty$, is now often known as 'Osgood's criterion'.

‡ Lindelöf's paper is primarily concerned with successive approximations, but his proof of uniqueness is independent of the theory of successive approximations.

for all $t \in J$, then $\psi = 0$.

The proof is simple. Let $\eta = \min\{\alpha, 1/(2k)\}$, let M be the supremum of ψ on $[t_0, t_0 + \eta]$, and let ρ be a point of this interval where $\psi(\rho) = M$. Then (12) gives

$$M = \psi(\rho) \leq k \int_{t_0}^{\rho} \psi(s)ds \leq kM(\rho - t_0) \leq \tfrac{1}{2}M,$$

whence $M = 0$. If $\eta = \alpha$, this is the result, while if $\eta < \alpha$ we repeat the argument with t_0 replaced by $t_0 + \eta$, and so on.

We note in passing that this argument easily extends to give the following result.

(G) *If $J = [t_0, t_0 + \alpha], h : J \to [0, \infty[$ and $\psi : J \to \mathbf{R}$ are continuous, and*

$$\psi(t) \leq \int_{t_0}^{t} h(s)\psi(s)ds \tag{13}$$

for all $t \in J$, then $\psi \leq 0$.

In fact, let $k = \sup h$, and let σ be the greatest element of J such that $\psi(t) \leq 0$ for all $t \in [t_0, \sigma]$ (note that (13) implies that $\psi(t_0) \leq 0$, so that σ exists). If $\sigma < t_0 + \alpha$, and ψ attains its supremum M on $[\sigma, \sigma + 1/(2k)] \cap J$ at a point ρ of this set, then (13) gives

$$M = \psi(\rho) \leq \int_{t_0}^{\rho} h(s)\psi(s)ds \leq \int_{\sigma}^{\rho} h(s)\psi(s)ds \leq kM(\rho - \sigma) \leq \tfrac{1}{2}M.$$

Hence $M \leq 0$ and this contradicts the definition of σ.

The next integral inequality to be discovered was the following lemma, which occurs in a paper by Gronwall (1919) on the differentiability of the solutions of a differential equation with respect to a parameter.

(H) *If $J = [0, \alpha], \phi : J \to [0, \infty[$ is continuous, $a \geq 0, k \geq 0$, and*

$$\phi(t) \leq at + k \int_{0}^{t} \phi(s)ds$$

for all $t \in J$, then $\phi(t) \leq ate^{kt} (t \in J)$.

Gronwall's argument employs an idea similar to that used by Lindelöf: if $\phi(t) = e^{kt}\psi(t)$, and ψ attains its supremum M on $[0, t]$ at a point ρ of this interval, then

$$e^{k\rho}M = \phi(\rho) \leq a\rho + kM \int_{0}^{\rho} e^{ks}ds = a\rho + M(e^{k\rho} - 1),$$

and therefore $M \leq a\rho \leq at$.

It should be noted that if in (2.6.2) (which is nowadays associated with Gronwall's name) we take $g(t) = at, h(t) = k$, we obtain a stronger result than (H), namely that $\phi(t) \le a(e^{kt} - 1)/k$, and, moreover, with the hypothesis that $a \ge 0$ removed. This stronger result is implicit in (G), for it follows immediately from (G) with $h(t) = k$, $\psi(t) = \phi(t) - a(e^{kt} - 1)/k$.

Some further integral inequalities are implicit in the proofs of various uniqueness criteria given in the 1920s (for example, Perron, 1928a). However, little interest seems to have been shown in such inequalities until the discovery by Bellman (1943) of the following special case of (2.6.2. Corollary) and his striking applications of this result in stability theory.

(I) *If* $J = [0, \alpha]$, $h, \phi : J \to [0, \infty[$ *are continuous, $c > 0$, and*

$$\phi(t) \le c + \int_0^t h(s)\phi(s)ds \qquad (14)$$

for all $t \in J$, then $\phi(t) \le c \exp(\int_0^t h(s)ds)$.

To prove this, Bellman sets $H(t) = \int_0^t h(s)\phi(s)ds$. Then $H'(t) = h(t)\phi(t) \le h(t)(c + H(t))$, whence $\log(c + H(t)) - \log c \le \int_0^t h(s)ds$, and this and (14) together give the result.

The general case of (2.6.2. Corollary) seems to be due to Weyl (1946), who reduced the result to that of (G) and then argued by induction to show that if ψ satisfies (12) with $t = t_0$, and $M = \sup \psi$, then

$$\psi(t) \le \frac{M}{n!} \int_0^t (t - u)^{n-1} h(u)du \qquad (n = 1, 2, \ldots),$$

so that $\psi \le 0$. The proofs of (2.6.2) and (2.6.2. Corollary) given in the text are due to Levinson (1949) and La Salle (1949).

The whole group of inequalities is now often known by the collective title 'Gronwall's inequality', and references to recent work can be found in W. Walter (1970, pp. 14–16, 43–4).† A number of generalizations of Gronwall's inequality are covered by (2.6.3), which is due to Viswanatham (1952). The proof of (2.6.3) in the text, which employs the idea of Levinson's proof of Bellman's inequality, is taken from Hartman (1964, p. 24).

Exercise 2.14

1 Let $h : \mathbf{R} \times \mathbf{R} \to [0, 1]$ be a continuous function of compact support which takes the value 1 in a neighbourhood of $(0, 0)$. Find continuous functions $g_1, g_2 : \mathbf{R} \to \mathbf{R}$ such that, if $f : \mathbf{R} \times \mathbf{R} \to \mathbf{R}$ is defined by

$$f(t, y) = \begin{cases} g_1(t)h(t, y) & (t \le y), \\ g_2(t)h(t, y) & (t > y), \end{cases}$$

† Gollwitzer (1969) should be added to these references.

then there is no function ϕ defined on any interval I containing 0 such that, for $t \in I$,

$$\phi(t) = \int_0^t f(s, \phi(s)) ds.$$

[Since the function f described above is Riemann integrable with respect to each of its variables taken separately, this shows that a result asserted by la Vallée Poussin (1893) is false (see p. 159).]

3

The Fréchet differential

Throughout this chapter, X, Y and Z will denote normed spaces over either the real or the complex field. In contexts in which two or more spaces appear (for example, in expressions like $\mathscr{L}(X, Y)$) they will be understood to be over the same field, although occasional use will be made of the elementary fact that a linear space over the complex field can be considered as a linear space over the real field. Completeness is required only in some later sections, and will be stated as a hypothesis whenever needed.

3.1 The Fréchet differential of a function

A function f from a set $A \subseteq X$ into Y is said to be *Fréchet differentiable*† *at* x_0 if x_0 is an interior point of A and there exists $T \in \mathscr{L}(X, Y)$ such that

$$\lim_{x \to x_0} \frac{f(x) - f(x_0) - T(x - x_0)}{\|x - x_0\|} = 0. \tag{1}$$

This limit relation is, of course, equivalent to the statement that

$$f(x) = f(x_0) + T(x - x_0) + o(\|x - x_0\|) \tag{2}$$

as $x \to x_0$.

We observe that the linear term on the right of (2) is of larger order than the last term except when $T = 0$, for if $T(x - x_0)/\|x - x_0\| \to 0$, then for any non-zero $z \in X$

$$Tz = \|z\| T(tz)/\|tz\| = \|z\| T(tz + x_0 - x_0)/\|tz + x_0 - x_0\| \to 0$$

as $t \to 0+$, so that $T = 0$. The statement that f is Fréchet differentiable at x_0 therefore means that it is possible to approximate to $f(x)$ in the neighbourhood of x_0 by an expression of the form $f(x_0) + T(x - x_0)$, with $T \in \mathscr{L}(X, Y)$, to the degree of approximation prescribed by (2).

This same argument shows also that, if a T satisfying (1) or (2) exists,

† Other forms of differentiability will be considered in Chapter 4.

then it is unique, for if (1) or (2) is satisfied with $T = T_1, T_2$, and $U = T_1 - T_2$, then $U(x - x_0)/\|x - x_0\| \to 0$ as $x \to x_0$, so that $U = 0$.[†]

If f is Fréchet differentiable at x_0, then the (unique) function $T \in \mathscr{L}(X, Y)$ determined by (1) or (2) is called the *Fréchet differential of* f *at* x_0, and we denote it by $df(x_0)$. The function $x \mapsto df(x)$, whose domain is the set of interior points x of A at which $df(x)$ exists, is called the *Fréchet differential of* f, and we denote it by df. Thus df is a function from a subset of A into $\mathscr{L}(X, Y)$ (df may, of course, be the empty function).

In the sequel, when we say that a function f is Fréchet differentiable at x, or that the Fréchet differential $df(x)$ of f at x exists, we take it as understood that x is an interior point of the domain of f.

If E is an open set in X, a function f which is Fréchet differentiable at each point of E is said to be *Fréchet differentiable on* E. We say also that f is C^1 on an open set $E \subseteq X$ if df is continuous on E.

If f is Fréchet differentiable at x_0, then f is obviously also Fréchet differentiable at x_0 when the norms on X and Y are replaced by equivalent norms. In particular, if X and Y are infinite-dimensional, then the property of Fréchet differentiability is independent of the choice of norms on X and Y.

It is also evident that if f and g are functions from subsets of X into Y which agree on a neighbourhood of x_0, then f is Fréchet differentiable at x_0 if and only if g is Fréchet differentiable at x_0, and in this case $df(x_0) = dg(x_0)$.

The Fréchet differential of f at x_0 is sometimes called the *Fréchet derivative* of f at x_0, and is then denoted by $f'(x_0)$. This notation gives rise to some obvious ambiguity when f is a function of a real variable (see the footnote to Example (*a*) below), and we shall therefore avoid it.

We note that it is often more convenient to write the limit relation (1) as

$$\lim_{h \to 0} \frac{f(x_0 + h) - f(x_0) - Th}{\|h\|} = 0,$$

and similarly for (2).

The Fréchet differentiability of f at x_0 obviously implies that f is continuous at x_0, and indeed we have a stronger result (cf. (1.1.2)).

(3.1.1) *Let* $A \subseteq X$, *and let* $f : A \to Y$ *be Fréchet differentiable at* x_0. *Then for each* $\varepsilon > 0$ *there exists a neighbourhood* U *of* x_0 *in* X, *contained in* A,

[†] We note that it is not possible to extend the definition of the Fréchet differentiability of f to a non-isolated point x_0 of A which is not an interior point of A, since the limit relation (1) may then fail to define a *unique* T (see also the footnote to Example (*a*) below).

such that for all $x \in U$,

$$\| f(x) - f(x_0) \| \leq (\| df(x_0) \| + \varepsilon) \| x - x_0 \|.$$

Let $T = df(x_0)$, and let $\varepsilon > 0$. Then we can find a neighbourhood U of x_0 in X, contained in A, such that for all $x \in U$

$$\| f(x) - f(x_0) - T(x - x_0) \| \leq \varepsilon \| x - x_0 \|,$$

whence also

$$\| f(x) - f(x_0) \| \leq \| T(x - x_0) \| + \varepsilon \| x - x_0 \|$$
$$\leq (\| T \| + \varepsilon) \| x - x_0 \|.$$

Example (a)

A function ϕ from a set $A \subseteq \mathbf{R}$ into the real normed space Y is Fréchet differentiable at a point t_0 if and only if t_0 is an interior point of A and ϕ has a derivative at t_0, and then $d\phi(t_0)$ is the function $h \mapsto h\phi'(t_0)$ from $\mathbf{R}$ into Y.

To prove the 'only if', let ϕ be Fréchet differentiable at t_0. Then t_0 is an interior point of A, and

$$\lim_{t \to t_0} \frac{\phi(t) - \phi(t_0) - T(t - t_0)}{|t - t_0|} = 0,$$

where $T = d\phi(t_0)$. Since any linear function from $\mathbf{R}$ into Y is of the form $h \mapsto h\eta$ for some $\eta \in Y$, we can find $\eta \in Y$ such that $Th = h\eta$ for all $h \in \mathbf{R}$. Hence

$$\lim_{t \to t_0} \frac{\phi(t) - \phi(t_0) - (t - t_0)\eta}{|t - t_0|} = 0,$$

and this implies that $\phi'(t_0)$ exists and is equal to η (cf. (1.1.1) (iv)). The 'if' is proved by reversing the argument.†

Example (b)

If f is a constant function from a set $A \subseteq X$ into Y, then f is Fréchet differentiable at each interior point x_0 of A, and $df(x_0)$ is the zero function from X into Y (so that df is constant).

† This result shows that the apparent ambiguity in the use of the symbol $\phi'(t_0)$ to denote both the derivative of ϕ at t_0 and the Fréchet differential (derivative) of ϕ at t_0 effectively disappears if we identify the linear map $h \mapsto h\eta$ from $\mathbf{R}$ into Y with the element η of Y (i.e. we simply use the isomorphism between $\mathscr{L}(\mathbf{R}, Y)$ and Y). However, difficulties may still arise from the fact that the derivative $\phi'(t_0)$ can exist at a point which is not an interior point of the domain of ϕ; in particular, a function $\phi : [a,b] \mapsto Y$ which is differentiable on $[a,b]$ in the sense used in Chapters 1 and 2 is not Fréchet differentiable on $[a,b]$, since it is not meaningful to speak of the Fréchet differentiability of ϕ at a and b.

Example (c)

If $f \in \mathscr{L}(X, Y)$, then f is Fréchet differentiable at each $x_0 \in X$ and $df(x_0) = f$ (so that df is constant). More generally, if f is the restriction of a function g of $\mathscr{L}(X, Y)$ to a set $A \subseteq X$, then f is Fréchet differentiable at each interior point x_0 of A, and $df(x_0) = g$.

Example (d)

Let X_1, X_2 be normed spaces, let $X = X_1 \times X_2$, and let $F \in \mathscr{L}(X_1, X_2; Y)$. Then F is Fréchet differentiable at each point $x = (x_1, x_2) \in X$, and $dF(x)$ is the function $T \in \mathscr{L}(X, Y)$ given by

$$T(h_1, h_2) = F(h_1, x_2) + F(x_1, h_2) \qquad ((h_1, h_2) \in X). \tag{3}$$

Moreover, dF is a continuous linear function from X into $\mathscr{L}(X, Y)$.

First, the function T defined by (3) obviously belongs to $\mathscr{L}(X, Y)$. Further, if $x, h \in X$ and $x = (x_1, x_2), h = (h_1, h_2)$, then

$F(x + h) - F(x) - Th$

$\qquad = F(x_1 + h_1, x_2 + h_2) - F(x_1, x_2) - F(h_1, x_2) - F(x_1, h_2)$

$\qquad = F(h_1, h_2).$

Hence

$\|F(x + h) - F(x) - Th\| \le \|F\| \, \|h_1\| \, \|h_2\|$

$\qquad\qquad \le \tfrac{1}{2} \|F\| (\|h_1\|^2 + \|h_2\|^2) = \tfrac{1}{2} \|F\| \, \|h\|^2,$

and this implies that $dF(x)$ exists and is equal to T.

Next, dF is linear, for if $x = (x_1, x_2)$, $x' = (x_1', x_2')$, and α is a scalar, then for all $h = (h_1, h_2)$

$dF(x + x')h = F(h_1, x_2 + x_2') + F(x_1 + x_1', h_2)$

$\qquad\qquad = F(h_1, x_2) + F(h_1, x_2') + F(x_1, h_2) + F(x_1', h_2)$

$\qquad\qquad = dF(x)h + dF(x')h,$

and

$dF(\alpha x)h = F(h_1, \alpha x_2) + F(\alpha x_1, h_2) = \alpha F(h_1, x_2) + \alpha F(x_1, h_2) = \alpha dF(x)h,$

so that $dF(x + x') = dF(x) + dF(x')$ and $dF(\alpha x) = \alpha dF(x)$. Further, by Cauchy's inequality,

$\|dF(x)h\| \le \|F(h_1, x_2)\| + \|F(x_1, h_2)\|$

$\qquad\quad \le \|F\| \, \|h_1\| \, \|x_2\| + \|F\| \, \|x_1\| \, \|h_2\|$

$\qquad\quad \le \|F\| (\|x_1\|^2 + \|x_2\|^2)^{1/2} (\|h_1\|^2 + \|h_2\|^2)^{1/2}$

$\qquad\quad = \|F\| \, \|x\| \, \|h\|,$

and therefore $\|dF(x)\| \le \|F\| \|x\|$, so that dF is continuous and $\|dF\| \le \|F\|$.

Example (e)

More generally, let $X_1, X_2, \ldots, X_p$ be normed spaces, let $X = X_1 \times X_2 \times \ldots \times X_p$, and let $F \in \mathscr{L}(X_1, \ldots, X_p; Y)$. Then F is Fréchet differentiable at each point $x \in X$ and $dF(x)$ is the function $T \in \mathscr{L}(X, Y)$ given by

$$T(h_1, \ldots, h_p) = F(h_1, x_2, \ldots, x_p) + F(x_1, h_2, x_3, \ldots, x_p)$$
$$+ \ldots + F(x_1, \ldots, x_{p-1}, h_p).$$

Moreover, dF is linear and continuous.

Here $F(x + h) - F(x) - Th$ is a sum of expressions of the form $F(z_1, \ldots, z_p)$, where each z_i is either x_i or h_i, and $z_i = h_i$ for at least two i. The statement therefore follows by obvious modifications of the argument in Example (d).

The next result gives some equivalent formulations of the property of differentiability.

(3.1.2) *Let f be a function from a set $A \subseteq X$ into Y, let x_0 be an interior point of A, let $T \in \mathscr{L}(X, Y)$, and let*

$$R(h) = f(x_0 + h) - f(x_0) - Th \qquad (x_0 + h \in A).$$

Then the following statements are equivalent:
(i) *f is Fréchet differentiable at x_0, with $df(x_0) = T$ (i.e. $R(h)/\|h\| \to 0$ as $h \to 0$);*
(ii) *for each bounded set $E \subseteq X$, $R(th)/t \to 0$ as $t \to 0$ in $\mathbf{R}$, uniformly for h in E;*
(iii) *for each sequence (t_n) in $\mathbf{R} \backslash \{0\}$ converging to 0 and for each bounded sequence (h_n) in X, $R(t_n h_n)/t_n \to 0$ as $n \to \infty$.*
Further, if X is finite-dimensional, then each of (i)–(iii) is equivalent to
(iv) *for each compact set $E \subseteq X$, $R(th)/t \to 0$ as $t \to 0$ in $\mathbf{R}$, uniformly for h in E.*

First, (i) implies (ii). To prove this, suppose that (i) holds, let E be a bounded set in X, let M be a positive number such that E is contained in the ball $\{x \in X : \|x\| \le M\}$, and let $\varepsilon > 0$. Then we can find $\delta > 0$ such that $\|R(h)\| \le \varepsilon \|h\|/M$ whenever $\|h\| \le \delta$. If now $h \in E$ and $|t| \le \delta/M$, then $\|th\| \le \delta$, and therefore

$$\|R(th)\| \le \varepsilon \|th\|/M = \varepsilon |t| \|h\|/M \le \varepsilon |t|,$$

i.e. $R(th)/t \to 0$ as $t \to 0$, uniformly for h in E.

Next, (ii) implies (iii), for if (h_n) is a bounded sequence in X we have only to take E in (ii) to be the set of points h_n $(n = 1, 2, \ldots)$, and we obtain (iii).

To prove that (iii) implies (i), suppose that (i) is false. Then $R(h)/\|h\| \nrightarrow 0$ as $h \to 0$, and hence we can find a sequence (k_n) in $X \backslash \{0\}$ tending to 0 such that $R(k_n)/\|k_n\| \nrightarrow 0$ as $n \to \infty$. If now $t_n = \|k_n\|$ and $h_n = k_n/\|k_n\|$, then the sequence (h_n) is bounded, $t_n \to 0$, and

$$R(t_n h_n)/t_n = R(k_n)/\|k_n\| \nrightarrow 0,$$

so that (iii) does not hold. Hence (i), (ii) and (iii) are equivalent.

We observe next that (ii) obviously implies (iv), and hence to prove the last part it is enough to show that if $\dim X < \infty$ then (iv) implies (ii). This, however, is immediate, since the closure of a bounded set in a finite-dimensional space is compact.†

We consider now some properties of differentiable functions.

(3.1.3) (i) *If $A \subseteq X$, and $f : A \to Y$ is Fréchet differentiable at x_0, then for each scalar α the function αf is Fréchet differentiable at x_0 and its differential there is $\alpha df(x_0)$.*

(ii) *If $A, B \subseteq X$, and $f : A \to Y$ and $g : B \to Y$ are Fréchet differentiable at x_0, then $f + g$ is Fréchet differentiable at x_0 and its differential there is $df(x_0) + dg(x_0)$.*

These two results follow immediately from the definition of differentiability (or from (3.1.2)(iii)). Together they show that if $\mathfrak{X}$ is the set of functions from a set $A \subseteq X$ into Y which are Fréchet differentiable at a point x_0 of A, then $\mathfrak{X}$ is a vector space under the operations of addition of functions and multiplication of functions by scalars. Moreover, the function $f \mapsto df(x_0)$ is a linear function from $\mathfrak{X}$ into $\mathscr{L}(X, Y)$.

Example (f)

Let $T \in \mathscr{L}(X, Y)$, let $c \in Y$, and let $F : X \to Y$ be given by $F(x) = Tx + c$. Then F is the sum of T and a constant function, whence F is Fréchet differentiable at each $x \in X$ and $dF(x) = T + 0 = T$.

(3.1.4) (The chain rule) *Let $A \subseteq X$, $B \subseteq Y$, let $f : A \to Y$ be Fréchet differentiable at x_0, and let $g : B \to Z$ be Fréchet differentiable at the*

---

† It will be shown in §4.2, Example (d), p. 266 that (iv) is not equivalent to (i)–(iii) when $\dim X = \infty$.

point $y_0 = f(x_0)$. *Then the function* $g \circ f$ *is Fréchet differentiable at* x_0, *and* $d(g \circ f)(x_0) = dg(y_0) \circ df(x_0)$.

The theorem asserts that if we approximate to f, g, and $g \circ f$ as in (2) (near x_0, y_0, and x_0, respectively), then the linear part of the approximation to the composite $g \circ f$ is the composite of the linear parts of the approximations to g and f.

To prove the theorem, we observe first that x_0 and y_0 are interior points of A and B respectively, and hence, as in the remark following the proof of (1.1.6), x_0 is an interior point of the domain of $g \circ f$.

Now let $T = df(x_0)$, let $U = dg(y_0)$, let (h_n) be a bounded sequence in X, let (t_n) be a sequence in $\mathbf{R} \setminus \{0\}$ converging to 0, and let

$$\mu_n = g(f(x_0 + t_n h_n)) - g(f(x_0)) - U(T(t_n h_n)).$$

Since $t_n h_n \to 0$ as $n \to \infty$, μ_n is defined for all sufficiently large n, and it is therefore enough (by (3.1.2)(iii)) to prove that $\mu_n / t_n \to 0$.

Let

$$R(h) = f(x_0 + h) - f(x_0) - Th \qquad (x_0 + h \in A),$$
$$S(k) = g(y_0 + k) - g(y_0) - Uk \qquad (y_0 + k \in B),$$

and let k_n be given by

$$t_n k_n = f(x_0 + t_n h_n) - f(x_0).$$

Then $k_n = Th_n + R(t_n h_n)/t_n$, and since $R(t_n h_n)/t_n \to 0$ as $n \to \infty$ (by (3.1.2)(iii) applied to f), the sequence (k_n) is bounded. Further,

$$\mu_n/t_n = (g(y_0 + t_n k_n) - g(y_0))/t_n - UTh_n$$
$$= S(t_n k_n)/t_n + Uk_n - UTh_n$$
$$= S(t_n k_n)/t_n + U(R(t_n h_n)/t_n),$$

and since $S(t_n k_n)/t_n \to 0$ as $n \to \infty$ (by (3.1.2)(iii) applied to g), and U is continuous, $\mu_n/t_n \to 0$, as required.

We state as corollaries some important special cases of the chain rule.

(3.1.4. Corollary 1) *Let* $A \subseteq X$, *let* $f : A \to Y$ *be Fréchet differentiable at* x_0, *and let* $g \in \mathscr{L}(Y, Z)$. *Then* $g \circ f$ *is Fréchet differentiable at* x_0, *and* $d(g \circ f)(x_0) = g \circ df(x_0)$.

(3.1.4. Corollary 2) *Let* $f \in \mathscr{L}(X, Y)$ *let* $x_0 \in X$, *let* $B \subseteq Y$, *and let* $g : B \to Z$ *be Fréchet differentiable at the point* $y_0 = f(x_0)$. *Then* $g \circ f$ *is Fréchet differentiable at* x_0, *and* $d(g \circ f)(x_0) = dg(y_0) \circ f$.

These are immediate consequences of the main theorem and Example (c).

(3.1.4. Corollary 3) *Let f be a function from a set $A \subseteq X$ into Y which is Fréchet differentiable at x_0, and let ϕ be a function from a neighbourhood of a point t_0 in $\mathbf{R}$ into X which takes the value x_0 at t_0 and which has a derivative at t_0. Then the function $\psi = f \circ \phi$ has a derivative at t_0 equal to $df(x_0)\phi'(t_0)$.*

First, let X, Y be real. Then ϕ is Fréchet differentiable at t_0, and $d\phi(t_0)h = h\phi'(t_0)$ for all $h \in \mathbf{R}$ (Example (a)). Hence ψ is Fréchet differentiable at t_0, and for all $h \in \mathbf{R}$

$$d\psi(t_0)h = df(x_0)(d\phi(t_0)h) = df(x_0)(h\phi'(t_0)) = h\,df(x_0)\phi'(t_0),$$

and (again by Example (a)) this gives the result. If X, Y are complex, then ψ cannot be Fréchet differentiable (for there are no complex linear maps from $\mathbf{R}$ to Y). However, if we consider X, Y as spaces over the real field, f is still Fréchet differentiable and its differential is $df(x_0)$ considered as a real-linear map. We may apply the first part of this proof to obtain our conclusion.

Example (g)

Let $Y_1, \ldots, Y_m$ be normed spaces, let $Y = Y_1 \times Y_2 \times \ldots \times Y_m$, let f be a function from a set $A \subseteq X$ into Y, and for $i = 1, \ldots, m$ let $f_i : A \to Y_i$ be the ith component of f, given by

$$f(x) = (f_1(x), \ldots, f_m(x)) \qquad (x \in A).$$

Then f is Fréchet differentiable at a point $x_0 \in A$ if and only if its components $f_1, \ldots, f_m$ are Fréchet differentiable at x_0, and $df(x_0)$ is the function in $\mathscr{L}(X, Y)$ with components $df_1(x_0), \ldots, df_m(x_0)$.

To prove this, we observe that if $P_i \in \mathscr{L}(Y, Y_i)$ is the projection map given by $P_i(y_1, \ldots, y_m) = y_i$, then

$$f_i = P_i \circ f.$$

Also, if $I_i \in \mathscr{L}(Y_i, Y)$ is the insertion map whose value at the point y_i of Y_i is the point of Y with ith coordinate y_i and all other coordinates 0, then

$$f = \sum_{i=1}^{m} I_i \circ f_i.$$

It therefore follows from (3.1.4. Corollary 1) and (3.1.3)(ii) that f is Fréchet differentiable at x_0 if and only if $f_1, \ldots, f_m$ are Fréchet differentiable at x_0. Moreover,

$$df_i(x_0) = P_i \circ df(x_0) \quad \text{and} \quad df(x_0) = \sum_{i=1}^{m} I_i \circ df_i(x_0),$$

so that $df(x_0)$ has components $df_1(x_0), \ldots, df_m(x_0)$.

We note also that if $\mathscr{P}_i : \mathscr{L}(X, Y) \to \mathscr{L}(X, Y_i)$ and $\mathscr{I}_i : \mathscr{L}(X, Y_i) \to \mathscr{L}(X, Y)$ are the continuous linear functions given by

$$\mathscr{P}_i(T) = P_i \circ T \text{ and } \mathscr{I}_i(T_i) = I_i \circ T_i,$$

then

$$df_i = \mathscr{P}_i \circ df \text{ and } df = \sum_{i=1}^{m} \mathscr{I}_i \circ df_i.$$

Example (h)

Let $Y_1, \ldots, Y_m, Z$ be normed spaces, let $Y = Y_1 \times Y_2 \times \ldots \times Y_m$, let $G \in \mathscr{L}(Y_1, \ldots, Y_m; Z)$, and let f be a function from a set $A \subseteq X$ into Y which is Fréchet differentiable at x_0. Then the function $G \circ f$ is Fréchet differentiable at x_0. Moreover, if the components of f are $f_1, \ldots, f_m$, and $T_i = df_i(x_0)(i = 1, \ldots, m)$, then $d(G \circ f)(x_0)$ is the function

$$h \mapsto G(T_1(h), f_2(x_0), \ldots, f_m(x_0)) + G(f_1(x_0), T_2(h), f_3(x_0), \ldots, f_m(x_0))$$
$$+ \ldots + G(f_1(x_0), \ldots, f_{m-1}(x_0), T_m(h)).$$

This follows immediately from the chain rule (3.1.4) and Examples (g) and (e).

We conclude this section with an analogue of (1.1.7) for Fréchet differentiable functions.

(3.1.5) *Let f be a one-to-one function from a set $A \subseteq X$ into Y, and let x_0 be an interior point of A such that*
  (i) *f is Fréchet differentiable at x_0,*
  (ii) *the function $T = df(x_0)$ belongs to $\mathscr{L}\mathscr{H}(X, Y)$,*
  (iii) *the point $y_0 = f(x_0)$ is an interior point of $f(A)$,*
  (iv) *f^{-1} is continuous at y_0.*
Then f^{-1} is Fréchet differentiable at y_0, and $df^{-1}(y_0) = T^{-1}$.

Let (k_n) be a bounded sequence in Y, and let (t_n) be a sequence in $\mathbf{R} \backslash \{0\}$ converging to 0. Then $y_0 + t_n k_n$ belongs to $f(A)$ for all sufficiently large n, say $n \geq n_0$, and for all such n let h_n be given by

$$t_n h_n = f^{-1}(y_0 + t_n k_n) - f^{-1}(y_0),$$

so that $t_n k_n = f(x_0 + t_n h_n) - f(x_0)$. Then

$$\frac{f^{-1}(y_0 + t_n k_n) - f^{-1}(y_0) - T^{-1}(t_n k_n)}{t_n} = \frac{t_n h_n - T^{-1}(f(x_0 + t_n h_n) - f(x_0))}{t_n}$$

$$= -T^{-1}\left(\frac{f(x_0 + t_n h_n) - f(x_0) - T(t_n h_n)}{t_n} \right),$$

and since T^{-1} is continuous, it is enough, by (3.1.2)(iii), to show that the sequence (h_n) is bounded in X.

We observe first that $\| T^{-1}y \| \le \| T^{-1} \| \, \| y \|$ for all $y \in Y$, and therefore $c\| x \| \le \| Tx \|$ for all $x \in X$, where $c = 1/\| T^{-1} \|$. Next, since f^{-1} is continuous at $y_0, t_n h_n \to 0$ as $n \to \infty$. Since f is differentiable at x_0, we can therefore find $n_1 \ge n_0$ such that

$$\| f(x_0 + t_n h_n) - f(x_0) - T(t_n h_n) \| \le \tfrac{1}{2}c \| t_n h_n \|$$

for all $n \ge n_1$, and then for such n we have

$$\| t_n k_n \| = \| f(x_0 + t_n h_n) - f(x_0) \| \ge \| T(t_n h_n) \| - \tfrac{1}{2}c \| t_n h_n \| \ge \tfrac{1}{2}c \| t_n h_n \|.$$

Hence $\| k_n \| \ge \tfrac{1}{2}c \| h_n \|$ for all $n \ge n_1$, so that (h_n) is bounded.

Exercises 3.1

1 Let f be a function from a set $A \subseteq X$ into Y and let x_0 be an interior point of A. Prove that f is Fréchet differentiable at x_0 with $df(x_0) = T$ if and only if
$$\frac{f(x_0 + th) - f(x_0) - T(th)}{t} \to 0$$
as $t \to 0$ in $\mathbf{R}$, uniformly for h in the unit sphere $\{h \in X : \| h \| = 1\}$.

2 Prove that if $F \in \mathscr{L}_p(X; Y)$ and $G : X \to Y$ is given by $G(x) = F(x, x, \dots, x)$, then G is Fréchet differentiable at each $x \in X$ and for all $h \in X$
$$dG(x)h = F(h, x, \dots, x) + F(x, h, x, \dots, x) + \dots + F(x, \dots, x, h).$$
Deduce that if F is symmetric (i.e. $F(x_1, \dots, x_p) = F(x_{\sigma(1)}, \dots, x_{\sigma(p)})$ for every permutation σ of the integers $1, \dots, p$), then
$$dG(x)h = pF(x, \dots, x, h).$$

3 Prove that if X is a real inner product space,† and $G(x) = \| x \|^2$, then G is Fréchet differentiable at each $x \in X$ and $dF(x)h = 2\langle x, h \rangle$ for all $h \in X$. Deduce that $x \mapsto \| x \|$ is Fréchet differentiable at each $x \ne 0$ and that its differential there is $h \mapsto \langle x, h \rangle / \| x \|$.

4 Let f be a function from a set $A \subseteq X$ into Y which is Fréchet differentiable at x_0, and let ϕ be a function from a set $B \subseteq \mathbf{R}$ into X such that $\phi(t_0) = x_0$ and that $\phi'_+(t_0)$ exists. Prove that the function $\psi = f \circ \phi$ has a right-hand derivative at t_0 equal to $df(x_0)\phi'_+(t_0)$.

5 Show that if X is a real Banach space and $Y = \mathbf{R}$ then the result of Exercise 4 continues to hold if ϕ'_+ is replaced by the weak right-hand derivative ϕ'_{w+}.
[Hint. Use the uniform boundedness principle to show that if (t_n) is a sequence of points of $B \cap \,]t_0, \infty[$ converging to t_0 and $h_n = (\phi(t_n) - \phi(t_0))/(t_n - t_0)$, then the sequence (h_n) is bounded.]

6 Let A, B be open sets in X, Y, respectively, and let f be a homeomorphism of A onto B which is Fréchet differentiable on A. Prove that f^{-1} is Fréchet differentiable on B if and only if $df(x) \in \mathscr{LH}(X, Y)$ for all $x \in A$.

7 (*Euler's theorem on homogeneous functions*) Let $A \subseteq X$ and let σ be a non-zero real

† Since a norm is real-valued, it cannot be differentiable over a complex normed space.

number. A function $f : A \to Y$ is said to be *positive homogeneous of degree σ* if
(i) whenever $x \in A$ then $tx \in A$ for all $t > 0$,
(ii) $f(tx) = t^\sigma f(x)$ for all $x \in A$ and $t > 0$.
Prove that if A has the property (i), then a function $f : A \to Y$ is positive homogeneous of degree σ if and only if $df(x)x = \sigma f(x)$ for all $x \in A$.
    [Hint. Let $x \in X$, and consider the function $t \mapsto f(tx)$ on $]0, \infty[$.]
8 Let A be an open convex set in the real space X, and let $f : A \to \mathbf{R}$ be Fréchet differentiable. Prove that f is convex if and only if

$$f(x) \geq f(x_0) + df(x_0)(x - x_0)$$

for all $x, x_0 \in A$.
    [Note that the result implies that if $df(x_0) = 0$, then $f(x) \geq f(x_0)$ for all $x \in A$ (cf. (3.6.4)).
    Hint. To prove the 'if', let $a, b \in A, 0 < t < 1, \sigma(t) = (1 - t)a + tb$, and apply the condition with $x_0 = \sigma(t), x = a, b$.]

3.2 Mean value inequalities for Fréchet differentiable functions

A mean value inequality for functions of a vector variable is an easy consequence of (3.1.4. Corollary 3) and the results of §1.6.

(3.2.1) *Let a, b be distinct points of X, let S be the closed line segment in X with endpoints a, b, and let f be a function from a subset of X containing S into Y which is continuous on S and Fréchet differentiable at nearly every point of S. Then there exist uncountably many $c \in S$ for which*

$$\| f(b) - f(a) \| \leq \| df(c)(b - a) \|. \qquad (1)$$

Moreover, either (1) holds with strict inequality for uncountably many $c \in S$, or (1) holds for all $c \in S$ for which $df(c)$ exists (and with equality nearly everywhere).

    Let ϕ be the homeomorphism of the interval $[0, 1]$ onto S given by

$$\phi(t) = (1 - t)a + tb \qquad (0 \leq t \leq 1),$$

and let

$$\psi(t) = f(\phi(t)) = f((1 - t)a + tb) \qquad (0 \leq t \leq 1).$$

Then ψ is continuous on $[0, 1]$, and, for nearly all $t \in]0, 1[$, ψ has a derivative at t equal to $df(x)\phi'(t) = df(x)(b - a)$, where $x = \phi(t)$, using (3.1.4. Corollary 3). Since $\psi(0) = f(a), \psi(1) = f(b)$, the result follows from (1.6.3. Corollary).

(3.2.2) *Let $M \geq 0$, let C be a closed convex set in X with a non-empty interior C°, and let $f : C \to Y$ be a continuous function such that, for nearly all $x \in C^\circ$, the Fréchet differential $df(x)$ exists and satisfies $\| df(x) \| \leq M$.*

Then f is M-Lipschitzian on C, i.e. for all $a, b \in C$

$$\| f(b) - f(a) \| \le M \| b - a \|. \tag{2}$$

The inequality (2) is trivial when $a = b$, so that we may assume a and b distinct. If $a, b \in C^\circ$, then (2) follows from (1), for C° is convex (A.3.1), so that the closed segment S in X with endpoints a, b lies in C°, and

$$\| df(c)(b - a) \| \le \| df(c) \| \, \| b - a \| \le M \| b - a \|$$

for nearly all $c \in S$. Next, if $a, b \in C$, then a, b belong to the closure of C° (A.3.1). Hence we can find sequences $(a_n), (b_n)$ of points of C° converging to a, b, and, by the case already proved,

$$\| f(b_n) - f(a_n) \| \le M \| b_n - a_n \| \tag{3}$$

for all n. Since f is continuous at a and b, $f(a_n) \to f(a)$ and $f(b_n) \to f(b)$ as $n \to \infty$, and hence on taking the limit in (3) we obtain (2).

(3.2.2. Corollary) *Let A be a connected open set in X, and let $f : A \to Y$ be a continuous function which has a Fréchet differential $df(x)$ equal to 0 at nearly every point x of A. Then f is constant.*

Let $a \in A$, and let $E = \{ x \in A : f(x) = f(a) \}$. By the main theorem, E is open. Since f is continuous, E is also closed in A, and since E is non-empty, $E = A$.

Exercises 3.2

1 Let $f : X \to Y$ be a function such that df is constant on X. Prove that $f - f(0) \in \mathscr{L}(X, Y)$.

2 Let $\phi : [a, b] \to Y$ be a continuous function possessing a right-hand derivative nearly everywhere in $[a, b[$. Prove that there exist uncountably many $\xi \in [a, b[$ for which

$$\left\| \phi(b) - \phi(a) - (b - a)\phi'(a) \right\| \le \tfrac{1}{2}(b - a)^2 \left\| \frac{\phi'_+(\xi) - \phi'_+(a)}{\xi - a} \right\|.$$

Deduce that if A is an open convex set in X and $f : A \to Y$ is a Fréchet differentiable function such that df is K-Lipschitzian on A, then for all $x, x' \in A$

$$\| f(x') - f(x) - df(x)(x' - x) \| \le \tfrac{1}{2} K \| x' - x \|^2.$$

Prove that if $T \in \mathscr{L}(X, Y)$ then for all $x, x' \in A$

$$\| f(x') - f(x) - T(x' - x) \| \le \tfrac{1}{2} K \| x' - x \|^2 + \| T - df(x) \| \, \| x' - x \|.$$

[Hint. For the first part apply (1.6.6) with $\psi(t) = (t - a)^2$ and ϕ replaced by $t \mapsto \phi(t) + (t - a)\phi'(a)$. For the second part, argue as in the proof of (3.2.1).]

3 Let a, b be distinct points of X, let S be the closed line segment in X with endpoints a, b, and let f be a function from a subset of X containing S into Y which is continuous on S and Fréchet differentiable at nearly every point of S. Prove that if E is a closed convex set in Y and $df(c)w \in E$ for nearly every $c \in S$, where $w = (b - a)/\| b - a \|$, then $(f(b) - f(a))/\| b - a \| \in E$.

4 Let X, Y be Banach spaces, let A be an open connected set in X, and let (f_n) be a sequence of Fréchet differentiable functions from A into Y. Suppose further that
   (i) there exists $c \in A$ such that the sequence $(f_n(c))$ converges,
   (ii) for each $x_0 \in A$ there is an open ball $B(x_0)$ in A with centre x_0 such that the sequence $(df_n(x))$ converges uniformly on $B(x_0)$.
Prove that for each $x_0 \in A$ the sequence $(f_n(x))$ converges uniformly on $B(x_0)$, and that if $f(x) = \lim_{n \to \infty} f_n(x)$ $(x \in A)$, then f is Fréchet differentiable on A and $df(x) = \lim_{n \to \infty} df_n(x)$ for all $x \in A$.

5 Let X be a normed space and Y a Banach space, let A be an open ball in X with centre x_0, let $B = A \backslash \{x_0\}$, and let $f : B \to Y$ be Fréchet differentiable. Prove that
   (i) if $\|df(x)\| \le M$ for all $x \in B$, then f has an M-Lipschitzian extension to A,
   (ii) if $df(x) \to l$ as $x \to x_0$, then f has a Fréchet differentiable extension $\bar{f}$ to A, and $d\bar{f}(x_0) = l$.

3.3 The partial Fréchet differentials of a function with domain in a product space

Let $X_1, \ldots, X_n$ be normed spaces, where $n \ge 2$, let $X = X_1 \times X_2 \times \ldots \times X_n$, and for $j = 1, \ldots, n$ let I_j be the insertion map from X_j into X which maps the point x_j of X_j to the point of X with jth coordinate x_j and all other coordinates 0. Clearly $I_j \in \mathcal{L}(X_j, X)$. Also, if $x_0 = (x_{10}, \ldots, x_{n0}) \in X$, then the function $x_j \mapsto x_0 + I_j(x_j - x_{j0})$ maps the point x_j of X_j to the point of X with jth coordinate x_j and all other coordinates equal to those of x_0.

Now let $A \subseteq X$, let $f : A \to Y$, let $x_0 \in A$, and let j be an integer such that $1 \le j \le n$. Then the function $x_j \mapsto f(x_0 + I_j(x_j - x_{j0}))$, with domain the set of $x_j \in X_j$ for which $x_0 + I_j(x_j - x_{j0}) \in A$, is the function which is obtained by fixing all the arguments of f except the jth and which takes the value $f(x_0)$ at $x_j = x_{j0}$. If this function has a Fréchet differential at $x_j = x_{j0}$, we call this differential the *partial Fréchet differential of f at x_0 with respect to the jth coordinate*, and we denote it by $d_j f(x_0)$ (thus $d_j f(x_0)$ is obtained by taking the differential with respect to the jth coordinate, keeping all other coordinates fixed). Further, the function $x \mapsto d_j f(x_0)$ whose domain is the set of points x of A at which $d_j f(x)$ exists is called the *partial Fréchet differential of f with respect to the jth coordinate*, and we denote it by $d_j f$.†

By writing $h_j = x_j - x_{j0}$, we see from the definition of the Fréchet differential in §3.1 that f has a partial differential at x_0 with respect to the jth coordinate if and only if 0 is an interior point of the set of $h_j \in X_j$

---

† Note that $d_j f(x) \in \mathcal{L}(X_j, Y)$, so that $d_j f$ is a function from a subset of A into $\mathcal{L}(X_j, Y)$.

such that $x_0 + I_j h_j \in A$ and there exists $T_j \in \mathcal{L}(X_j, Y)$ such that

$$\frac{f(x_0 + I_j h_j) - f(x_0) - T_j h_j}{\|h_j\|} \to 0 \text{ in } Y \text{ as } h_j \to 0,$$

and then $d_j f(x_0) = T_j$. In other words, the partial differential $d_j f(x_0)$ exists and is equal to T_j if and only if the function $h_j \mapsto f(x_0 + I_j h_j)$ has a differential at $h_j = 0$ equal to T_j.

We prove now

(3.3.1) *Let* $X_1, \ldots, X_n$ *be normed spaces, where* $n \geq 2$, *let* $X = X_1 \times X_2 \times \ldots \times X_n$, *let* $A \subseteq X$, *and let* $f : A \to Y$ *be Fréchet differentiable at the point* x_0 *of A. Then f has a partial Fréchet differential at* x_0 *with respect to each coordinate, and, moreover,*

$$d_j f(x_0) = df(x_0) \circ I_j \qquad (j = 1, \ldots, n), \tag{1}$$

and

$$df(x_0)h = \sum_{j=1}^{n} d_j f(x_0) h_j \qquad (h = (h_1, \ldots, h_n) \in X). \tag{2}$$

Further, if df is continuous at x_0, *then so is each partial differential* $d_j f$.

By §3.1, Example (f), p. 171, the function $h_j \mapsto x_0 + I_j h_j$ from X_j into X is differentiable at each $h_j \in X_j$ with differential there equal to I_j. It therefore follows from the chain rule (3.1.4) that each partial differential $d_j f(x_0)$ exists, and that (1) holds. Hence $d_j f(x_0) h_j = df(x_0)(0, \ldots, 0, h_j, 0, \ldots, 0)$ for all $h_j \in X_j$, and this implies (2). Further, $d_j f = F_j \circ df$, where $F_j : \mathcal{L}(X, Y) \to \mathcal{L}(X_j, Y)$ is given by $F_j(T) = T \circ I_j$, and since F_j is clearly (linear and) continuous, the continuity of df at x_0 implies that of $d_j f$.

The converse of (3.3.1) is false, i.e. the existence of the partial Fréchet differentials of f at x_0 does not imply the Fréchet differentiability of f at x_0, nor indeed even the continuity of f there. For example, if $f : \mathbf{R}^2 \to \mathbf{R}$ is given by

$$f(x_1, x_2) = \frac{2x_1 x_2}{x_1^2 + x_2^2} \quad ((x_1, x_2) \neq (0, 0)), \quad f(0, 0) = 0,$$

then f takes the value 0 everywhere on the axes, so that the partial differentials of f at the origin exist and are equal to the zero function from $\mathbf{R}$ into itself. On the other hand, $f(t, t) = 1$ for all real $t \neq 0$, so that f is not continuous at the origin.

The following theorem provides a sufficient condition for the Fréchet differentiability of f.

(3.3.2) *Let* $X_1, \ldots, X_n, X$ *be as before, let* $A \subseteq X$, *and let* $f : A \to Y$ *be a* *function whose partial Fréchet differentials $d_1 f, \ldots, d_n f$ are defined on a neighbourhood of x_0 and are continuous at x_0. Then f is Fréchet differentiable at x_0.*

By hypothesis, x_0 is an interior point of A. Further, by (2), the differential of f at x_0, if it exists, is the function

$$(h_1, \ldots, h_n) \mapsto \sum_{j=1}^{n} d_j f(x_0) h_j.$$

We therefore have to prove that if $h = (h_1, \ldots, h_n)$ and

$$R(h) = f(x_0 + h) - f(x_0) - \sum_{j=1}^{n} d_j f(x_0) h_j,$$

then $R(h)/\|h\| \to 0$ as $h \to 0$.

Given $\varepsilon > 0$, we can find an open ball B in X with centre x_0, contained in A, such that for all $x \in B$

$$\|d_j f(x) - d_j f(x_0)\| \leq \varepsilon \qquad (j = 1, \ldots, n).$$

Let $h = (h_1, \ldots, h_n)$ be a point such that $x_0 + h \in B$, let $z_0 = x_0$, and for $j = 1, \ldots, n$ let

$$z_j = (x_{10} + h_1, \ldots, x_{j0} + h_j, x_{j+1,0}, \ldots, x_{n0}) = x_0 + \sum_{i=1}^{j} I_i h_i.$$

Then each $z_j \in B$, $z_n = x_0 + h$, and

$$f(x_0 + h) - f(x_0) = \sum_{j=1}^{n} (f(z_j) - f(z_{j-1})) = \sum_{j=1}^{n} (f(z_{j-1} + I_j h_j) - f(z_{j-1})),$$

so that

$$R(h) = \sum_{j=1}^{n} (f(z_{j-1} + I_j h_j) - f(z_{j-1}) - d_j f(x_0) h_j).$$

To estimate the sum on the right we apply the mean value inequality in the form (3.2.1) to the function

$$k \mapsto f(z_{j-1} + I_j k) - f(z_{j-1}) - d_j f(x_0) k$$

on the closed segment S in X_j with endpoints $0, h_j$. The differential of this function at the point k is

$$d_j f(z_{j-1} + I_j k) - d_j f(x_0),$$

and since $z_{j-1} + I_j k \in B$ for all k on S, the norm of this differential does not exceed ε for all such k. Hence

$$\|f(z_{j-1} + I_j h_j) - f(z_{j-1}) - d_j f(x_0) h_j\| \leq \varepsilon \|h_j\|,$$

so that

$$\|R(h)\| \leq \varepsilon \sum_{j=1}^{n} \|h_j\| \leq \varepsilon n^{1/2} \|h\|,$$

i.e. $R(h)/\|h\| \to 0$ as $h \to 0$.

(3.3.3) *Let* $X_1, \ldots, X_n$, X *be as before, and let* A *be an open subset of* X. *Then* $f : A \to Y$ *is* C^1 *on* A *if and only if all the partial differentials* $d_1 f, \ldots, d_n f$ *are continuous on* A.

The 'only if' follows directly from the last part of (3.3.1). To prove the 'if', suppose that $d_1 f, \ldots, d_n f$ are continuous on A. Then, by (3.3.2), f is differentiable at each point x of A. Further, by (2),

$$df(x) = \sum_{j=1}^{n} d_j f(x) \circ P_j,$$

where $P_j \in \mathscr{L}(X, X_j)$ is the projection map $(x_1, \ldots, x_n) \mapsto x_j$, and an argument similar to that used to prove the last part of (3.3.1) now shows that df is continuous on A.

Exercises 3.3

1   Let $X_1, \ldots, X_n$ be normed spaces, where $n \geq 2$, let $X = X_1 \times \ldots \times X_n$, let A be an open convex set in X, and let $f : A \to Y$ be a function whose partial differentials $d_1 f, \ldots, d_n f$ are defined and bounded on A. Prove that f is Lipschitzian on A.

2   Let X_1, X_2 be normed spaces, let $X = X_1 \times X_2$, let A be a convex set in X, and let $f : A \to Y$ be a continuous function whose partial differential $d_2 f$ is defined and zero on A. Prove that, for all $(x_1, x_2) \in A$, $f(x_1, x_2) = g(x_1)$, where g is a continuous function whose domain is the set of first coordinates of points of A.

3.4   The partial derivatives of a function with domain in $\mathbf{R}^n$

Let f be a function from a set $A \subseteq \mathbf{R}^n$ into Y, where $n \geq 2$, let $x_0 = (x_{10}, \ldots, x_{n0}) \in A$, let $\{e_1, \ldots, e_n\}$ be the natural basis in $\mathbf{R}^n$, and let j be an integer such that $1 \leq j \leq n$. Then the function $x_j \mapsto f(x_0 + (x_j - x_{j0})e_j)$, with domain the set of real x_j for which $x_0 + (x_j - x_{j0})e_j \in A$, is the function which is obtained by fixing all the arguments of f except the jth and which takes the value $f(x_0)$ at $x_j = x_{j0}$. If this function has a derivative at $x_j = x_{j0}$ we call this derivative the *partial derivative of* f *with respect to the* j*th coordinate*, and we denote it by

$$D_j f(x_0), \quad D_j f(x_{10}, \ldots, x_{n0}), \quad \text{or} \quad \left[\frac{\partial}{\partial x_j} f(x_1, \ldots, x_n)\right]_{x_0} \quad (1)$$

(thus $D_j f(x_0)$ is obtained by taking the derivative with respect to the jth coordinate, keeping all other coordinates fixed). Further, the function $x \mapsto D_j f(x)$, whose domain is the set of points $x \in A$ at which $D_j f(x)$ exists, is called the *partial derivative of* f *with respect to the* j*th coordinate*, and we denote it by $D_j f$ ($D_j f$ may, of course, be the empty function). The last notation in (1) will normally be used when f is given by some specific formula. We also write $\partial F(x_1, \ldots, x_n)/\partial x_j$ for the value of the partial derivative at the point $(x_1, \ldots, x_n)$.†

By writing $t = x_j - x_{j0}$, we see that

$$D_j f(x_0) = \left[\frac{d}{dt} f(x_0 + t e_j) \right]_{t=0}$$

whenever either side exists, that is to say that $D_j f(x_0)$ exists and is equal to η if and only if 0 is a non-isolated point of the set of real t for which $x_0 + t e_j \in A$ and $(f(x_0 + t e_j) - f(x_0))/t \to \eta$ as $t \to 0$. (It should be noted that the insertion map I_j of §3.3, which is here a map from $\mathbf{R}$ into $\mathbf{R}^n$, is given by $I_j t = t e_j$.)

We remark that, by §3.1, Example (a) (p. 168), a function f from a set $A \subseteq \mathbf{R}^n$ into Y has a partial Fréchet differential with respect to the jth coordinate at the point x_0 of A if and only if 0 is an interior point of the set of real t for which $x_0 + t e_j \in A$ and f has a partial derivative $D_j f(x_0)$ at x_0, and then $d_j f(x_0)$ is the function $h \mapsto h D_j f(x_0)$ from $\mathbf{R}$ into Y (so that $D_j f(x_0)$ is the value of $d_j f(x_0)$ at e_j). By combining this remark with (3.3.1), we obtain immediately the following result.

(3.4.1) *Let* $A \subseteq \mathbf{R}^n$, *where* $n \geq 2$, *and let* $f: A \to Y$ *be Fréchet differentiable at the point* x_0. *Then for* $j = 1, \ldots, n$ *the partial derivative* $D_j f(x_0)$ *exists and is equal to* $df(x_0)e_j$, *and for all* $h = (h_1, \ldots, h_n) \in \mathbf{R}^n$

$$df(x_0)h = \sum_{j=1}^{n} h_j D_j f(x_0).$$

If in addition df *is continuous at* x_0, *so is each partial derivative* $D_j f$.

If Y is the product of real normed spaces $Y_1, \ldots, Y_m$, and $f_1, \ldots, f_m$ are the components of f, then, by §1.1, Example (c) (p. 4), f has a partial derivative at x_0 with respect to the jth coordinate if and only if each of its components has a partial derivative at x_0 with respect to this coordinate, and in this case $D_j f(x_0)$ is the point $(D_j f_1(x_0), \ldots, D_j f_m(x_0))$ of Y. In the important case where $Y = \mathbf{R}^m$, the components of f and their partial derivatives are real-valued functions, and the following result shows that these partial derivatives provide a simple means for computing the matrix of the linear function $df(x_0)$.

(3.4.2) *Let* $A \subseteq \mathbf{R}^n$, *where* $n \geq 2$, *let* $f: A \to \mathbf{R}^m$ *be Fréchet differentiable*

---

† Those authors who use the notation $f'(x_0)$ for the Fréchet differential of f at x_0 normally use the notations in (1) for both the partial Fréchet differential of f at x_0 and the partial derivative, and, as before, rely on the isomorphism between $\mathscr{L}(\mathbf{R}, Y)$ and Y to avoid confusion.

at the point x_0, and let $f_1, \ldots, f_m$ be the components of f. Then for $i = 1, \ldots, m$ and $j = 1, \ldots, n$ the partial derivative $D_j f_i(x_0)$ exists, and the $m \times n$ matrix $[D_j f_i(x_0)]$ is the matrix of the linear function $df(x_0): \mathbf{R}^n \to \mathbf{R}^m$ with respect to the natural bases in $\mathbf{R}^n$ and $\mathbf{R}^m$ (so that $df(x_0)$ is the function $(x_1, \ldots, x_n) \mapsto (y_1, \ldots, y_m)$ given by

$$y_i = \sum_{j=1}^{n} x_j D_j f_i(x_0) \qquad (i = 1, \ldots, m)).$$

Further,

$$\|df(x_0)\| \le \left(\sum_{i=1}^{m} \sum_{j=1}^{n} (D_j f_i(x_0))^2 \right)^{1/2} \le n^{1/2} \|df(x_0)\|.$$

By (3.4.1) and the remark above, each partial derivative $D_j f_i(x_0)$ exists, and $D_j f(x_0)$ is the point $(D_j f_1(x_0), \ldots, D_j f_m(x_0))$ of $\mathbf{R}^m$. Since $D_j f(x_0) = df(x_0)e_j$, and since the elements in the jth column of the matrix of $df(x_0)$ are the coordinates of $df(x_0)e_j$, the matrix of $df(x_0)$ is therefore $[D_j f_i(x_0)]$. The inequalities for $\|df(x_0)\|$ are easy to obtain.

The $m \times n$ matrix $[D_j f_i(x_0)]$ in (3.4.2) is called the *Jacobian* of f at x_0. When $m = n$, the matrix is square, and the determinant of this matrix is then called the *Jacobian determinant* of f at x_0; in this case the Jacobian itself is often denoted by the symbol $\partial(f_1, \ldots, f_n)/\partial(x_1, \ldots, x_n)$.

Since the matrix of the composite $U \circ T$ of two linear maps U, T is the product of the matrices of U and T, the chain rule (3.1.4) gives the following result concerning the Jacobian of the composite $g \circ f$ of two differentiable functions. This result enables us to calculate the partial derivatives of the composite in terms of those of f and g, and the relation (3) below is the usual form of the chain rule given in elementary textbooks on the calculus.

(3.4.3) *Let $A \subseteq \mathbf{R}^n, B \subseteq \mathbf{R}^m$, let $f: A \to \mathbf{R}^m$ be Fréchet differentiable at x_0, let $g: B \to \mathbf{R}^l$ be Fréchet differentiable at the point $y_0 = f(x_0)$, let $H = g \circ f$, and let the components of f, g, H be $f_1, \ldots, f_m, g_1, \ldots, g_l, H_1, \ldots, H_l$. Then the Jacobians of f, g, H satisfy the relation*

$$[D_j H_k(x_0)] = [D_i g_k(y_0)][D_j f_i(x_0)] \tag{2}$$

(*so that*

$$D_j H_k(x_0) = \sum_{i=1}^{m} D_i g_k(y_0) D_j f_i(x_0) \qquad (k = 1, \ldots, l; j = 1, \ldots, n)). \tag{3}$$

Further, if $l = m = n$, the Jacobian determinants of f, g, H satisfy the relation

$$\det[D_j H_k(x_0)] = \det[D_i g_k(y_0)] \det[D_j f_i(x_0)]. \tag{4}$$

If $n = 1$ or $m = 1$, the appropriate partial derivatives in (2), (3) and (4) are to be replaced by ordinary derivatives.

The formula (3) is often written in a more easily remembered form involving a 'dependent variable' notation; thus with

$$y_i = f_i(x_1, \ldots, x_n), \ z_k = g_k(y_1, \ldots, y_m) = H_k(x_1, \ldots, x_n),$$

the formula (3) is written as

$$\frac{\partial z_k}{\partial x_j} = \frac{\partial z_k}{\partial y_1}\frac{\partial y_1}{\partial x_j} + \frac{\partial z_k}{\partial y_2}\frac{\partial y_2}{\partial x_j} + \cdots + \frac{\partial z_k}{\partial y_m}\frac{\partial y_m}{\partial x_j}.$$

There is a good deal of ambiguity in the use of the various symbols here, but it does in fact cause remarkably few difficulties, and these can always be avoided by recourse to the form (3).

It is easy to see from the example following (3.3.1) that the converse of (3.4.1) is false, i.e. the existence of the partial derivatives of f at x_0 does not imply the Fréchet differentiability of f at x_0 (nor even the continuity of f there). The following theorem provides a useful sufficient condition for the Fréchet differentiability of a function with domain in $\mathbf{R}^n$.

(3.4.4) *Let $A \subseteq \mathbf{R}^n$, where $n \geq 2$, and let $f : A \to Y$ be a function whose partial derivatives $D_1 f, \ldots, D_n f$ are defined on a neighbourhood of x_0 and are continuous at x_0. Then f is Fréchet differentiable at x_0.*

In fact, it is easy to see from the remark preceding (3.4.1) that the hypotheses here imply that the partial differentials $d_1 f, \ldots, d_n f$ are defined on a neighbourhood of x_0 and are continuous at x_0. The result therefore follows from (3.3.2).

We note also the analogue of (3.3.3) for partial derivatives.

(3.4.5) *Let A be an open subset of $\mathbf{R}^n$, where $n \geq 2$. Then $f : A \to Y$ is C^1 on A if and only if all the partial derivatives $D_1 f, \ldots, D_n f$ are continuous on A.*

The partial derivatives $D_1 f, \ldots, D_n f$ of a function from a set $A \subseteq \mathbf{R}^n$ into Y are themselves functions from subsets of $\mathbf{R}^n$ into Y, and we can therefore form their partial derivatives. If j, k are integers such that $1 \leq j \leq n$, $1 \leq k \leq n$, we define the *partial derivative of f of the second order with respect to the jth and kth coordinates*, denoted by $D_k D_j f$, to be the partial derivative of the function $D_j f$ with respect to the kth coordinate, so that $D_k D_j f = D_k(D_j f)$ ($D_k D_j f$ may, of course, be the empty function, and is certainly so if $D_j f$ is empty). The value $D_k D_j f(x_0)$ of the partial derivative $D_k D_j f$ at a point x_0 of its domain will be denoted also by $[\partial^2 f(x_1,\ldots,x_n)/\partial x_k \partial x_j]_{x_0}$. We also write $\partial^2 f(x_1,\ldots,x_n)/\partial x_k \partial x_j$ or simply $\partial^2 f(x_1,\ldots,x_n)/\partial x_k \partial x_j$ for the value of $D_k D_j f$ at the point $(x_1,\ldots,x_n)$. When $j = k$, we write D_j^2 in place of $D_k D_j$, and $\partial^2/\partial x_j^2$ in place of $\partial^2/\partial x_k \partial x_j$.

The partial derivatives of f of higher order are defined inductively, the partial derivatives of f of order $r + 1$ being the partial derivatives of the first order of the partial derivatives of f of order r. It follows immediately from this that if r, p are positive integers, then the partial derivatives of order q of the partial derivatives of f of order r are the partial derivatives of f of order $q + r$.

The connection between the existence and continuity of the partial derivatives of f of order r and the existence of the rth differential of f is considered in the next section.

Exercise 3.4

1 Let Z_n^+ denote the set of n-tuples of non-negative integers, with the usual addition and subtraction for elements of $\mathbf{R}^n$, and with the partial ordering relation $\alpha \geq \beta$ defined to mean that $\alpha_j \geq \beta_j$ for $j = 1,\ldots,n$, where α_j, β_j are the jth coordinates of α, β. For any $\alpha = (\alpha_1,\ldots,\alpha_n) \in Z_n^+$ we define

$$\alpha! = \prod_{j=1}^{n} \alpha_j!.$$

Further, for any $x = (x_1,\ldots,x_n) \in \mathbf{R}^n$ we define x^α to be the product

$$x^\alpha = x_1^{\alpha_1} x_2^{\alpha_2} \ldots x_n^{\alpha_n},$$

and we define the partial differential operator D^α by

$$D^\alpha = D_1^{\alpha_1} D_2^{\alpha_2} \ldots D_n^{\alpha_n}.$$

(i)   Verify that $D^\beta x^\alpha = (\alpha!/(\alpha - \beta)!)x^{\alpha-\beta}$ if $\alpha \geq \beta$ and is 0 otherwise.

(ii) Any real polynomial P on $\mathbf{R}^n$ can be uniquely expressed as $P(x) = \Sigma a_\alpha x^\alpha$, where $a_\alpha \in \mathbf{R}$. Prove that $a_\alpha = D^\alpha P(0)/\alpha!$, and hence deduce that for all $x, y \in \mathbf{R}^n$

$$P(x + y) = \sum D^\alpha P(y) x^\alpha/\alpha!.$$

(iii) Prove that

$$(x + y)^\beta = \sum_{\alpha \leq \beta} \frac{\beta!}{(\beta - \alpha)!\alpha!} x^\alpha y^{\beta-\alpha}.$$

3.5 Fréchet differentials of higher order

Let f be a function from a set $A \subseteq X$ into Y. The Fréchet differential df of f is a function from a subset of A into the normed space $\mathscr{L}(X, Y)$, and we define the *second Fréchet differential of f*, denoted by $d^2 f$, to be the Fréchet differential of df (d^2f may be the empty function, and is certainly so if df is the empty function). Further, if d^2f is defined at a point $x_0 \in X$, we say that f is *twice Fréchet differentiable at* x_0, and we call the value of d^2f at x_0, which we denote by $d^2f(x_0)$, the *second Fréchet differential of f at* x_0. In other words, f is twice Fréchet differentiable at x_0 if and only if df is defined on a neighbourhood of x_0 in X and there exists a continuous linear function $T: X \to \mathscr{L}(X, Y)$ such that

$$\lim_{x \to x_0} \frac{df(x) - df(x_0) - T(x - x_0)}{\|x - x_0\|} = 0,$$

and then $d^2f(x_0) = T$. We stress that $d^2f(x_0)$ (if it exists) belongs to $\mathscr{L}(X, \mathscr{L}(X, Y))$; thus for each $h \in X$ the value $d^2f(x_0)h$ of $d^2f(x_0)$ at h belongs to $\mathscr{L}(X, Y)$, and for each $k \in X$ the value $(d^2f(x_0)h)k$ of $d^2f(x_0)h$ at k is a point of Y.†

Generally, the *rth Fréchet differential d^rf of f* is defined inductively by the formulae

$$d^1f = df, \quad d^rf = d(d^{r-1}f) \qquad (r = 2, 3, \ldots).$$

Further, if d^rf is defined at the point $x_0 \in X$, we say that f is *r-times Fréchet differentiable at x_0*, and we call the value $d^rf(x_0)$ of d^rf there the *rth Fréchet differential of f at x_0*.

It is obvious that if x_0 is a point such that $d^rf(x_0)$ exists for some r, then either $d^rf(x_0)$ exists for all positive integers r, or there exists a positive integer p such that $d^rf(x_0)$ exists for $1 \le r \le p$ and does not exist for any $r > p$. It is also obvious that, for any positive integers q, r, the qth differential of d^rf is $d^{q+r}f$.

It should be noted that if $r \ge 2$ then the existence of the rth differential of f at x_0 implies that x_0 is an interior point of the domain of $d^{r-1}f$, and hence x_0 is an interior point of the domains of $f, df, \ldots, d^{r-1}f$.

If E is an open set in X, a function which is r-times Fréchet differentiable at each point of E is said to be *r-times Fréchet differentiable on E*. Further, f is said to be *C^r on E* if d^rf is continuous on E. We say also that f is *C^∞ on E* if f is r-times Fréchet differentiable on E for every positive integer r (by (3.1.1), this implies that f is C^r on E for every r). If $E = A$, we say simply that f is *C^r or C^∞*.

Example (a)

If f is a constant function from a set $A \subseteq X$ into Y, and the interior A° of A is non-empty, then $df(x)$ exists for each $x \in A^\circ$ and is the zero function

---

† We shall see shortly that we can simplify this notation by using the isometric isomorphism between $\mathscr{L}(X, \mathscr{L}(X, Y))$ and the space $\mathscr{L}_2(X; Y)$ of continuous bilinear maps from $X \times X$ into Y.

from X into Y. Hence df is the constant function from A° into $\mathscr{L}(X,Y)$ which maps each point of A° to the zero function $0:X \to Y$. It follows that d^2f is the constant function from $A^\circ(=(A^\circ)^\circ)$ into $\mathscr{L}(X,\mathscr{L}(X,Y))$ which maps each point of A° to the zero function $0:X \to \mathscr{L}(X,Y)$, and generally $d^r f$ is the constant function which maps each point of A° to the zero function $0:X \to \mathscr{L}(X,\mathscr{L}(X,\ldots,\mathscr{L}(X,Y))\ldots)$. Hence f is C^∞ on A°.

Example (b)

If $f\in\mathscr{L}(X,Y)$, then $df(x)=f$ for all $x\in X$, so that df is the constant function from X into $\mathscr{L}(X,Y)$ which maps each point of X to the element f of $\mathscr{L}(X,Y)$. Hence the differentials $df,d^2f,d^3f,\ldots$ are all constant, so that f is C^∞.

Example (c)

Let $X_1,\ldots,X_p$ be normed spaces, let $X=X_1\times X_2\times\ldots\times X_p$, and let $F\in\mathscr{L}(X_1,\ldots,X_p,Y)$. By §3.1, Example (*e*), $dF\in\mathscr{L}(X,\mathscr{L}(X,Y))$, and therefore, by Example (*b*)(p. 170), the differentials d^2F, d^3F, … are all constant, and F is C^∞.

Example (d)

Let $G:\mathscr{L}(X,Y)\times\mathscr{L}(Y,Z)\to\mathscr{L}(X,Z)$ be given by $G(T,U)=U\circ T$ then G is obviously bilinear, and it is continuous, since $\|G(T,U)\|\le\|U\|\|T\|$. Hence G is C^∞, by Example (*c*).

We consider now the analogues of (3.1.3) and the chain rule (3.1.4) for higher order differentiability.

(3.5.1) (i) *If $f:A\to Y$ is r-times Fréchet differentiable at x_0, where $A\subseteq X$, and α is a scalar, then αf is r-times Fréchet differentiable at x_0.*

(ii) *If $f:A\to Y$ and $g:B\to Y$ are r-times Fréchet differentiable at x_0, where $A,B\subseteq X$, then $f+g$ is r-times Fréchet differentiable at x_0.*

The results (i) and (ii) hold also if r-times Fréchet differentiability is replaced throughout by the condition that the functions are C^r on an open set E.

Consider the proof of (ii) for r-times differentiability. The case $r=1$ follows from (3.1.3)(ii), and we complete the proof by induction on r. Let r be a positive integer such that (ii) holds for every pair of functions r-times differentiable at x_0, and let f,g be $(r+1)$-times differentiable at x_0. We prove that $f+g$ is $(r+1)$-times differentiable at x_0.

Since f, g are $(r + 1)$-times differentiable at x_0, they are differentiable on some open neighbourhood V of x_0, and we may obviously replace f and g by their restrictions to V, so that V is the domain of $f + g$. By (3.1.3)(ii), we then have

$$d(f + g) = df + dg. \dagger$$

Since df and dg are r-times differentiable at x_0, the induction hypothesis implies that $d(f + g)$ is r-times differentiable at x_0, whence $f + g$ is $(r + 1)$-times differentiable at x_0.

A similar argument proves (i) and the corresponding C^r results.

To deal with the extension of the chain rule, we have to prove first the special case corresponding to (3.1.4. Corollary 1).

(3.5.2) *Let f be a function from a set $A \subseteq X$ into Y, and let $g \in \mathscr{L}(Y, Z)$. Then*

(i) *if f is r-times Fréchet differentiable at x_0, so is $g \circ f$,*

(ii) *if f is C^r on an open set E, so is $g \circ f$.*

Consider the proof of (i). We argue by induction on r, the case $r = 1$ being part of (3.1.4. Corollary 1). Let r be a positive integer such that (i) holds for every pair of functions satisfying the stated hypotheses, let f be a function $(r + 1)$-times differentiable at x_0, and let $g \in \mathscr{L}(Y, Z)$. Then f is differentiable on some open neighbourhood V of x_0, and we may obviously replace f by its restriction to V, so that V is the domain of $g \circ f$. By (3.1.4. Corollary 1), $d(g \circ f)(x) = g \circ df(x)$ for all $x \in V$, and therefore $d(g \circ f) = \mu(df)$, where $\mu : \mathscr{L}(X, Y) \to \mathscr{L}(X, Z)$ is given by $\mu(T) = g \circ T$. The function μ is linear (since g is linear) and continuous (since $\| \mu(T) \| \leq \| g \| \, \| T \|$), and hence is C^∞ (Example (b)). Since df is r-times differentiable at x_0, the induction hypothesis therefore implies that $d(g \circ f)$ is r-times differentiable. whence $g \circ f$ is $(r + 1)$-times differentiable.

The proof of (ii) is similar.

(3.5.2. Corollary) *Let $Y_1, \ldots, Y_m$ be normed spaces, let $Y = Y_1 \times Y_2 \times \ldots \times Y_m$, let f be a function from a set $A \subseteq X$ into Y, and let $f_i : A \to Y_i$ ($i = 1, \ldots, m$) be the ith component of f. Then*

(i) *f is r-times Fréchet differentiable at x_0 if and only if $f_1, \ldots, f_m$ are r-times Fréchet differentiable at x_0,*

(ii) *f is C^r on an open set E if and only if $f_1, \ldots, f_m$ are C^r on E.*

---

† This relation is not true in general, since $d(f + g)$ may be defined at points where df and dg are undefined.

This follows immediately from the theorem and the formulae of §3.1, Example (*g*) (p. 173).

(3.5.3) *Let* $A \subseteq X$, $B \subseteq Y$, *let* $f : A \to Y$ *be r-times Fréchet differentiable at the point* x_0, *and let* $g : B \to Z$ *be r-times Fréchet differentiable at the point* $y_0 = f(x_0)$. *Then* $g \circ f$ *is r-times Fréchet differentiable at* x_0.

The result holds also if r-times differentiability is replaced throughout by the property that the functions are C^r *on appropriate open sets.*

Consider the proof for *r*-times differentiability. The case $r = 1$ follows from the chain rule (3.1.4), and we again complete the proof by induction on *r*. Let *r* be a positive integer such that the result holds for every pair of functions satisfying the stated hypotheses, and let *f*, *g* be a pair of functions satisfying the hypotheses with *r* replaced by $r + 1$. We show that $g \circ f$ is $(r + 1)$-times differentiable at x_0.

As before, *f* is differentiable on an open neighbourhood *V* of x_0, and *g* is differentiable on an open neighbourhood *W* of y_0. Moreover, since *f* is continuous at x_0 and $y_0 = f(x_0)$, we may suppose that $f(V) \subseteq W$. Further, we may obviously replace *f* by its restriction to *V*, and then *V* is the domain of $g \circ f$.

By (3.1.4), $d(g \circ f)(x) = dg(y) \circ df(x)$ for all $x \in V$, where $y = f(x)$. Let $w = dg \circ f$, so that *w* maps *V* into $\mathscr{L}(Y, Z)$, and $dg(y) = w(x)$. Then

$$d(g \circ f)(x) = w(x) \circ df(x),$$

and hence

$$d(g \circ f) = G \circ F,$$

where $F : V \to \mathscr{L}(X, Y) \times \mathscr{L}(Y, Z)$ is given by $f(x) = (df(x), w(x))$, and $G : \mathscr{L}(X, Y) \times \mathscr{L}(Y, Z) \to \mathscr{L}(X, Z)$ is given by $G(T, U) = U \circ T$. By the induction hypothesis, *w* is *r*-times differentiable at x_0, and since *df* is also *r*-times differentiable at x_0, so is *F*, by (3.5.2. Corollary). Since *G* is C^∞ (Example (*d*)), the induction hypothesis now implies that $d(g \circ f)$ is *r*-times differentiable at x_0, whence $g \circ f$ is $(r + 1)$-times differentiable there.

The proof for the C^r case is similar.

In the further study of higher order Fréchet differentiability, it is convenient to regard the *r*th differential $d^r f(x_0)$ of *f* at x_0 as a multilinear function.

Consider first the case $r = 2$. We have shown in (A.2.6) that the space $\mathscr{L}(X, \mathscr{L}(X, Y))$ is isometrically isomorphic with the space $\mathscr{L}_2(X; Y)$ of continuous bilinear functions, so that these two spaces can be naturally identified; this is done by identifying the function $T \in \mathscr{L}(X, \mathscr{L}(X, Y))$

with the function $(h, k) \mapsto (Th)k$, which is continuous and bilinear. Hence if f is twice Fréchet differentiable at x_0, we can regard the element $(d^2f(x_0)h)k$ of Y (i.e. the value of the function $d^2f(x_0)h : X \to Y$ at k) as the value at the point (h, k) of $X \times X$ of a continuous bilinear function from $X \times X$ into Y. We use the same symbol $d^2f(x_0)$ for this continuous bilinear function, so that $d^2f(x_0)(h, k)$ is equal to $(d^2f(x_0)h)k$.

The following result makes the bilinear nature of $d^2f(x_0)$ clearer.

(3.5.4) *Let f be a function from a set $A \subseteq X$ into Y, and let x_0 be an interior point of A. In order that f is twice Fréchet differentiable at x_0, it is necessary and sufficient that there exists $F \in \mathscr{L}_2(X; Y)$ with the property that for each $\varepsilon > 0$ we can find $\delta > 0$ such that*

$$\| df(x_0 + h)k - df(x_0)k - F(h, k) \| \le \varepsilon \| h \| \, \| k \| \qquad (1)$$

whenever $\| h \| \le \delta$ and $k \in X$, and then $d^2f(x_0)(h, k) = F(h, k)$.

If f is twice-differentiable at x_0, then for each $\varepsilon > 0$ we can find $\delta > 0$ such that

$$\| df(x_0 + h) - df(x_0) - d^2f(x_0)h \| \le \varepsilon \| h \|$$

whenever $\| h \| \le \delta$, and this trivially implies (1). Conversely, suppose that the condition (1) holds, and let $F^*(h)$ be the function $k \mapsto F(h, k)$. Then $F^* \in \mathscr{L}(X, \mathscr{L}(X, Y))$, and, by (1),

$$\| df(x_0 + h) - df(x_0) - F^*(h) \| \le \varepsilon \| h \|$$

whenever $\| h \| \le \delta$, so that df is differentiable at x_0, and $d^2f(x_0) = d(df)(x_0) = F^*$.

We prove now that $d^2f(x_0)$, regarded as a bilinear function, is symmetric; this is the extension to Fréchet differentials of the 'mixed derivative theorem'.

(3.5.5) *If f is a function from a set $A \subseteq X$ into Y which is twice Fréchet differentiable at x_0, then $d^2f(x_0)$ is symmetric, i.e. for all $h, k \in X$*

$$d^2f(x_0)(h, k) = d^2f(x_0)(k, h).$$

Let G be the function given by

$$G(h, k) = f(x_0 + h + k) - f(x_0 + h) - f(x_0 + k) + f(x_0) - d^2f(x_0)(h, k),$$

where $x_0 + h + k, x_0 + h, x_0 + k \in A$. Since x_0 is an interior point of A, the domain of G is obviously a neighbourhood of the origin in $X \times X$, and the crux of the proof lies in showing that

$$G(h, k) / \| (h, k) \|^2 \to 0 \text{ in } Y \text{ as } (h, k) \to (0, 0). \qquad (2)$$

Let $\varepsilon > 0$. Since f is twice-differentiable at x_0, we can find $\delta > 0$ such that $df(x_0 + h)$ is defined and satisfies

$$\| df(x_0 + h) - df(x_0) - d^2f(x_0)h \| \leq \varepsilon \| h \| \tag{3}$$

whenever $\| h \| \leq 2\delta$. Then, if B is the open ball in X with centre 0 and radius δ, the set $B \times B$ is clearly contained in the domain of G. Let $h \in B$, and let $\Phi_h : B \to Y$ be given by

$$\begin{aligned} \Phi_h(k) &= f(x_0 + h + k) - f(x_0 + k) - d^2f(x_0)(h, k) \\ &= f(x_0 + h + k) - f(x_0 + k) - (d^2f(x_0)h)k. \end{aligned}$$

Then for all $h, k \in B$ we have

$$G(h, k) = \Phi_h(k) - \Phi_h(0). \tag{4}$$

We observe now that, by the chain rule (3.1.4) and §3.1, Example (f) (p. 171) (note that $d^2f(x_0)h \in \mathscr{L}(X, Y)$),

$$\begin{aligned} d\Phi_h(k) &= df(x_0 + h + k) - df(x_0 + k) - d^2f(x_0)h \\ &= (df(x_0 + h + k) - df(x_0) - d^2f(x_0)(h + k)) \\ &\quad - (df(x_0 + k) - df(x_0) - d^2f(x_0)k), \end{aligned}$$

and therefore, by (3),

$$\| d\Phi_h(k) \| \leq \varepsilon(\| h + k \| + \| k \|) \leq 3\varepsilon \| (h, k) \|$$

for all $h, k \in B$. Hence, by (4) and the mean value inequality (3.2.2),

$$\| G(h, k) \| \leq 3\varepsilon \| (h, k) \| \, \| k \| \leq 3\varepsilon \| (h, k) \|^2,$$

again for all $h, k \in B$, and this gives (2).

It is now easy to complete the proof of the theorem. Interchanging h and k in (2), and noting that $\| (k, h) \| = \| (h, k) \|$, we see that

$$G(k, h) / \| (h, k) \|^2 \to 0 \text{ in } Y \text{ as } (h, k) \to (0, 0). \tag{5}$$

Let

$$H(h, k) = d^2f(x_0)(h, k) - d^2f(x_0)(k, h).$$

Then H is evidently bilinear and continuous, and, by (2) and (5),

$$H(h, k) / \| (h, k) \|^2 \to 0 \text{ as } (h, k) \to (0, 0).$$

Hence for every $(h, k) \neq (0, 0)$,

$$H(th, tk) / \| (th, tk) \|^2 \to 0 \text{ as } t \to 0 \text{ in } \mathbf{R},$$

and since this last expression is equal to $H(h, k) / \| (h, k) \|^2$, we deduce that $H = 0$.

Suppose now that $r \geq 3$. By (A.2.6), we can identify the space $\mathscr{L}(X, \mathscr{L}(X, \ldots, \mathscr{L}(X, Y)) \ldots)$ with the space $\mathscr{L}_r(X; Y)$ of continuous multilinear functions from X^r into Y, and hence we can regard the element

$(\dots((d^r f(x_0)h_1)h_2)\dots)h_r$ of Y as the value of a continuous multilinear function from X^r into Y at the point $(h_1,\dots,h_r)$ of X^r. As before, we use the same symbol $d^r f(x_0)$ for this multilinear function, so that $d^r f(x_0)$ $(h_1,\dots,h_r) = (\dots((d^r f(x_0)h_1)\dots)h_r$.

We show now that this multilinear function is symmetric, and for this we require the following lemma.

(3.5.6) **Lemma.** *If f is r-times Fréchet differentiable at x_0, where $r \geq 2$, then for all $k_1,\dots,k_{r-1}$ the function $x \mapsto d^{r-1}f(x)(k_1,\dots,k_{r-1})$ is Fréchet differentiable at x_0, and its differential there is $h \mapsto d^r f(x_0)(h,k_1,\dots,k_{r-1})$.*

Let $k_1,\dots,k_{r-1} \in X$. Then the given function is the composite $w \circ d^{r-1}f$, where $w: \mathscr{L}_{r-1}(X;Y) \to Y$ is the evaluation map given by $w(F) = F(k_1,\dots,k_{r-1})$. Clearly w is linear and continuous, and hence, by (3.1.4. Corollary 1), $w \circ d^r f$ is Fréchet differentiable at x_0, and its differential there is $w \circ d^r f(x_0)$, i.e. it is the function

$$h \mapsto w(d^r f(x_0)h) = d^r f(x_0)(h,k_1,\dots,k_{r-1}).$$

(3.5.7) *If f is a function from a set $A \subseteq X$ into Y which is r-times Fréchet differentiable at x_0, then $d^r f(x_0)$ is symmetric, i.e. if σ is a permutation of the numbers $1,\dots,r$, then for all $h_1,\dots,h_r \in X$*

$$d^r f(x_0)(h_{\sigma(1)},\dots,h_{\sigma(r)}) = d^r f(x_0)(h_1,\dots,h_r). \tag{6}$$

We prove this by induction on r, the case $r = 2$ being (3.5.5). Let r be an integer greater than 2 such that the result holds for all functions that are $(r-1)$-times differentiable, and let f be a function which is r-times differentiable at x_0. By the lemma (3.5.6), the function $x \mapsto d^{r-2}f(x)$ $(h_3,\dots,h_r)$ is twice-differentiable at x_0, and its second differential there is the function $(h_1,h_2) \mapsto d^r f(x_0)(h_1,h_2,h_3,\dots,h_r)$. It therefore follows from (3.5.5) that

$$d^r f(x_0)(h_2,h_1,h_3,\dots,h_r) = d^r f(x_0)(h_1,h_2,h_3,\dots,h_r).$$

Next, since f is $(r-1)$-times differentiable on a neighbourhood V of x_0, the induction hypothesis shows that if τ is a permutation of the numbers $2,\dots,r$, then for all $x \in V$ and all $h_2,\dots,h_r \in X$

$$d^{r-1}f(x)(h_{\tau(2)},\dots,h_{\tau(r)}) = d^{r-1}f(x)(h_2,\dots,h_r).$$

Hence, again by the lemma, for all $h_1 \in X$ we have

$$d^r f(x_0)(h_1,h_{\tau(2)},\dots,h_{\tau(r)}) = d^r f(x_0)(h_1,h_2,\dots,h_r).$$

We have thus shown that (6) holds (a) when σ interchanges 1 and 2 only, and (b) when σ leaves 1 fixed. Since every permutation σ is a product of permutations of the types in (a) and (b), the relation (b) therefore holds for all σ, and this completes the proof.

We turn now to the special case of a function from a set $A \subseteq \mathbf{R}^n$ into Y. The case $n = 1$ is particularly simple, for a function ϕ from a set $A \subseteq \mathbf{R}$ into Y is r-times Fréchet differentiable at a point t_0 if and only if ϕ has an $(r-1)$th derivative at each point of some neighbourhood of t_0 and has an rth derivative at t_0, and then $d^r\phi(t_0)$ is the function $(h_1, \ldots, h_r) \mapsto h_1 h_2 \ldots h_r \phi^{(r)}(t_0)$ from $\mathbf{R}^r$ into Y.

For $n \geq 2$, the relation between the differentiability of a function and that of its partial derivatives is less simple, and we begin by proving an extension of (3.4.1). As in §3.4, we denote the natural basis vectors in $\mathbf{R}^n$ by $e_1, \ldots, e_n$.

(3.5.8) *Let $A \subseteq \mathbf{R}^n$, where $n \geq 2$, let $f : A \to Y$ be r-times Fréchet differentiable at x_0, and let $j_1, \ldots, j_r$ be integers between 1 and n inclusive. Then the partial derivative $D_{j_1} D_{j_2} \ldots D_{j_r} f(x_0)$ exists and is equal to $d^r f(x_0)(e_{j_1}, e_{j_2}, \ldots, e_{j_r})$.*

By (3.4.1), the result is true for $r = 1$. Suppose then that r is a positive integer for which the result holds, let f be a function from a subset of $\mathbf{R}^n$ into Y which is $(r+1)$-times differentiable at x_0, and let $j, j_1, \ldots, j_r$ be integers between 1 and n inclusive. Then f is r-times differentiable on some open neighbourhood V of x_0, and hence, by the induction hypothesis, $D_{j_1} \ldots D_{j_r} f(x)$ exists and is equal to $d^r f(x)(e_{j_1}, \ldots, e_{j_r})$ for all $x \in V$. It therefore follows that

$$D_j D_{j_1} \ldots D_{j_r} f(x_0) = \left[\frac{d}{dt} D_{j_1} \ldots D_{j_r} f(x_0 + t e_j) \right]_{t=0}$$
$$= \left[\frac{d}{dt} d^r f(x_0 + t e_j)(e_{j_1}, \ldots, e_{j_r}) \right]_{t=0}.$$

To compute this last derivative, we use (3.1.4. Corollary 3). By (3.5.6), the function $x \mapsto d^r f(x)(e_{j_1}, \ldots, e_{j_r})$ is differentiable at x_0, and its differential there is $h \mapsto d^{r+1} f(x_0)(h, e_{j_1}, \ldots, e_{j_r})$. Since the derivative of $t \mapsto x_0 + t e_j$ at 0 is e_j, we obtain that

$$\left[\frac{d}{dt} d^r f(x_0 + t e_j)(e_{j_1}, \ldots, e_{j_r}) \right]_{t=0} = d^{r+1} f(x_0)(e_j, e_{j_1}, \ldots, e_{j_r}),$$

and this completes the proof.

We deduce immediately the following corollaries.

(3.5.8. Corollary 1) *Under the hypotheses of (3.5.8),*

$$d^r f(x_0)(h_1, h_2, \ldots, h_r) = \sum h_{j_1 1} h_{j_2 2} \ldots h_{j_r r} D_{j_1} D_{j_2} \ldots D_{j_r} f(x_0),$$

where $h_i = (h_{1i}, h_{2i}, \ldots, h_{ni}) \in \mathbf{R}^n$ $(i = 1, \ldots, r)$, and the summation is taken over all distinct r-tuples $(j_1, \ldots, j_r)$ of integers between 1 and n inclusive.

(3.5.8. Corollary 2) *Under the hypotheses of* (3.5.8),

$$D_{k_1} D_{k_2} \cdots D_{k_r} f(x_0) = D_{j_1} D_{j_2} \cdots D_{j_r} f(x_0)$$

for every permutation $k_1, \ldots, k_r$ of $j_1, \ldots, j_r$.

(3.5.9) *Let f be a function from a set $A \subseteq \mathbf{R}^n$ into Y, where $n \geq 2$, and let r be an integer such that $r \geq 2$. Then f is r-times Fréchet differentiable at x_0 if and only if all the partial derivatives of f of order $r - 2$ are Fréchet differentiable on an open neighbourhood V of x_0 and all the partial derivatives of f of order $r - 1$ (which are necessarily defined on V) are Fréchet differentiable at x_0.*

Here the 'only if' follows immediately from (3.5.8) and (3.5.6), so that it remains to prove the 'if'. We give the proof for the case $r = 2$, the argument for the general case being similar.

Suppose then that f is differentiable on an open neighbourhood V of x_0 and that the partial derivatives $D_1 f, \ldots, D_n f$ of f of order 1 are differentiable at x_0. By (3.4.1), $D_1 f, \ldots, D_n f$ are defined on V, and for each $x \in V$ and for each j we have $D_j f(x) = df(x)e_j$. We therefore have to prove that if the functions $x \mapsto df(x)e_j$ $(j = 1, \ldots, n)$ are differentiable at x_0, then df is differentiable at x_0..

Let $\varepsilon > 0$. Then given the differentiability of the functions $x \mapsto df(x)e_j$ at x_0, for each j we can find $T_j \in \mathscr{L}(\mathbf{R}^n, Y)$ and $\delta_j > 0$ such that

$$\| df(x_0 + h)e_j - df(x_0)e_j - T_j h \| \leq \varepsilon \| h \|$$

whenever $\| h \| \leq \delta_j$. If now $k = (k_1, \ldots, k_n) \in \mathbf{R}^n$, then $k = \sum_{j=1}^n k_j e_j$, and therefore, by the linearity of $df(x_0 + h)$ and $df(x_0)$,

$$\left\| df(x_0 + h)k - df(x_0)k - \sum_{j=1}^n k_j T_j h \right\|$$

$$= \left\| \sum_{j=1}^n k_j (df(x_0 + h)e_j - df(x_0)e_j - T_j h) \right\| \leq \varepsilon \| h \| \sum_{j=1}^n |k_j| \leq \varepsilon n^{1/2} \| h \| \| k \|$$

whenever $\| h \| \leq \min \delta_j$. Since the function

$$(h, k) \mapsto \sum_{j=1}^n k_j T_j h$$

is clearly bilinear and continuous, the result therefore follows from (3.5.4).

(3.5.10) *Let f be a function from an open subset A of $\mathbf{R}^n$ into Y, where $n \geq 2$,* *and let p,r be integers such that $0 \leq p < r$. Then the following statements are equivalent*:

(i) f *is* C^r *on* A;

(ii) $d^p f$ *is* C^{r-p} *on* A;

(iii) *all the partial derivatives of f of order r are continuous on A*;

(iv) *all the partial derivatives of f of order p are* C^{r-p} *on* A.

Since $d^r f = d^{r-p}(d^p f)$, (i) and (ii) are obviously equivalent. Also (i) implies (iii), by (3.5.8). Conversely, (iii) implies (i), for if all the partial derivatives of f of order r are continuous on A, then, by (3.4.5), all the partial derivatives of f of order $r-1$ are differentiable on A, and are therefore continuous on A. Hence, again by (3.4.5), the partial derivatives of f of order $r-2$ are differentiable on A, and therefore $d^r f$ is defined on A, by (3.5.9). Moreover, by the formula of (3.5.8. Corollary 1), $d^r f$ is continuous on A, so that f is C^r on A. Next, (iii) implies (iv), for if (iii) holds, then all the partial derivatives of order $r-p$ of the partial derivatives of f of order p are continuous, and (since (iii) implies (i)) this implies (iv). Finally, since (i) implies (iii), we see that (iv) implies (iii), and this completes the proof.

Partial differentials of higher order can be defined in a similar manner to partial derivatives of higher order. Thus let $X_1, \ldots, X_n$ be normed spaces, let $X = X_1 \times X_2 \times \ldots \times X_n$, let f be a function from a set $A \subseteq X$ into Y, and let j, k be integers such that $1 \leq j \leq n$, $1 \leq k \leq n$. Then we define the *partial Fréchet differential of f of the second order with respect to the jth and kth coordinates*, denoted by $d_k d_j f$, to be the partial Fréchet differential of the function $d_j f$ with respect to the kth coordinate, so that $d_k d_j f = d_k(d_j f)$ ($d_k d_j f$ may be the empty function, and is certainly so if $d_j f$ is empty). The value $d_k d_j f(x_0)$ of $d_k d_j f$ at a point x_0 of its domain is an element of $\mathscr{L}(X_k, \mathscr{L}(X_j, Y))$, and, by (A.2.6), can be identified with a continuous bilinear function from $X_k \times X_j$ into Y, i.e. with an element of $\mathscr{L}(X_k, X_j; Y)$. As before, we use the same symbol $d_k d_j f(x_0)$ for this bilinear function, so that $d_k d_j f(x_0)(h_k, h_j) = (d_k d_j f(x_0) h_k) h_j$. When $j = k$ we write d_j^2 in place of $d_k d_j$.

The partial Fréchet differentials of f of higher order are defined inductively, the partial differentials of f of order $r+1$ being the partial differentials of order 1 of the partial differentials of f of order r. If $j_1, \ldots, j_r$ are integers between 1 and n inclusive, then $d_{j_1} d_{j_2} \ldots d_{j_r} f(x_0)$, if it exists, can be identified with an element of $\mathscr{L}(X_{j_1}, X_{j_2}, \ldots, X_{j_r}; Y)$; in particular, $d_j^r f(x_0)$ can be identified with an element of $\mathscr{L}_r(X_j; Y)$.

It is not difficult to obtain analogues of (3.5.8–10) for partial differentials, and these are given as Exercises 3.5.7–9.

Exercises 3.5

1 Let X be a real Hilbert space, let $A \subseteq X$, and let $f : A \to \mathbf{R}$ be a function which is Fréchet differentiable at x_0. Then there exists a unique element $\bar{x}_0 \in X$ such that $df(x_0)h = \langle \bar{x}_0, h \rangle$ for all $h \in X$. This element $\bar{x}_0$ is called the *gradient of f* at x_0, and we denote it by $\nabla f(x_0)$. Equivalently, if I is the isometric isomorphism of X' onto X such that $u(h) = \langle Iu, h \rangle$ for all $u \in X'$ and $h \in X$, then $\nabla f(x_0) = I(df(x_0))$, so that $\nabla f = I \circ df$ (note that ∇f maps a subset of A into X, while df maps the same subset of A into X').

   Prove that
   (i) if $df(x_0) \neq 0$, then the supremum $\| df(x_0) \|$ of $df(x_0)h$ on the unit sphere $\{ h \in Y : \| h \| = 1 \}$ is attained at and only at $h = \nabla f(x_0) / \| df(x_0) \|$,
   (ii) if $X = \mathbf{R}^n$, then $\nabla f(x_0) = (D_1 f(x_0), \ldots, D_n f(x_0))$.
   Show further that
   (iii) f is twice Fréchet differentiable at x_0 if and only if ∇f is Fréchet differentiable at x_0, and that if $Hf = d(\nabla f)$, then for all $h, k \in X$
$$d^2 f(x_0)(h, k) = \langle Hf(x_0)h, k \rangle,$$
   (iv) if $X = \mathbf{R}^n$, then the matrix of $Hf(x_0)$ with respect to the natural basis in $\mathbf{R}^n$ is $[D_i D_j f(x_0)]$ (this matrix is called the *Hessian* of f).
   [Hint. For (ii), note that if $X = \mathbf{R}^n$ then $Iu = (u(e_1), \ldots, u(e_n))$, where $e_1, \ldots, e_n$ is the natural basis for $\mathbf{R}^n$.]

2 (*Leibniz's formula for the differentiation of a product*) With the notation of Exercise 3.4.1, prove that if $\beta = (\beta_1, \ldots, \beta_n) \in Z_n^+$, $|\beta| = \beta_1 + \ldots + \beta_n$, and u, v are $|\beta|$-times Fréchet differentiable real-valued functions on a set $A \subseteq \mathbf{R}^n$, then for all $x \in A$
$$D^\beta (u(x)v(x)) = \sum_{\alpha \leq \beta} \frac{\beta!}{(\beta - \alpha)! \alpha!} (D^\alpha u(x))(D^{\beta - \alpha} v(x)).$$

3 Let f be a function from a subset A of $\mathbf{R}^n$ into the real normed space Y which is twice Fréchet differentiable at x_0. Prove that for $j = 1, \ldots, n$
$$d(D_j f)(x_0) = D_j (df)(x_0).$$

4 Let $X = X_1 \times X_2$, where X_1, X_2 are normed spaces, and let G be a function from an open neighbourhood of 0 in X into Y such that
   (i) $G(h, k) / (\| h \| \| k \|) \to 0$ as $(h, k) \to (0, 0)$ in $(X_1 \backslash \{0\}) \times (X_2 \backslash \{0\})$,
   (ii) for all sufficiently small $k \neq 0$ there exists $L_k \in \mathcal{L}(X_1, Y)$ such that
$$\frac{G(h, k) - L_k h}{\| h \| \| k \|} \to 0 \quad \text{as } h \to 0.$$
   Prove that $L_k / \| k \| \to 0$ in $\mathcal{L}(X_1, Y)$ as $k \to 0$.

5 (*Inversion of mixed partial differentials*) Let $X = X_1 \times X_2$, where X_1, X_2 are normed spaces, and let $Q = (X_1 \backslash \{0\}) \times (X_2 \backslash \{0\})$. Let also $x = (x_1, x_2) \in X$, let W be an open ball in X with centre 0, and let f be a function from a subset of X into Y such that $d_1 f$ exists on $x + W$.
   (i) Prove that if $d_2 d_1 f(x_1 + h, y_1 + k)$ exists for all $(h, k) \in Q \cap W$ and tends to a limit B in $\mathcal{L}(X_2, \mathcal{L}(X_1, Y))$ ($\approx \mathcal{L}(X_2, X_1 ; Y)$) as $(h, k) \to (0, 0)$ in Q, then $d_2 d_1 f(x)$ exists and is equal to B.
   (ii) Prove further that if, in addition to the hypotheses of (i), $d_2 f(x_1 + h, x_2)$ exists for all sufficiently small $h \in X_1$, then $d_1 d_2 f(x)$ exists, and for all $(h, k) \in X$

$$d_1 d_2 f(x)(h, k) = d_2 d_1 f(x)(k, h).$$

[This result shows in particular that if $X = \mathbf{R}^2$, $D_1 f$ and $D_2 D_1 f$ exist on an open neighbourhood of x, $D_2 D_1 f$ is continuous at x, and $D_2 f$ exists on an open segment through x parallel to the x_1-axis, then $D_1 D_2 f(x)$ exists and is equal to $D_2 D_1 f(x)$.

Hint. (i) By a repeated application of (3.2.1) show that if $G(h, k) = f(x_1 + h, x_2 + k) - f(x_1 + h, x_2) - f(x_1, x_2 + k) + f(x_1, x_2) - B(k, h)$,   then $G(h, k)/(\|h\| \|k\|) \to 0$ as $(h, k) \to (0, 0)$ in Q; note that

$$G(h, k) = F_k(x_1 + h) - F_k(h) - B(k, h),$$

where $F_k(z) = f(z, x_2 + k) - f(z, x_2)$. Then use Exercise 4 with

$$L_k h = d_1 f(x_1, x_2 + k)h - d_1 f(x_1, x_2)h - B(k, h).$$

(ii) Use Exercise 4 for an $L_h \in \mathcal{L}(X_2, Y)$.]

6 Let $f : \mathbf{R}^2 \to \mathbf{R}$ be given by

$$f(x, y) = \frac{xy(x^2 - y^2)}{x^2 + y^2} \quad ((x, y) \neq (0, 0)), \quad f(0, 0) = 0.$$

Prove that $D_1 D_2 f(x, y)$ and $D_2 D_1 f(x, y)$ exist for all $(x, y) \in \mathbf{R}^2$, and that $D_1 D_2 f(0, 0) \neq D_2 D_1 f(0, 0)$.

7 Let $X_1, \dots, X_n$ be normed spaces, where $n \geq 2$, let $X = X_1 \times X_2 \times \dots \times X_n$, and for $j = 1, \dots, n$ let $I_j : X_j \to X$ be the insertion map defined in §3.3. Let also $A \subseteq X$, and let $f : A \to Y$ be r-times Fréchet differentiable at x_0.

Prove that

(i) if $j_1, \dots, j_r$ are integers between 1 and n inclusive, then the partial differential $d_{j_1} d_{j_2} \dots d_{j_r} f(x_0)$ exists and is equal to $d^r f(x_0) \circ I$, where I is the continuous linear map from $X_{j_1} \times \dots \times X_{j_r}$ into X^r given by

$$I(h_{j_1}, \dots, h_{j_r}) = (I_{j_1} h_{j_1}, \dots, I_{j_r} h_{j_r}),$$

(ii) for every permutation $k_1, \dots, k_r$ of $j_1, \dots, j_r$,

$$d_{k_1} d_{k_2} \dots d_{k_r} f(x_0) = d_{j_1} d_{j_2} \dots d_{j_r} f(x_0),$$

(iii) for all $h_1, \dots, h_r \in X$,

$$d^r f(x_0)(h_1, \dots, h_r) = \sum d_{j_1} d_{j_2} \dots d_{j_r} f(x_0) (h_{1 j_1}, h_{2 j_2}, \dots, h_{r j_r}),$$

where the sum on the right is taken over all distinct r-tuples $j_1, \dots, j_r$ of integers between 1 and n inclusive, and $h_i = (h_{i1}, \dots, h_{in})$ $(i = 1, \dots, r)$.

8 Let $X_1, \dots, X_n$ be normed spaces, where $n \geq 2$, let $X = X_1 \times X_2 \times \dots \times X_n$, and let r be an integer greater than 1. Prove that a function f from a subset of X into Y is r-times Fréchet differentiable at x_0 if and only if all the partial differentials of f of order $r - 2$ are Fréchet differentiable on an open neighbourhood V of x_0 and all the partial differentials of f of order $r - 1$ (which are necessarily defined on V) are Fréchet differentiable at x_0.

9 Let $X_1, \dots, X_n$ be normed spaces, where $n \geq 2$, let $X = X_1 \times X_2 \times \dots \times X_n$, let f be a function from an open subset A of X into Y, and let p, r be integers such that $0 \leq p < r$. Prove that the following statements are equivalent:

(i) f is C^r on A;

(ii) all the partial differentials of f of order r are continuous on A;

(iii) all the partial differentials of f of order p are C^{r-p} on A.

3.6 Taylor's theorem for Fréchet differentiable functions

We can obtain various versions of Taylor's theorem for Fréchet differentiable functions by using the results of §1.8 together with the following lemma.

(3.6.1) Lemma. *Let $x_0, h \in X$, let f be a function from a set $A \subseteq X$ into Y, and let $\psi(t) = f(x_0 + th)$ $(t \in \mathbf{R}, x_0 + th \in A)$. If f is r-times Fréchet differentiable at the point $z_0 = x_0 + t_0 h$, where $t_0 \in \mathbf{R}$, then ψ has an rth derivative at t_0 equal to $d^r f(z_0)(h)^r$, where $(h)^r$ denotes the point of X^r with each coordinate equal to h.*

The proof is similar to that of (3.5.8). By (3.1.4. Corollary 3), the result is true for $r = 1$. Let r be a positive integer for which the result holds, and let f be $(r + 1)$-times differentiable at z_0. Then the set E of real t such that f is r-times differentiable at $x_0 + th$ is a neighbourhood of t_0 in $\mathbf{R}$, and, by the induction hypothesis, $\psi^{(r)}(t)$ exists and is equal to $d^r f(z_0)(h)^r$ for all $t \in E$. The restriction of $\psi^{(r)}$ to E is therefore the composite of the functions $x \mapsto d^r f(x)(h)^r$ and $t \mapsto x_0 + th$. By (3.5.6), the first of these two functions is differentiable at $x = z_0$, and its differential there is $k \mapsto d^{r+1} f(z_0)(k, h, \ldots, h)$, while the second function has derivative h. Hence, again by (3.1.4. Corollary 3),

$$\psi^{(r+1)}(t_0) = \left[\frac{d}{dt}\psi^{(r)}(t)\right]_{t=t_0} = d^{r+1}f(z_0)(h)^{r+1},$$

and the proof of the induction step is complete.

(3.6.2) *If f is a function from a set $A \subseteq X$ into Y which is r-times Fréchet differentiable at the point x_0, then*

$$f(x_0 + h) = f(x_0) + df(x_0)h + \frac{1}{2!}d^2f(x_0)(h)^2 + \ldots + \frac{1}{r!}d^rf(x_0)(h)^r + o(\|h\|^r)$$

as $h \to 0$ in X.

Since f is r-times differentiable at x_0, we can find an open ball B in X with centre x_0 and radius η such that f is $(r-1)$-times differentiable on B. Let $x_0 + h \in B$, where $h \neq 0$, and let ψ be defined as in the lemma. Since $x_0 + th \in B$ for all $t \in [0, 1]$, the domain of ψ contains the interval $[0, 1]$, and ψ has an $(r-1)$th derivative at each point of this interval and an rth derivative at 0. By (1.8.1) with $a = 0$, $b = 1$, and q the norm function on Y, there exists (uncountably many) $\xi \in]0, 1[$ for which

$$r! \left\| f(x_0 + h) - f(x_0) - df(x_0)h - \ldots - \frac{1}{r!} d^rf(x_0)(h)^r \right\|$$

$$\leq \left\| \frac{d^{r-1}f(x_0 + \xi h)(h)^{r-1} - d^{r-1}f(x_0)(h)^{r-1}}{\xi} - d^rf(x_0)(h)^r \right\|. \quad (1)$$

But since f is r-times differentiable at x_0, given $\varepsilon > 0$ we can find δ satisfying $0 < \delta < \eta$ such that

$$\| d^{r-1}f(x_0 + k) - d^{r-1}f(x_0) - d^r f(x_0)k \| \le \varepsilon \| k \|$$

whenever $\| k \| \le \delta$. This implies that the expression on the right of (1) does not exceed $\varepsilon \| h \|^r$ whenever $\| h \| \le \delta$, and this is the required result.

(3.6.3) *Let $p > 0$, let f be a function from a subset of X into Y, and suppose that f is $(r - 1)$-times Fréchet differentiable at each point of the closed segment S in X with endpoints $x_0, x_0 + h$, and is r-times Fréchet differentiable at nearly every point of S. Then there exist uncountably many $\xi \in [0,1]$ such that*

$$\Big\| f(x_0 + h) - f(x_0) - df(x_0)h - \frac{1}{2!}d^2 f(x_0)(h)^2 - \ldots$$

$$- \frac{1}{(r-1)!}d^{r-1}f(x_0)(h)^{r-1} \Big\|$$

$$\le \frac{(1 - \xi)^{r-p}}{p(r-1)!} \| d^r f(x_0 + \xi h)(h)^r \|.$$

This is an immediate consequence of the lemma (3.6.1) and (1.8.3).

We end this section with some applications of (3.6.2) to local minima and maxima.

Let f be a real-valued function with domain $A \subseteq X$.† We say that a point $x_0 \in A$ is a *local minimum* of f if there exists a neighbourhood V of x_0 in X such that $f(x) \ge f(x_0)$ for all $x \in (A \cap V)$; if strict inequality holds except at x_0, then x_0 is a *strict local minimum*. A *local maximum* is defined similarly.

(3.6.4) *If f is a real-valued function with domain $A \subseteq X$, x_0 is a local minimum of f, and f is Fréchet differentiable at x_0, then $df(x_0) = 0$. If in addition f is twice Fréchet differentiable at x_0, then also $d^2 f(x_0)(h)^2 \ge 0$ for all $h \in X$.*

To prove the first part, suppose on the contrary that there exists $h \in X$ such that $df(x_0)h \ne 0$, and let $\psi(t) = f(x_0 + th)$ $(t \in \mathbf{R}, x_0 + th \in A)$. Then $\psi'(0) = df(x_0)h \ne 0$, and by (1.3.1) this contradicts the hypothesis that x_0 is a local minimum of f.

Next, suppose that f is twice differentiable at x_0, that $df(x_0) = 0$, and that there exists $h \in X$ such that $d^2 f(x_0)(h)^2 = -\alpha < 0$. By (3.6.2)

$$f(x_0 + th) = f(x_0) - \tfrac{1}{2}t^2\alpha + o(t^2)$$

† The requirement that f should be differentiable then implies that X is over the real field.

as $t \to 0$, so that $f(x_0 + th) < f(x_0)$ for all sufficiently small non-zero t, and again this contradicts the hypothesis that x_0 is a local minimum of f.

It is easy to see that the condition that $d^2f(x_0)(h)^2 \geq 0$ does not imply that x_0 is a local minimum of f. However, in this direction we have:

(3.6.5) *If f is a real-valued function which is twice Fréchet differentiable at x_0, $df(x_0) = 0$, and there exists $c > 0$ such that $d^2f(x_0)(h)^2 \geq c\|h\|^2$ for all $h \in X$, then x_0 is a strict local minimum of f.*
    By (3.6.2),

$$f(x_0 + h) = f(x_0) + \tfrac{1}{2}d^2f(x_0)(h)^2 + o(\|h\|^2)$$
$$\geq f(x_0) + \|h\|^2(\tfrac{1}{2}c + o(1)),$$

and this obviously implies that x_0 is a strict local minimum of f.

(3.6.5. Corollary) *Let X be finite-dimensional, let f be a real-valued function with domain $A \subseteq X$ which is twice Fréchet differentiable at x_0, and suppose that $df(x_0) = 0$ and that $d^2f(x_0)(h)^2 > 0$ for all non-zero $h \in X$. Then x_0 is a strict local minimum of f.*
    Here the unit sphere S in X is compact. Since $x \mapsto d^2f(x_0)(x)^2$ is continuous, it is bounded below on S, and attains its infimum c there, so that $c > 0$. If now h is a non-zero element of X and $x = h/\|h\|$, then

$$d^2f(x_0)(h)^2 = \|h\|^2 d^2f(x_0)(x)^2 \geq c\|h\|^2,$$

and the result therefore follows from the main theorem.

If $X = \mathbf{R}^n$, then by (3.5.8. Corollary 1), $d^2f(x_0)(h)^2 > 0$ for all $h \neq 0$ if and only if for all $h = (h_1, \ldots, h_n) \neq 0$

$$\sum_{i,j=1}^{n} h_i h_j D_i D_j f(x_0) > 0.$$

Exercises 3.6

1 Let $p > 0$, let f be a function from a subset of X into $\mathbf{R}$, and suppose that f is $(r - 1)$-times Fréchet differentiable at each point of the closed segment S in X with endpoints $x_0, x_0 + h$ and is r-times Fréchet differentiable at nearly every point of S. Prove that there exist uncountably many $\xi \in [0, 1]$ such that

$$f(x_0 + h) = f(x_0) + df(x_0)h + \frac{1}{2!}d^2f(x_0)(h)^2 + \ldots + \frac{1}{(r-1)!}d^{r-1}f(x_0)(h)^{r-1}$$
$$+ \frac{(1 - \xi)^{r-p}}{p(r-1)!}d^r f(x_0 + \xi h)(h)^r.$$

2 Formulate and prove a result analogous to (3.6.3) in which the norm function is replaced by a quasi-convex function.

3 Let X be a normed space and Y a Banach space, let A be an open subset of X, and let $f : A \to Y$ be C^r. Prove that if the segment joining x_0 and $x_0 + h$ is in A, then

$$f(x_0 + h) = f(x_0) + df(x_0)h + \frac{1}{2!}d^2f(x_0)(h)^2 + \ldots + \frac{1}{(r-1)!}d^{r-1}f(x_0)(h)^{r-1}$$

$$+ \frac{1}{(r-1)!}\left(\int_0^1 (1-t)^{r-1}d^rf(x+th)dt \right)(h)^r.$$

[Hint. Cf. Exercises 1.9.3, 4.]

4 Let A be an open convex set in the real normed space X, and let $f : A \to \mathbf{R}$ be twice Fréchet differentiable on A. Prove that f is convex if and only if $d^2f(x)(h)^2 \geq 0$ for all $x \in A$ and $h \in X$.

[Hint. Use Exercise 3.1.8.]

5 Let X be a real Hilbert space, let $A \in \mathcal{L}(X, X)$ be self-adjoint, and let $F : X \to \mathbf{R}$ be given by $F(x) = \langle Ax, x \rangle$. Prove that F is Fréchet differentiable at each $x \in X$ and that $dF(x)h = 2\langle Ax, h \rangle$ $(h \in X)$. Show also that $F(x) > 0$ for all non-zero $x \in X$ if and only if A is one-to-one.

[Hint. For the 'if' use the fact that if there exists a non-zero x_0 with $F(x_0) = 0$ then F has a minimum at x_0.]

3.7 The inverse function theorem

Let $A \subseteq X$, and let $f : A \to Y$ be Fréchet differentiable at the point x_0. Then the function

$$x \mapsto f(x_0) + df(x_0)(x - x_0) \qquad (x \in X) \tag{1}$$

provides an approximation to f in the vicinity of x_0, in the sense explained in §3.1. If $df(x_0) \in \mathcal{LH}(X, Y)$, then the function (1) is a homeomorphism of X onto Y, and it is natural to ask whether the restriction of f to some neighbourhood V of x_0 is a homeomorphism of V onto $f(V)$. The answer is in general negative, as is easily seen by considering the function defined in the footnote to (1.3.1). However, the result in question is true if X, Y are complete, f is differentiable on some neighbourhood of x_0, and df is continuous at x_0. Moreover, under these conditions further differentiability properties of f are reflected in properties of the inverse of the restriction. This result, (3.7.3) below, is among the most important theorems of the differential calculus.

We prove first two lemmas.

(3.7.1) Lemma. *Let X, Y be Banach spaces, let $T \in \mathcal{LH}(X, Y)$, and let $M = \| T^{-1} \|$. Let also $0 < \varepsilon < 1$, let A be an open set in X, and let $f : A \to Y$ be a function such that*

$$\| f(x) - f(x') - T(x - x') \| \leq \frac{\varepsilon}{M}\| x - x' \| \tag{2}$$

for all $x, x' \in A$. *Then* f *is a homeomorphism of* A *onto* $f(A)$, *and* $f(A)$ *is open in* Y. *Further, if* $A = X$ *then* $f(A) = Y$.

Since $\|z\|/M \leq \|Tz\| \leq \|T\| \|z\|$ for all $z \in X$, we deduce from (2) that for all $x, x' \in A$

$$(1 - \varepsilon)\|x - x'\|/M \leq \|f(x) - f(x')\| \leq (\|T\| + \varepsilon/M)\|x - x'\|. \quad (3)$$

The right-hand inequality here implies that f is continuous, and the left-hand inequality implies that f is one-to-one and that f^{-1} is continuous, so that f is a homeomorphism of A onto $f(A)$.

To prove that $f(A)$ is open in Y, let $\bar{y} \in f(A)$, and let $\bar{x}$ be the point of A such that $f(\bar{x}) = \bar{y}$. Let also C be a closed ball in X with centre $\bar{x}$ contained in A, let α be the radius of C, let D be the closed ball in Y with centre $\bar{y}$ and radius $\alpha(1 - \varepsilon)/M$, let $y \in D$, and for x in C let $h(x) = x - T^{-1}(f(x) - y)$. Then h maps C into itself, since

$$\|h(x) - \bar{x}\| = \|T^{-1}(y - \bar{y} + f(\bar{x}) - f(x) - T(\bar{x} - x))\|$$
$$\leq M\|y - \bar{y}\| + \varepsilon\|x - \bar{x}\| \leq (1 - \varepsilon)\alpha + \varepsilon\alpha = \alpha.$$

Further, h is a contraction with constant ε, for if $x, x' \in C$ then

$$\|h(x) - h(x')\| = \|T^{-1}(f(x') - f(x) - T(x' - x))\| \leq \varepsilon\|x - x'\|.$$

It follows now from the contraction mapping principle (A.1.6) that there exists a unique $x \in C$ such that $h(x) = x$, i.e. such that $f(x) = y$. Hence $D \subseteq f(C) \subseteq f(A)$, so that $f(A)$ is open in Y. Further, if $A = X$ we can take α to be arbitrarily large, whence $f(A) = Y$.

In the special case where $f \in \mathscr{L}(X, Y)$ we have the following useful result.

(3.7.1. **Corollary 1**) *Let* X, Y *be Banach spaces, let* $T \in \mathscr{LH}(X, Y)$, *let* $M = \|T^{-1}\|$, *and let* S *be an element of* $\mathscr{L}(X, Y)$ *such that* $\|S - T\| < 1/M$. *Then* $S \in \mathscr{LH}(X, Y)$ *and*

$$\|S^{-1}\| \leq M/(1 - M\|S - T\|). \quad (4)$$

Taking $A = X$, $f = S$ in the main theorem, we see that (1) holds with $\varepsilon = M\|S - T\|$, and therefore $S \in \mathscr{LH}(X, Y)$. Further, the left-hand inequality in (3) implies that $\|S^{-1}\| \leq M/(1 - \varepsilon)$, and this gives (4).

By further specialization, we have also

(3.7.1 **Corollary 2**) *Let* X *be a Banach space, let* L *be an element of* $\mathscr{L}(X, X)$ *such that* $\|L\| < 1$, *and let* 1_X *be the identity function on* X. *Then* $1_X + L \in \mathscr{LH}(X, X)$,

$$\|(1_X + L)^{-1}\| \leq 1/(1 - \|L\|),$$

and

$$\|(1_X + L)^{-1} - 1_X + L\| \le \|L\|^2/(1 - \|L\|). \qquad (5)$$

The first two statements are immediate consequences of Corollary 1 with $Y = X, S = 1_X + L, T = 1_X$. To prove (5) we observe that

$$(1_X + L)(1_X - L) = 1_X - L^2$$

(where $L^2 = L \circ L$). Hence

$$(1_X + L)^{-1} - 1_X + L = (1_X + L)^{-1}(1_X - (1_X - L^2)),$$

and since $\|L^2\| \le \|L\|^2$, this gives (5).

(3.7.2) **Lemma.** *Let X, Y be Banach spaces, and let $\mathscr{H}$ denote the set $\mathscr{LH}(X, Y)$. Then $\mathscr{H}$ is open in $\mathscr{L}(X, Y)$. Further, if $\mathscr{H}$ is non-empty, and I is the function $T \mapsto T^{-1}$ from $\mathscr{H}$ into $\mathscr{L}(Y, X)$, then I is C^∞ on $\mathscr{H}$, and for $T \in \mathscr{H}$, $dI(T)$ is the function $U \mapsto -T^{-1} \circ U \circ T^{-1}$ from $\mathscr{L}(X, Y)$ into $\mathscr{L}(Y, X)$.*

Let $T \in \mathscr{H}$. Then, by (3.7.1. Corollary 1), $\mathscr{H}$ contains the open ball B in $\mathscr{L}(X, Y)$ with centre T and radius $1/\|T^{-1}\|$, so that $\mathscr{H}$ is open in $\mathscr{L}(X, Y)$. Further, if U is an element of $\mathscr{L}(X, Y)$ such that $T + U \in B$, then $\|T^{-1} \circ U\| \le \|T^{-1}\| \|U\| < 1$, so that $1_X + T^{-1} \circ U \in \mathscr{LH}(X, X)$. Since

$$(T + U)^{-1} - T^{-1} + T^{-1} \circ U \circ T^{-1}$$
$$= (1_X + T^{-1} \circ U)^{-1} \circ T^{-1} - T^{-1} + T^{-1} \circ U \circ T^{-1}$$
$$= ((1_X + T^{-1} \circ U)^{-1} - 1_X + T^{-1} \circ U) \circ T^{-1},$$

it follows from (5) that

$$\|(T + U)^{-1} - T^{-1} + T^{-1} \circ U \circ T^{-1}\|$$
$$\le \|T^{-1}\|^3 \|U\|^2/(1 - \|T^{-1}\| \|U\|),$$

and this implies that I is differentiable at T and that $dI(T)U = -T^{-1} \circ U \circ T^{-1}$.

Now let $\mathscr{V} = \mathscr{L}(X, Y)$, $\mathscr{W} = \mathscr{L}(Y, X)$. Then $dI = F \circ G \circ I$, where $G: \mathscr{W} \to \mathscr{W} \times \mathscr{W}$ is given by $G(S) = (S, S)$, and $F: \mathscr{W} \times \mathscr{W} \to \mathscr{L}(\mathscr{V}, \mathscr{W})$ is the function whose value at (S_1, S_2) is the function $U \mapsto -S_1 \circ U \circ S_2$. Clearly G is linear and continuous, and F is bilinear and continuous, so that F and G are C^∞. By (3.5.3), $F \circ G$ is therefore C^∞, and a further application of (3.5.3) now shows that if I is r-times differentiable on $\mathscr{H}$, then so is dI, whence I is $(r + 1)$-times differentiable on $\mathscr{H}$. Hence I is C^∞ on $\mathscr{H}$, and this completes the proof.

(3.7.3) (The inverse function theorem) *Let X, Y be Banach spaces, let f be* *a function from a set $A \subseteq X$ into Y, let x_0 be an interior point of A, and suppose that the Fréchet differential df of f is defined on a neighbourhood of x_0 and is continuous at x_0, and that $df(x_0) \in \mathscr{L}\mathscr{H}(X, Y)$. Then we can find an open neighbourhood V of x_0 in X, contained in A, such that*

(i) *for each $x \in V$ the function f is Fréchet differentiable at x, and $df(x) \in \mathscr{L}\mathscr{H}(X, Y)$,*

(ii) *the restriction F of f to V is a homeomorphism of V onto $f(V)$, and $f(V)$ is open in Y,*

(iii) *the inverse F^{-1} of F is Fréchet differentiable at each point $y \in f(V)$, and $dF^{-1}(y) = df(x)^{-1}$, where $x = F^{-1}(y)$,*

(iv) *dF^{-1} is continuous at the point $y_0 = f(x_0)$,*

(v) *if f is r-times Fréchet differentiable at a point $x \in V$, where $r \geq 2$, then F^{-1} is r-times Fréchet differentiable at $y = f(x)$,*

(vi) *if f is C^r on an open set $E \subseteq V$, F^{-1} is C^r on $f(E)$.*

Let $K = 1/\|df(x_0)^{-1}\|$. Since df is defined on a neighbourhood of x_0 and is continuous at x_0, we can find an open ball V in X with centre x_0 such that $df(x)$ exists and satisfies $\|df(x) - df(x_0)\| \leq \frac{1}{2}K$ for all $x \in V$. By (3.7.1. Corollary 1), this obviously implies (i). Further, by the mean value inequality (3.2.2) applied to $f - df(x_0)$, we have

$$\|f(x) - f(x') - df(x_0)(x - x')\| \leq \tfrac{1}{2}K\|x - x'\|$$

for all $x, x' \in V$, and by (3.7.1), this implies (ii).

Next, (iii) is an immediate consequence of (i), (ii), and (3.1.5). Moreover, by the formula of (iii),

$$dF^{-1} = I \circ df \circ F^{-1}, \tag{6}$$

where I is the function $T \mapsto T^{-1}$ of (3.7.2). Since F^{-1} is continuous at y_0, df is continuous at x_0, and I is C^∞, it follows that dF^{-1} is continuous at y_0, and this proves (iv). There remain now only (v) and (vi), and these follow from (6) by an obvious induction argument.

If $X = \mathbf{R}^n$, then the dimension of Y is necessarily n. In particular, if $Y = \mathbf{R}^n$, and $f_1, \ldots, f_n$ are the components of f, the condition that $df(x_0)$ is a linear homeomorphism of X onto Y is equivalent to the condition that the Jacobian matrix $[D_j f_i(x_0)]$ of f at x_0 is non-singular.

Exercises 3.7

1 Let X be a Hilbert space, and let $\alpha > 0$. Prove that
(i) if T is a continuous linear function from X into itself such that $\langle Th, h \rangle \geq \alpha\|h\|^2$ for all $h \in X$, then $T \in \mathscr{L}\mathscr{H}(X, X)$,

(ii) if f is a continuous function from X into itself such that

$$\langle f(b) - f(a), b - a \rangle \geq \alpha \| b - a \|^2 \qquad (*)$$

for all $a, b \in X$, then f is a homeomorphism of X onto a closed subset of X,
(iii) if f is a C^1 function from X into itself such that $\langle df(x)h, h \rangle \geq \alpha \| h \|^2$ for all $x, h \in X$, then $(*)$ holds for all $a, b \in X$, f is a homeomorphism of X onto itself, and f^{-1} is C^1.

[Hints. (i) Use the given condition to show that T is one-to-one and that T^{-1} is continuous (at 0), and hence deduce that the range Z of T is closed in X. There then exists a unit vector x such that $\langle z, x \rangle = 0$ for all $z \in Z$, and this gives a contradiction.
(ii) To prove $(*)$ apply the mean value theorem to

$$t \mapsto \langle f(\sigma(t)), b - a \rangle - \alpha \langle \sigma(t), b - a \rangle,$$

where $\sigma(t) = (1 - t)a + tb$. Then, by (ii), f is a homeomorphism of X onto a closed subset R of X. Now use (i) and the inverse function theorem to show that R is open.]
2 Give an alternative proof of (3.7.1. Corollary 2) by showing that if

$$S_n = \sum_{\nu=0}^{n} (-1)^\nu L^\nu \qquad (n = 0, 1, \ldots),$$

where $L^0 = 1_X, L^1 = L$, and $L^\nu = L \circ L^{\nu-1}$, then (S_n) is a Cauchy sequence in $\mathscr{L}(X, X)$ and its limit S satisfies the relation $(1_X + L) \circ S = 1_X$. Deduce the result of (3.7.1. Corollary 1).

3.8 The implicit function theorem

Let X, Y be Banach spaces, let $Z = X \times Y$, let $A \subseteq Z$, and let $F : A \to Y$ be Fréchet differentiable at the point $z_0 = (x_0, y_0)$. Let also $F(z_0) = c$, and consider the equation

$$F(x, y) = c \qquad (1)$$

for (x, y) near (x_0, y_0). Since the function

$$(x, y) \mapsto c + dF(z_0)(x - x_0, y - y_0)$$

provides an approximation to F near z_0, a first approximation to the equation (1) is

$$dF(z_0)(x - x_0, y - y_0) = 0,$$

or equivalently (cf. (3.3.1)),

$$d_1 F(z_0)(x - x_0) + d_2 F(z_0)(y - y_0) = 0. \qquad (2)$$

This last equation certainly has a solution for y in terms of x if $d_2 F(z_0)$ is a linear homeomorphism of Y onto itself, and we might expect that this condition, together with the continuity of dF at z_0,† will imply the existence in some neighbourhood of x_0 of a unique differentiable solution $y = f(x)$ of the equation (1) such that $y_0 = f(x_0)$. The implicit function theorem shows that these conditions do in fact suffice.

† This extra condition is the analogue of that in the inverse function theorem.

(3.8.1) (The implicit function theorem) *Let X, Y be Banach spaces, let $Z = X \times Y$, let A be an open set in Z, and let $F : A \to Y$ be Fréchet differentiable on A. Let also $z_0 = (x_0, y_0) \in A$, let $F(z_0) = c$, and suppose that the Fréchet differential dF of F is continuous at z_0 and that the partial differential $d_2 F(z_0)$ of F at z_0 belongs to $\mathscr{L}\mathscr{H}(Y, Y)$. Then there exist an open neighbourhood V of x_0 in X, an open neighbourhood W of z_0 in Z, and a Fréchet differentiable function $f : V \to Y$, such that $f(x_0) = y_0$ and that, for each $x \in V$,*

$$(x, f(x)) \in W, \quad F(x, f(x)) = c,$$

and $y = f(x)$ is the only solution of the equation $F(x, y) = c$ for which $(x, f(x)) \in W$. Further, $d_2 F(z) \in \mathscr{L}\mathscr{H}(Y, Y)$ for all $z \in W$,

$$df(x) = -d_2 F(\bar{z})^{-1} \circ d_1 F(\bar{z})$$

for all $x \in V$, where $\bar{z} = (x, f(x))$, and df is continuous at x_0. Moreover, if F is r-times Fréchet differentiable at $(x, f(x))$, where $x \in V$ and $r \geq 2$, then f is r-times Fréchet differentiable at x, and if F is C^r on A, where $r \geq 1$, then f is C^r on V. Also, f is unique in the sense that if V^* is any connected open neighbourhood of x_0 contained in V and $f^* : V^* \to Y$ is a continuous function such that $f^*(x_0) = y_0$ and that $F(x, f^*(x)) = c$ for all $x \in V^*$, then f^* agrees with f on V^*.

We require the following lemma.

(3.8.2) Lemma. *Let X, Y be normed spaces, let $Z = X \times Y$, let $T_1 \in \mathscr{L}(X, Y)$, $T_2 \in \mathscr{L}(Y, Y)$, and $T \in \mathscr{L}(Z, Z)$ be given by*

$$T(h, k) = (h, T_1 h + T_2 k) \qquad ((h, k) \in Z).$$

Then $T \in \mathscr{L}\mathscr{H}(Z, Z)$ if and only if $T_2 \in \mathscr{L}\mathscr{H}(Y, Y)$.

Suppose first that $T_2 \in \mathscr{L}\mathscr{H}(Y, Y)$. Then T is one-to-one and onto Z, for if $(\bar{h}, \bar{k}) \in Z$, the equation $T(h, k) = (\bar{h}, \bar{k})$ is equivalent to the pair of equations $h = \bar{h}, T_1 h + T_2 k = \bar{k}$, and this pair of equations has the unique solutions

$$h = \bar{h}, k = T_2^{-1}(\bar{k} - T_1 h) = T_2^{-1}\bar{k} - T_2^{-1}T_1\bar{h}. \tag{3}$$

Since the function $(\bar{h}, \bar{k}) \mapsto (h, k)$ defined by the formulae (3) is clearly continuous on Z, it follows that $T \in \mathscr{L}\mathscr{H}(Z, Z)$.

Conversely, suppose that $T \in \mathscr{L}\mathscr{H}(Z, Z)$. Since T preserves the first coordinate of its argument, the function $k \mapsto T(0, k) = (0, T_2 k)$ is a linear homeomorphism of Y onto the subspace $\{0\} \times Y$ of Z, whence $T_2 \in \mathscr{L}\mathscr{H}(Y, Y)$.

Consider now the proof of (3.8.1). Our argument depends on the follow-

ing observation. If G is a function of the form
$$G(z) = G(x,y) = (x, F(x,y)), \tag{4}$$
and G is one-to-one, then, since G preserves the first coordinate of its argument, so does G^{-1}, and therefore G^{-1} is of the form
$$G^{-1}(x,y) = (x, H(x,y)). \tag{5}$$
It follows that for all (x,y) in the domain of G^{-1} we have
$$(x,y) = G(G^{-1}(x,y)) = G(x, H(x,y)) = (x, F(x, H(x,y))),$$
and on equating the second coordinates we deduce that
$$F(x, H(x,y)) = y. \tag{6}$$
In particular, this gives
$$F(x, H(x,c)) = c, \tag{7}$$
so that $y = H(x,c)$ is a solution of the equation $F(x,y) = c$.

We have assumed here that G is one-to-one, and this will not in general be true if F is the function of the theorem and the domain of G is taken to be the set A. However, the hypotheses of the theorem ensure that the function G given by (4) satisfies the conditions of the inverse function theorem at z_0, and hence we can find an open neighbourhood W of z_0 in Z such that the restriction of G to W is one-to-one. The preceding argument can then be applied to this restriction, and the required function f is given by $f(x) = H(x,c)$.

The full argument largely consists of patient verification, and the details are as follows. Let $G : A \to Z$ be given by (4). Then $G(z_0) = (x_0, c)$. Also, by §3.1, Examples (c) and (g) (pp. 169, 173), G is differentiable on A, and for each $z \in A$ and all $(h, k) \in Z$
$$dG(z)(h, k) = (h, dF(z)(h, k)).$$
In particular, this implies that dG is continuous at z_0. Further, since
$$dF(z)(h, k) = d_1 F(z)h + d_2 F(z)k,$$
and $d_2 F(z_0) \in \mathscr{LH}(Y, Y)$, it follows from the lemma (3.8.2) that $dG(z_0) \in \mathscr{LH}(Z, Z)$. Moreover, if F is r-times differentiable at any point $z \in A$, where $r \geq 2$, or is C^r on A, where $r \geq 1$, then by (3.5.2. Corollary), G is respectively r-times differentiable at z or C^r on A.

We have now shown that G satisfies the conditions of the inverse function theorem at z_0, and hence we can find an open neighbourhood W of z_0, contained in A, such that

(a) for each $z \in W$, $dG(z) \in \mathscr{LH}(Z, Z)$, whence (by (3.8.2)) $d_2 F(z) \in \mathscr{LH}(Y, Y)$,

(b) the restriction of G to W is a homeomorphism of W onto the set $U = G(W)$, and U is open in Z,

(c) the inverse of this restriction, which for simplicity we denote by G^{-1}, is differentiable at each $\bar{z} \in U$, and dG^{-1} is continuous at the point $G(z_0) = (x_0, c)$,

(d) if G is r-times differentiable at a point $z \in W$, where $r \geq 2$, or is C^r on A, where $r \geq 1$, G^{-1} is respectively r-times differentiable at $\bar{z} = G(z)$ or C^r on U.

Next, let $H : U \to Y$ be given by (5), so that $H(x_0, c) = y_0$, and H satisfies (6) for all $(x, y) \in U$. By (c), H is differentiable on U, and dH is continuous at (x_0, c). Also, by (d), if F is r-times differentiable at a point $z \in W$, or is C^r on A, then H is respectively r-times differentiable at $\bar{z} = G(z)$ or C^r on U.

Now let $V = \{x \in X : (x, c) \in U\}$, and let $f : V \to Y$ be given by $f(x) = H(x, c)$. Since U is open in Z, V is open in X, and obviously $x_0 \in V, f(x_0) = H(x_0, c) = y_0$, and $(x, f(x)) \in W$ and $F(x, f(x)) = c$ for all $x \in V$. Moreover, for each $x \in V$, $y = f(x)$ is the only solution $F(x, y) = c$ for which $(x, y) \in W$, for if $(x, y), (x, y') \in W$ and $F(x, y) = F(x, y')$, then $G(x, y) = G(x, y')$, whence $y = y'$, since the restriction of G to W is one-to-one.

Next, since f is the composite of H and the function $x \mapsto (x, c)$ (which is the sum of the function $x \mapsto (x, 0)$ in $\mathscr{L}(X, Z)$ and the constant function $x \mapsto (0, c)$), f is differentiable on V and df is continuous at x_0, and f has the higher differentiability properties specified in the theorem. Further, since $F(x, f(x)) = c$ for all $x \in V$, the chain rule gives

$$d_1 F(z) + d_2 F(z) \circ df(x) = 0,$$

where $z = (x, f(x))$, and therefore $df(x) = -d_2 F(z)^{-1} \circ d_1 F(z)$.

It remains only to prove that f is unique in the sense stated. Let V^* be a connected open neighbourhood of x_0 contained in V, let $f^* : V^* \to Y$ be a continuous function such that $f^*(x_0) = y_0$ and that $F(x, f^*(x)) = c$ for all $x \in V^*$, and let $E = \{x \in V^* : f^*(x) = f(x)\}$. We have to show that $E = V^*$, and since V^* is connected, and E is non-empty and closed in V^* (note that $x_0 \in E$ and that f and f^* are continuous), it is enough to show that E is open. Let $x^* \in E$. Then $(x^*, f^*(x^*)) = (x^*, f(x^*)) \in W$, and since W is open in Z and the function $x \mapsto (x, f^*(x))$ is continuous on V^*, we can find an open neighbourhood V' of x^* contained in V^* such that $(x, f^*(x)) \in W$ for all $x \in V'$. By what we have already proved, $f^*(x) = f(x)$ for all $x \in V'$, whence $V' \subseteq E$, and E is open.

If $X = \mathbf{R}^n$, $Y = \mathbf{R}^m$, and F has components $F_1, \ldots, F_m$, then $dF(z_0)$ is the function from $\mathbf{R}^{n+m}$ into $\mathbf{R}^m$ whose ith component $(i = 1, \ldots, m)$ is the function

$$(x_1, \ldots, x_n, y_1, \ldots, y_m) \mapsto \sum_{j=1}^{n} x_j D_j F_i(z_0) + \sum_{k=1}^{m} y_k D_{n+k} F_i(z_0)$$

(cf. (3.4.1)). Hence $d_1 F(z_0)$ and $d_2 F(z_0)$ are the functions from $\mathbf{R}^n$ and $\mathbf{R}^m$ into $\mathbf{R}^m$ whose ith components are the functions

$$(x_1, \ldots, x_n) \mapsto \sum_{j=1}^{n} x_j D_j F_i(z_0) \quad \text{and} \quad (y_1, \ldots, y_m) \mapsto \sum_{k=1}^{m} y_k D_{n+k} F_i(z_0).$$

The condition in the implicit function theorem that $d_2 F(z_0)$ is a linear homeomorphism of Y ($= \mathbf{R}^m$) onto itself is therefore equivalent to the condition that the $m \times m$ matrix

$$\begin{bmatrix} D_{n+1} F_1(z_0) & \cdots & D_{n+m} F_1(z_0) \\ D_{n+1} F_2(z_0) & \cdots & D_{n+m} F_2(z_0) \\ \cdots & \cdots & \cdots \\ D_{n+1} F_m(z_0) & \cdots & D_{n+m} F_m(z_0) \end{bmatrix}$$

is non-singular.

The next result is a rather sophisticated extension of (3.8.1).

(3.8.3) *Let X, Y be Banach spaces, let $Z = X \times Y$, let A be an open set in Z, and let $F : A \to Y$ be C^1 on A. Let also K be a compact set in X, and let $f : K \to Y$ be a continuous function with graph in A such that $F(x, f(x)) = 0$ and $d_2 F(x, f(x)) \in \mathscr{L}\mathscr{H}(Y, Y)$ for all $x \in K$. Then there exist an open set B in X containing K, an open subset A_1 of A containing the graph of f, and a C^1 function $\bar{f} : B \to Y$ with graph in A_1 and agreeing with f on K, such that $F(x, \bar{f}(x)) = 0$ for all $x \in B$, and that if $x \in B$ then $y = \bar{f}(x)$ is the only solution of $F(x, y) = 0$ for which $(x, y) \in A_1$. Moreover, if F is r-times Fréchet differentiable or C^r on A, where $r \geq 2$, then $\bar{f}$ is r-times Fréchet differentiable or C^r on B.*†

Let K_0 be the graph of f (so that K_0 is a compact subset of A). We show first that there exists an open subset A_0 of A containing K_0, and a positive number η, such that if $(x, y), (x, y') \in A_0$, $\|y - y'\| < \eta$, and $F(x, y) = F(x, y')$, then $y = y'$.

For each $w \in K$ let $p(w) = 1/\|d_2 F(w, f(w))^{-1}\|$, and let $\Gamma(w, \varepsilon)$ be the open ball in Z with centre at the point $(w, f(w))$ of K_0 and radius $\varepsilon > 0$. Then, since A is open and $d_2 F$ is continuous on A, for each $w \in K$ we can find $\varepsilon_w > 0$ such that $\Gamma(w, \varepsilon_w) \subseteq A$ and that for all $(x, y) \in \Gamma(w, \varepsilon_w)$

$$\|d_2 F(x, y) - d_2 F(w, f(w))\| \leq \tfrac{1}{2} p(w).$$

---

† It is tempting to try to prove (3.8.3) by covering K with a finite set of neighbourhoods $V_1, \ldots, V_N$ of points $x_1, \ldots, x_N$ on each of which the equation $F(x, y) = 0$ has a solution in accordance with (3.8.1), and by piecing these solutions together to form the extension $\bar{f}$. However, the solutions corresponding to overlapping V_i may not agree on the entire overlap, and it seems difficult to employ (3.8.1) directly in this way.

By the mean value inequality (3.2.2) applied to the function $y \mapsto F(x, y) - d_2 F(w, f(w))y$, it follows that for all $(x, y), (x, y') \in \Gamma(w, \varepsilon_w)$

$$\| F(x, y) - F(x, y') - d_2 F(w, f(w))(y - y') \| \leq \tfrac{1}{2} p(w) \| y - y' \|,$$

and as in the first part of the proof of (3.7.1) this implies that

$$\tfrac{1}{2} p(w) \| y - y' \| \leq \| F(x, y) - F(x, y') \|,$$

so that if $F(x, y) = F(x, y')$, then $y = y'$.

Next, the open balls $\Gamma(w, \tfrac{1}{2}\varepsilon_w)$ $(w \in K)$ cover K_0, and hence we can find a finite number of them, say for $w = w_1, \ldots, w_N$, which cover K_0. Let A_0 be their union, and let $\eta = \tfrac{1}{2} \min \{\varepsilon_{w_1}, \ldots, \varepsilon_{w_N}\}$. Then if (x, y), $(x, y') \in A_0$ and $\| y - y' \| < \eta$, the points $(x, y), (x, y')$ belong the *same* ball $\Gamma(w_i, \varepsilon_{w_i})$, and therefore A_0 has the required property.

We observe now that since f is uniformly continuous on K, we can find $\delta > 0$ such that $\| f(x) - f(x')) \| < \tfrac{1}{3}\eta$ whenever $x, x' \in K$ and $\| x - x' \| < \delta$. We now apply (3.8.1) to each point of K_0; thus for each $x_0 \in K$ we can find an open ball $V(x_0)$ in X with centre x_0 and radius less than $\tfrac{1}{2}\delta$, and a C^1 function $f_0 : V(x_0) \to Y$ such that $f_0(x_0) = f(x_0)$ and that, for all $x \in V(x_0)$,

$$(x, f_0(x)) \in A_0, \quad F(x, f_0(x)) = 0, \quad \text{and} \quad \| f_0(x) - f_0(x_0) \| < \tfrac{1}{3}\eta.$$

The functions f_0 and f agree on $V(x_0) \cap K$, for if $x \in V(x_0) \cap K$, then $(x, f(x)), (x, f_0(x)) \in A_0$, and

$$\| f(x) - f_0(x) \| \leq \| f(x) - f(x_0) \| + \| f(x_0) - f_0(x) \|$$
$$= \| f(x) - f(x_0) \| + \| f_0(x_0) - f_0(x) \| < \tfrac{2}{3}\eta,$$

so that $f_0(x) = f(x)$. Further, if $x_0, x_1 \in K$ and $V(x_0)$ meets $V(x_1)$, then f_0 and f_1 agree on $V(x_0) \cap V(x_1)$, for in this case $\| x_1 - x_0 \| < \delta$, so that $\| f_0(x_0) - f_1(x_1) \| = \| f(x_0) - f(x_1) \| < \tfrac{1}{3}\eta$. Hence if $x \in V(x_0) \cap V(x_1)$ then $\| f_0(x) - f_1(x) \| \leq \| f_0(x) - f_0(x_0) \| + \| f_0(x_0) - f_1(x_1) \| + \| f_1(x_1) - f_1(x) \| < \eta$, and since $(x, f_0(x)), (x, f_1(x)) \in A_0$, it follows that $f_0(x) = f_1(x)$.

Now let B be the union of the balls $V(x_0)$ for all $x_0 \in K$, let $\bar{f} : B \to Y$ be given by $\bar{f}(x) = f_0(x)$ whenever $x \in V(x_0)$, and let A_1 be the set of $(x, y) \in A_0$ such that $x \in B$ and $\| y - \bar{f}(x) \| < \eta$. Then $B, \bar{f}, A_1$ have the required properties.

Exercises 3.8

1 Let X be a real Banach space, let A be an open set in X, let $a \in X$, and let $f : A \to X$ be a Fréchet differentiable function such that df is continuous at a. Prove that there is an open neighbourhood V of 0 in $\mathbf{R}$ and a function $\phi : V \to X$ such that $y = \phi(x)$ is the only continuous solution of the equation

$$y = a + xf(y)$$

valid on V. Prove that ϕ is differentiable on V, and evaluate $\phi'(0)$.

2 Let X be a real Banach space, let A be an open set in X containing 0, and let $f : A \to \mathbf{R}$ be a Fréchet differentiable function such that df is continuous at 0. Prove that there exists an open neighbourhood V of $(0,0)$ in $X \times X$ and a function $\phi : V \to \mathbf{R}$ such that $u = \phi(x, y)$ is the only continuous solution of the equation

$$u = f(ux + y)$$

valid on V. Prove that ϕ is Fréchet differentiable on V and that

$$d_1\phi(x, y) = \phi(x, y)d_2\phi(x, y) \qquad ((x, y) \in V).$$

3 (*Another implicit function theorem*) Let X be a normed space and Y a Banach space, let $x_0 \in X, y_0 \in Y$, let V be an open neighbourhood of x_0 in X and W a closed ball in Y with centre y_0, and let $F : V \times W \to Y$ be a continuous function such that $F(x_0, y_0) = y_0$ and that $F(x, y)$ is K-Lipschitzian in y, where $0 < K < 1$. Prove that there exists an open neighbourhood U of x_0 in V and a continuous function $f : U \to Y$ such that $y_0 = f(x_0)$ and that, for each $x \in U, y = f(x)$ is the only solution of the equation $F(x, y) = y$.

[Hint. Use the contraction mapping principle.]

3.9 Examples of Fréchet differentiable functions

The examples of Fréchet differentiable functions given so far have all been of a relatively simple nature. In general, however, Fréchet differentials can be difficult to calculate, and in this section we give some less simple examples of differentiable functions; these examples will be used in §§3.10, 11.

(3.9.1) *Let Y, Z be normed spaces, let J be a compact interval in $\mathbf{R}$, let A be an open set in the metric space $J \times Y$,† and let $f : A \to Z$ be a continuous function whose partial differential $d_2 f$ is continuous on A. Let also $\mathscr{E}$ be the set of functions in $C(J, Y)$ whose graphs are contained in A, and let $G : \mathscr{E} \to C(J, Z)$ be given by*

$$G(\phi)(t) = f(t, \phi(t)) \qquad (\phi \in \mathscr{E}, t \in J).$$

Then $\mathscr{E}$ is open in $C(J, Y)$, G is C^1 on $\mathscr{E}$, and for each $\phi \in \mathscr{E}$ and all $\eta \in C(J, Y)$

$$(dG(\phi)\eta)(t) = d_2 f(t, \phi(t))\eta(t) \qquad (t \in J).‡ \tag{1}$$

If in addition $d_2^2 f, \ldots, d_2^r f$ are defined and continuous on A, then G is C^r on $\mathscr{E}$, and for each $\phi \in \mathscr{E}$ and all $\eta_1, \ldots, \eta_r \in C(J, Y)$

$$d^r G(\phi)(\eta_1, \ldots, \eta_r)(t) = d_2^r f(t, \phi(t))(\eta_1(t), \ldots, \eta_r(t)) \qquad (t \in J).* \tag{2}$$

† Thus A is the intersection of $J \times Y$ with an open set in $\mathbf{R} \times Y$.
‡ The expression on the left in (1) is the value at t of the function $dG(\phi)\eta$ in $C(J, Z)$ (note that $dG(\phi) \in \mathscr{L}(C(J, Y), C(J, Z))$), while the expression on the right is the value at $\eta(t)$ of the function $d_2 f(t, \phi(t))$ in $\mathscr{L}(Y, Z)$.
* Note that $d^r G(\phi) \in \mathscr{L}_r(C(J, Y), C(J, Z))$, while $d_2^r f(t, \phi(t)) \in \mathscr{L}_r(Y; Z)$.

To prove that $\mathscr{E}$ is open in $C(J, Y)$, let $\phi \in \mathscr{E}$. Then the set $\{(t, \phi(t)) : t \in J\}$ is contained in A, and since this set is compact it lies at a positive distance δ from the complement of A in $J \times Y$. If now ψ belongs to the open ball in $C(J, Y)$ with centre ϕ and radius δ, then

$$\| (t, \psi(t)) - (t, \phi(t)) \| = \| \psi(t) - \phi(t) \| \le \| \psi - \phi \| < \delta$$

for all $t \in J$, so that the graph of ψ lies in A. Hence $\psi \in \mathscr{E}$, so that $\mathscr{E}$ is open.

We observe next that if $\phi \in \mathscr{E}$, $\eta \in C(J, Y)$, and $\omega_\eta : J \to Z$ is given by

$$\omega_\eta(t) = d_2 f(t, \phi(t)) \eta(t),$$

then $\omega_\eta \in C(J, Z)$. In fact, the function $t \mapsto d_2 f(t, \phi(t))$ from J into $\mathscr{L}(Y, Z)$ is continuous, since it is the composite of $d_2 f$ and $t \mapsto (t, \phi(t))$; hence ω_η is continuous, since it is the composite of the function $t \mapsto (d_2 f(t, \phi(t)), \eta(t))$ from J into $\mathscr{L}(Y, Z) \times Y$ and the function $(L, y) \mapsto Ly$ from $\mathscr{L}(Y, Z) \times Y$ into Z, and both of these functions are continuous (the second is bilinear). Further, if $\Omega(\eta) = \omega_\eta$, then $\Omega \in \mathscr{L}(C(J, Y), C(J, Z))$, for Ω is obviously linear, and for all $t \in J$

$$\| \Omega(\eta)(t) \| = \| \omega_\eta(t) \| \le \| d_2 f(t, \phi(t)) \| \, \| \eta(t) \|,$$

so that

$$\| \Omega(\eta) \| \le \sup_{t \in J} \| d_2 f(t, \phi(t)) \| \, \| \eta \|$$

(note that the supremum on the right is the supremum of a continuous function over a compact interval).

We show now that G is Fréchet differentiable at ϕ and that $dG(\phi) = \Omega$. For each η such that $\phi + \eta \in \mathscr{E}$ and for each $t \in J$ let

$$R(\eta, t) = G(\phi + \eta)(t) - G(\phi)(t) - \Omega(\eta)(t)$$
$$= f(t, \phi(t) + \eta(t)) - f(t, \phi(t)) - d_2 f(t, \phi(t)) \eta(t).$$

Then

$$R(\eta, t) = \Phi_t(\eta(t)), \tag{3}$$

where

$$\Phi_t(y) = f(t, \phi(t) + y) - f(t, \phi(t)) - d_2 f(t, \phi(t)) y \qquad ((t, \phi(t) + y) \in A).$$

To estimate the expression on the right of (3), we use the mean value inequality. It is evident that

$$d\Phi_t(y) = d_2 f(t, \phi(t) + y) - d_2 f(t, \phi(t)).$$

Further, an argument closely similar to that of (2.2.2) shows that $d\Phi_t(y) \to 0$ in $\mathscr{L}(Y, Z)$ as $y \to 0$, uniformly for t in J, that is, given $\varepsilon > 0$ we can find $\delta_1 > 0$ such that $\| d\Phi_t(y) \| \le \varepsilon$ whenever $\| y \| \le \delta_1$ and $t \in J$. Hence, by the mean value inequality (3.2.2),

$$\| \Phi_t(y) \| = \| \Phi_t(y) - \Phi_t(0) \| \le \varepsilon \| y \|$$

whenever $\|y\| \le \delta_1$ and $t \in J$. In particular,

$$\|R(\eta, t)\| = \|\Phi_t(\eta(t))\| \le \varepsilon \|\eta(t)\| \le \varepsilon \|\eta\|$$

whenever $\|\eta\| \le \delta_1$, and on taking the supremum of $\|R(\eta, t)\|$ for t in J we obtain that

$$\|G(\phi + \eta) - G(\phi) - \Omega(\eta)\| \le \varepsilon \|\eta\|$$

whenever $\|\eta\| \le \delta_1$, so that $dG(\phi)$ exists and is equal to Ω. Further, dG is continuous on $\mathscr{E}$, for if $\phi, \psi \in \mathscr{E}, \eta \in C(J, Y)$, and $t \in J$, then

$$\|(dG(\phi)\eta)(t) - (dG(\psi)\eta)(t)\| = \|d_2 f(t, \phi(t))\eta(t) - d_2 f(t, \psi(t))\eta(t)\|$$
$$\le \|d_2 f(t, \phi(t)) - d_2 f(t, \psi(t))\| \, \|\eta\|.$$

Hence

$$\|dG(\phi) - dG(\psi)\| \le \sup_{t \in J} \|d_2 f(t, \phi(t)) - d_2 f(t, \psi(t))\|,$$

and a repetition of the argument of (2.2.2) shows that the expression on the right here tends to 0 as $\psi \to \phi$ in $C(J, Y)$.

To prove the last part of the theorem we use induction on r. Let r be an integer greater than 1 such that the result holds with r replaced by $r-1$, let $f, d_2 f, \ldots, d_2^r f$ be defined and continuous on r, let $\phi, \phi + \eta \in \mathscr{E}$ and let $\eta_1, \ldots, \eta_{r-1} \in C(J, Y)$. Then if $\Omega_r(\eta, \eta_1, \ldots, \eta_{r-1})$ is the function from J into Z given by

$$\Omega_r(\eta, \eta_1, \ldots, \eta_{r-1})(t) = d_2^r f(t, \phi(t))(\eta(t), \eta_1(t), \ldots, \eta_{r-1}(t)),$$

we see exactly as above that $\Omega_r(\eta, \eta_1, \ldots, \eta_{r-1}) \in C(J, Z)$, and that the function Ω_r on $(C(J, Y))^r$ thus defined belongs to $\mathscr{L}_r(C(J, Y), C(J, Z))$. Further, by the induction hypothesis,

$$\|d^{r-1}G(\phi + \eta)(\eta_1, \ldots, \eta_{r-1})(t) - d^{r-1}G(\phi)(\eta_1, \ldots, \eta_{r-1})(t)$$
$$- \Omega_r(\eta, \eta_1, \ldots, \eta_{r-1})(t)\|$$
$$= \|d_2^{r-1}f(t, \phi(t) + \eta(t))(\eta_1(t), \ldots, \eta_{r-1}(t)) -$$
$$d_2^{r-1}f(t, \phi(t))(\eta_1(t), \ldots, \eta_{r-1}(t)) - d_2^r f(t, \phi(t))(\eta(t), \eta_1(t), \ldots, \eta_{r-1}(t))\|$$
$$\le \|d_2^{r-1}f(t, \phi(t) + \eta(t)) - d_2^{r-1}f(t, \phi(t)) - d_2^r f(t, \phi(t))\eta(t)\|$$
$$\times \|\eta_1\| \cdots \|\eta_{r-1}\|.$$

To prove (2), it is therefore enough (by an obvious extension of (3.5.4) for general r) to prove that given $\varepsilon > 0$ we can find $\delta_2 > 0$ such that

$$\|d_2^{r-1}f(t, \phi(t) + \eta(t)) - d_2^{r-1}f(t, \phi(t)) - d_2^r f(t, \phi(t))\eta(t)\| \le \varepsilon \|\eta\|$$

whenever $\|\eta\| \le \delta_2$. This, however, follows by an argument entirely similar to that for the case $r = 1$, and we omit the details. Finally, we see from (2) that if $d_2^r f$ is continuous on A, then $d^r G$ is continuous on $\mathscr{E}$, and this completes the proof.

(3.9.2) *Let Y be a normed space and Z a Banach space, let J be a compact interval in* **R**, *let A be an open set in the metric space J × Y, and let f : A → Z be a continuous function whose partial differential $d_2 f$ is continuous on A. Let also $\mathscr{E}_1$ be the set of functions in $C^1(J, Y)$ whose graphs are contained in A, let $t_0 \in J$, and let $H : \mathscr{E}_1 \to C^1(J, Z)$ be given by*

$$H(\phi)(t) = \int_{t_0}^{t} f(s, \phi(s))ds \qquad (\phi \in \mathscr{E}_1, t \in J).$$

Then $\mathscr{E}_1$ is open in $C^1(J, Y)$, H is C^1 on $\mathscr{E}_1$, and for each $\phi \in \mathscr{E}_1$ and all $\eta \in C^1(J, Y)$

$$(dH(\phi)\eta)(t) = \int_{t_0}^{t} d_2 f(s, \phi(s))\eta(s)ds.$$

If in addition $d_2^2 f, \ldots, d_2^r f$ are defined and continuous on A, then H is C^r on $\mathscr{E}_1$ and for each $\phi \in \mathscr{E}_1$ and all $\eta_1, \ldots, \eta_r \in C^1(J, Y)$

$$d^r H(\phi)(\eta_1, \ldots, \eta_r)(t) = \int_{t_0}^{t} d_2^r f(s, \phi(s))(\eta_1(s), \ldots, \eta_r(s))ds.$$

Here $H = T \circ G \circ I$, where G is the function of (3.9.1), I is the inclusion map $\phi \mapsto \phi$ from $C^1(J, Y)$ into $C(J, Y)$, and $T : C(J, Z) \to C^1(J, Z)$ is given by

$$T\chi(t) = \int_{t_0}^{t} \chi(s)ds \qquad (t \in J). \tag{4}$$

Since T and I are continuous and linear (note that $\|I\phi\| = \|\phi\| \le \|\phi\|_1$, where $\| \|$ and $\| \|_1$ are the norms on $C(J, Y)$ and $C^1(J, Y)$), the result therefore follows from (3.9.1), (3.1.4. Corollaries 1, 2), and (3.5.2).

(3.9.3) *Let Y be a normed space and Z a Banach space, let J be a compact interval in* **R**, *let A be an open set in the metric space J × Y × Y, and let f : A → Z be a continuous function whose partial differentials $d_2 f$ and $d_3 f$ are continuous on A. Let also $\mathscr{F}$ be the set of functions $\phi \in C^1(J, Y)$ such that $(t, \phi(t), \phi'(t)) \in A$ for all $t \in J$, let $t_0 \in J$, and let $K : \mathscr{F} \to C^1(J, Z)$ be given by*

$$K(\phi)(t) = \int_{t_0}^{t} f(s, \phi(s), \phi'(s))ds \qquad (\phi \in \mathscr{F}, t \in J).$$

Then $\mathscr{F}$ is open in $C^1(J, Y)$, K is C^1 on $\mathscr{F}$, and for each $\phi \in \mathscr{F}$ and all $n \in C^1(J, Y)$

$$(dK(\phi)\eta)(t) = \int_{t_0}^{t} (d_2 f(s, \phi(s), \phi'(s))\eta(s) + d_3 f(s, \phi(s), \phi'(s))\eta'(s))ds. \tag{5}$$

If in addition all the partial differentials $d_2^m d_3^n f$, where $2 \le m + n \le r$, are defined and continuous on A, then K is C^r on $\mathscr{F}$.

Let A_0 be the set A regarded as a subset of $J \times Y^2$ (so that A_0 is open in $J \times Y^2$), and let $f_0 : A_0 \to Z$ be given by

$$f_0(t, (y_1, y_2)) = f(t, y_1, y_2) \qquad ((t, y_1, y_2) \in A).$$

Since $d_2 f$ and $d_3 f$ are defined and continuous on A, it follows from (3.3.2) and (3.3.1) that $d_2 f_0$ is defined and continuous on A_0, and that for all $(t, y_1, y_2) \in A$ and $(h_1, h_2) \in Y^2$

$$d_2 f_0(t, (y_1, y_2))(h_1, h_2) = d_2 f(t, y_1, y_2) h_1 + d_3 f(t, y_1, y_2) h_2. \qquad (6)$$

Now let $\mathscr{E}_0$ be the set of functions in $C(J, Y^2)$ whose graphs are contained in A_0, and let $G_0 : \mathscr{E}_0 \to C(J, Z)$ be given by

$$G_0(\Phi)(t) = f_0(t, \Phi_0(t)) \qquad (\Phi \in \mathscr{E}_0, t \in J).$$

Then, by (3.9.1), $\mathscr{E}_0$ is open in $C(J, Y^2)$ and G_0 is C^1 on $\mathscr{E}_0$. Further, by (1) and (6), for each $\Phi \in \mathscr{E}_0$ and all $\Psi \in C(J, Y^2)$ we have

$$(d_2 G_0(\Phi)\Psi)(t) = d_2 f(t, \Phi_1(t), \Phi_2(t))\Psi_1(t) + d_3 f(t, \Phi_1(t), \Phi_2(t))\Psi_2(t),$$

where Φ_1, Φ_2 and Ψ_1, Ψ_2 are the components of Φ and Ψ. Moreover, if the additional hypothesis on the higher differentials of f is satisfied, then G_0 is clearly C^r on $\mathscr{E}_0$. Since $K = T \circ G_0 \circ B$, where T is defined by (4), and B is the function in $\mathscr{L}(C^1(J, Y), C(J, Y^2))$ given by $B\phi(t) = (\phi(t), \phi'(t))$, the result therefore follows from (3.1.4. Corollaries 1, 2) and (3.5.2).

By taking $t_0 = a$, and composing K with the continuous linear function $\psi \mapsto \psi(b)$ from $C^1(J, Z)$ into Z, we obtain exactly similar results for the function $F : \mathscr{F} \to Z$ given by

$$F(\phi)(t) = \int_a^b f(s, \phi(s), \phi'(s)) ds,$$

the formula (5) here being replaced by

$$dF(\phi)\eta = \int_a^b (d_2 f(\Phi(s))\eta(s) + d_3 f(\Phi(s))\eta'(s)) ds,$$

where $\Phi(s) = (s, \phi(s), \phi'(s))$ and $\eta \in C^1(J, Y)$. We note also for future use that the value of the second differential $d^2 F(\phi)$ at the point (η, η) is given by

$$d^2 F(\phi)(\eta)^2 = \int_a^b (d_2^2 f(\Phi(s))(\eta(s))^2 + 2 d_2 d_3 f(\Phi(s))(\eta(s), \eta'(s)) +$$
$$d_3^2 f(\Phi(s))(\eta'(s))^2) ds.$$

Exercises 3.9

1 Let Y, Z be normed spaces, let J be a compact interval in $\mathbf{R}$, let A be an open set in the metric space $J \times Y$, and let $f : A \to Z$ be a continuous function whose partial

differential d_2f is continuous on A. Let also $R(J, Y)$ and $R(J, Z)$ be the spaces of regulated functions on J with values in Y and Z, with the sup norm (see Exercise 1.9.7), let $\mathscr{E}$ be the set of functions in $R(J, Y)$ such that the closures of their graphs are contained in A, and let $G : \mathscr{E} \to R(J, Z)$ be given by

$$G(\phi)(t) = f(t, \phi(t)) \qquad (\phi \in \mathscr{E}, t \in J).$$

Prove that $\mathscr{E}$ is open in $R(J, Y)$, G is C^1 on $\mathscr{E}$, and for each $\phi \in \mathscr{E}$ and all $\eta \in R(J, Y)$

$$(dG(\phi)\eta)(t) = d_2 f(t, \phi(t))\eta(t).$$

2   Let Y be a real normed space, and let $J = [a, b]$. A function $\phi : J \to Y$ is said to be *piecewise smooth* if it is continuous and there exists a partition $\{t_0, \ldots, t_n\}$ of J such that ϕ' is defined and continuous on each subinterval $]t_i, t_{i+1}[$ and possesses right-hand limits at $t_0, \ldots, t_{n-1}$ and left-hand limits at $t_1, \ldots, t_n$ (thus, by (1.7.4), ϕ' is defined except at those points $t_i (i = 1, \ldots, n - 1)$ where $\phi'_-(t_i), \phi'_+(t_i)$ are unequal). The set $PS(J, Y)$ of piecewise smooth functions from J into Y forms a normed space with the norm defined by

$$\| \phi \| = \sup_{t \in J} \| \phi(t) \| + \sup_{t \in K} \| \phi'(t) \|,$$

where K is the domain of ϕ' (in fact $PS(J, Y)$ is a linear subspace of the normed space considered in Exercise 1.7.1. It should be noted that if $\phi \in PS(J, Y)$, then ϕ' has a unique extension ϕ^* to J that is continuous at a and left continuous on $]a, b]$, and

$$\| \phi \| = \sup_{t \in J} \| \phi(t) \| + \sup_{t \in J} \| \phi^*(t) \|.)$$

Now let A be an open set in the metric space $J \times Y \times Y$, and let $f : A \to \mathbf{R}$ be a continuous function whose partial differentials d_2f and d_3f are continuous on A. Let also $\mathscr{F}$ be the set of $\phi \in PS(J, Y)$ such that the closure of the set $\{(t, \phi(t), \phi'(t)) : t \in J\}$ is contained in A, and let $F : \mathscr{F} \to \mathbf{R}$ be given by

$$F(\phi)(t) = \int_a^b f(t, \phi(t), \phi'(t))dt \qquad (\phi \in \mathscr{F}).$$

Prove that $\mathscr{F}$ is open in $PS(J, Y)$, F is C^1 on $\mathscr{F}$, and for each $\phi \in \mathscr{F}$ and all $\eta \in PS(J, Y)$

$$dF(\phi)\eta = \int_a^b (d_2 f(t, \phi(t), \phi'(t))\eta(t) + d_3 f(t, \phi(t), \phi'(t))\eta'(t))dt.$$

[Hint. Nothing in the exercise is changed if we replace ϕ' by ϕ^* and η' by η^*, and for this altered version the argument of (3.9.3) applies, using Exercise 1. Note that the function $\phi \mapsto (\phi, \phi^*)$ from $PS(J, Y)$ into $R(J, Y^2)$ is linear and continuous.

A similar result could be obtained with $PS(J, Y)$ replaced by the space W of functions from J into Y that are the primitives of regulated functions, considered as a linear subspace of the normed space of Exercise 1.7.1. However, the set of $\phi \in W$ such that the closure of the set $\{(t, \phi(t), \phi'(t)) : t \in J\}$ is contained in A does not seem to be necessarily open in W, and we therefore have to alter the hypotheses accordingly.]

3   (*Continuity and differentiability of integrals involving a parameter*) Let $J = [a, b]$, let X be a normed space and Y a Banach space, let A be an open set in X, let $f : J \times A \to Y$ be continuous, and let

$$g(x) = \int_a^b f(t, x)dt \qquad (x \in A).$$

Prove that
(i) g is continuous on A,

(ii) if in addition $d_2 f, \ldots, d_2^r f$ are defined and continuous on $J \times A$, where $r \geq 1$, then g is C^r on A, and for all $x \in A$

$$d^r g(x) = \int_a^b d_2^r f(t,x) dt.$$

[Hint. Use (2.2.2). The proof of (ii) is by induction on r.]

4 Let $J = [a,b]$, let X be a real normed space and Y a real Banach space, let A be an open set in X, and let $f : J \times A \to Y$ be a continuous function such that $d_2 f$ is continuous on $J \times A$. Let also $\alpha, \beta : A \to J$ be C^1, and let

$$G(x) = \int_{\alpha(x)}^{\beta(x)} f(s,x) ds \qquad (x \in A).$$

Prove that G is C^1 on A and that for all $x \in A$ and $h \in X$

$$dG(x)h = (d\beta(x)h)f(\beta(x),x) - (d\alpha(x)h)f(\alpha(x),x) + \left(\int_{\alpha(x)}^{\beta(x)} d_2 f(t,x) dt \right) h.$$

5 Let X be a normed space and Y a Banach space, let A be an open ball in X with centre 0, let $f : A \to Y$ be C^r, where $r \geq 1$, and let

$$g(x) = \int_0^1 \phi(t) f(tx) dt \qquad (x \in A),$$

where $\phi : [0,1] \to \mathbf{R}$ is continuous. Prove that g is C^r on A and that

$$d^r g(x) = \int_0^1 t^r \phi(t) d^r f(tx) dt \qquad (x \in A).$$

6 Let Y be a real Banach space, let A be an open ball in $\mathbf{R}^n$ with centre 0, and let $f : A \to Y$ be a C^r function such that $df(0) = 0$. Prove that if $a_{ij} = \frac{1}{2} D_i D_j f(0)$ $(i,j = 1, \ldots, n)$, there exist C^{r-2} functions $g_{ij} : A \to Y$ such that $g_{ij} = g_{ji}, g_{ij}(0) = a_{ij}$, and

$$f(x) = \sum_{i,j=1}^n x_i x_j g_{ij}(x) \qquad (x = (x_1, \ldots, x_n) \in A).$$

[Hint. Use Exercise 3.6.3 and Exercise 5.]

3.10 Higher order differentiability of solutions of differential equations; differentiability with respect to the initial conditions and parameters

Let Y be a real Banach space, let E be an open set in $\mathbf{R} \times Y$, and let $f : E \to Y$ be a continuous function whose partial differential $d_2 f$ is continuous on E. Since $d_2 f$ is locally bounded, it follows directly from the mean value inequality (3.2.2) that $f(t,y)$ is locally Lipschitzian in y in E. Hence, by (2.7.6), for each $(t_0, y_0) \in E$ the differential equation $y' = f(t,y)$ has a unique solution ϕ taking the value y_0 at t_0 and having maximal (open) interval of existence, I say. Moreover, since f is continuous, ϕ is C^1 on I. We prove now:

(3.10.1) *If in addition to the above conditions f is C^r on E, where $r \geq 1$, then ϕ is C^{r+1} on I.*

Since ϕ is C^1 on I, so is the function $t \mapsto (t, \phi(t))$. Since $\phi'(t) = f(t, \phi(t))$ for all $t \in I$, it therefore follows from the chain rule that if f is C^1 on E then ϕ' is C^1 on I, i.e. ϕ is C^2 on I. This proves the case $r = 1$, and the proof is completed by induction on r, using a similar argument.

We turn next to differentiability of the solution with respect to the initial conditions, and we consider first the case where the initial value y_0 varies, the initial point t_0 being kept fixed.

(3.10.2) *Let Y be a real Banach space, let E be an open set in $\mathbf{R} \times Y$, let $f : E \to Y$ be a continuous function whose partial differential $d_2 f$ is continuous on E, let $t_0 \in \mathbf{R}$, let $S = \{y_0 \in Y : (t_0, y_0) \in E\} \neq \varnothing$, and for each $y_0 \in S$ let $t \mapsto \phi(t, y_0)$ be the solution of the equation $y' = f(t, y)$ taking the value y_0 at t_0 and having maximal interval of existence, $I(y_0)$, say. Then*
  (i) *the domain of the function ϕ, i.e. the set A of points (t, y_0) such that $y_0 \in S$ and $t \in I(y_0)$, is open in $\mathbf{R} \times Y$, and ϕ is C^1 on A,*
  (ii) *the function $t \mapsto d_2 \phi(t, y_0)$ is the solution ξ of the differential equation*
$$\xi' = d_2 f(t, \phi(t, y_0)) \circ \xi \tag{1}$$
on $I(y_0)$ such that $\xi(t_0) = 1_Y$,
  (iii) *the function $y_0 \mapsto \partial \phi(t, y_0)/\partial t$ is C^1 for each t. Further, if $d_2^2 f, \ldots, d_2^r f$ are defined and continuous on E, where $r \geq 2$, then the functions $y_0 \mapsto \phi(t, y_0)$ and $y_0 \mapsto \partial \phi(t, y_0)/\partial t$ are C^r.*

Our arguments prove also the following companion result.

(3.10.3) *Suppose that the conditions of (3.10.2) are satisfied, let $\bar{y}_0 \in S$, and let J be a compact subinterval of $I(\bar{y}_0)$ containing t_0 as an interior point. Then there exist an open neighbourhood V of $\bar{y}_0$ in Y and a C^1 function $\Phi : V \to C^1(J, Y)$ such that, for each $y_0 \in V$, $J \subseteq I(y_0)$ and $\Phi(y_0)$ is the restriction of the solution $t \mapsto \phi(t, y_0)$ to J. Further, if $d_2^2 f, \ldots, d_2^r f$ are defined and continuous on E, then Φ is C^r.†*

We note that if the differential equation $y' = f(t, y)$ and the initial condition that $y = y_0$ when $t = t_0$ are written in terms of ϕ, i.e.

$$\frac{\partial \phi}{\partial t}(t, y_0) = f(t, \phi(t, y_0)), \qquad \phi(t_0, y_0) = y_0,$$

† Of these two results (3.10.3) is the stronger kind, since it says that the solution, as a function-valued map, is C^1 in the initial conditions; it implies the statements of (3.10.2) for $t \in J$. However, $C^1(I(y_0), Y)$ may not consist of bounded functions, and $I(y_0)$ may vary with y_0. Therefore in general we have only the weaker results of (3.10.2).

then the result of (3.10.2)(ii) is formally obtained by differentiating both of these relations with respect to y_0. In fact, formal differentiation gives

$$d_2\left(\frac{\partial\phi}{\partial t}\right)(t,y_0) = d_2 f(t,\phi(t,y_0))\, d_2\phi(t,y_0) \quad \text{and} \quad d_2\phi(t_0,y_0) = 1_Y,$$

and on interchanging the order of the differentiations on the left in the first equation we obtain the result of (3.10.2)(ii). Unfortunately, this formal argument does not provide the basis for a proof.

Before we do prove (3.10.2, 3), we consider some corollaries.

If $Y = \mathbf{R}^m$, $\{e_1,\dots,e_m\}$ is the natural basis for $\mathbf{R}^m$, and ξ is the function of (3.10.2)(ii), then for $i = 1,\dots,m$,

$$\frac{\partial}{\partial y_i}\phi(t,y_0) = d_2\phi(t,y_0)e_i = \xi(t)e_i.$$

It therefore follows from (2.8.3)(i) that the function $t\mapsto \partial\phi(t,y_0)/\partial y_i$ is the solution η of the linear differential equation

$$\eta' = d_2 f(t,\phi(t,y_0))\eta \tag{2}$$

on $I(y_0)$ such that $\eta(t_0) = e_i$.

In the general case the function ξ of (3.10.2)(ii) is a fundamental kernel for the equation (2) (cf. (2.8.4); note that ξ takes the value 1_Y at t_0). We therefore have the following corollary of (3.10.2).

(3.10.2. Corollary) *If the conditions of* (3.10.2)† *are satisfied, then* $d_2\phi(t,y_0)\in\mathscr{L}\mathscr{H}(Y,Y)$ *for all* $(t,y_0)\in A$.

From this we deduce in turn the following result.

(3.10.4) (The rectification theorem) *If the conditions of* (3.10.2) *are satisfied, then the function* $G : A \to E$ *given by*

$$G(t,y_0) = (t,\phi(t,y_0))$$

is a homeomorphism of A onto an open subset B of E, and both G and G^{-1} are C^1.

The set B consists of the graphs of the solutions $t\mapsto\phi(t,y_0)$ for all $y_0\in S$, and since obviously

$$G^{-1}(t,\phi(t,y_0)) = (t,y_0)$$

for all $(t,y_0)\in A$, the theorem asserts that *there exists a C^1 homeomorphism of B onto A, with a C^1 inverse, which maps the graphs of the solutions onto line segments in $\mathbf{R}\times Y$ parallel to the t-axis.*

† Here and in (3.10.4) below the extra conditions in (3.10.2) (iii) are excluded.

To prove (3.10.4), we observe first that G is one-to-one, for if $G(t, y_0) = G(t', y_0')$, then $t = t'$ and $\phi(t', y_0) = \phi(t, y_0) = \phi(t', y_0')$, so that $y_0 = y_0'$, by the uniqueness of the solution ϕ. Further, G is C^1 (by (3.10.2)(i)), and for all $(t, y_0) \in A$ and $(h, k) \in \mathbf{R} \times Y$

$$dG(t, y_0)(h, k) = (h, d_1\phi(t, y_0)h + d_2\phi(t, y_0)k).$$

Since $d_2\phi(t, y_0) \in \mathscr{L}\mathscr{H}(Y, Y)$, it follows from (3.8.2) that $dG(t, y_0) \in \mathscr{L}\mathscr{H}(\mathbf{R} \times Y, \mathbf{R} \times Y)$. Hence, by applying the inverse function theorem (3.7.3) to G at each point of A, we deduce that G maps A onto an open subset B of E, and that G^{-1} is C^1 on B.

(3.10.4. Corollary) (Existence of a first integral) *Let G and B be defined as in* (3.10.4) *and let* $\Psi : B \to Y$ *be given by*

$$G^{-1}(t, y) = (t, \Psi(t, y)).$$

Then Ψ is C^1, and for each $(t, y) \in B$, $\Psi(t, y) = y_0$ if and only if $y = \phi(t, y_0)$.

In other words, the level surfaces of Ψ (i.e. the sets $\{(t, y) \in B : \Psi(t, y) = y_0\}$) are the graphs of the solutions $t \mapsto \phi(t, y_0)$ of $y' = f(t, y)$. A function Ψ with this property is called a *first integral* of the equation.

The corollary is almost immediate, for Ψ is obviously C^1. Also $\Psi(t, y) = y_0$ if and only if $G^{-1}(t, y) = (t, y_0)$, and therefore if and only if $(t, y) = G(t, y_0) = (t, \phi(t, y_0))$.

Consider now the proofs of (3.10.2, 3). By (2.10.1. Corollary), the set A is open in $\mathbf{R} \times Y$, and indeed the proof of that result shows that if $(t, \bar{y}_0) \in A$, then there exist a compact subinterval J of $I(\bar{y}_0)$ containing both t_0 and t as interior points, and an open neighbourhood W of $\bar{y}_0$ in Y, such that $J \times W \subseteq A$. In the remainder of the proof, it is obviously enough to confine ourselves to those y_0 that belong to W, and we may assume also that the corresponding solutions $t \mapsto \phi(t, y_0)$ are restricted to J.

Let $\mathscr{E}_1$ be the set of functions in $C^1(J, Y)$ whose graphs lie in E, and let $F : W \times \mathscr{E}_1 \to C^1(J, Y)$ be given by

$$F(y_0, \psi)(t) = y_0 + \int_{t_0}^t f(s, \psi(s))ds - \psi(t) \qquad (t \in J).$$

By (2.2.1), $F(y_0, \psi) = 0$ if and only if ψ is the solution of $y' = f(t, y)$ on J such that $\psi(t_0) = y_0$, i.e. if and only if $\psi(t) = \phi(t, y_0)$ for all $t \in J$. In particular, we see that $F(\bar{y}_0, \bar{\phi}) = 0$, where $\bar{\phi}(t) = \phi(t, \bar{y}_0) (t \in J)$.

We prove now that F satisfies the conditions of the implicit function theorem (3.8.1). By (3.9.2), $\mathscr{E}_1$ is open in $C^1(J, Y)$, d_2F is continuous, and for each $(y_0, \psi) \in W \times \mathscr{E}_1$ and all $\eta \in C^1(J, Y)$

$$(d_2F(y_0, \psi)\eta)(t) = \int_{t_0}^t d_2f(s, \psi(s))\eta(s)ds - \eta(t).$$

Moreover, if $d_2^2 f, \ldots, d_2^r f$ are defined and continuous on E, then $d_2 F$ is C^{r-1} (note that $d_2 F$ is independent of y_0). It is also easy to see that for each $(y_0, \psi) \in W \times \mathscr{E}_1$ and all $h \in Y$, $d_1 F(y_0, \psi)h$ is the constant function $t \mapsto h$ in $C^1(J, Y)$, so that $d_1 F$ is constant and is therefore C^∞. From (3.3.3) we therefore deduce that F is C^1 and that if $d_2^2 f, \ldots, d_2^r f$ are defined and continuous on E then F is C^r.

Next, for each $(y_0, \psi) \in W \times \mathscr{E}_1$ the function $d_2 F(y_0, \psi)$ is a linear homeomorphism of $C^1(J, Y)$ onto itself, for if $\chi \in C^1(J, Y)$ the equation $d_2 F(y_0, \psi)\eta = \chi$ is equivalent to the integral equation

$$\int_{t_0}^t d_2 f(s, \psi(s))\eta(s)ds - \eta(t) = \chi(t),$$

and this is in turn equivalent to the differential equation

$$\eta'(t) = d_2 f(t, \psi(t))\eta(t) - \chi'(t) \qquad (3)$$

subject to the initial condition that

$$\eta(t_0) = -\chi(t_0). \qquad (4)$$

Since $t \mapsto d_2 f(t, \psi(t))$ is continuous on J, the differential equation (3) is linear, and therefore, by (2.8.1), it has a unique solution η in $C^1(J, Y)$ satisfying (4). Hence $d_2 F(y_0, \psi)$ is a one-to-one continuous linear function from $C^1(J, Y)$ onto itself, and this implies that it is a linear homeomorphism, by the open mapping theorem.

It now follows from the implicit function theorem that there exist an open neighbourhood V of $\bar{y}_0$ in W, and a C^1 function $\Phi : V \to C^1(J, Y)$ satisfying $\Phi(\bar{y}_0) = \bar{\phi}$, such that for all $y_0 \in V$,
(a) $F(y_0, \Phi(y_0)) = 0$, i.e. $\Phi(y_0)$ is the solution of $y' = f(t, y)$ on J taking the value y_0 at t_0, so that $\Phi(y_0)(t) = \phi(t, y_0)$ $(t \in J)$,
(b) $d\Phi(y_0) = -d_2 F(y_0, \Phi(y_0))^{-1} \circ d_1 F(y_0, \Phi(y_0))$.
Further,
(c) if $d_2^2 f, \ldots, d_2^r f$ are defined and continuous on E, then Φ is C^r.
We have thus proved (3.10.3), and we now deduce the results of (3.10.2).

Let $t \in J$, and let ζ_t, ζ_t' be the functions in $\mathscr{L}(C^1(J, Y), Y)$ given by $\zeta_t \psi = \psi(t), \zeta_t' \psi = \psi'(t)$. Since the functions $y_0 \mapsto \phi(t, y_0)$ and $y_0 \mapsto \partial \phi(t, y_0)/\partial t$ are the composites $\zeta_t \circ \Phi$ and $\zeta_t' \circ \Phi$, we deduce immediately that these two functions are C^1, and that they are C^r if $d_2^2 f, \ldots, d_2^r f$ are continuous on E. Moreover, $d_2 \phi(t, y_0) = \zeta_t \circ d\Phi(y_0)$, so that for all $h \in Y$

$$d_2 \phi(t, y_0)h = (d\Phi(y_0)h)(t).$$

But, by (b), $d\Phi(y_0)h$ is the inverse image by $d_2 F(y_0, \Phi(y_0))$ of the constant function $t \mapsto -h$, and hence by (3) and (4) it is the solution η of the linear differential equation

$$\eta' = d_2 f(t, \phi(t, y_0))\eta$$

on J satisfying $\eta(t_0) = h$. If now ξ is the solution of the equation (1) such that $\xi(t_0) = 1_Y$, then $\eta(t) = \xi(t)h$. Hence $d_2\phi(t, y_0)h = (d\Phi(y_0)h)(t) = \xi(t)h$ for all $h \in Y$, so that $d_2\phi(t, y_0) = \xi(t)$.

To complete the proof we have to show that ϕ is C^1 on A. By (2.10.1), ϕ is continuous on A, and since

$$\frac{\partial}{\partial t}\phi(t, y_0) = f(t, \phi(t, y_0)),$$

it follows that $\partial\phi/\partial t$ is continuous on A. Also, by a further application of (2.10.1) to the equation (1), we deduce that $d_2\phi$ is continuous on A. Hence ϕ is C^1 on A, by (3.3.3).

We now consider variations in both the initial value y_0 and the initial point t_0, and we prove the following result, which contains (3.10.2).

(3.10.5) *Let* Y *be a real Banach space, let* E *be an open set in* $\mathbf{R} \times Y$, *let* $f : E \to Y$ *be a continuous function whose partial differential* $d_2 f$ *is continuous on* E, *and for each* $(t_0, y_0) \in E$ *let* $t \mapsto \phi(t, t_0, y_0)$ *be the solution of the equation* $y' = f(t, y)$ *taking the value* y_0 *at* t_0 *and having maximal interval of existence* $I(t_0, y_0)$, *say. Then*

(i)   *the domain of the function* ϕ, *i.e. the set* A *of points* (t, t_0, y_0) *such that* $(t_0, y_0) \in E$ *and* $t \in I(t_0, y_0)$, *is open in* $\mathbf{R} \times \mathbf{R} \times Y$, *and* ϕ *is* C^1 *on* A,

(ii)   *the function* $t \mapsto d_3\phi(t, t_0, y_0)$ *is the solution of the differential equation*

$$\xi' = d_2 f(t, \phi(t, t_0, y_0)) \circ \xi$$

on $I(t_0, y_0)$ *such that* $\xi(t_0) = 1_Y$,

(iii) $\partial\phi(t, t_0, y_0)/\partial t_0 = -d_3\phi(t, t_0, y_0)f(t_0, y_0)$,

(iv)   *the function* $y_0 \mapsto \partial\phi(t, t_0, y_0)/\partial t$ *is* C^1 *for each* t, t_0. *Further, if* $d_2^2 f, \dots,$ $d_2^r f$ *are defined and continuous on* E, *then the functions* $y_0 \mapsto \phi(t, t_0, y_0)$ *and* $y_0 \mapsto \partial\phi(t, t_0, y_0)/\partial t$ *are* C^r, *and the function* $y_0 \mapsto \partial\phi(t, t_0, y_0)/\partial t_0$ *is* C^{r-1}.

That A is open follows from (2.10.1. Corollary). Also (ii) and the results in (iv) concerning the functions ϕ and $\partial\phi/\partial t$ follow immediately from (3.10.2)(ii) and (iii). Further, the results of (ii) and (iii), together with the formula

$$\frac{\partial}{\partial t}\phi(t, t_0, y_0) = f(t, \phi(t, t_0, y_0)),$$

imply that ϕ is C^1 on A, and (iii) also implies the result in (iv) concerning $\partial\phi/\partial t_0$. It therefore remains to prove (iii).

Let $(t, t_0, y_0) \in A$. Then for all sufficiently small real τ we have

$$\phi(t, t_0 + \tau, y_0) = \phi(t, t_0, y_\tau),$$

where $y_\tau = \phi(t_0, t_0 + \tau, y_0)$. Clearly $y_\tau \to y_0$ as $\tau \to 0$, and hence we can write

$$\phi(t, t_0 + \tau, y_0) - \phi(t, t_0, y_0) + \tau d_3\phi(t, t_0, y_0) f(t_0, y_0)$$
$$= \phi(t, t_0, y_\tau) - \phi(t, t_0, y_0) + \tau d_3\phi(t, t_0, y_0) f(t_0, y_0)$$
$$= d_3\phi(t, t_0, y_0)(y_\tau - y_0 + \tau f(t_0, y_0)) + o(\| y_\tau - y_0 \|). \tag{5}$$

Since $y_0 = \phi(t_0 + \tau, t_0 + \tau, y_0)$, the mean value inequality (1.6.3. Corollary) gives

$$\| y_\tau - y_0 + \tau f(t_0, y_0) \| = \| \phi(t_0, t_0 + \tau, y_0) - \phi(t_0 + \tau, t_0 + \tau, y_0) + \tau f(t_0, y_0) \|$$
$$\leq |\tau| \| - D_1\phi(t_0 + \theta\tau, t_0 + \tau, y_0) + f(t_0, y_0) \|$$
$$= |\tau| \| - f(t_0 + \theta\tau, \phi(t_0 + \theta\tau, t_0 + \tau, y_0)) - f(t_0, y_0) \|$$

for some $\theta \in]0, 1[$, and here the expression on the right is $o(|\tau|)$ as $\tau \to 0$. Hence the expression on the right of (5) is also $o(|\tau|)$ as $\tau \to 0$, so that $\partial\phi(t, t_0, y_0)/\partial t_0$ exists and satisfies (iii).

If f is C^r on E, where $r \geq 1$, we have a stronger result, namely:

(3.10.6) *If Y is a real Banach space, E is an open set in $\mathbf{R} \times Y$, and $f : E \to Y$ is C^r, where $r \geq 1$, then the function ϕ of (3.10.5) is C^r on A.*

This follows from the formulae of (3.10.5)(ii) and (iii) by an easy induction argument, and we omit the details.

We complete this section by dealing with differentiation with respect to a parameter.

(3.10.7) (Differentiation with respect to a parameter) *Let Y, Z be real Banach spaces, let E be an open set in $\mathbf{R} \times Y \times Z$, let $f : E \to Y$ be a continuous function whose partial differentials $d_2 f$ and $d_3 f$ are continuous on E, and for each $(t_0, y_0, z_0) \in E$ let $t \mapsto \phi(t, t_0, y_0, z_0)$ be the solution of the equation $y' = f(t, y, z_0)$ taking the value y_0 at t_0 and having maximal interval of existence, $I(t_0, y_0, z_0)$, say. Then*

(i) *the domain of the function ϕ, i.e. the set A of points (t, t_0, y_0, z_0) such that $(t_0, y_0, z_0) \in E$ and $t \in I(t_0, y_0, z_0)$, is open in $\mathbf{R} \times \mathbf{R} \times Y \times Z$, and ϕ is C^1 on A,*

(ii) *the function $t \mapsto d_3\phi(t, t_0, y_0, z_0)$ is the solution ξ of the differential equation*

$$\xi' = d_2 f(t, \phi(t, t_0, y_0, z_0), z_0) \circ \xi$$

on $I(t_0, y_0, z_0)$ such that $\xi(t_0) = 1_Y$,

(iii) $\partial\phi(t, t_0, y_0, z_0)/\partial t_0 = - d_3\phi(t, t_0, y_0, z_0) f(t_0, y_0, z_0),$

(iv) *the function* $t \mapsto d_4\phi(t,t_0,y_0,z_0)$ *is the solution* ζ *of the differential equation*

$$\zeta' = d_2 f(t,\phi(t,t_0,y_0,z_0),z_0)\circ\zeta + d_3 f(t,\phi(t,t_0,y_0,z_0),z_0)$$

on $I(t_0,y_0,z_0)$ *such that* $\zeta(t_0) = 0$,

(v) *if* f *is* C^r *on* E, ϕ *is* C^r *on* A.

By (2.10.1. Corollary), A is open. Also (ii) and (iii) follow from (3.10.5)(ii) and (iii), and (ii), (iii), and (iv) together imply that ϕ is C^1 on A and that ϕ is C^r if f is C^r. It is therefore enough to prove (iv).

As remarked in §2.10, the equation $y' = f(t,y,z_0)$ subject to the initial condition that $y = y_0$ when $t = t_0$ is equivalent to the pair of equations

$$y' = f(t,y,z), \qquad z' = 0 \tag{6}$$

subject to the initial conditions that $y = y_0$ and $z = z_0$ when $t = t_0$. Hence if $w = (y,z)$ and $g(t,w) = (f(t,y,z),0)$, then the equations (6) subject to the stated conditions are equivalent to the equation $w' = g(t,w)$ subject to the condition that $w = w_0 = (y_0,z_0)$ when $t = t_0$. We apply (3.10.5) to this last equation.

If $t \mapsto \psi(t,t_0,w_0)$ is the solution of $w' = g(t,w)$ taking the value w_0 at t_0, then, by (3.10.5), ψ is C^1, and $d_3\psi(t,t_0,w_0) = \theta(t)$, where θ is the solution of

$$\theta' = d_2 g(t,\psi(t,t_0,w_0))\circ\theta, \qquad \theta(t_0) = 1_{Y \times Z}.$$

Since $\psi(t,t_0,w_0) = (\phi(t,t_0,y_0,z_0),z_0)$ this implies that ϕ is C^1, and that

$$(d_4\phi(t,t_0,y_0,z_0)k,k) = d_3\psi(t,t_0,w_0)(0,k) = \theta(t)(0,k),$$

i.e. the expression on the left is the solution of

$$\lambda' = d_2 g(t,\psi(t,t_0,w_0))\lambda, \qquad \lambda(t_0) = (0,k).$$

Writing $\lambda(t) = (\mu(t),\nu(t))$, and noting that for all $(a,b)\in Y \times Z$

$$d_2 g(t,w)(a,b) = (d_2 f(t,y,z)a + d_3 f(t,y,z)b, 0),$$

we deduce that $\nu(t) = k$ and that

$$\mu'(t) = d_2 f(t,\phi(t,t_0,y_0,z_0),z_0)\mu(t) + d_3 f(t,\phi(t,t_0,y_0,z_0),z_0)k. \tag{7}$$

If now ζ satisfies the equation in (iv), then $t \mapsto \zeta(t)k$ satisfies (7), whence $\mu(t) = \zeta(t)k$, i.e. $d_4\phi(t,t_0,y_0,z_0)k = \zeta(t)k$ for all $k\in Z$, and this gives the result.

3.11 Applications of Fréchet differentiation to the calculus of variations

The calculus of variations deals with the problem of determining the minima of certain real-valued functions, usually called *functionals*, whose domains are sets of functions.† In this section we consider the

---

† Since real-valued functions are not differentiable over complex spaces, our hypotheses will imply that the field of scalars is also **R**.

simplest such problem, which concerns a functional F of the form

$$F(\phi) = \int_a^b f(t, \phi(t), \phi'(t))dt, \tag{1}$$

and where we have to find the function (or functions) ϕ, continuously differentiable on the interval $[a, b]$ and taking given values at a and b, for which $F(\phi)$ is a minimum.

Historically, the first problem of this type was posed by John Bernoulli in 1696, namely to find the curve in a vertical plane joining two given points P, Q such that the time taken by a heavy particle to slide along the curve from P to Q under the influence of gravity is a minimum (the curve with this property is called the *brachistochrone*). To find the functional F in this case, let the curve have equation $y = \phi(x)$, where the positive direction of the y-axis is vertically downwards, let $P = (a, c), Q = (b, d)$, where $a < b, c < d$, and let u be the (initial) speed of the particle at P. Then the speed v of the particle along the curve at (x, y) is given by

$$v^2 - u^2 = 2g(y - c).$$

Hence, writing $\alpha = c - u^2/2g$ and denoting arc length by s, we see that the time taken by the particle to slide from P to Q is

$$\int \frac{ds}{(2g(y - \alpha))^{1/2}} = \int_a^b \left(\frac{1 + y'^2}{2g(y - \alpha)} \right)^{1/2} dx,$$

so that we have to find the ϕ that minimizes

$$F(\phi) = \int_a^b \left(\frac{1 + (\phi'(x))^2}{2g(\phi(x) - \alpha)} \right)^{1/2} dx$$

and takes the values c, d at a, b, respectively.

We now reformulate the problem of minimizing the general functional F given by (1). Let Y be a real normed space, let $J = [a, b]$, let A be an open set in the metric space $J \times Y \times Y$,† let $f : A \to \mathbf{R}$ be a continuous function whose partial differentials $d_2 f$ and $d_3 f$ are continuous on A, and let F be defined by (1), where the domain $\mathscr{F}$ of F is the set of $\phi \in C^1(J, Y)$ such that $(t, \phi(t), \phi'(t)) \in A$ for all $t \in J$. By the remark following (3.9.3), $\mathscr{F}$ is open in $C^1(J, Y)$, F is C^1 on $\mathscr{F}$, and for all $\phi \in \mathscr{F}$ and $\psi \in C^1(J, Y)$ we have

$$dF(\phi)\psi = \int_a^b (d_2 f(t, \phi(t), \phi'(t))\psi(t) + d_3 f(t, \phi(t), \phi'(t))\psi'(t))dt.$$

Next, let c, d be given elements of Y, let $\mathscr{E}$ be the set of $\phi \in \mathscr{F}$ such that $\phi(a) = c, \phi(b) = d$, and let $F_{\mathscr{E}}$ be the restriction of F to $\mathscr{E}$. We wish to solve

† i.e. the intersection of $J \times Y \times Y$ with an open set in $\mathbf{R} \times Y \times Y$.

the problem **(P)**: *to determine necessary conditions for a function $\phi \in \mathscr{E}$ to be a local minimum of $F_{\mathscr{E}}$.*

Let $C_0^1(J, Y)$ be the closed linear subspace of $C^1(J, Y)$ consisting of those $\psi \in C^1(J, Y)$ such that $\psi(a) = \psi(b) = 0$, let $\phi \in \mathscr{E}$, and let $G(\psi) = F_{\mathscr{E}}(\phi + \psi)$ $(\phi + \psi \in \mathscr{E})$. Then the domain of G is obviously an open set in $C_0^1(J, Y)$, and ϕ is a local minimum of $F_{\mathscr{E}}$ if and only if 0 is a local minimum of G. Since the differential of G at 0 is clearly equal to the restriction of $dF(\phi)$ to $C_0^1(J, Y)$, we deduce from (3.6.4) that if ϕ is a local minimum of $F_{\mathscr{E}}$, then $dF(\phi)\psi = 0$ for all $\psi \in C_0^1(J, Y)$. Hence we have proved:

(3.11.1) *Let A be an open set in the metric space $J \times Y \times Y$, let $f : A \to \mathbf{R}$ be a continuous function whose partial differentials $d_2 f$ and $d_3 f$ are continuous on A, and let*

$$F(\phi) = \int_a^b f(t, \phi(t), \phi'(t))dt,$$

where the domain $\mathscr{F}$ of F is the set of $\phi \in C^1(J, Y)$ such that $(t, \phi(t), \phi'(t)) \in A$ for all $t \in J$. Let also $c, d \in Y$, let $\mathscr{E}$ be the set of $\phi \in \mathscr{F}$ such that $\phi(a) = c$, $\phi(b) = d$, and let $F_{\mathscr{E}}$ be the restriction of F to $\mathscr{E}$. If ϕ is a local minimum of $F_{\mathscr{E}}$, then for all $\psi \in C_0^1(J, Y)$

$$\int_a^b (d_2 f(t, \phi(t), \phi'(t))\psi(t) + d_3 f(t, \phi(t), \phi'(t))\psi'(t))dt = 0. \tag{2}$$

Suppose next that in addition to the hypotheses of (3.11.1) the partial differentials $d_2^2 f$, $d_2 d_3 f$, and $d_3^2 f$ are defined and continuous on A. Then F is twice Fréchet differentiable on $\mathscr{F}$, and for all $\phi \in \mathscr{F}$ and $\psi \in C^1(J, Y)$

$$d^2 F(\phi)(\psi)^2 = \int_a^b (d_2^2 f(\Phi(t))(\psi(t))^2 + 2d_2 d_3 f(\Phi(t))(\psi(t), \psi'(t))$$
$$+ d_3^2 f(\Phi(t))(\psi'(t))^2)dt,$$

where $\Phi(t) = (t, \phi(t), \phi'(t))$. The second differential of G at 0 is then the restriction of $d^2 F(\phi)$ to $C_0^1(J, Y) \times C_0^1(J, Y)$, and therefore, by (3.6.4), a necessary condition for ϕ to be a local minimum of $F_{\mathscr{E}}$ is that $d^2 F(\phi)(\psi)^2 \geq 0$ for all $\psi \in C_0^1(J, Y)$. Hence we have:

(3.11.2) *Suppose that in addition to the hypotheses of (3.11.1) the partial differentials $d_2^2 f, d_2 d_3 f$, and $d_3^2 f$ are defined and continuous on A. Then for all $\psi \in C_0^1(J, Y)$*

$$\int_a^b (d_2^2 f(\Phi(t))(\psi(t))^2 + 2d_2 d_3 f(\Phi(t))(\psi(t), \psi'(t)) + d_3^2 f(\Phi(t))(\psi'(t))^2)dt \geq 0,$$

$$\tag{3}$$

where $\Phi(t) = (t, \phi(t), \phi'(t))$.

(3.11.2. Corollary) *If the hypotheses of* (3.11.2) *are satisfied and* $t \mapsto d_2 d_3 f(\Phi(t))$ *has a derivative at each point of J, then for all* $\psi \in C_0^1(J, Y)$

$$\int_a^b (P(\psi'(t))^2 + Q(\psi(t))^2) dt \geq 0, \tag{4}$$

where $P = d_3^2 f(\Phi(t))$ *and* $Q = d_2^2 f(\Phi(t)) - d(d_2 d_3 f(\Phi(t)))/dt$.

This follows easily from (3), for if $R(t) = d_2 d_3 f(\Phi(t))$, then

$$\frac{d}{dt}(R(t)(\psi(t))^2) = R'(t)(\psi(t))^2 + 2R(t)(\psi(t), \psi'(t)),$$

so that (since $\psi(a) = \psi(b) = 0$)

$$2 \int_a^b R(t)(\psi(t), \psi'(t)) dt = - \int_a^b R'(t)(\psi(t))^2 dt.$$

We prove next three lemmas which enable us to express the conditions (2) and (4) in a simpler form. Here, as later, Y' is the dual space of Y.

(3.11.3) Lemma. *Let* $v : J \to Y'$ *be continuous. Then*

$$\int_a^b v(t)\psi'(t) dt = 0 \tag{5}$$

for all $\psi \in C_0^1(J, Y)$ *if and only if* v *is constant.*

The proof of the 'if' is trivial. To prove the 'only if', observe that if $\psi \in C_0^1(J, \mathbf{R})$ and $y \in Y$ then $t \mapsto \psi(t)y$ is in $C_0^1(J, Y)$. Hence for any $k \in Y'$, $y \in Y$, and $\psi \in C_0^1(J, \mathbf{R})$,

$$\int_a^b (v(t) - k)y\psi'(t) dt = 0.$$

Take

$$k = \int_a^b v(t) dt / (b - a) \quad \text{and} \quad \psi(t) = \int_a^t (v(s) - k)y \, ds.$$

Then $\psi \in C_0^1(J, \mathbf{R})$, and the condition gives

$$\int_a^b ((v(t) - k)y)^2 dt = 0,$$

so that $v(t) = k$ for all t.

(3.11.4) Lemma. *Let* $v, w : J \to Y'$ *be continuous. Then*

$$\int_a^b (v(t)\psi(t) + w(t)\psi'(t)) dt = 0 \tag{6}$$

for all $\psi \in C_0^1(J, Y)$ *if and only if* $t \mapsto w(t) - \int_a^t v(s) ds$ *is constant on J.*

Let $V(t) = \int_a^t v(s) ds$ $(t \in J)$. Integrating the first term in the integral by parts and remembering that $\psi(a) = \psi(b) = 0$ we get

$$\int_a^b (v(t)\psi(t) + w(t)\psi'(t))dt = \int_a^b (-V(t)\psi'(t) + w(t)\psi'(t))dt.$$

The result therefore follows from (3.11.3).

(3.11.5) Lemma. *Let* $P, Q : J \to \mathscr{L}_2(Y; \mathbf{R})$ *be continuous functions such that*

$$\int_a^b (P(t)(\psi'(t))^2 + Q(t)(\psi(t))^2)dt \geq 0$$

for all $\psi \in C_0^1(J, Y)$. *Then* $P(t)(y)^2 \geq 0$ *for all* $t \in J$ *and* $y \in Y$.

Suppose that the result is false. Then we can find $t_0 \in J^\circ$ and $y \in Y$ such that $P(t_0)(y)^2 = -\alpha < 0$. Since P is continuous, it follows that for all sufficiently small positive ε the interval $I = [t_0 - \varepsilon, t_0 + \varepsilon]$ is contained in J and $P(t)(y)^2 < -\frac{1}{2}\alpha$ for all $t \in I$. Let $\chi : J \to \mathbf{R}$ be the function taking the value 0 outside I and given in that interval by

$$\chi(t) = \frac{2\varepsilon^{1/2}}{\pi} \cos^2 \left(\frac{\pi(t - t_0)}{2\varepsilon} \right).$$

Then $|\chi(t)|^2 \leq 4\varepsilon/\pi$ for all t, and $\int_{t_0-\varepsilon}^{t_0+\varepsilon} (\chi'(t))^2 dt = 1$. Taking $\psi(t) = \chi(t)y$, we therefore have

$$-\frac{\alpha}{2} = -\frac{\alpha}{2} \int_{t_0-\varepsilon}^{t_0+\varepsilon} (\chi'(t))^2 dt \geq \int_{t_0-\varepsilon}^{t_0+\varepsilon} P(t)(\chi'(t)y)^2 dt$$

$$= \int_a^b P(t)(\chi'(t)y)^2 dt \geq -\int_a^b Q(t)(\chi(t)y)^2 dt \geq -\frac{4\varepsilon}{\pi} \sup(Q(t)(y)^2)$$

for all sufficiently small ε, and this gives a contradiction.

From (3.11.1) and (3.11.4) we now have:

(3.11.6) *Under the hypotheses of* (3.11.1) *there exists* $k \in Y'$ *such that for all* $t \in J$

$$d_3 f(\Phi(t)) = \int_a^t d_2 f(\Phi(s))ds + k,$$

where $\Phi(t) = (t, \phi(t), \phi'(t))$. *Hence also the function* $t \mapsto d_3 f(\Phi(t))$ *has a derivative at each* $t \in J$ *and*

$$\frac{d}{dt} d_3 f(\Phi(t)) = d_2 f(\Phi(t)). \tag{7}$$

The differential equation (7) is called the *Euler equation* for the problem

(P) (p. 226), and any solution ϕ of this equation is called an *extremal of the functional F*. Theorem (3.11.6) asserts that any local minimum of $F_{\mathscr{E}}$ is an extremal of F. However, an extremal ϕ satisfying the boundary conditions $\phi(a) = c, \phi(b) = d$ is not necessarily a local minimum (or local maximum) of $F_{\mathscr{E}}$ (see Exercise 3.11.4). The problem of determining sufficient conditions for an extremal to be a local minimum of $F_{\mathscr{E}}$ is more difficult, and we do not consider it here.

Example (a)

If f does not depend explicitly on y, the functional F is of the form

$$F(\phi) = \int_a^b g(s, \phi'(s))ds,$$

and in this case the Euler equation (7) becomes

$$\frac{d}{dt}d_2 g(t, \phi'(t)) = 0,$$

whence

$$d_2 g(t, \phi'(t)) = \alpha,$$

where α is a constant. If $Y = \mathbf{R}$ this last equation is equivalent to the first-order equation

$$D_2 g(t, \phi'(t)) = \alpha,$$

while if $Y = \mathbf{R}^m$ it is equivalent to the system of equations

$$D_i g(t, \phi'_1(t), \dots, \phi'_m(t)) = \alpha_i \qquad (i = 2, \dots, m+1),$$

where $\phi_1, \dots, \phi_m$ are the components of ϕ. For instance, if $Y = \mathbf{R}$ and $g(t, y') = t^{-3}y'^2$, then the Euler equation becomes

$$2t^{-3}y' = \alpha,$$

with the solution

$$y = \tfrac{1}{8}\alpha t^4 + \beta.$$

So far we have considered only the consequences of (3.11.1). Similarly, from (3.11.2. Corollary) and (3.11.5) we have:

(3.11.7) *Under the hypotheses of* (3.11.2. Corollary),

$$d_3^2 f(t, \phi(t), \phi'(t))(h)^2 \geq 0 \tag{8}$$

for all $t \in J$ *and* $h \in Y$.

If $Y = \mathbf{R}^m$ and we write $z = (t, y, y') = (t, y_1, \dots, y_m, y'_1, \dots, y'_m)$, the condition (8) is equivalent to the condition that for all $t \in J$ and $h = (h_1, \ldots, h_m) \in \mathbf{R}^m$

$$\sum_{i,j=1}^{m} h_i h_j \frac{\partial^2 f}{\partial y_i' \partial y_j'} \geq 0, \qquad (9)$$

where the partial derivatives are evaluated at $(t, \phi(t), \phi'(t))$.

To make further progress we have to impose extra conditions on f, and we now suppose that the domain of f is an open set E in $\mathbf{R} \times Y \times Y$, that f is C^r on E, where $r \geq 2$, and that $d_3^2 f(t, y, y') \in \mathscr{L}\mathscr{H}(Y, Y')$ for all $(t, y, y') \in E$.† Thus among other things our hypotheses require that Y is linearly homeomorphic to its dual space Y'. This last requirement is, of course, satisfied when Y is a Hilbert space,‡ and in this particular case we can reformulate our results in terms of gradients.

The following theorem shows that under these extra hypotheses each extremal inherits the differentiability properties of f.

(3.11.8) *Let E be an open set in $\mathbf{R} \times Y \times Y$, let $f : E \to \mathbf{R}$ be C^r, where $r \geq 2$, and suppose that $d_3^2 f(t, y, y') \in \mathscr{L}\mathscr{H}(Y, Y')$ for all $(t, y, y') \in E$. Let also F be defined as in (3.11.1) (where now A is the intersection of E and $J \times Y \times Y$), and let $\phi \in C^1(J, Y)$ be an extremal of F. Then $\phi \in C^r(J, Y)$ and for all $t \in J$*

$$d_1 d_3 f(\Phi(t))1 + d_2 d_3 f(\Phi(t))\phi'(t) + d_3^2 f(\Phi(t))\phi''(t) = d_2 f(\Phi(t)), \quad (10)$$

where $\Phi(t) = (t, \phi(t), \phi'(t))$.*

Let t be a given point of J, let s be a point of J distinct from t, let $P = \Phi(t)$, $Q = \Phi(s)$, and let $\delta(s) = (\phi'(s) - \phi'(t))/(s - t)$. Since $d_3 f$ is differentiable at P, we can write

$$\begin{aligned} d_3 f(Q) - d_3 f(P) = &\, d_1 d_3 f(P)(s - t) + d_2 d_3 f(P)(\phi(s) - \phi(t)) \\ &+ d_3^2 f(P)(\phi'(s) - \phi'(t)) \\ &+ \omega(s)(|s - t| + \|\phi(s) - \phi(t)\| + \|\phi'(s) - \phi'(t)\|), \end{aligned}$$

where $\omega(s) \to 0$ as $s \to t$. By (3.11.6), the expression

$$(d_3 f(Q) - d_3 f(P))/(s - t)$$

† It should be stressed that $y' \in Y$ (so that $d_3 f(t, y, y') \in Y'$).

‡ The condition that Y is linearly homeomorphic to Y' does not imply that there is an equivalent norm on Y under which Y is an inner product space. This is easily seen by taking Y to be the product $X \times X'$, where X is a reflexive normed space which does not have an equivalent norm under which it is an inner product space.

* In the formula (10) the second differentials are to be regarded as linear rather than bilinear mappings; thus $d_1 d_3 f(\Phi(t)) \in \mathscr{L}(\mathbf{R}, \mathscr{L}(Y, \mathbf{R}))$, and its value at $1 \in \mathbf{R}$ is $d_1 d_3 f(\Phi(t))1 \in \mathscr{L}(Y, \mathbf{R}) = Y'$.

tends to the limit $d_2 f(P)$ as $s \to t$, and therefore

$$d_3^2 f(P)\delta(s) + \omega(s)\|\delta(s)\| \to d_2 f(P) - d_1 d_3 f(P)1 - d_2 d_3 f(P)\phi'(t)$$

as $s \to t$. Since the norm of the expression on the left here is greater than or equal to $(K - \|\omega(s)\|)\|\delta(s)\|$, where $1/K = \|d_3^2 f(P)^{-1}\|$, this implies first that $\delta(s)$ is bounded, and then that $\delta(s)$ tends to the limit

$$d_3^2 f(P)^{-1}(d_2 f(P) - d_1 d_3 f(P)1 - d_2 d_3 f(P)\phi'(t)) \tag{11}$$

as $s \to t$. Hence $\phi''(t)$ exists and is equal to the expression in (11), and this gives (10) and shows also that $\phi'' \in C^{r-2}(J, Y)$, whence $\phi \in C^r(J, Y)$.

The condition that f is C^r can obviously be weakened. For instance, to ensure that ϕ'' exists and satisfies (10) on J, it is enough that $d_2 f$ and $d_3 f$ exist on E and $d_3 f$ is differentiable on E.

We obtain the following corollary by calculating the derivative on the left side of (12) and evaluating (10) at the point $\phi'(t)$.

(3.11.8. Corollary) *Under the hypotheses of* (3.11.8)

$$\frac{d}{dt}(f(\Phi(t)) - d_3 f(\Phi(t))\phi'(t)) = D_1 f(\Phi(t)). \tag{12}$$

Suppose now that Y is a real Hilbert space. Then the partial gradients $\nabla_2 f(t, y, y')$ and $\nabla_3 f(t, y, y')$ of f at the point (t, y, y') of E (cf. Exercise 3.5.1) are the (unique) elements of Y such that for all $h \in Y$

$$d_2 f(t, y, y')h = \langle \nabla_2 f(t, y, y'), h \rangle \quad \text{and} \quad d_3 f(t, y, y')h = \langle \nabla_2 f(t, y, y'), h \rangle.$$

Equivalently, if I is the isometric isomorphism of Y' onto Y such that $u(h) = \langle Iu, h \rangle$ for all $u \in Y'$ and $h \in Y$, then

$$\nabla_2 f(t, y, y') = I(d_2 f(t, y, y')) \quad \text{and} \quad \nabla_3 f(t, y, y') = I(d_3 f(t, y, y')).$$

We thus have $\nabla_2 f = I \circ d_2 f$ and $\nabla_3 f = I \circ d_3 f$, whence it follows by the chain rule that

$$d_2 \nabla_3 f(t, y, y') = I \circ d_2 d_3 f(t, y, y') \quad \text{and} \quad d_3 \nabla_3 f(t, y, y') = I \circ d_3^2 f(t, y, y').$$

From the second relation here we see that $d_3^2 f(t, y, y') \in \mathscr{LH}(Y, Y')$ if and only if $d_3 \nabla_3 f(t, y, y') \in \mathscr{LH}(Y, Y)$. In particular, when $Y = \mathbf{R}^m$ the matrix of $d_3 \nabla_3 f(t, y, y')$ with respect to the natural basis in $\mathbf{R}^m$ is

$$\left[\frac{\partial^2 f(z)}{\partial y_i' \partial y_j} \right] \tag{13}$$

where $z = (t, y, y') = (t, y_1, \ldots, y_m, y_1', \ldots, y_m')$ (cf. Exercise 3.5.1), so that $d_3^2 f(t, y, y') \in \mathscr{LH}(Y, Y')$ if and only if this matrix (13) is non-singular. We note in passing that by Exercise 3.6.5 this last condition is satisfied if and only if the condition (9) holds with strict inequality for all non-zero $h \in \mathbf{R}^m$.

Next, by applying I to both sides of the equation (10), we see that (10) is equivalent to the equation

$$D_1 \nabla_3 f(\Phi(t)) + d_2 \nabla_3 f(\Phi(t)) \phi'(t) + d_3 \nabla_3 f(\Phi(t)) \phi''(t) = \nabla_2 f(\Phi(t)).$$

In particular, when $Y = \mathbf{R}^m$ the equation (10) is equivalent to the system of equations

$$\frac{\partial^2 f(z)}{\partial t \partial y_i'} + \sum_{j=1}^{m} \frac{\partial^2 f(z)}{\partial y_i \partial y_j'} \phi_j'(t) + \sum_{j=1}^{m} \frac{\partial^2 f(z)}{\partial y_i' \partial y_j'} \phi_j''(t) = \frac{\partial f(z)}{\partial y_i} \qquad (i = 1, \ldots, m),$$

where $\phi_1, \ldots, \phi_m$ are the components of ϕ and $z = (t, \phi_1(t), \ldots, \phi_m(t), \phi_1'(t), \ldots, \phi_m'(t))$. When $m = 1$, this reduces to the single equation

$$\frac{\partial^2 f(z)}{\partial t \partial y'} + \frac{\partial^2 f(z)}{\partial y \partial y'} \phi'(t) + \frac{\partial^2 f(z)}{\partial y'^2} \phi''(t) = \frac{\partial f(z)}{\partial y}.$$

In the same way, when Y is a Hilbert space the equation (12) of (3.11.8. Corollary) can be written as

$$\frac{d}{dt}(f(\Phi(t)) - \langle \nabla_3 f(\Phi(t)), \phi'(t) \rangle) = D_1 f(\Phi(t)), \tag{14}$$

with the corresponding interpretation when $Y = \mathbf{R}^m$.

Example (b)

Let Y be a Hilbert space, let g be a C^1 function from an open set E in $\mathbf{R} \times Y$ into $]0, \infty[$, and let $f : E \times Y \to \mathbf{R}$ be given by

$$f(t, y, y') = g(t, y)(1 + \|y'\|^2)^{1/2}$$

(in this case F represents the integral of $g(t, y)$ taken with respect to arc length $ds = (1 + \|y'\|^2)^{1/2} dt)$.

If $L = (1 + \|y'\|^2)^{1/2}$ and $z = (t, y, y') \in E \times Y$, then

$$d_3 f(z) h = g(t, y) L^{-1} \langle y', h \rangle$$

for all $h \in Y$, so that

$$\nabla_3 f(z) = g(t, y) L^{-1} y',$$

and therefore

$$d_3 \nabla_3 f(z) h = g(t, y)(L^{-1} h - L^{-3} \langle y', h \rangle y').$$

To prove that $d_3^2 f(z) \in \mathscr{LH}(Y, Y')$, or equivalently, that $d_3 \nabla_3 f(z) \in \mathscr{LH}(Y, Y)$, we have to show that if $w \in Y$ there exists a unique $h \in Y$ such that

$$g(t, y)(L^{-1} h - L^{-3} \langle y', h \rangle y') = w. \tag{15}$$

First, if such an h exists, it is unique, for taking the inner product of both sides of (15) with respect to y' we get

$$\langle y', h \rangle = L^3 \langle w, y' \rangle / g(t, y), \tag{16}$$

and therefore

$$h = L(w + \langle w, y' \rangle y') / g(t, y). \tag{17}$$

Conversely, if h is given by (17), then $\langle y', h \rangle$ satisfies (16) and therefore (15) holds. Hence $d_3^2 f(z) \in \mathscr{LH}(Y, Y')$, as required.

We can now apply (3.11.8. Corollary) in the form (14), whence we deduce that the Euler equation for F can be written in the form

$$\frac{d}{dt}((1 + \| \phi'(t) \|^2)^{-1/2} g(t, \phi(t))) = (1 + \| \phi'(t) \|^2)^{1/2} D_1 g(t, \phi(t)).$$

In particular, if g is independent of t, this gives

$$g(\phi(t)) = \alpha(1 + \| \phi'(t) \|^2)^{1/2}, \tag{18}$$

where α is a constant.

For instance, if $Y = \mathbf{R}$ and $g(y) = y$, then (18) becomes

$$y = \alpha(1 + y'^2)^{1/2},$$

and under the condition that $y > 0$ this equation is easily integrated by separation of variables to give

$$y = \alpha \cosh\left(\frac{t - \beta}{\alpha}\right). \tag{19}$$

In this particular case, $2\pi F(\phi)$ is the area of the surface of revolution generated by rotating the curve $y = \phi(t)$ about the t-axis (Fig. 3.1), so that here the problem **(P)** (p. 226) is that of finding the curve which generates the minimum area for given a, b, c, d. The extremal curve given by (19) is a catenary and it can be shown that the family of such catenaries which pass

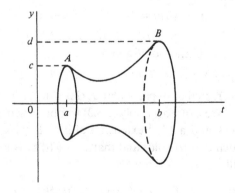

Figure 3.1

through the initial point $A = (a, c)$ has an envelope C lying above the t-axis. If the point $B = (b, d)$ lies above this envelope, there are two catenaries of the family through B, and the upper one gives the minimum area. If B lies on the envelope, there is exactly one catenary of the family through B, but this catenary gives an area of revolution bigger than the total area of the two end discs, and the minimum is not attained for a ϕ in $C^1(J, \mathbf{R})$. Finally, if B lies below the envelope, there is no catenary of the family through B, and again the minimum is not attained.

To conclude this section we consider the transformation of the Euler equation (7) by the use of the canonical coordinates. We prove first:

(3.11.9) *Let Y be a real Banach space, let E be an open set in $\mathbf{R} \times Y \times Y$, let $f : E \to \mathbf{R}$ be C^r, where $r \geq 2$, and suppose that $d_3^2 f(t, y, y') \in \mathscr{L}\mathscr{H}(Y, Y')$ for all $(t, y, y') \in E$. Let also F be defined as in (3.11.1) (where again A is the intersection of E and $J \times Y \times Y$), let $\phi : J \to Y$ be an extremal for F, let $\theta(t) = d_3 f(t, \phi(t), \phi'(t))$, and let*

$$K = \{(t, \phi(t), \theta(t)) : t \in J\}, \quad L = \{(t, \phi(t), \phi'(t), \theta(t)) : t \in J\}.$$

Then there exist an open set B in $\mathbf{R} \times Y \times Y'$ containing K, an open set A_1 in $E \times Y'$ containing L, and a C^{r-1} function $g : B \to Y$ such that
(i) *given $t \in \mathbf{R}$, $y \in Y$, $u \in Y'$ for which $(t, y, u) \in B$, the only point $y' \in Y$ for which both $u = d_3 f(t, y, y')$ and $(t, y, y', u) \in A_1$ is given by $y' = g(t, y, u)$,*
(ii) *$y = \phi(t), u = \theta(t)$ satisfy on J the system of differential equations*

$$y' = g(t, y, u), \quad u' = d_2 f(t, y, g(t, y, u)). \qquad (20)$$

It is important to note here that g is defined globally, not just locally.

Let T be a linear homeomorphism of Y' onto Y (for instance the inverse of $d_3^2 f(t, y, y')$ for some fixed point (t, y, y') of E), and let $G : E \times Y' \to Y$ be given by

$$G(t, y, y', u) = T(d_3 f(t, y, y') - u).$$

Define $g : K \to Y$ by $g(t, \phi(t), \theta(t)) = \phi'(t)$ for $t \in J$. Then G vanishes on the graph L of g by definition of θ, and

$$d_3 G(t, y, y', u) = T \circ d_3^2 f(t, y, y') \in \mathscr{L}\mathscr{H}(Y, Y)$$

for $(t, y, y', u) \in E \times Y$, so that in particular $d_3 G$ is a linear homeomorphism at the points of L. We can now apply (3.8.3) to obtain sets B, A_1 with the properties specified, and a C^{r-1} extension of g to B, which we continue to denote by g, such that (i) holds and that $y' = \phi'(t)$ satisfies the equation $y' = \bar{g}(t, y, u)$ for all $(t, y, u) \in K$. Since also

$$\theta'(t) = \frac{d}{dt} d_3 f(t, \phi(t), \phi'(t)) = d_2 f(t, \phi(t), \phi'(t)),$$

(using the Euler equation (7)) it follows that $y = \phi(t), u = \theta(t)$ satisfy the equations (20) on J, and this completes the proof.

The *Hamiltonian corresponding to the functional F* is the function $\hat{H}: B \to \mathbf{R}$ given by

$$\hat{H}(t,y,u) = u(y') - f(t,y,y'),$$

where $y' = g(t,y,u)$. Since $u = d_3 f(t,y,y')$, we have

$$d_1\hat{H}(t,y,u) = u \circ d_1 g(t,y,u) - d_1 f(t,y,y') - d_3 f(t,y,y') \circ d_1 g(t,y,u)$$
$$= -d_1 f(t,y,y'),$$
$$d_2\hat{H}(t,y,u) = u \circ d_2 g(t,y,u) - d_2 f(t,y,y') - d_3 f(t,y,y') \circ d_2 g(t,y,u)$$
$$= -d_2 f(t,y,y'),$$

and for all $v \in Y'$

$$d_3\hat{H}(t,y,u)v = u(d_3 g(t,y,u)v) + v(g(t,y,u)) - d_3 f(t,y,y')(d_3 g(t,y,u)v)$$
$$= v(g(t,y,u)). \tag{21}$$

It follows from these formulae that $d_1\hat{H}, d_2\hat{H}$, and $d_3\hat{H}$ are C^{r-1} on B, so that $\hat{H}$ is C^r.

We note that $d_3\hat{H}(t,y,u) \in Y''$; moreover, if we embed Y in Y'' by the canonical mapping $y \mapsto w$, where $w(v) = v(y)$ $(v \in Y')$, then (21) asserts that $g(t,y,u)$ is identified with $d_3\hat{H}(t,y,u)$. Hence with this identification the pair of equations (20) can be written in the form

$$y' = d_3\hat{H}(t,y,u), \quad u' = -d_2\hat{H}(t,y,u). \tag{22}$$

We observe also that the equation

$$\frac{d}{dt}(f(t,y,y') - d_3 f(t,y,y')y') = D_1 f(t,y,y'),$$

which holds along any extremal (cf. (3.11.8. Corollary)), takes the form

$$\frac{d\hat{H}}{dt} = D_1\hat{H}.$$

In particular, if f is independent of t, then so is $\hat{H}$, whence $d\hat{H}/dt = 0$, so that $\hat{H}$ is constant on the set K.

When Y is a real Hilbert space, it is convenient to replace the variable u in $\hat{H}$ by a variable p taking values in Y and to reformulate our results in terms of gradients. As before, let I be the isometry of Y' onto Y such that $u(y) = \langle Iu, y \rangle$ for all $u \in Y'$ and $y \in Y$, and let

$$P(t,y,p) = g(t,y,I^{-1}p), \quad H(t,y,p) = \hat{H}(t,y,I^{-1}p),$$

where the domain of P and H is the set C of points $(t,y,p) \in \mathbf{R} \times Y \times Y$ such that $(t,y,I^{-1}p) \in B$. Then P is C^{r-1}, H is C^r, and

$$H(t,y,p) = \langle p,y' \rangle - f(t,y,y'), \tag{23}$$

where $y' = P(t, y, p)$. Clearly

$$p = Iu = I(d_3 f(t, y, y')) = \nabla_3 f(t, y, y') \tag{24}$$

whenever $(t, y, p) \in C$, and

$$d_2 H(t, y, p) = d_2 \hat{H}(t, y, u) = - d_2 f(t, y, y').$$

Also, by the chain rule and (21), for all $h \in Y$ we have

$$d_3 H(t, y, p)h = d_3 \hat{H}(t, y, I^{-1}p)(I^{-1}h) = (I^{-1}h)(P(t, y, p)) = \langle P(t, y, p), h \rangle,$$

so that

$$P(t, y, p) = \nabla_3 H(t, y, p).$$

Hence the pair of equations (20) (cf. (22)) can be written in the form

$$y' = \nabla_3 H(t, y, p), \quad p' = - \nabla_2 H(t, y, p). \tag{25}$$

In particular, if $Y = \mathbf{R}^m, y = (y_1, \ldots, y_m), y' = (y'_1, \ldots, y'_m)$, and $p = (p_1, \ldots, p_m)$, then by (24)

$$P_i = \frac{\partial f}{\partial y'_i} \qquad (i = 1, \ldots, m),$$

and by (25) the pair of equations (20) is equivalent to the system of equations

$$y'_i = \frac{\partial H}{\partial p_i}, \quad p'_i = - \frac{\partial H}{\partial y_i} \qquad (i = 1, \ldots, m).$$

The function H defined by (23) is again called the *Hamiltonian corresponding to the functional F*, the variables t, y, p are called the *canonical coordinates*, and the equations (25) are called the *Euler equations in canonical form*.

Example (c)

If $Y = \mathbf{R}$ and $f(t, y, y') = t^2 y'^2 + 6y^2 \ (t > 0)$, then

$$p = \frac{\partial f}{\partial y'} = 2t^2 y',$$

so that $y' = p/2t^2$. Hence the Hamiltonian is given by

$$H(t, y, p) = py' - (t^2 y'^2 + 6y^2) = p^2/4t^2 - 6y^2,$$

and the Euler equations in canonical form are

$$y' = \frac{\partial H}{\partial p} = p/2t^2, \quad p' = - \frac{\partial H}{\partial y} = 12y.$$

Example (d)

Suppose that we have a system of n particles in $\mathbf{R}^3$, where the ith particle has mass m_i, and that when the ith particle is at the point (x_i, y_i, z_i) the potential energy of the system is $U(t, x_1, y_1, z_1, \ldots, x_n, y_n, z_n)$, so that the external force acting on the ith particle has components

$$X_i = -\frac{\partial U}{\partial x_i}, \; Y_i = -\frac{\partial U}{\partial y_i}, \; Z_i = -\frac{\partial U}{\partial z_i}.$$

Suppose further that at time t_0 the system is in some given position, and that at a subsequent time t the position of the ith particle is $(x_i(t), y_i(t), z_i(t))$. The kinetic energy of the system is then

$$T = \tfrac{1}{2} \sum_{i=1}^{n} m_i(x_i'^2 + y_i'^2 + z_i'^2),$$

and we consider the *Lagrangian* $L = T - U$ of the system. Hamilton's *Principle of Least Action* states that the motion of the system during any sufficiently short time interval $[t_0, t_1]$ is described by those functions x_i, y_i, z_i for which the functional

$$\int_{t_0}^{t_1} L\,dt \tag{26}$$

(called the *action*) is a minimum.

We confine ourselves here to showing that the motion is an extremal for the functional (26). In fact, the Euler equations for this functional are

$$\frac{d}{dt}\frac{\partial L}{\partial x_i'} - \frac{\partial L}{\partial x_i} = 0, \quad \frac{d}{dt}\frac{\partial L}{\partial y_i'} - \frac{\partial L}{\partial y_i} = 0, \quad \frac{d}{dt}\frac{\partial L}{\partial z_i'} - \frac{\partial L}{\partial z_i} = 0,$$

i.e.

$$\frac{d}{dt}(m_i x_i') - \frac{\partial U}{\partial x_i} = 0, \quad \frac{d}{dt}(m_i y_i') - \frac{\partial U}{\partial y_i} = 0, \quad \frac{d}{dt}(m z_i') - \frac{\partial U}{\partial z_i} = 0,$$

i.e.

$$m_i x_i'' - X_i = 0, \quad m_i y_i'' - Y_i = 0, \quad m_i z_i'' - Z_i = 0,$$

and these are Newton's equations of motion for the system.

The canonical variables p_i, q_i, r_i corresponding to the functional (26) are given by

$$p_i = \frac{\partial L}{\partial x_i'} = m_i x_i', \quad q_i = m_i y_i', \quad r_i = m_i z_i',$$

(see (24)) and thus are the components of the momentum of the particle m_i. Hence the Hamiltonian is given by

$$H(t, x_1, y_1, z_1, \ldots, x_n, y_n, z_n, p_1, q_1, r_1, \ldots, p_n, q_n, r_n)$$

$$= \sum_{i=1}^{n} (x_i' p_i + y_i' q_i + z_i' r_i) - L = 2T - (T - U) = T + U,$$

and this last expression is the total energy of the system. In particular, if the system is conservative, i.e. if U does not depend on t, then L does not depend on t, so that H is constant along each extremal, i.e. the total energy is constant during the motion.

Exercises 3.11

1 Find the extremals for the functional

$$F(\phi) = \int_a^b f(t, \phi(t), \phi'(t)) dt$$

when
(i) $Y = \mathbf{R}$ and $f(t, y, y') = y' \cos t + y'^2$,
(ii) $Y = \mathbf{R}$ and $f(t, y, y') = t^2 y'^2 + 6y^2 + 12y$ $(t > 0)$,
(iii) $Y = \mathbf{R}^2$ and $f(t, y_1, y_2, y'_1, y'_2) = t^3 y'^3_1 y'_2$,
(iv) $Y = \mathbf{R}^2$ and $f(t, y_1, y_2, y'_1, y'_2) = (y_1 + y'_2)(y'_1 + y_2)$.

2 Let $f :]0, \infty[\times \mathbf{R} \to \mathbf{R}$ be given by $f(y, y') = y'(1 + y'^2)^{1/2}$. By using the substitution $y' = \tan \theta$ integrate the Euler equation for the functional

$$F(\phi) = \int_a^b f(\phi(t), \phi'(t)) dt$$

(see Example (b) above), and obtain the solution in the parametric form

$$y = \beta \sec^{1/r}\theta, \quad t = \alpha + \beta r^{-1} \int_0^\theta \sec^{1/r} u \, du.$$

Show that if $r = -1$ the solution is a circular arc and that if $r = -\frac{1}{2}$ it is a cycloid.

3 Use the method of Exercise 2 to find the extremals for the brachistochrone problem.

4 Let $J = [a, b]$, let $F : C^1(J, \mathbf{R}) \to \mathbf{R}$ be given by

$$F(\phi) = \int_a^b ((\phi(t))^2 - (\phi'(t))^2) dt,$$

and let $c, d \in \mathbf{R}$. Prove that
(i) if $b - a \not\equiv 0 \pmod{\pi}$, there is a unique extremal ϕ for F such that $\phi(a) = c$ and $\phi(b) = d$,
(ii) if $b - a \equiv 0 \pmod{2\pi}$, there is no extremal ϕ for F such that $\phi(a) = c$ and $\phi(b) = d$ unless $c = d$, and then there is an infinity of such extremals,
(iii) if $b - a \equiv \pi \pmod{2\pi}$, there is no extremal ϕ for F such that $\phi(a) = c$ and $\phi(b) = d$ unless $c = -d$, and then there is an infinity of such extremals.
Show further that if ϕ is an extremal for F and $\psi \in C_0^1(J, Y)$, then

$$F(\phi + \psi) = F(\phi) + F(\psi),$$

and hence deduce that ϕ is not a local minimum of F on the set of functions in $C^1(J, \mathbf{R})$ taking the values c, d at a, b.

[Hint. For the last part, use a function ψ similar to the χ in the proof of (3.11.5).
It should be mentioned that if $b - a > \pi$ the extremals ϕ for F are neither local minima nor local maxima of F. That ϕ is not a local maximum can be seen by considering a function ψ which differs arbitrarily little from the function χ whose values on $[a, a + c]$, $[a + c, a + 2c]$, $[a + 2c, b]$ are $\sin(t - a)$, $\sin(a + 2c - t)$, 0 respectively, where $\pi < 2c < b - a$.]

5 Find the Hamiltonian corresponding to the functional

$$\int_{-t}^{t} (\phi(t))^2 (1 - \phi'(t))^2 \, dt$$

(where $Y = \mathbf{R}$), and express the Euler equation in canonical form.

6 Let $J = [a, b]$, let Y be a real normed space, let $PS_0(J, Y)$ be the space of all piecewise smooth functions $\psi : J \to Y$ (see Exercise 3.9.2) such that $\psi(a) = \psi(b) = 0$, and let $v : J \to Y'$ be regulated. Prove that

$$\int_a^b v(t)\psi'(t)dt = 0$$

for all $\psi \in PS_0(J, Y)$ if and only if v is constant on its set of points of continuity.

[The same result holds if $PS_0(J, Y)$ is replaced by the set of primitives ψ of regulated functions from J into Y such that $\psi(a) = \psi(b) = 0$.]

7 Let $J = [a, b]$, let Y be a real normed space, and let $PS(J, Y)$ be the normed space of piecewise smooth functions from J into Y defined in Exercise 3.9.2. Let also A be an open set in the metric space $J \times Y \times Y$, let $f : A \to \mathbf{R}$ be a continuous function whose partial differentials $d_2 f$ and $d_3 f$ are continuous on A, let $\mathscr{F}$ be the set of $\phi \in PS(J, Y)$ such that the closure of the set $\{(t, \phi(t), \phi'(t)) : t \in J\}$ is contained in A (so that $\mathscr{F}$ is open in $PS(J, Y)$, by Exercise 3.9.2), and let $F : \mathscr{F} \to \mathbf{R}$ be given by

$$F(\phi) = \int_a^b f(t, \phi(t), \phi'(t))dt.$$

Suppose further that c, d are given elements of Y, let $\mathscr{E}$ be the set of $\phi \in \mathscr{F}$ such that $\phi(a) = c, \phi(b) = d$, and let $F_{\mathscr{E}}$ be the restriction of F to $\mathscr{E}$. Prove that if ϕ is a local minimum of $F_{\mathscr{E}}$, then there exists $k \in Y'$ such that

$$d_3 f(t, \phi(t), \phi'(t)) = \int_a^t d_2 f(s, \phi(s), \phi'(s))ds + k$$

at each point of continuity of $t \mapsto d_3 f(t, \phi(t), \phi'(t))$.

[Since a, b are points of continuity of ϕ', they are points of continuity of $t \mapsto d_3 f(t, \phi(t), \phi'(t))$, so that the above relation holds in particular when $t = a, b$. If t is a point of discontinuity of ϕ', the relation gives

$$d_3 f(t, \phi(t), \phi'(t -)) = d_3 f(t, \phi(t), \phi'(t +)) = \int_a^t d_2 f(s, \phi(s), \phi'(s))ds + k.]$$

8 Let $J = [0, 1]$. Prove that there are no extremals ϕ in $C^1(J, Y)$ (or $PS(J, Y)$) for the functional

$$F(\phi) = \int_0^1 t^2 (\phi'(t))^2 dt$$

such that $\phi(0) = 0, \phi(1) = 1$. By considering the function $\phi \in PS(J, Y)$ given by

$$\phi(t) = t/\varepsilon \quad (0 \le t \le \varepsilon), \quad \phi(t) = 1 \quad (\varepsilon \le t \le 1),$$

where $0 < \varepsilon \le 1$, show that the infimum of $F(\phi)$ taken over all $\phi \in PS(J, Y)$ satisfying $\phi(0) = 0, \phi(1) = 1$ is 0 (and is unattained).

9 Let $F_{\mathscr{E}}$ be defined as in Exercise 7, and suppose that $\phi \in \mathscr{E}$ is a local minimum of $F_{\mathscr{E}}$. Prove that if

$$E(t, y, y', u) = f(t, y, u) - f(t, y, y') - d_3 f(t, y, y')(u - y'),$$

then $E(t, \phi(t), \phi'(t), u) \geq 0$ whenever t is a point of continuity of ϕ' and $(t, \phi(t), u) \in A$.

[This necessary condition for ϕ to be a local minimum is known as *Weierstrass's condition*. It tells us that at each point $(t, \phi(t), u) \in A$ the surface with equation $z = f(t, y, u)$ lies above the hyperplane tangent to this surface at the point $(t, \phi(t), \phi'(t))$).

Hint. Let t_0 be a point of $[a, b[$ at which ϕ' is continuous, let $y_0 = \phi(t_0), u_0 = \phi'(t_0)$, let $u \in Y$, let $t_0 < t_0 + \delta \leq b, 0 < \varepsilon < 1, \eta = \varepsilon/(1 - \varepsilon)$, and let

$$\psi(t) = \begin{cases} \phi(t) & (a \leq t \leq t_0, t_0 + \delta \leq t \leq b), \\ \phi(t) + (t - t_0)(u - u_0) & (t_0 < t \leq t_0 + \varepsilon\delta), \\ \phi(t) + \eta(t_0 + \delta - t)(u - u_0) & (t_0 + \varepsilon\delta < t < t_0 + \delta). \end{cases}$$

Clearly $\psi \in \mathscr{E}$ for small enough δ, and

$$0 \leq F_{\mathscr{E}}(\psi) - F_{\mathscr{E}}(\phi) = \int_{t_0}^{t_0 + \varepsilon\delta} + \int_{t_0 + \varepsilon\delta}^{t_0 + \delta} (f(t, \psi(t), \psi'(t)) - f(t, \phi(t), \phi'(t))) dt = I_1 + I_2.$$

Prove that as $\delta \to 0 +$

$$I_1/(\varepsilon\delta) \to f(t_0, y_0, u) - f(t_0, y_0, u_0)$$

and

$$I_2/(\varepsilon\delta) \to (f(t_0, y_0, u_0 - \eta(u - u_0)) - f(t_0, y_0, u_0))/\eta.$$

The result is then obtained by making $\varepsilon \to 0 +$.]

10 By using the Weierstrass condition in Exercise 9, obtain an alternative proof of the result of (3.11.8). What hypotheses are required?

3.12 Newton's method for the solution of the equation $f(x) = 0$

Let A be an open set in X, let $f : A \to Y$ be a Fréchet differentiable function, and consider the equation $f(x) = 0$. If x_0 is a point of A near a root of this equation, then a first approximation to the equation is the linear equation

$$f(x_0) + df(x_0)(x - x_0) = 0,$$

and this has the solution

$$x_1 = x_0 - df(x_0)^{-1} f(x_0),$$

provided that the inverse $df(x_0)^{-1}$ exists. Continuing in this manner, starting from the initial approximation x_0, we obtain points $x_1, x_2, \ldots$ given by

$$x_{n+1} = x_n - df(x_n)^{-1} f(x_n) \qquad (n = 0, 1, \ldots), \tag{1}$$

x_{n+1} being defined so long as $x_1, \ldots, x_n \in A$; in effect, the x_n are successive approximations for the equation

$$x = x - df(x)^{-1} f(x).$$

It is intuitive that if we start from a point x_0 for which $f(x_0)$ is sufficiently small, and df does not vary too much near x_0, then the recurrence relation (1) will define a sequence (x_n) that converges to a root x^* of the equation

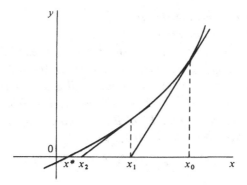

Figure 3.2

$f(x) = 0$. This is particularly transparent in the case of a real-valued function f of a real variable, for here the formula (1) becomes

$$x_{n+1} = x_n - f(x_n)/f'(x_n),$$

so that x_{n+1} is the abscissa of the point where the tangent to the graph of f at x_n meets the x-axis (Fig. 3.2; this special case was considered by Newton, and the sequence (x_n) given by (1) is usually known as a Newton sequence for the equation $f(x) = 0$).

An alternative possibility is to consider the recurrence relation

$$x_{n+1} = x_n - df(x_0)^{-1}f(x_n) \qquad (n = 0, 1, \ldots), \tag{2}$$

which defines successive approximations for the solution of the equation

$$x = x - df(x_0)^{-1}f(x).$$

More generally, we can consider a recurrence relation of the form

$$x_{n+1} = x_n - T_n^{-1}f(x_n), \tag{3}$$

where, for each n for which x_n is defined, T_n is a linear homeomorphism of X onto Y, possibly depending on $x_0, \ldots, x_n$. Sequences (x_n) defined by recurrence relations of the forms (1),(2),(3) are of considerable use in both numerical and theoretical work, and in this section we prove several theorems giving conditions for convergence of (x_n) to a root x^* of the equation $f(x) = 0$, and also estimates for the difference $\| x^* - x_n \|$.

We begin with an important theorem concerning the basic Newton sequence defined by (1).

(3.12.1) (Kantorovich's theorem) *Let X, Y be Banach spaces, let A be an open convex set in X, let $x_0 \in A$, let $K > 0, M > 0$, and let $f : A \to Y$ be* *a Fréchet differentiable function such that*

(i) *df is K-Lipschitzian on A,*

(ii) *df*(x_0) *is a linear homeomorphism of X onto Y, and* $\| df(x_0)^{-1} \| \leq M$,

(iii) $\| df(x_0)^{-1} f(x_0) \| \leq h\rho$, *where* $\rho = 1/KM$ *and* $0 < h \leq \frac{1}{2}$.

Suppose further that A contains the closed ball B in X with centre x_0 *and radius* $t^* = \rho(1 - (1 - 2h)^{1/2})$. *Then the recurrence relation* (1) *defines a sequence* (x_n) *in the interior* B° *of B converging to a point* x^* *of B such that* $f(x^*) = 0$. *Moreover, for* $n = 0, 1, \ldots,$

$$\| x^* - x_n \| \leq \rho 2^{-n}(1 - (1 - 2h)^{1/2})^{2^n} \tag{4}$$

and

$$\| x^* - x_{n+1} \| \leq \| x_{n+1} - x_n \|. \tag{5}$$

Also, if $h < \frac{1}{2}$ *then*

$$\| x^* - x_{n+1} \| \leq \frac{\| x_{n+1} - x_n \|^2}{\rho(1 - 2h)^{1/2}}. \tag{6}$$

Our proof makes use of the Newton sequence for an auxiliary function, and we treat this separately in the following lemma.

(3.12.2) Lemma. *Let* $\rho > 0, 0 < h \leq \frac{1}{2}$, *let*

$$q(t) = t^2 - 2\rho t + 2h\rho^2, \quad Q(t) = t - q(t)/q'(t),$$

and let t^* *be the smaller root of the equation* $q(t) = 0$, *i.e.* $t^* = \rho(1 - (1 - 2h)^{1/2})$. *Then the recurrence relation*

$$t_0 = 0, \quad t_{n+1} = Q(t_n) \quad (n = 0, 1, \ldots),$$

defines a sequence (t_n) *which is strictly increasing to the limit* t^*, *and for* $n = 1, 2, \ldots$

$$t_{n+1} - t_n = \frac{(t_n - t_{n-1})^2}{2(\rho - t_n)}. \tag{7}$$

Further, for all n,

$$t^* - t_n \leq \rho 2^{-n}(1 - (1 - 2h)^{1/2})^{2^n} \tag{8}$$

and

$$t_{n+1} - t_n \leq \frac{1}{2}(t_n - t_{n-1}). \tag{9}$$

Since $q(t^*) = 0$, we have

$$t^* - Q(t) = t^* - t + \frac{(q(t) - q(t^*))}{q'(t)} = \frac{(t^* - t)^2}{2(\rho - t)}$$

whenever $t \neq \rho$, and since $t^* \leq \rho$, this implies that if $t_n < t^*$, then $t_{n+1} = Q(t_n) < t^*$. Since $t_0 < t^*$, it follows that t_n is defined and satisfies $t_n < t^*$ for all n, and that

$$t^* - t_{n+1} = \frac{(t^* - t_n)^2}{2(\rho - t_n)}. \tag{10}$$

Next,

$$t_{n+1} - t_n = Q(t_n) - t_n = \frac{t_n^2 - 2\rho t_n + 2h\rho^2}{2(\rho - t_n)}. \tag{11}$$

This gives

$$-2\rho t_{n+1} + 2h\rho^2 = -2t_{n+1}t_n + t_n^2,$$

and on replacing n in this last equation by $n-1$ and combining the resulting equation with (11) we obtain (7). The sequence (t_n) is therefore strictly increasing and bounded above by t^*, and hence it converges to a limit $\tau \le t^*$. Clearly $Q(\tau) = \tau$, so that $q(\tau) = 0$, and therefore $\tau = t^*$, as required.

Next, by (11),

$$t_{n+1} = t_n + \tfrac{1}{2}(\rho - t_n) - \frac{\rho^2(1 - 2h)}{2(\rho - t_n)} \le \tfrac{1}{2}(\rho + t_n),$$

and since $t_0 = 0$ an easy induction argument shows that $t_n \le \rho(1 - 2^{-n})$. From (10) we therefore have

$$t^* - t_{n+1} \le \rho^{-1}2^{n-1}(t^* - t_n)^2 \qquad (n = 0, 1, \ldots),$$

and since $t^* - t_0 = t^*$, induction on n gives (8). Also, by (7),

$$t_{n+1} - t_n \le \rho^{-1}2^{n-1}(t_n - t_{n-1})^2 \qquad (n = 1, 2, \ldots),$$

and since $t_1 - t_0 \le \tfrac{1}{2}\rho$, this gives $t_n - t_{n-1} \le \rho 2^{-n}$. Hence

$$t_{n+1} - t_n \le \tfrac{1}{2}(t_n - t_{n-1}) \qquad (n = 1, 2, \ldots),$$

and this is (9). We note that if $h = \tfrac{1}{2}$ then $t_{n+1} = \tfrac{1}{2}(\rho + t_n)$, and therefore $t_n = \rho(1 - 2^{-n})$, so that there is equality in (8) and (9).

We observe also that the sequence (t_n) is completely determined by the recurrence relation (7) together with the conditions that $t_0 = 0$ and $t_1 = h\rho$.

Consider now the proof of (3.12.1). If $x \in A$ and $\|x - x_0\| < \rho$, then, by (i), $\|df(x) - df(x_0)\| < 1/M$, and hence, by (3.7.1. Corollary 1), $df(x)$ is a linear homeomorphism of X onto Y and

$$\|df(x)^{-1}\| \le \frac{\|df(x_0)^{-1}\|}{1 - \|df(x_0)^{-1}\| \, \|df(x) - df(x_0)\|} \le \frac{M}{1 - MK\|x - x_0\|}. \tag{12}$$

We show now that the recurrence relation (1) defines a sequence (x_n) such that $\|x_{n+1} - x_n\| \le t_{n+1} - t_n$ for all n. Clearly x_1 is defined and

$\|x_1 - x_0\| \le h\rho = t_1 = t_1 - t_0$. Let n be a positive integer such that $x_1, \ldots, x_n$ are defined and that

$$\|x_{r+1} - x_r\| \le t_{r+1} - t_r \tag{13}$$

for $r = 0, \ldots, n-1$. Then $\|x_r - x_0\| \le t_r < t^* \le \rho$ for $r = 0, \ldots, n$, so that $x_1, \ldots, x_n \in B^\circ$, and x_{n+1} is defined. By (1) with n replaced by $n-1$,

$$f(x_{n-1}) + df(x_{n-1})(x_n - x_{n-1}) = 0,$$

and therefore

$$\|x_{n+1} - x_n\| = \|df(x_n)^{-1} f(x_n)\| \le \|df(x_n)^{-1}\| \, \|f(x_n)\|$$
$$= \|df(x_n)^{-1}\| \, \|f(x_n) - f(x_{n-1}) - df(x_{n-1})(x_n - x_{n-1})\|.$$

Hence by (12), Exercise 3.2.2, (13) and (7),

$$\|x_{n+1} - x_n\| \le \frac{MK\|x_n - x_{n-1}\|^2}{2(1 - MK\|x_n - x_0\|)} \le \frac{(t_n - t_{n-1})^2}{2(\rho - t_n)} = t_{n+1} - t_n, \tag{14}$$

so that (13) holds for $r = n$, and therefore for all r. In particular, this implies that $x_n \in B^\circ$ for all n, and that (x_n) is a Cauchy sequence in X. Hence (x_n) converges to a point x^* of B, and since (by (i))

$$\|df(x_n)\| \le \|df(x_0)\| + K\|x_n - x_0\| \le \|df(x_0)\| + Kt^*,$$

x^* is a solution of the equation $f(x) = 0$.

Clearly (4) now follows from (8). Also, since (13) holds for all r, (14) holds for all n. In particular, from the first inequality in (14), (13) and (9), we have

$$\|x_{n+1} - x_n\| \le \frac{(t_n - t_{n-1})\|x_n - x_{n-1}\|}{2(\rho - t_n)}$$
$$= \frac{(t_{n+1} - t_n)\|x_n - x_{n-1}\|}{(t_n - t_{n-1})} \le \tfrac{1}{2}\|x_n - x_{n-1}\|,$$

and this gives (5). It gives also $\|x^* - x_{n+1}\| \le 2\|x_{n+2} - x_{n+1}\|$, and since $\rho - t_n \ge \rho - t^* = \rho(1 - 2h)^{1/2}$, it follows from the first inequality in (14) that if $h < \tfrac{1}{2}$ then

$$\|x^* - x_{n+1}\| \le \frac{2\|x_{n+1} - x_n\|^2}{2(\rho - t_n)} \le \frac{\|x_{n+1} - x_n\|^2}{\rho(1 - 2h)^{1/2}},$$

and this is (6).

It is by no means obvious from this proof how one obtains the particular auxiliary function q from which the sequence (t_n) is derived. In the next theorem we pull the curtain aside a little, and indicate how an appropriate auxiliary function can be determined in certain cases – the result of (3.12.1) is included as a special case. In this theorem we deal with the general recurrence relation

$$x_{n+1} = x_n - T_n^{-1}f(x_n), \qquad (15)$$

where x_{n+1} is defined so long as $x_0, \ldots, x_n$ belong to the domain of f, and where, for each n for which x_n is defined, T_n is a linear homeomorphism of X onto Y, possibly depending on $x_0, \ldots, x_n$. We consider not only the existence of a root x^* and the convergence of the sequence (x_n) to it, but also the uniqueness of this root and the convergence of sequences of successive approximations starting from some initial approximation different from x_0.

(3.12.3) *Let X, Y be Banach spaces, let A be an open convex set in X, let $f : A \to Y$ be a Fréchet differentiable function such that df is K-Lipschitzian on A, where $K > 0$, let $x_0 \in A$, and let $T_0, T_1, \ldots$ be the linear homeomorphisms of X onto Y associated with the points $x_0, x_1, \ldots$ determined by the recurrence relation (15). Let also c, a_0, δ be positive numbers, let*

$$p(t) = \tfrac{1}{2}Kt^2 - ct + a_0\delta \qquad (t \in \mathbf{R}),$$

and suppose that

(i) *the equation $p(t) = 0$ has positive roots t^* and $t^{**} \geq t^*$,*

(ii) *A contains the closed ball B in X with centre x_0 and radius t^*,*

(iii) *$\| T_0 \| \leq 1/a_0$ and $\| T_0 - df(x_0) \| \leq a_0 - c$,*

(iv) *$\| x_1 - x_0 \| = \| T_0^{-1}f(x_0) \| \leq \delta$.*

Suppose further that $a : \{0, 1, \ldots\} \times [0, t^*[\to]0, \infty [$ is a bounded function such that $a(0, 0) = a_0$, let the sequence (t_n) be given by

$$t_0 = 0, \ t_{n+1} = t_n + p(t_n)/a_n \qquad (n = 0, 1, \ldots),$$

where $a_n = a(n, t_n)$, and suppose that

(v) *for each positive integer n for which $t_n < t^*$ and for which the recurrence relation (15) determines a point x_n satisfying $\| x_n - x_0 \| \leq t_n$, we have*

$$\| T_n^{-1} \| \leq 1/a_n \quad and \quad \| T_n - df(x_n) \| \leq a_n + Kt_n - c.$$

Then the sequence (t_n) is strictly increasing to the limit t^*, the recurrence relation (15) defines a sequence (x_n) in B° converging to a point x^* such that $f(x^*) = 0$, and $\| x_{n+1} - x_n \| \leq t_{n+1} - t_n$ for all n (so that $\| x^* - x_n \| \leq t^* - t_n$). Moreover, if

(vi) *C is the closed ball in X with centre x_0 and radius s_0, where either $s_0 = t^* = t^{**}$ or $t^* \leq s_0 < t^{**}$ and $C \subseteq A$,†*

then x^* is the only solution of the equation $f(x) = 0$ in C, and for each $y_0 \in C$ the recurrence relation

$$y_{n+1} = y_n - T_n^{-1}f(y_n)$$

defines a sequence (y_n) which converges to x^*.

† If $s_0 = t^*$, then $C = B$ and this condition is redundant.

Although this result is stated for a given p, in applications we have essentially to select the a_n and c to satisfy (iii) and (v), and then to choose the number δ to satisfy (i). We can then determine an appropriate auxiliary function p from the data.

Suppose first the conditions (i)–(v) are satisfied. By (iii), $a_0 \geq c$, and this implies that $t_1 = \delta < t^*$ (for it implies that $p(\delta) \geq \frac{1}{2} K \delta^2 > 0$, so that either $\delta < t^*$ or $\delta > t^{**}$, and in the latter case we have $t^* t^{**} = 2a_0 \delta / K > 2ct^{**}/K$, so that $t^* > 2c/K = t^* + t^{**}$, which is impossible). Hence $0 = t_0 < t_1 < t^*$ and $\|x_1 - x_0\| \leq \delta = t_1 - t_0$.

Now let n be a positive integer such that $t_0 < t_1 < \ldots < t_n < t^*$, that $x_1, \ldots, x_n$ are defined, and that

$$\|x_{r+1} - x_r\| \leq t_{r+1} - t_r \tag{16}$$

for $r = 0, \ldots, n - 1$. We show that $t_n < t_{n+1} < t^*$, that x_{n+1} is defined, and that (16) holds for $r = n$.

By our hypotheses on n, we have $\|x_r - x_0\| \leq t_r < t^*$ for $r = 1, \ldots, n$, so that $x_1, \ldots, x_n \in B^\circ$, and x_{n+1} is defined. Further, by (v), $a_n + Kt_n - c \geq 0$. It therefore follows that

$$t^* - t_{n+1} = a_n^{-1}(a_n(t^* - t_n) - p(t_n)) = a_n^{-1}(a_n(t^* - t_n) + p(t^*) - p(t_n))$$
$$= a_n^{-1}(t^* - t_n)(a_n + \tfrac{1}{2}K(t^* + t_n) - c) > 0.$$

Moreover, $t_{n+1} - t_n = a_n^{-1} p(t_n) > 0$ (since $t_n < t^*$), so that $t_n < t_{n+1} < t^*$.

We observe next that (15) with n replaced by $n - 1$ gives

$$f(x_{n-1}) + T_{n-1}(x_n - x_{n-1}) = 0,$$

and therefore, by (v),

$$\|x_{n+1} - x_n\| = \|T_n^{-1} f(x_n)\| \leq a_n^{-1} \|f(x_n)\|$$
$$= a_n^{-1} \|f(x_n) - f(x_{n-1}) - T_{n-1}(x_n - x_{n-1})\|.$$

Hence, by Exercise 3.2.2, (16) and (v),

$$a_n \|x_{n+1} - x_n\| \leq \tfrac{1}{2} K \|x_n - x_{n-1}\|^2 + \|T_{n-1} - df(x_{n-1})\| \|x_n - x_{n-1}\|$$
$$\leq \tfrac{1}{2} K(t_n - t_{n-1})^2 + (a_{n-1} + Kt_{n-1} - c)(t_n - t_{n-1})$$
$$= p(t_n) - p(t_{n-1}) + a_{n-1}(t_n - t_{n-1})$$
$$= p(t_n) = a_n(t_{n+1} - t_n), \tag{17}$$

so that (16) holds for $r = n$.

It now follows by induction that the sequence (t_n) is strictly increasing and bounded above by t^*, that (x_n) is defined, and that (16) holds for all n, so that $x_n \in B^\circ$ for all n. Hence (t_n) converges to a limit $\tau \leq t^*$, and since the sequence (a_n) is bounded, $p(\tau) = 0$, so that $\tau = t^*$. Moreover, (x_n) is clearly a Cauchy sequence in X, and therefore converges to a point

$x^* \in B$. Since also

$$\|T_n\| \le \|T_n - df(x_n)\| + \|df(x_n) - df(x_0)\| + \|df(x_0)\|$$
$$\le a_n + Kt_n - c + K\|x_n - x_0\| + \|df(x_0)\|$$
$$\le a_n + 2Kt^* + \|df(x_0)\|,$$

the sequence $(\|T_n\|)$ is bounded, and hence $f(x^*) = 0$. This completes the proof of the first part of the theorem.

Suppose next that in addition the condition (vi) is satisfied, and let the sequence (s_n) be given by

$$s_{n+1} = s_n + a_n^{-1}p(s_n) \qquad (n = 0, 1, \ldots),$$

where s_0 is defined as in (vi). We show first that this sequence (s_n) decreases to the limit t^*. If $s_0 = t^*$, then $s_n = t^*$ for all n, so that we may suppose $t^* < s_0 < t^{**}$. If now $t^* < s_n < t^{**}$, then $s_{n+1} - s_n < 0$, and

$$s_{n+1} - t^* = a_n^{-1}(a_n(s_n - t^*) + p(s_n) - p(t^*))$$
$$= a_n^{-1}(s_n - t^*)(a_n + \tfrac{1}{2}K(s_n + t^*) - c) > 0.$$

Hence (s_n) is decreasing and bounded below by t^*, and therefore it converges to a limit which is clearly t^*.

Now let $y_0 \in C$, and let

$$y_{n+1} = y_n - T_n^{-1}f(y_n) \qquad (n = 0, 1, \ldots),$$

y_{n+1} being defined so long as $y_0, \ldots, y_n \in A$. We prove that this recurrence relation defines a sequence (y_n) in C such that $\|y_n - x_n\| \le s_n - t_n$ for all n.

Obviously $\|y_0 - x_0\| \le s_0 = s_0 - t_0$. Suppose then that n is a nonnegative integer such that $y_0, \ldots, y_n$ are defined and that

$$\|y_r - x_r\| \le s_r - t_r \tag{18}$$

for $r = 0, \ldots, n$. Then $y_n \in C$, so that y_{n+1} is defined. Further, by Exercise 3.2.2,

$$\|y_{n+1} - x_{n+1}\| = \|y_n - x_n - T_n^{-1}(f(y_n) - f(x_n))\|$$
$$\le \|T_n^{-1}\| \, \|T_n(y_n - x_n) - f(y_n) + f(x_n)\|$$
$$\le a_n^{-1}(\tfrac{1}{2}K\|y_n - x_n\|^2 + \|T_n - df(x_n)\| \, \|y_n - x_n\|)$$
$$\le a_n^{-1}(s_n - t_n)(\tfrac{1}{2}K(s_n - t_n) + (a_n + Kt_n - c))$$
$$= a_n^{-1}(s_n - t_n)(a_n + \tfrac{1}{2}K(s_n + t_n) - c)$$
$$= s_n - t_n + a_n^{-1}(p(s_n) - p(t_n))$$
$$= s_{n+1} - t_{n+1},$$

so that (18) holds for $r = n + 1$. Hence (18) holds for all r and, since $s_n - t_n \to 0$, it follows that $y_n \to x^*$. But if y_0 is a solution of the equation $f(x) = 0$ in C, then $y_n = y_0$ for all n, so that $y_n \to y_0$. Hence x^* is the only solution of $f(x) = 0$ in C, and this completes the proof.

We note that from the first inequality in (17) we have also

$$\|x_{n+1} - x_n\| \le \left(\frac{t_{n+1} - t_n}{t_n - t_{n-1}}\right) \|x_n - x_{n-1}\| ; \tag{19}$$

this can sometimes be used to derive an estimate of the form

$$\|x^* - x_{n+1}\| \le \gamma \|x_{n+1} - x_n\|$$

(cf. (5)).

We note also that the proof of (3.12.3) still applies if we replace (v) by

(v)′ *for each positive integer n for which* $t_n < t^*$ *and for which the recurrence relation* (15) *determines points* $x_1, \ldots, x_n$ *such that* $\sum_{j=0}^{n-1} \|x_{j+1} - x_j\| \le t_n$ *we have*

$$\|T_n^{-1}\| \le 1/a_n \quad and \quad \|T_n - df(x_n)\| \le a_n + Kt_n - c.$$

In particular, if (v)′ is satisfied for $n = m$, and for $m \le r \le m + k$ we take $T_r = T_m$, then (v)′ is satisfied for $n = r$ with $a_r = a_m$ provided that x_r is defined. In other words, we can use T_m for k iterations, and then calculate a new T_{m+k}.

In the case of Kantorovich's theorem (3.12.1), we have $T_n = df(x_n)$, and hence, by (12),

$$\|T_n^{-1}\| = \|df(x_n)^{-1}\| \le \frac{M}{1 - MK\|x_n - x_0\|} \le \frac{M}{1 - MKt_n}$$

whenever $\|x_n - x_0\| < \rho$ and $\|x_n - x_0\| \le t_n$. Also the condition

$$\|T_n - df(x_n)\| \le a_n + Kt_n - c$$

is satisfied provided that $a_n + Kt_n - c \ge 0$. If we now take $a(n, t) = 1/M - Kt$, so that $a_n = 1/M - Kt_n$, and take $c = 1/M$ and $\delta = h\rho$, where $0 < h \le \frac{1}{2}$, then the function p of (3.12.3) is given by

$$p(t) = \tfrac{1}{2}Kt^2 - \frac{t}{M} + \frac{h}{KM^2},$$

and the roots of $p(t) = 0$ are $t^* = \rho(1 - (1 - 2h)^{1/2})$ and $t^{**} = \rho(1 + (1 - 2h)^{1/2})$, where $\rho = 1/KM$. Since $t^* \le \rho$, $a(n, t) > 0$ whenever $0 \le t < t^*$, and $\|T_n^{-1}\| \le 1/a_n$ whenever $t_n < t^*$ and $\|x_n - x_0\| \le t_n$, so that (iii) and (v) hold with this choice of a, c, δ. The sequence (t_n) of (3.12.3) is identical to that of the lemma (3.12.2), and hence the existence part of (3.12.3) gives (3.12.1). Further, from the uniqueness part of (3.12.3) we have:

(3.12.3. Corollary) *Under the conditions of* (3.12.1) *the point* x^* *is the only solution of* $f(x) = 0$ *in* B. *Moreover, if* $h < \frac{1}{2}$ *and* A *contains the closed ball* C *in* X *with centre* x_0 *and radius* s_0, *where* $t^* < s_0 < t^{**}$, *then* x^* *is the only solution of* $f(x) = 0$ *in* C.

A further application of (3.12.3) is given in Exercise 1 below.

Exercises 3.12

1 Let X, Y be Banach spaces, let A be an open convex set in X, and let $f : A \to Y$ be a
Fréchet differentiable function such that df is K-Lipschitzian on A, where $K > 0$.
Let also $x_0 \in A$, let $M > 0$, $0 \le \mu < 1$, and let (T_n) be a sequence of linear homeomorphisms of X onto Y such that
  (i) $\| T_n^{-1} \| \le M$ and $\| T_n - df(x_0) \| \le \mu/M$ for all n,
  (ii) $\| T_0^{-1} f(x_0) \| \le h\rho(1 - \mu)^2$, where $\rho = 1/KM$ and $0 < h \le \tfrac{1}{2}$.
Prove that if A contains the closed ball B in X with centre x_0 and radius $t^* = \rho(1 - \mu)$
$\times (1 - (1 - 2h)^{1/2})$, then the recurrence relation
$$x_{n+1} = x_n - T_n^{-1} f(x_n)$$
defines a sequence (x_n) in B° converging to a point x^* of B for which $f(x^*) = 0$, and
that for $n = 0, 1, \ldots$,
$$\| x^* - x_n \| \le \alpha^n t^* \quad \text{and} \quad \| x^* - x_{n+1} \| \le \alpha \| x_{n+1} - x_n \| /(1 - \alpha),$$
where $\alpha = 1 - (1 - \mu)(1 - 2h)^{1/2}$. Prove that if $t^{**} = \rho(1 - \mu)(1 + (1 - 2h)^{1/2})$, C is the
closed ball in X with centre x_0 and radius s_0, where either $s_0 = t^* = t^{**}$ (when $h = \tfrac{1}{2}$)
or $t^* \le s_0 < t^{**}$ (when $h < \tfrac{1}{2}$) and $C \subseteq A$, then x^* is the only solution of the equation
$f(x) = 0$ in C, and for each $y_0 \in C$ the recurrence relation
$$y_{n+1} = y_n - T_n^{-1} f(y_n) \qquad (n = 0, 1, \ldots)$$
defines a sequence (y_n) in C which converges to x^*.
  [When $\mu = 0$, the sequence (x_n) is the modified Newton sequence defined by (2).
Hint. Apply (3.12.3) with $a(n, t) = 1/M$, $c = (1 - \mu)/M$, $\delta = h\rho(1 - \mu)^2$, so that
$$p(t) = \tfrac{1}{2} K(t^2 - 2\rho(1 - \mu)t + 2h\rho^2(1 - \mu)^2)$$
and
$$t_{n+1} = t_n + Mp(t_n) \qquad (n = 0, 1, \ldots).]$$

2 Let X be a Banach space, let A be an open convex set in X, and let $g : A \to X$ be a
Fréchet differentiable function such that dg is K-Lipschitzian on A, where $K > 0$.
Let also $x_0 \in A$, let $M > 0$, $0 \le \mu < 1$, $0 < h \le \tfrac{1}{2}$, let
$$t^* = (1 - \mu)(1 - (1 - 2h)^{1/2})/K, \quad t^{**} = (1 - \mu)(1 + (1 - 2h)^{1/2})/K,$$
and suppose that
  (i) A contains the closed ball C in X with centre x_0 and radius s_0, where either
$s_0 = t^* = t^{**}$ (when $h = \tfrac{1}{2}$) or $t^* \le s_0 < t^{**}$ (when $h < \tfrac{1}{2}$),
  (ii) $\| dg(x_0) \| \le \mu$,
  (iii) $\| x_0 - g(x_0) \| \le h(1 - \mu)^2/K$.
Prove that g has a unique fixed point x^* in C, and that for each $y_0 \in C$ the recurrence
relation $y_{n+1} = g(y_n)$ $(n = 0, 1, \ldots)$ defines a sequence (y_n) which converges to x^*.
  [Hint. Apply Exercise 1 with $f(x) = x - g(x)$, $T_n = 1_X$, $M = 1$.]

3 Let X, Y be Banach spaces, let $M > 0$, and let (T_n) be a sequence of linear homeo-
morphisms from X onto Y such that $\| T_n^{-1} \| \le M$ for all n. Let also B be the closed
ball in X with centre x_0 and radius $\delta > 0$, let $f : B \to Y$ be a continuous function
which is Fréchet differentiable in the interior B° of B, and suppose that
  (i) $\| df(x) - df(x_0) \| \le \tfrac{1}{4} M^{-1}$ for all $x \in B^\circ$,
  (ii) $\| T_0^{-1} f(x_0) \| \le \tfrac{1}{2} \delta$,   (iii) $\| T_0^{-1} f(x_0) \| \le \tfrac{1}{2} \delta$.
Prove that the sequence (x_n) given by
$$x_{n+1} = x_n - T_n^{-1} f(x_n) \qquad (n = 0, 1, \ldots)$$

is defined and converges to a point x^* of B for which $f(x^*) = 0$, and that for all n

$$\|x^* - x_{n+1}\| \le \|x_{n+1} - x_n\| \le 2^{-n-1}\delta.$$

Prove that $x = x^*$ is the only solution of the equation in B.

[Hint. Note that

$$f(x_n) = f(x_n) - f(x_{n-1}) - T_{n-1}(x_n - x_{n-1})$$
$$= f(x_n) - f(x_{n-1}) - df(x_0)(x_n - x_{n-1}) - (T_{n-1} - df(x_0)(x_n - x_{n-1})),$$

and that, by (i), for all $x, x' \in X$

$$\|f(x) - f(x') - df(x_0)(x - x')\| \le \tfrac{1}{4}M^{-1}\|x - x'\|.$$

For the last part, use this last relation together with the fact that $\|df(x_0)^{-1}\| \le 4M/3$.]

4 Let $J = [0, \alpha]$, let $\phi : J \to \mathbf{R}$ be continuous, let $S = \{(t, y) \in J \times \mathbf{R} : |y - \phi(t)| < \sigma\}$, and let $f : S \to \mathbf{R}$ be a continuous function such that the partial derivative $D_2 f$ of f is continuous on S and that $D_2 f(t, y)$ is K-Lipschitzian in y in S, where $K > 0$. Prove that if

(i) $\left|\int_0^t D_2 f(s, \phi(s))ds\right| \le L$ for all $t \in J$,

(ii) $\left|\phi(t) - \int_0^t f(s, \phi(s))ds\right| \le h\varepsilon^{-4L}/(\alpha^3 K)$, where $0 < h \le \tfrac{1}{2}$,

(iii) $t^* = \rho(1 - (1 - 2h)^{1/2})$, $t^{**} = \rho(1 + (1 - 2h)^{1/2})$, where $\rho = e^{-2L}/\alpha^2 K$, and either $s_0 = t^* = t^{**} < \sigma$ or $t^* \le s_0 < t^{**}$ and $s_0 < \sigma$,

then the differential equation $y' = f(t, y)$ has a unique solution ψ satisfying $\psi(0) = 0$ such that $|\psi(t) - \phi(t)| \le s_0$ for all $t \in J$.

[Hint. Let $\mathscr{S}$ be the open subset of $C(J, \mathbf{R})$ consisting of those functions in $C(J, \mathbf{R})$ whose graphs lie in S, and let $P : \mathscr{S} \to C^1(J, Y)$ be given by

$$P(\psi)(t) = \psi(t) - \int_0^t f(s, \psi(s))ds \qquad (\psi \in \mathscr{S}).$$

Then (see (3.9.2))

$$(dP(\psi)\eta)(t) = \eta(t) - \int_0^t D_2 f(s, \psi(s))\eta(s)ds \qquad (\psi \in \mathscr{S}, \eta \in C(J, \mathbf{R})),$$

and dP is αK-Lipschitzian on $\mathscr{S}$. Prove that if $\chi \in C^1(J, \mathbf{R})$, then $dP(\phi)\eta = \chi$ if and only if η is the solution of the linear equation

$$\eta'(t) = D_2 f(t, \phi(t))\eta(t) + \chi'(t)$$

on J such that $\eta(0) = 0$. Deduce that $dP(\phi)$ is a linear homeomorphism of $C(J, Y)$ onto $C^1(J, Y)$, and by solving this last equation explicitly show that $\|dP(\psi)^{-1}\| \le \alpha e^{2L}$. Then apply (3.12.1).]

4

The Gâteaux and Hadamard variations and differentials

In this chapter X, Y and Z again denote normed spaces over either the real or the complex field.

4.1 The Gâteaux variation and the Gâteaux differential

Let f be a function from a set $A \subseteq X$ into Y, let $x_0 \in A$, and let $h \in X$. We say that *f has a Gâteaux variation at x_0 for the increment h* if the function $t \mapsto f(x_0 + th)$ is defined on the interval $[0, \eta[$ for some $\eta > 0$ and has a right-hand derivative at $t = 0$. The value of this derivative is then called the *Gâteaux variation of f at x_0 for the increment h*, and we denote it by $\delta f(x_0)h$.† Thus

$$\delta f(x_0)h = \lim_{t \to 0+} (f(x_0 + th) - f(x_0))/t$$

whenever the limit exists. Further, the function $h \mapsto \delta f(x_0)h$ with domain the set of $h \in X$ for which $\delta f(x_0)h$ exists is called the *Gâteaux variation of f at x_0*, and we denote it by $\delta f(x_0)$.‡ Obviously $\delta f(x_0)0$ exists whenever $x_0 \in A$, and is equal to 0.

The Gâteaux variation $\delta f(x_0)$ need not be linear (see Example (*d*) below), but we have the result that if $\delta f(x_0)h$ *exists and* $\alpha \geq 0$, *then* $\delta f(x_0)(\alpha h)$ *exists and is equal to* $\alpha \delta f(x_0)h$. This is obvious when $\alpha = 0$, while if $\alpha > 0$

$$\delta f(x_0)(\alpha h) = \lim_{t \to 0+} \frac{f(x_0 + \alpha th) - f(x_0)}{t}$$

$$= \alpha \lim_{t \to 0+} \frac{f(x_0 + \alpha th) - f(x_0)}{\alpha t} = \alpha \delta f(x_0)h.$$

† We use the same bracketing convention as with $df(x_0)$.
‡ There is no really standard notation and terminology for the Gâteaux variation. Some authors speak of the 'Gâteaux variation of f at x_0 for the increment h' only when it exists for all $h \in X$, and it is more usual to employ the ordinary derivative in the definition than the right-hand derivative. The most common notations for the Gâteaux variation are $\delta f(x_0; h)$ and $Vf(x_0; h)$.

The following examples illustrate the range of possible behaviour of the function $\delta f(x_0)$.

Example (a)

Let ϕ be a function from a set $A \subseteq \mathbf{R}$ into Y, and let $t_0 \in A$. If $h > 0$, then ϕ has a Gâteaux variation at t_0 for the increment h if and only if A contains an interval of the form $[t_0, t_0 + \eta[$, where $\eta > 0$, and ϕ has a right-hand derivative $\phi'_+(t_0)$, and then $\delta\phi(t_0)h = h\phi'_+(t_0)$. Similarly, if $h < 0$, then ϕ has a Gâteaux variation at t_0 for the increment h if and only if A contains an interval of the form $]t_0 - \eta, t_0]$ and ϕ has a left-hand derivative $\phi'_-(t_0)$, and then $\delta\phi(t_0)h = h\phi'_-(t_0)$. Thus $\delta\phi(t_0)h$ exists for all $h \in \mathbf{R}$ if and only if t_0 is an interior point of A and both $\phi'_+(t_0)$ and $\phi'_-(t_0)$ exist.

There is, of course, no correspondence between the existence of $\phi'_\pm(t_0)$ and that of $\delta\phi(t_0)0$, since the latter exists whenever $t_0 \in A$.

Example (b)

If f is a function from a set $A \subseteq X$ into Y which is Fréchet differentiable at x_0, then the Gâteaux variation $\delta f(x_0)h$ exists and is equal to $df(x_0)h$ for all $h \in X$, so that the function $\delta f(x_0)$ belongs to $\mathscr{L}(X, Y)$. This follows directly from either Exercise 3.1.4 or (3.1.2).

Example (c)

If $f : \mathbf{R}^2 \to \mathbf{R}$ is given by
$$f(x_1, x_2) = 1 \text{ if } x_2 = x_1^2 \neq 0, \quad f(x_1, x_2) = 0 \text{ otherwise,}$$
then $\delta f(0)h$ exists and is equal to 0 for all $h \in \mathbf{R}^2$, so that the function $\delta f(0)$ belongs to $\mathscr{L}(\mathbf{R}^2, \mathbf{R})$. However, f is discontinuous at the origin, and is therefore not Fréchet differentiable there.

Example (d)

Let f be the (continuous) function from $\mathbf{R}^2$ into $\mathbf{R}$ given by
$$f(x_1, x_2) = x_1 \text{ if } |x_2| \geq x_1^2, \quad f(x_1, x_2) = |x_2|/x_1 \text{ otherwise.}$$
Then f takes the value 0 everywhere on the axes, so that $\delta f(0)h$ exists and is equal to 0 when h lies on either axis. Also, if $h = (h_1, h_2)$ does not lie on either axis, then
$$(f(th_1, th_2) - f(0, 0))/t = h_1 \quad \text{whenever} \quad 0 < t \leq |h_2|/h_1^2,$$

so that for such h we have $\delta f(0)h = h_1$. Hence $\delta f(0)$ is non-linear, and is discontinuous at each point of the h_1-axis other than the origin.

Example (e)

If p is a convex function on a convex set $A \subseteq X$, and x_0 is an internal point of A, then $\delta p(x_0)h$ exists for all $h \in X$ and $\delta p(x_0)$ is sublinear.

To prove this we note first that if $h \in X$ then the function $t \mapsto p(x_0 + th)$ is convex on a neighbourhood of 0 in $\mathbf{R}$, and therefore the limit

$$\delta p(x_0)h = \lim_{t \to 0+} (p(x_0 + th) - p(x_0))/t$$

exists (cf. §1.1, Example (g)). It remains to prove that $\delta p(x_0)$ is sublinear, and since $\delta p(x_0)(\alpha h) = \alpha \delta p(x_0)h$ for all $\alpha \geq 0$, it is enough to show that $\delta p(x_0)$ is convex. Let $h, k \in X$, let $0 < \sigma < 1$, and let $l = \sigma h + (1 - \sigma)k$. Then for all sufficiently small positive t, $p(x_0 + tl) = p(\sigma(x_0 + th) + (1 - \sigma)(x_0 + tk)) \leq \sigma p(x_0 + th) + (1 - \sigma)p(x_0 + tk)$, and this trivially implies that

$$\delta p(x_0)l \leq \sigma \delta p(x_0)h + (1 - \sigma)\delta p(x_0)k.$$

Theorem (3.1.3) applies without change to the Gâteaux variation, i.e. we have:

(4.1.1) (i) *If f is a function from a set $A \subseteq X$ into Y which has a Gâteaux variation at x_0 for the increment h, then so does αf for every scalar α and $\delta(\alpha f)(x_0)h = \alpha \delta f(x_0)h$.*
(ii) *If f, g are functions from sets $A, B \subseteq X$ into Y which have Gâteaux variations at x_0 for the increment h, then so does $f + g$, and $\delta(f + g)(x_0)h = \delta f(x_0)h + \delta g(x_0)h$.*

The following version of the chain rule is an immediate consequence of Exercise 3.1.4.

(4.1.2) *Let f be a function from a set $A \subseteq X$ into Y which has a Gâteaux variation at x_0 for the increment h, and let g be a function from a set $B \subseteq Y$ into Z which is Fréchet differentiable at the point $y_0 = f(x_0)$. Then $g \circ f$ has a Gâteaux variation at x_0 for the increment h, equal to $dg(y_0)(\delta f(x_0)h)$. In particular, if $g \in \mathscr{L}(Y, Z)$, then the Gâteaux variation of $g \circ f$ at x_0 for the increment h is $g(\delta f(x_0)h)$.*

In (4.1.2) we cannot replace the condition that g is Fréchet differentiable at y_0 by the condition that g has a Gâteaux variation $\delta g(y_0)k$ for every increment k, even if $\delta g(y_0) \in \mathcal{L}(Y, Z)$. For example, if $g : \mathbf{R}^2 \to \mathbf{R}$ is the function of Example (c) above, so that

$$g(x_1, x_2) = 1 \text{ if } x_2 = x_1^2 \neq 0, \quad g(x_1, x_2) = 0 \text{ otherwise,}$$

and $f : \mathbf{R} \to \mathbf{R}^2$ is given by $f(t) = (t, t^2)$, then $\delta g(0) \in \mathcal{L}(\mathbf{R}^2, \mathbf{R})$ and $\delta f(0) \in \mathcal{L}(\mathbf{R}, \mathbf{R}^2)$, but $g(f(t)) = 1$ except when $t = 0$, so that the Gâteaux variation of $g \circ f$ at 0 exists only for the increment 0. This example shows also that we cannot interchange the Gâteaux and Fréchet conditions in (4.1.2).

It is useful to note that if f is defined on the closed line segment with endpoints $x_0, x_0 + h$, and $\xi \in [0, 1[$, then the function $t \mapsto f(x_0 + th)$ has a right-hand derivative at ξ equal to l if and only if $\delta f(x_0 + \xi h)h$ exists and is equal to l (for

$$\delta f(x_0 + \xi h)h = \lim_{t \to 0+} \frac{f(x_0 + \xi h + th) - f(x_0 + \xi h)}{t}$$

$$= \lim_{s \to \xi+} \frac{f(x_0 + sh) - f(x_0 + \xi h)}{s - \xi}$$

whenever either limit exists). In particular, by combining this remark with (1.6.3. Corollary), we obtain the following mean value inequality for Gâteaux variations.

(4.1.3) *Let* $x_0, x_0 + h \in X$, *let* S *be the closed line segment in* X *with endpoints* $x_0, x_0 + h$, *and let* $f : S \to Y$ *be a continuous function such that* $\delta f(x_0 + th)h$ *exists for nearly all* $t \in]0, 1[$. *Then there exist uncountably many* $\xi \in]0, 1[$ *for which*

$$\| f(x_0 + h) - f(x_0) \| \leq \| \delta f(x_0 + \xi h)h \|. \tag{1}$$

Moreover, either (1) *holds with strict inequality for uncountably many* ξ, *or it holds whenever* $\delta f(x_0 + \xi h)h$ *exists (and with equality for nearly all* ξ).

We give next a simple sufficient condition for the linearity of the Gâteaux variation $\delta f(x_0)$ when the normed spaces involved are real. We observe that since $\delta f(x_0)(\alpha h) = \alpha \delta f(x_0)h$ for every $\alpha \geq 0$, to prove the linearity of $\delta f(x_0)$ over the real field it is enough to show that $\delta f(x_0)(h + k) = \delta f(x_0)h + \delta f(x_0)k$ for all h, k.

(4.1.4) *If* X, Y *are real normed spaces and* f *is a function from a set* $A \subseteq X$ *into* Y *such that*

(i) *for each* x *in some neighbourhood of* x_0 *the Gâteaux variation* $\delta f(x)h$ *exists for all* $h \in X$,

(ii) *for each $h \in X$ the function $x \mapsto \delta f(x)h$ is continuous at x_0,*
then the function $\delta f(x_0)$ is linear.

This result is a direct consequence of the preceding remark and the following lemma.

(4.1.5) **Lemma.** *Let f be a function from a set $A \subseteq X$ into Y, and let $x_0 \in A$. Let also $h, k \in X$ and suppose that $x \mapsto \delta f(x)h$ is defined on some neighbourhood V of x_0 and is continuous at x_0, and that $\delta f(x_0)k$ exists. Then $\delta f(x_0)(h + k)$ exists and is equal to $\delta f(x_0)h + \delta f(x_0)k$.*

If t is a point of $]0, \infty[$ for which $x_0 + th + tk$ and $x_0 + tk$ belong to A, then

$$f(x_0 + th + tk) - f(x_0) - t\delta f(x_0)h - t\delta f(x_0)k$$
$$= (f(x_0 + th + tk) - f(x_0 + tk) - t\delta f(x_0)h)$$
$$+ (f(x_0 + tk) - f(x_0) - t\delta f(x_0)k)$$
$$= P(t) + Q(t),$$

say, and here $Q(t)/t \to 0$ as $t \to 0+$, since $\delta f(x_0)k$ exists.

We have now to show that $P(t)/t \to 0$ as $t \to 0+$, and to do this we consider the function

$$s \mapsto f(x_0 + sh + tk) - f(x_0 + tk) - s\delta f(x_0)h \qquad (2)$$

on the interval $I = [0, t]$. For all sufficiently small t, the closed line segment in X joining the points $x_0 + th + tk$ and $x_0 + tk$ lies in V, and hence, by the remark preceding (4.1.3), for all such t the right-hand derivative of the function (2) at the point s of I is

$$\delta f(x_0 + sh + tk)h - \delta f(x_0)h.$$

By the mean value inequality (1.6.3. Corollary) applied to the function (2) on I, we deduce that for all sufficiently small t there exist uncountably many $\xi \in I$ for which

$$\| P(t) \| \leq |t| \, \| \delta f(x_0 + \xi h + tk)h - \delta f(x_0)h \|,$$

and since the function $x \mapsto \delta f(x)h$ is continuous at x_0, this implies that $P(t)/t \to 0$ as $t \to 0+$.

We now define Gâteaux differentiability. Let f be a function from a set $A \subseteq X$ into Y and let x_0 be an interior point of A. We say that f is *Gâteaux differentiable at x_0* if the Gâteaux variation $\delta f(x_0)h$ exists for all $h \in X$ and the function $\delta f(x_0)$ belongs to $\mathscr{L}(X, Y)$. This function $\delta f(x_0)$ is then called the *Gâteaux differential of f at x_0*.† Further, the function

---

† As with the Gâteaux variation, there is no really standard notation for the Gâteaux differential of f at x_0; perhaps the most common notation is $Df(x_0)$.

$x \mapsto \delta f(x)$ whose domain is the set of interior points of A at which f is Gâteaux differentiable is called the *Gâteaux differential* of f, and we denote it by δf.

If f is Fréchet differentiable at x_0, then by Example (*b*) above, f is Gâteaux differentiable at x_0 and $\delta f(x_0) = df(x_0)$. The converse is true when $X = \mathbf{R}$ (and Y is real), for if ϕ is a function from a set $A \subseteq \mathbf{R}$ into Y, and t_0 is an interior point of A, then the Fréchet differentiability of ϕ at t_0 and the Gâteaux differentiability of ϕ at t_0 are each equivalent to the existence of the derivative $\phi'(t_0)$, and then $d\phi(t_0)h = \delta\phi(t_0)h = h\phi'(t_0)$ for all $h \in X$ (see Example (*a*) above and §3.1, Example (*a*) (p. 168)). If $\dim X \geq 2$, then Gâteaux differentiability does not imply Fréchet differentiability; this is easily seen from Example (*c*). This same example shows also that when $\dim X \geq 2$ the Gâteaux differentiability of f at x_0 does not imply the continuity of f there.

When $X = \mathbf{R}^n$, where $n \geq 2$, the Gâteaux differentiability of f at x_0 implies the existence of the partial derivatives of f at x_0, and in fact $D_j f(x_0) = \delta f(x_0)e_j$ $(j = 1, \ldots, n)$, where $e_1, \ldots, e_n$ is the natural basis in $\mathbf{R}^n$. Hence if $Y = \mathbf{R}^m$, and $f_1, \ldots, f_m$ are the components of f, then the matrix of the Gâteaux differential $\delta f(x_0)$ with respect to the natural bases in $\mathbf{R}^n$ and $\mathbf{R}^m$ is the Jacobian matrix $[D_j f_i(x_0)]$.

It should be noted also that if $X = \mathbf{R}^n$, then the existence of the partial derivatives $D_j f(x_0)$ of f at an interior point x_0 of its domain does not imply that f is Gâteaux differentiable at x_0. For instance, if f is the function of Example (*d*), then $D_1 f(0,0) = D_2 f(0,0) = 0$, but f is not Gâteaux differentiable at $(0, 0)$.

The analogue of (3.1.2) for the Gâteaux differential is that f is Gâteaux differentiable at an interior point x_0 of its domain if and only if there exists $T \in \mathscr{L}(X, Y)$ such that for each $h \in X$

$$(f(x_0 + th) - f(x_0) - T(th))/t \to 0 \text{ as } t \to 0 \text{ in } \mathbf{R}\backslash\{0\}, \qquad (3)$$

and then $\delta f(x_0) = T$. The argument used in §3.1 to prove the uniqueness of the Fréchet differential shows that if there exists $T \in \mathscr{L}(X, Y)$ such that (3) holds for all $h \in X$, then T is unique.

By (4.1.1) the Gâteaux differential $\delta f(x_0)$ has the same linearity properties (as a function of f) as the Fréchet differential (cf. (3.1.3)). Also, by (4.1.2), the chain rule here takes the form that *if f is Gâteaux differentiable at x_0 and g is Fréchet differentiable at the point $y_0 = f(x_0)$, then $g \circ f$ is Gâteaux differentiable at x_0 and its Gâteaux differential there is $dg(y_0) \circ \delta f(x_0)$.†* In particular, if $g \in \mathscr{L}(Y, Z)$, then the Gâteaux differential of

---

† We shall see in (4.2.5) that the Fréchet differentiability of g here can be replaced by the weaker property of Hadamard differentiability.

$g \circ f$ at x_0 is $g \circ \delta f(x_0)$. When $X = \mathbf{R}^n$, $Y = \mathbf{R}^m$, $Z = \mathbf{R}^l$, there is also a direct analogue of (3.4.3) involving the Jacobian matrices of f and g.

We note in passing that the particular case of the above chain rule where $g \in \mathscr{L}(Y, Z)$ implies that the result of §3.1, Example (*g*) (p. 173), holds for Gâteaux differentiable functions.

By virtue of (4.1.3), the result of (3.2.2) extends to Gâteaux differentiability; more precisely, we have:

(4.1.6) *Let $M \geq 0$, let C be a closed convex set in X with a non-empty interior C°, and let $f : C \to Y$ be a continuous function such that, for nearly all $x \in C^\circ$ the Gâteaux differential $\delta f(x)$ exists and satisfies $\| \delta f(x) \| \leq M$. Then f is M-Lipschitzian on C, i.e. for all $a, b \in C$*

$$\| f(b) - f(a) \| \leq M \| b - a \|.$$

There is a simpler variant of (4.1.6) which is frequently useful, namely:

(4.1.7) *If C is an open convex set in X, $f : C \to Y$ is Gâteaux differentiable on C, and $\| \delta f(x) \| \leq M$ for all $x \in C$, then f is M-Lipschitzian on C.*

To prove this, we observe that if $a, b \in C$ and S is the closed segment joining a and b, then the Gâteaux differentiability of f implies that f is continuous on S. We can therefore apply (4.1.3) to f on S to obtain the result.

(4.1.7. Corollary 1) *If f is a function from a set $A \subseteq X$ into Y whose Gâteaux differential δf is defined on a neighbourhood of x_0 and is continuous at x_0, then f is Fréchet differentiable at x_0.*

Given $\varepsilon > 0$, we can find an open ball B with centre x_0 contained in A such that $\delta f(x)$ is defined and satisfies $\| \delta f(x) - \delta f(x_0) \| \leq \varepsilon$ for all $x \in B$. By (4.1.7) applied to the function

$$x \mapsto f(x) - f(x_0) - \delta f(x_0)(x - x_0) \qquad (x \in B),$$

we deduce that for all $x \in B$

$$\| f(x) - f(x_0) - \delta f(x_0)(x - x_0) \| \leq \varepsilon \| x - x_0 \|$$

whence f is Fréchet differentiable at x_0.

(4.1.7. Corollary 2) *If f is a function from an open set $E \subseteq X$ into Y, then f is C^1 on E if and only if its Gâteaux differential δf is continuous on E.*

If X is a product space, say $X = X_1 \times \ldots \times X_n$, we can define partial Gâteaux differentials of f exactly as in the case of Fréchet differentials discussed in §3.3. Thus for $j = 1, \ldots, n$ let I_j be the insertion map from

X_j into X which maps the point x_j of X_j to the point of X with the jth co-ordinate x_j and all other coordinates 0. Then the *partial Gâteaux differential of a function* $f : A \to Y$ *at the point* $x_0 = (x_{10}, \ldots, x_{n0})$ *of A with respect to the jth coordinate* is the Gâteaux differential of the function $x_j \mapsto f(x_0 + I_j(x_j - x_{j0}))$ at x_{j0}, provided that this latter differential exists. By analogy with §3.3, we denote this partial differential by $\delta_j f(x_0)$. It is readily verified that the results of (3.3.1, 2) hold for partial Gâteaux differentials.

It is also possible to define Gâteaux differentials of higher order. However, we make no use of such differentials, and we do not consider them further.

We mention finally the upper and lower Gâteaux variations $\bar{\delta} f(x_0)h$ and $\underline{\delta} f(x_0)h$ of a real-valued function f. These are respectively the upper and lower right-hand Dini derivatives of the function $t \mapsto f(x_0 + th)$ at $t = 0$, i.e.

$$\bar{\delta} f(x_0)h = \lim_{t \to 0+} \sup (f(x_0 + th) - f(x_0))/t,$$

$$\underline{\delta} f(x_0)h = \lim_{t \to 0+} \inf (f(x_0 + th) - f(x_0))/t$$

(the values $\pm \infty$ being allowed). These upper and lower variations will be used in §4.6.

Exercises 4.1

1 Let X be real and let f be a function from a convex set $A \subseteq X$ into $\mathbf{R}$. Prove that if for each $x \in A$ the Gâteaux variation $\delta f(x)$ of f is defined and convex on X, and

$$f(x) \geq f(x_0) + \delta f(x_0)(x - x_0)$$

for all $x, x_0 \in A$, then f is convex.
   [Hint. Cf. Exercise 3.1.8.]

2 Let f be a function from a set $A \subseteq X$ into Y which has a Gâteaux variation at x_0 for all increments h, and suppose that $\|f(x_0)\| \geq \|f(x)\|$ for all $x \in A$. Prove that, for all $h \in X$,

$$\|f(x_0)\| \leq \|f(x_0) + \delta f(x_0)h\|.$$

4.2 The Hadamard variation and the Hadamard differential

The notion of differentiability introduced in this section is intermediate between those of Fréchet and Gâteaux. We consider first the corresponding variation, and we begin by defining an auxiliary concept, the set of vectors directed into a set at a point.

Let $E \subseteq X$, and let $x_0 \in X$. We say that a vector $h \in X$ is *directed into E at x_0* if there exist a neighbourhood W of h in X and a positive number ε

such that $x_0 + tk \in E$ whenever $k \in W$ and $0 < t < \varepsilon$. It is easy to see that the set of vectors directed into E at x_0 is an open cone in X with vertex 0, and we denote this cone by $K_d(E;x_0)$.† For example, we show in (4.5.3) that if E is non-empty and convex, and $x_0 \in \bar{E}$, then

$$K_d(E;x_0) = \{h \in X : \text{there exists } \lambda > 0 \text{ for which } x_0 + \lambda h \in E^\circ\}.$$

The following statements are readily verified:

(a) if $K_d(E;x_0) \neq \varnothing$, then $x_0 \in \bar{E}$ (so that if $x_0 \notin \bar{E}$ then $K_d(E;x_0) = \varnothing$,

(b) $0 \in K_d(E;x_0)$ if and only if $x_0 \in E^\circ$, and then $K_d(E;x_0) = X$,

(c) $h \in K_d(E;x_0)$ if and only if, for each sequence (h_n) in X converging to h and each sequence (t_n) of positive numbers converging to 0, the point $x_0 + t_n h_n$ belongs to E for all sufficiently large n.‡

Now let f be a function from a set $A \subseteq X$ into Y, and let $x_0 \in A$. We say that f has a *Hadamard variation at* x_0 *for the increment* h if h is directed into A at x_0 and there exists $l \in Y$ such that, for each sequence (h_n) in X converging to h and each sequence (t_n) of positive numbers converging to 0, we have

$$\lim_{n \to \infty} (f(x_0 + t_n h_n) - f(x_0))/t_n = l.$$

The element l is then called the *Hadamard variation of* f *at* x_0 *for the increment* h, and we denote it by $\partial f(x_0)h$. Further, the function $h \mapsto \partial f(x_0)h$ with domain the set of $h \in X$ for which $\partial f(x_0)h$ exists will be called the *Hadamard variation of* f *at* x_0, and we denote it by $\partial f(x_0)$.

It is obvious that if $\partial f(x_0)h$ exists and $\alpha > 0$, then $\partial f(x_0)(\alpha h)$ exists and is equal to $\alpha \partial f(x_0)h$.

By taking $h_n = h$ for all n we see that if f has a Hadamard variation at x_0 for the increment h, then f has a Gâteaux variation at x_0 for h and $\delta f(x_0)h = \partial f(x_0)h$. In particular, this implies that if $\partial f(x_0)0$ exists then it is equal to 0. In the opposite direction, the existence of the Gâteaux variation $\delta f(x_0)$ does not in general imply the existence of the Hadamard variation $\partial f(x_0)h$ (see Example (c) below), but we have the result that *if* f *has a Gâteaux variation at* x_0 *for* h *and is Lipschitzian in a neighbourhood of* x_0, *then* f *has a Hadamard variation at* x_0 *for* h. This is a trival consequence of the identity

$$\frac{f(x_0 + t_n h_n) - f(x_0)}{t_n} - \frac{f(x_0 + t_n h) - f(x_0)}{t_n} = \frac{f(x_0 + t_n h_n) - f(x_0 + t_n h)}{t_n}. \quad (1)$$

---

† $K_d(E;x_0)$ is often called the cone of *feasible directions* at x_0. For the purposes of this section it is enough to consider the case where $x_0 \in E$, but we require the more general definition in §4.5.

‡ For such sequences $(h_n),(t_n)$, the points $t_n h_n$ approach 0'tangentially' to the vector h (see §4.3).

Example (a)

Let ϕ be a function from a set $A \subseteq \mathbf{R}$ into Y, and let $t_0 \in A$. If $h > 0$, then ϕ has a Hadamard variation at t_0 for the increment h if and only if A contains an interval of the form $[t_0, t_0 + \eta[$, where $\eta > 0$, and ϕ has a right-hand derivative $\phi'_+(t_0)$, and then $\partial \phi(t_0)h = h\phi'_+(t_0)$. Similarly, if $h < 0$, then $\partial \phi(t_0)h$ exists if and only if A contains an interval of the form $]t_0 - \eta, t_0]$ and ϕ has a left-hand derivative $\phi'_-(t_0)$, and then $\partial \phi(t_0)h = h\phi'_-(t_0)$. Further, if t_0 is an interior point of A and both $\phi'_+(t_0)$ and $\phi'_-(t_0)$ exist, then $\partial \phi(t_0)0$ exists. Hence (as with the Gâteaux variation) $\partial \phi(t_0)h$ exists for all $h \in \mathbf{R}$ if and only if t_0 is an interior point of A and both $\phi'_+(t_0)$ and $\phi'_-(t_0)$ exist.

It should be noted that the existence of $\partial \phi(t_0)0$ does not imply the existence of $\phi'_\pm(t_0)$ (see 4.2.2) below).

Example (b)

If f is a function from a subset of X into Y which is Fréchet differentiable at a point x_0, then $\partial f(x_0)h$ exists for all $h \in X$ and is equal to $df(x_0)h$ (so that $\partial f(x_0) \in \mathcal{L}(X, Y)$). This follows easily from (3.1.2)(iii).

Example (c)

If $f : \mathbf{R}^2 \to \mathbf{R}$ is given by

$$f(x_1, x_2) = 1 \text{ if } x_2 = x_1^2 \neq 0, \quad f(x_1, x_2) = 0 \text{ otherwise,}$$

then f has a Gâteaux variation at $(0,0)$ which belongs to $\mathcal{L}(\mathbf{R}^2, \mathbf{R})$ (§4.1, Example (c)). On the other hand, the Hadamard variation of f at $(0,0)$ for the increment $h = (1,0)$ does not exist, for if (t_n) is a sequence of positive numbers converging to 0 and $h_n = (1, t_n)$, then $h_n \to h$, and $t_n h_n = (t_n, t_n^2)$, so that $(f(t_n h_n) - f(0))/t_n = 1/t_n$.

The following result gives an equivalent formulation of the definition of the Hadamard variation which is perhaps more illuminating, though less useful, than that above.

(4.2.1) *Let f be a function from a set $A \subseteq X$ into Y, let $x_0 \in A$, and let $h \in X$. Then the following statements are equivalent:*
(i) *the Hadamard variation $\partial f(x_0)h$ exists and is equal to l;*
(ii) *for each function ϕ from a subset B of $\mathbf{R}$ containing 0 into X such that $\phi(0) = x_0$, $\phi'_+(0) = h$, the function $f \circ \phi$ is defined on $B \cap [0, \eta[$ for some $\eta > 0$ and has a right-hand derivative at 0 equal to l.*

Suppose first that (i) holds, let ϕ be a function from a subset B of $\mathbf{R}$ containing 0 into X such that $\phi(0) = x_0, \phi'_+(0) = h$, let (t_n) be a sequence of positive numbers in B converging to 0, and let $h_n = (\phi(t_n) - \phi(0))/t_n$. Then $h_n \to h$ as $n \to \infty$, and since h is directed into A at x_0, it follows that $\phi(t_n) = x_0 + t_n h_n \in A$ for all sufficiently large n. Further, since $\partial f(x_0)h = l$,

$$\frac{f(\phi(t_n)) - f(\phi(0))}{t_n} = \frac{f(x_0 + t_n h_n) - f(x_0)}{t_n} \to l \quad \text{as} \quad n \to \infty.$$

This implies that $(f \circ \phi)'_+(0)$ exists and is equal to l. Moreover, it implies that $f \circ \phi$ is defined on $B \cap [0, \eta[$ for some $\eta > 0$, for otherwise we can find a sequence (t_n) of positive numbers in B converging to 0 such that, for each $n, \phi(t_n) \notin A$, and this gives a contradiction.

Conversely, suppose that (ii) holds, let (h_n) be a sequence in X converging to h, let (t_n) be a sequence of positive numbers converging to 0, and let $\phi : [0, \infty[\to X$ be given by

$$\phi(t_n) = x_0 + t_n h_n \ (n = 1, 2, \ldots), \qquad \phi(t) = x_0 + th \ \text{otherwise}.$$

Then $\phi(0) = x_0$ and $(\phi(t) - \phi(0))/t \to h$ as $t \to 0 +$, so that $\phi'_+(0) = h$. Hence $f \circ \phi$ is defined on the interval $[0, \eta[$ for some $\eta > 0$, and this implies that $x_0 + t_n h_n = \phi(t_n) \in A$ for all sufficiently large n, so that h is directed into A at x_0. Also

$$\frac{f(x_0 + t_n h_n) - f(x_0)}{t_n} = \frac{f(\phi(t_n)) - f(\phi(0))}{t_n} \to (f \circ \phi)'_+(0) = l$$

as $n \to \infty$, so that $\partial f(x_0)h = l$.

The position of the increment 0 in the theory is anomalous, as is shown by the following result.

(4.2.2) *Let f be a function from a set $A \subseteq X$ into Y, and let $x_0 \in A$. Then $\partial f(x_0)0$ exists if and only if x_0 is an interior point of A and there exist $K \geq 0$ and a neighbourhood V of x_0 in A such that*

$$\| f(x) - f(x_0) \| \leq K \| x - x_0 \| \tag{2}$$

for all $x \in V$. (In particular, if $\partial f(x_0)0$ exists then f is continuous at x_0.)

Suppose first that $\partial f(x_0)0$ exists. Then 0 is directed into A at x_0, and therefore $x_0 \in A^\circ$, by (b) (p. 259). Further, f is continuous at x_0, for if (x_n) is a sequence in $A \backslash \{x_0\}$ converging to x_0, and

$$t_n = \| x_n - x_0 \|^{1/2}, h_n = (x_n - x_0)/t_n,$$

then $h_n \to 0$ in $X, t_n \to 0$ in $]0, \infty[$, and therefore

$$f(x_n) - f(x_0) = t_n \frac{f(x_0 + t_n h_n) - f(x_0)}{t_n} \to 0 \cdot \partial f(x_0)0 = 0.$$

To prove the inequality (2), suppose on the contrary that no such K and V exist. Then we can find a sequence (k_n) in $X \backslash \{0\}$ converging to 0 such that

$$\|f(x_0 + k_n) - f(x_0)\| / \|k_n\| \to \infty \quad \text{as} \quad n \to \infty,$$

and obviously $\|f(x_0 + k_n) - f(x_0)\| > 0$ for all sufficiently large n. For such n let

$$t'_n = \|f(x_0 + k_n) - f(x_0)\|, \quad h'_n = k_n / t'_n.$$

Then $h'_n \to 0$ in X, $t'_n \to 0$ in $]0, \infty[$, and

$$\|f(x_0 + t'_n h'_n) - f(x_0)\| / t'_n = 1,$$

and again we obtain a contradiction, since $\partial f(x_0) 0 = 0$.

It remains to prove the 'if', and this is simple, for if (2) holds for all $x \in V$ then

$$\|f(x_0 + th) - f(x_0)\| / t \le K \|h\|$$

whenever $th \in V$ and $t > 0$.

Theorem (3.1.3) on the linearity (as a mapping defined on a vector space of functions) of the Fréchet differential applies without change to the Hadamard variation, just as it does for the Gâteaux variation (cf. (4.1.1)).

(4.2.3)(i) *If f is a function from a set $A \subseteq X$ into Y which has a Hadamard variation at x_0 for the increment h, then so does αf for every scalar α and $\partial(\alpha f)(x_0)h = \alpha \partial f(x_0)h$.*
(ii) *If f, g are functions from sets $A, B \subseteq X$ into Y which have Hadamard variations at x_0 for the increment h, then so does $f + g$, and $\partial(f + g)(x_0)h = \partial f(x_0)h + \partial g(x_0)h$.*

In contrast to the Gâteaux variation, the Hadamard variation obeys the chain rule, viz.

(4.2.4) *Let f be a function from a set $A \subseteq X$ into Y which has a Hadamard variation at the point x_0 for the increment h, and let g be a function from a set $B \subseteq Y$ into Z which has a Hadamard variation at the point $y_0 = f(x_0)$ for the increment $k = \partial f(x_0)h$. Then $g \circ f$ has a Hadamard variation at x_0 for the increment h, equal to $\partial g(y_0)k = \partial g(y_0)(\partial f(x_0)h)$.*

Let (h_n) be a sequence in X converging to h, and let (t_n) be a sequence of positive numbers converging to 0. Then $x_0 + t_n h_n \in A$ for all sufficiently large n, and if for such n we write $k_n = (f(x_0 + t_n h_n) - f(x_0))/t_n$, then $k_n \to \partial f(x_0)h = k$ as $n \to \infty$. This implies in turn that $f(x_0 + t_n h_n) =$

$y_0 + t_n k_n \in B$ for all sufficiently large n, whence h is directed into the domain of $g \circ f$ at x_0. Further,

$$\frac{g(f(x_0 + t_n h_n)) - g(f(x_0))}{t_n} = \frac{g(y_0 + t_n k_n) - g(y_0)}{t_n} \to \partial g(y_0)k$$

as $n \to \infty$, and this gives the result.

A similar but simpler proof gives the following companion result (cf. (4.1.2)).

(4.2.5) *Let f be a function from a set $A \subseteq X$ into Y which has a Gâteaux variation at the point x_0 for the increment h, and let g be a function from a set $B \subseteq Y$ into Z which has a Hadamard variation at the point $y_0 = f(x_0)$ for the increment $k = \delta f(x_0)h$. Then $g \circ f$ has a Gâteaux variation at x_0 for the increment h, equal to $\partial g(y_0) k = \partial g(y_0)(\delta f(x_0)h)$.*

The next group of results deal with the case where the variation $\partial f(x_0)h$ exists for all $h \in X$, and here there are marked contrasts with the Gâteaux variation (cf. §4.1, Examples (c), (d)).

(4.2.6) *Let f be a function from a set $A \subseteq X$ into Y, let $x_0 \in A$, and suppose that the Hadamard variation $\partial f(x_0)h$ exists for all $h \in X$. Then x_0 is an interior point of A, f is continuous at x_0 (and indeed (2) holds on some neighbourhood of x_0), and $\partial f(x_0)$ is continuous on X.*

The first two statements and the parenthesis follow from the existence of $\partial f(x_0)0$. To prove the last statement, suppose on the contrary that $\partial f(x_0)$ is not continuous at a point $k \in X$. Then we can find a positive number ε and a sequence (k_n) in X converging to k such that

$$\| \partial f(x_0)k_m - \partial f(x_0)k \| \geq \varepsilon$$

for all m. Further, since

$$\lim_{n \to \infty} (f(x_0 + t_n k_m) - f(x_0))/t_n = \partial f(x_0)k_m$$

for each m and every sequence (t_n) of positive numbers converging to 0, we can find a sequence (s_m) of positive numbers converging to 0 such that

$$\left\| \frac{f(x_0 + s_m k_m) - f(x_0)}{s_m} - \partial f(x_0)k \right\| \geq \tfrac{1}{2}\varepsilon$$

for all m, and this gives a contradiction, since the expression on the left tends to 0 as $m \to \infty$.

The most important example where the Hadamard variation is defined on X, but may be non-linear, is that of a convex function.

(4.2.7) *If p is a continuous convex function on a convex set $A \subseteq X$ and x_0 is an interior point of A, then $\partial p(x_0)h$ exists for all $h \in X$ and the function $\partial p(x_0)$ is (continuous and) sublinear.*

By §4.1, Example (e), the Gâteaux variation $\delta p(x_0)$ is defined and sublinear on X. Since p is Lipschitzian on a neighbourhood of x_0 (A.3.11.), it follows that the Hadamard variation $\partial p(x_0)$ is defined and sublinear on X.

By an obvious translation, we deduce from (4.2.1) the following corollary.

(4.2.7. Corollary 1) *Let p be a continuous convex function on a convex set $A \subseteq X$, let x_0 be an interior point of A, and let ϕ be a function from a set $B \subseteq \mathbf{R}$ into X taking the value x_0 at t_0 and having there a right-hand derivative $\phi'_+(t_0)$. Then $p \circ \phi$ is defined on $B \cap [t_0, t_0 + \eta[$ for some $\eta > 0$ and has a right-hand derivative at t_0 equal to $\partial p(x_0)\phi'_+(t_0)$.*

The hypotheses in (4.2.7) can be weakened. We recall (A.3.12) that if p is a lower semicontinuous convex function on a convex set $A \subseteq X$, and x_0 is an internal point of A, then x_0 is an interior point of A, and p is continuous on a neighbourhood of x_0. Moreover, it is clear from the definition of Gâteaux variation that if $\delta p(x_0)h$ exists for all $h \in X$, then x_0 is an internal point of A. Hence we have:

(4.2.7. Corollary 2) *Let p be a lower semicontinuous convex function on a convex set $A \subseteq X$, and suppose that p has a Gâteaux variation $\delta p(x_0)h$ at a point $x_0 \in A$ for all $h \in X$. Then x_0 is an interior point of A, p is continuous on a neighbourhood of x_0, $\partial p(x_0)h$ exists for all $h \in X$, and $\partial p(x_0)$ is continuous and sublinear.*

We now define Hadamard differentiability. Let f be a function from a set $A \subseteq X$ into Y, and let x_0 be an interior point of A. We say that f is *Hadamard differentiable* at x_0 if the Hadamard variation $\partial f(x_0)h$ exists for all $h \in X$ and the function $\partial f(x_0)$ belongs to $\mathscr{L}(X, Y)$. This function $\partial f(x_0)$ is then called the *Hadamard differential* of f at x_0. Further, the function $x \mapsto \partial f(x)$ whose domain is the set of interior points of A at which f is Hadamard differentiable is called the *Hadamard differential* of f, and we denote it by ∂f.

The following result is the analogue of (3.1.2) for Hadamard differentiability.

(4.2.8) *Let f be a function from a set $A \subseteq X$ into Y, let x_0 be an interior point of A, and let $T \in \mathscr{L}(X, Y)$. Then the following statements are equivalent*:

(i) *f is Hadamard differentiable at x_0, with $\partial f(x_0) = T$, i.e. if $h \in X$ then*

$$(f(x_0 + t_n h_n) - f(x_0))/t_n \to Th \quad as \quad n \to \infty \tag{3}$$

for every sequence (h_n) in X converging to h and every sequence (t_n) of positive numbers converging to 0;

(ii) *if $h \in X$ then (3) holds for every sequence (h_n) in X converging to h and every sequence (t_n) of non-zero real numbers converging to 0*;

(iii) *for each function ϕ from a neighbourhood of 0 in $\mathbf{R}$ into X which takes the value x_0 at 0 and has a derivative at 0, the function $f \circ \phi$ is defined on a neighbourhood of 0 and has a derivative at 0 equal to $T(\phi'(0))$*;

(iv) *for each compact set $E \subseteq X$,*

$$(f(x_0 + th) - f(x_0))/t \to Th$$

as $t \to 0$, uniformly for h in E.

Here the equivalence of (i) and (ii) is almost immediate. Also (i) implies (iii), by (4.2.1), and (iii) implies (ii), by an argument similar to that of (4.2.1). Next, (ii) implies (iv). To prove this, suppose that (ii) holds, and let

$$D(t,h) = (f(x_0 + th) - f(x_0))/t \qquad (x_0 + th \in A, t \neq 0).$$

Then $D(t, h) \to Th$ as $t \to 0$, for each $h \in X$, and we have to show that the convergence is uniform on any compact $E \subseteq X$. If this is false, we can find a positive number ε, a sequence (t_n) of non-zero real numbers converging to 0, and a sequence (h_n) in E, such that $\| D(t_n, h_n) - Th_n \| \geq \varepsilon$ for all n. Since E is compact, the sequence (h_n) has a subsequence (h_{n_m}) converging to a point h of E, and since T is continuous we have

$$\| D(t_{n_m}, h_{n_m}) - Th_{n_m} \| \to \| Th - Th \| = 0.$$

This gives a contradiction, and hence (ii) implies (iv).

We show finally that (iv) implies (ii). If (h_n) is a sequence in X converging to a point h, then the set consisting of h together with the points h_n is compact. Hence if (iv) holds, then $D(t, h_n) - Th_n \to 0$ as $t \to 0$, uniformly in n. It follows that if (t_n) is a sequence of non-zero real numbers converging to 0, then $D(t_n, h_n) - Th_n \to 0$ as $n \to \infty$, and since T is continuous, this implies that (ii) holds.

For a function ϕ of a real variable taking values in a real normed space, the notions of Fréchet, Hadamard and Gâteaux differentiability at a point coincide, and each is equivalent to the property that the derivative of ϕ exists when the point in question is an interior point of the domain of ϕ (cf. Example (*a*) and §4.1, Example (*a*)).

For a function f from a subset of X into Y, the Hadamard differentiability of f at x_0 obviously implies the Gâteaux differentiability of f there, and $\delta f(x_0) = \partial f(x_0)$. The converse is false when dim $X \geq 2$, as is readily seen by considering an example similar to the function of Example (c). However, if in addition f is Lipschitzian in a neighbourhood of x_0, then the Gâteaux differentiability of f at x_0 implies the Hadamard differentiability of f there (see p. 259).

If f is Fréchet differentiable at x_0, then it is Hadamard differentiable there, and $\partial f(x_0) = df(x_0)$ (cf. Example (b)). From (4.2.8)(iv) and (3.1.2)(iv) we see further that the converse is true when dim $X < \infty$, so that Fréchet and Hadamard differentiability are then equivalent. In fact, the following example shows that *Fréchet and Hadamard differentiability are equivalent if and only if* dim $X < \infty$.

Example (d)

Let dim $X = \infty$, let (w_n) be a sequence on the unit sphere of X with no convergent subsequence, let c be a unit vector in Y, and let $f : X \to Y$ be given by

$$f(w_m/m) = c/m \qquad (m = 1, 2, \ldots), \qquad f(x) = 0 \text{ otherwise.}$$

Then f is Hadamard differentiable at 0 with $\partial f(0) = 0$, but f is not Fréchet differentiable at 0.

To prove the first statement, let $h \in X$, let (h_n) be a sequence in X converging to h, and let (t_n) be a sequence of positive numbers converging to 0. If $f(t_n h_n) = 0$ for all sufficiently large n, then obviously $f(t_n h_n)/t_n \to 0$ as $n \to \infty$. In the contrary case, let (n_r) be the subsequence of the positive integers for which $f(t_{n_r} h_{n_r}) \neq 0$. We have to show that $f(t_{n_r} h_{n_r})/t_{n_r} \to 0$ as $r \to \infty$.

For each r we have $t_{n_r} h_{n_r} = w_{m_r}/m_r$ for some positive integer m_r, and since $t_{n_r} h_{n_r} \to 0, m_r \to \infty$ as $r \to \infty$. If $h \neq 0$ and for each non-zero x of X we write $x\hat{\;} = x/\|x\|$, then

$$w_{m_r} = w\hat{\;}_{m_r} = (m_r t_{n_r} h_{n_r})\hat{\;} = h\hat{\;}_{n_r} \to h\hat{\;}$$

as $r \to \infty$, and this contradicts our choice of the w_m. Hence $h = 0$, and therefore

$$\|f(t_{n_r} h_{n_r})/t_{n_r}\| = \|c/(t_{n_r} m_r)\| = \|w_{m_r}/(t_{n_r} m_r)\| = \|h_{n_r}\| \to 0,$$

as required.

It follows now that if f is Fréchet differentiable at 0, then $df(0) = 0$. This, however, is impossible, since the sequence (w_m) is bounded and $m(f(w_m/m) - f(0)) = c$ for all m (cf. (3.1.2)(iii)).

From (4.2.6) we see that if a function f is Hadamard differentiable at x_0 then it is continuous at x_0, and indeed there exist $K \geq 0$ and a neighbourhood V of x_0 such that

$$\| f(x) - f(x_0) \| \leq K \| x - x_0 \|$$

for all $x \in V$. The function we have just considered in Example (*d*) shows that in this inequality we cannot take K to be of the form $\| \partial f(x_0) \| + \varepsilon$ for arbitrarily small positive ε (cf. (3.1.1)).

By (4.2.3), the Hadamard differential has the same linearity properties as the Fréchet differential (cf. (3.1.3)). Further, by (4.2.4), the chain rule here takes the following form.

(4.2.9) *Let f be a function from a set $A \subseteq X$ into Y which is Hadamard differentiable at a point x_0, and let g be a function from a set $B \subseteq Y$ into Z which is Hadamard differentiable at the point $y_0 = f(x_0)$. Then $g \circ f$ is Hadamard differentiable at x_0, and $\partial(g \circ f) = \partial g(y_0) \circ \partial f(x_0)$.*

From (4.2.9) and (4.2.8) we shall deduce that *Hadamard differentiability is the most general form of differentiability for which the chain rule holds*, the phrase 'most general' here being used in the following sense.

It is reasonable to require of any definition of differentiability, say $\mathfrak{X}$-differentiability, that

(i) if f is a function from a set $A \subseteq X$ into Y, and x_0 is an interior point of A, then the $\mathfrak{X}$-differential of f at x_0, if it exists, is a unique continuous linear function from X into Y,

(ii) if ϕ is a function from a set $A \subseteq \mathbf{R}$ into Y considered as a normed space over $\mathbf{R}$ and t_0 is an interior point of A, then ϕ is $\mathfrak{X}$-differentiable at t_0 if and only if the derivative $\phi'(t_0)$ exists, and the $\mathfrak{X}$-differential of ϕ at t_0 is the function $h \mapsto h\phi'(t_0)$ from $\mathbf{R}$ into Y.

We shall show that *if the chain rule holds for $\mathfrak{X}$-differentiability* (i.e. the result of (4.2.9) holds with $\mathfrak{X}$-differentiability throughout) *then every function $\mathfrak{X}$-differentiable at a point is also Hadamard differentiable at that point and the two differentials are the same.*

To prove this, suppose that $\mathfrak{X}$-differentiability satisfies these conditions, let $f : A \to Y$ be $\mathfrak{X}$-differentiable at x_0, and let ϕ be a function from a neighbourhood of 0 in $\mathbf{R}$ into X such that $\phi(0) = x_0$ and that $\phi'(0)$ exists. Then, by applying the chain rule to f and ϕ, we see that f satisfies the criterion of (4.2.8)(iii), whence f is Hadamard differentable at x_0.

The mean value inequalities of (4.1.6, 7) and the results of (4.1.7, Corollaries 1, 2) obviously hold with the Hadamard differential in place of the Gâteaux differential. In particular, when we define the functions which are C^1 on an open set E, it is immaterial whether we use the Fréchet, Gâteaux, or Hadamard differentials.

Partial Hadamard differentials can be defined exactly as for Fréchet and Gâteaux differentials (§§3.3,4.1). We can also define Hadamard differentials of higher order, but since we make no use of these higher order differentials we do not pursue this point.

We next consider some results concerning the upper and lower Hadamard variations of a real-valued function.

Let f be a function from a set $A \subseteq X$ into $\mathbf{R}$, let $x_0 \in A$, and let h be directed into A at x_0. We define the *upper Hadamard variation of f at x_0* for the increment h by the relation

$$\bar{\partial}f(x_0)h = \sup(\limsup_{n \to \infty} (f(x_0 + t_n h_n) - f(x_0))/t_n),$$

where the supremum is taken over all sequences (h_n) in X converging to h and all sequences (t_n) of positive numbers converging to 0 (the values $\pm \infty$ being permitted). The *lower variation* $\underline{\partial}f(x_0)h$ is then defined to be the negative of the upper variation of $(-f)$. Obviously if f has a Hadamard variation at x_0 for the increment h, then $\bar{\partial}f(x_0)h = \underline{\partial}f(x_0)h = \partial f(x_0)h$.

We observe that there exist a sequence (h_n) in X converging to h and a sequence (t_n) of positive numbers converging to 0 such that

$$\bar{\partial}f(x_0)h = \lim_{n \to \infty} (f(x_0 + t_n h_n) - f(x_0))/t_n \qquad (4)$$

(so that $\bar{\partial}f(x_0)h$ is the greatest extended real number with this property). To prove this, we may suppose that $\bar{\partial}f(x_0)h > -\infty$, otherwise the result is trivial. Let (s_m) be a strictly increasing sequence of real numbers such that $s_m \to \bar{\partial}f(x_0)h$ as $m \to \infty$. Then for each positive integer m we can find sequences $(h_n^{(m)})$ and $(t_n^{(m)})$ in X and $]0, \infty[$ converging to h and 0 respectively such that for all m and n

$$(f(x_0 + t_n^{(m)} h_n^{(m)}) - f(x_0))/t_n^{(m)} > s_m.$$

If now for each m we choose n_m so large that $\| h_{n_m}^{(m)} - h \| < 1/m$ and $t_{n_m}^{(m)} < 1/m$, then $h_{n_m}^{(m)} \to h$ and $t_{n_m}^{(m)} \to 0$ as $m \to \infty$, and for all m

$$(f(x_0 + t_{n_m}^{(m)} h_{n_m}^{(m)}) - f(x_0))/t_{n_m}^{(m)} > s_m.$$

Hence

$$\limsup_{m \to \infty} \; (f(x_0 + t_{n_m}^{(m)} h_{n_m}^{(m)}) - f(x_0))/t_{n_m}^{(m)} \geq \bar{\partial}f(x_0)h,$$

and there is obviously equality here. On taking appropriate subsequences of $(h_{n_m}^{(m)})$ and $(t_{n_m}^{(m)})$ we therefore obtain the required result.

The following result is the analogue of (4.2.1).

(4.2.10) *Let f be a function from a set $A \subseteq X$ into $\mathbf{R}$, let $x_0 \in A$, and let h be directed into A at x_0. If ϕ is a function from a subset of $\mathbf{R}$ containing 0 into X such that $\phi(0) = x_0, \phi'_+(0) = h$, and $\psi = f \circ \phi$, then*

$$D^+\psi(0) \le \bar{\partial}f(x_0)h. \tag{5}$$

Moreover, there exists a function ϕ for which equality holds in (5).

Let (t_n) be a sequence of positive numbers converging to 0 such that $(\psi(t_n) - \psi(0))/t_n \to D^+\psi(0)$ as $n \to \infty$. Then $h_n = (\phi(t_n) - \phi(0))/t_n \to h$ as $n \to \infty$, whence

$$D^+\psi(0) = \lim_{n \to \infty} (\psi(t_n) - \psi(0))/t_n = \lim_{n \to \infty} (f(\phi(t_n)) - f(\phi(0)))/t_n$$

$$= \lim_{n \to \infty} (f(x_0 + t_n h_n) - f(x_0))/t_n \le \bar{\partial}f(x_0)h.$$

Next, let (h_n), (t_n) be sequences for which (4) holds, let $\phi_1 : \mathbf{R} \to X$ be given by

$$\phi_1(t_n) = x_0 + t_n h_n \quad (n = 1, 2, \ldots), \qquad \phi_1(t) = x_0 + th \text{ otherwise,}$$

and let $\psi_1 = f \circ \phi_1$. Then $\phi_1(0) = x_0, \phi'_1(0) = h$, and

$$(\psi_1(t_n) - \psi_1(0))/t_n = (f(x_0 + t_n h_n) - f(x_0))/t_n \to \bar{\partial}f(x_0)h$$

as $n \to \infty$, so that $D^+\psi_1(0) \ge \bar{\partial}f(x_0)h$. Hence $D^+\psi_1(0) = \bar{\partial}f(x_0)h$, by (4).

It is easily verified that if h is directed into A at x_0 and $\alpha > 0$, then $\bar{\partial}f(x_0)(\alpha h) = \alpha\bar{\partial}f(x_0)h$. In particular, this implies that if x_0 is an interior point of A (so that 0 is directed into A at x_0) then $\bar{\partial}f(x_0)0$ is one of $-\infty$, $0, \infty$. The following result, which is the analogue of the last part of (4.2.6), shows in particular that if $\bar{\partial}f(x_0)0 = -\infty$ then $\bar{\partial}f(x_0)h = -\infty$ for all h.

(4.2.11) *The upper Hadamard variation of a real-valued function at a point x_0 is upper semicontinuous.*

Let f be a real-valued function on a set $A \subseteq X$, and let $x_0 \in A$. The domain of the function $\bar{\partial}f(x_0)$ is then the set $K_d(A; x_0)$ of vectors directed into A at x_0, and is open in X. If $\bar{\partial}f(x_0)$ is not upper semicontinuous at a point $k \in K_d(A; x_0)$, we can find a real number $L > \bar{\partial}f(x_0)k$ and a sequence (k_m) of points of X converging to k such that $\bar{\partial}f(x_0)k_m > L$ for all m. By (4), we can then choose sequences $(h_n^{(m)})$ and $(t_n^{(m)})$ in X and $]0, \infty[$ converging to k_m and 0 respectively such that for all m and n

$$(f(x_0 + t_n^{(m)} h_n^{(m)}) - f(x_0))/t_n^{(m)} > L.$$

If now for each m we choose n_m so large that $\| h_{n_m}^{(m)} - k_m \| < 1/m$ and $t_{n_m}^{(m)} <$

$1/m$, then $h_{nm}^{(m)} \to k$ and $t_{nm}^{(m)} \to 0$ as $m \to \infty$, and

$$\limsup_{m \to \infty} \ (f(x_0 + t_{nm}^{(m)} h_{nm}^{(m)}) - f(x_0))/t_{nm}^{(m)} \geq L > \bar{\partial}f(x_0)k,$$

and this contradicts the definition of $\bar{\partial}f(x_0)k$.

(4.2.11. Corollary) *If $\bar{\partial}f(x_0)$ is convex,† then it is continuous and sublinear on X.*

Continuity is immediate. It implies that the relation $\alpha\bar{\partial}f(x_0)h = \bar{\partial}f(x_0)(\alpha h)$, which we know for $\alpha > 0$, holds also for $\alpha = 0$, and this relation together with convexity implies sublinearity.

It is obvious that if $\bar{\partial}f(x_0)0 < \infty$, then f is upper semicontinuous at x_0. In particular, this implies that if f is a convex function on a convex open set $A \subseteq X$, and $\bar{\partial}f(x_0)0 < \infty$ for some $x_0 \in A$, then f is continuous on A.

Now let X be a real normed space. We define the *subvariation* $\partial_* f(x_0)$ of the function $f : A \to \mathbf{R}$ at the interior point x_0 of A to be the set

$$\{u \in X' : u(h) \leq \bar{\partial}f(x_0)h \qquad (h \in X)\}.$$

(4.2.12) *Let X be real, let f be a function from a set $A \subseteq X$ into $\mathbf{R}$, let x_0 be an interior point of A, and let $u \in X'$. Then the following statements are equivalent:*

(i) $u \in \partial_* f(x_0)$;

(ii) *for each $h \in X$ there exist a sequence (h_n) in X converging to h and a sequence (t_n) of positive numbers converging to 0 such that*

$$\lim_{n \to \infty} (f(x_0 + t_n h_n) - f(x_0))/t_n \geq u(h).$$

If in addition f is lower semicontinuous and convex on A (and A is convex), then each of (i) and (ii) is equivalent to

(iii) $f(x_0 + h) - f(x_0) \geq u(h)$ *whenever* $x_0 + h \in A$.‡

Clearly (ii) implies (i), while the converse follows from (4).

Suppose now that f is lower semicontinuous and convex, so that the Hadamard variation $\partial f(x_0)h$ of f exists for all $h \in X$. If $x_0 + h \in A$, then

$$f(x_0 + h) - f(x_0) \geq (f(x_0 + th) - f(x_0))/t$$

for $0 < t \leq 1$, and on making $t \to 0+$ we obtain that

$$f(x_0 + h) - f(x_0) \geq \partial f(x_0)h.$$

† Recall that by our conventions a convex function is finite-valued.

‡ In books on convexity theory, the subvariation $\partial_* f(x_0)$ is defined (for convex f only) as the set of all $u \in X'$ that satisfy (iii). It is usually called the *subdifferential* of f at x_0, and denoted by $\partial f(x_0)$.

Hence if $u \in \partial_* f(x_0)$, then (iii) holds. Conversely, if (iii) holds, then for all $h \in X$ and all sufficiently small positive t we have

$$f(x_0 + th) - f(x_0) \geq u(th) = tu(h),$$

and this obviously implies that $\partial f(x_0)h \geq u(h)$, so that $u \in \partial_* f(x_0)$.

Example (e)

It is easy to see that if f is Hadamard differentiable at x_0, then the subvariation of f at x_0 consists of the single element $\partial f(x_0)$. The converse, that if $\partial_* f(x_0)$ consists of a single element then f is Hadamard differentiable at x_0, is false. For example, if $X = \mathbf{R}$, and

$$f(x) = -|x| \sin^2(1/x) \qquad (x \neq 0), \quad f(0) = 0,$$

then $\bar{\partial} f(0)h = 0$ for all $h \in \mathbf{R}$, so that $\partial_* f(0)$ consists of the zero function, while f is clearly not Hadamard differentiable at 0.

Example (f)

If $N(x) = \|x\|$ and $x_0 \neq 0$, then

$$\partial_* N(x_0) = \{u \in X' : \|u\| = 1, u(x_0) = \|x_0\|\}.$$

To prove this let $u \in \partial_* N(x_0)$. Then for all $h \in X$ we have

$$u(h) \leq N(x_0 + h) - N(x_0) = \|x_0 + h\| - \|x_0\| \leq \|h\|,$$

so that $\|u\| \leq 1$. Also, taking $h = -x_0$ we obtain that $u(-x_0) \leq -\|x_0\|$, so that $u(x_0) \geq \|x_0\|$. Since also $u(x_0) \leq \|x_0\|$, it follows that $u(x_0) = \|x_0\|$, and therefore $\|u\| = 1$. On the other hand, if $\|u\| = 1$ and $u(x_0) = \|x_0\|$, then

$$\|x_0\| + u(h) = u(x_0 + h) \leq \|x_0 + h\|,$$

so that $u \in \partial_* N(x_0)$.

For the exceptional case where $x_0 = 0$ we have trivially $\partial_* N(0) = \{u \in X' : \|u\| \leq 1\}$.

The subvariation $\partial_* f(x_0)$ is of most interest when the upper variation $\bar{\partial} f(x_0)$ is convex, for by (4.2.11. Corollary) $\bar{\partial} f(x_0)$ is continuous and sublinear. From (A.4.1. Corollary 2) we therefore have:

(4.2.13) *If the upper variation $\bar{\partial} f(x_0)$ of the function $f : A \to \mathbf{R}$ at x_0 is convex, then the subvariation $\partial_* f(x_0)$ is non-empty, and for all $h \in X$*

$$\bar{\partial} f(x_0)h = \sup \{u(h) : u \in \partial_* f(x_0)\}. \tag{6}$$

(4.2.13. Corollary) *If f has a Hadamard variation at x_0 which is convex, then f is Hadamard differentiable at x_0 if and only if the subvariation $\partial_* f(x_0)$ consists of a single element u, and in this case $u = \partial f(x_0)$.*

Here the 'if' follows from (6), since now $\overline{\partial} f(x_0) = \partial f(x_0)$. The 'only if' is a restatement of the first part of Example (*e*).

Example (g)

If $N(x) = \|x\|$, then N is Hadamard differentiable at a point $x_0 \neq 0$ if and only if the closed ball $B = \{x \in X : \|x\| \leq \|x_0\|\}$ has exactly one supporting hyperplane at x_0. In fact, by (4.2.13. Corollary) and Example (*e*), N is Hadamard differentiable at x_0 if and only if there exists exactly one $u \in X'$ such that $\|u\| = 1$ and $u(x_0) = \|x_0\|$. Obviously if u has these properties then $u(x) \leq \|u\| \, \|x_0\| = u(x_0)$ for all $x \in B$, so that $u(x) = u(x_0)$ is the equation of a supporting hyperplane to B at x_0. Conversely, let $u(x) = u(x_0)$ be the equation of a supporting hyperplane to B at x_0, so that $u(x) \leq u(x_0)$ for all $x \in B$. Since the hyperplane is unchanged if we multiply u by a positive constant, we may suppose that $\|u\| = 1$. The condition that $u(x) \leq u(x_0)$ for all $x \in B$ implies that $\pm u(x) = u(\pm x) \leq u(x_0)$ for all $x \in B$, so that also $|u(x)| \leq u(x_0)$.
Hence

$$1 = \|u\| = \sup_{\|x\| = \|x_0\|} |u(x)| / \|x\| \leq u(x_0)/\|x_0\| \leq \|u\| = 1,$$

so that $u(x_0) = \|x_0\|$. Hence there exists exactly one $u \in X'$ such that $\|u\| = 1$ and $u(x_0) = \|x_0\|$ if and only if B has exactly one supporting hyperplane at x_0.

A norm which is Hadamard differentiable at every point of $X \backslash \{0\}$ is said to be *smooth*.

Exercises 4.2

1 Let f be a function from a set $A \subseteq X$ into Y, let $x_0 \in A$, let h be directed into A at x_0, and let $l \in Y$. Prove that f has a Hadamard variation at x_0 for the increment h, equal to l, if and only if for each $\varepsilon > 0$ there exist a neighbourhood V of h in X and a positive number δ such that

$$\left\| \frac{f(x_0 + tk) - f(x_0)}{t} - l \right\| \leq \varepsilon$$

whenever $k \in V$ and $0 < t < \delta$.

2 Let f be a function from a set $A \subseteq X$ into Y, and let x_0 be an interior point of A. Prove that
   (i) if the Hadamard variation $\partial f(x_0)h$ exists for all $h \in X$, then, for each compact

set $E \subseteq X$,

$$(f(x_0 + th) - f(x_0))/t \to \partial f(x_0)h$$

as $t \to 0+$, uniformly for h in E,

(ii) if there exists a continuous function $T : X \to Y$ such that, for each compact set $E \subseteq X$,

$$(f(x_0 + th) - f(x_0))/t \to T(h)$$

as $t \to 0+$, uniformly for h in E, then $\partial f(x_0)h$ exists and is equal to $T(h)$ for all $h \in X$.

3 Give a direct proof of (4.2.7) without using the Lipschitzian property of p.

[Hint. Since $\delta p(x_0)$ is defined and convex on X, it is enough to prove that if $h \in X$, (h_n) is a sequence in X converging to h, and (t_n) is a sequence of positive numbers converging to 0, then

$$(p(x_0 + t_n h_n) - p(x_0 + t_n h))/t_n \to 0$$

as $n \to \infty$ (cf. (1)). Use the fact that if $y, y - z, y + z \in A$, then $t \mapsto p(y + tz)$ is convex on $[-1, 1]$, whence for $0 < t \le 1$

$$p(y) - p(y - z) \le (p(y + tz) - p(y))/t \le p(y + z) - p(y)$$

(these inequalities are particular cases of those used in §1.1, Example (g) (p. 6)).]

4 Let f be a function from a set $A \subseteq X$ into Y which has a Hadamard variation at the point x_0 for the increment h, let g be a function from a set $B \subseteq Y$ into $\mathbf{R}$, and let $k = \partial f(x_0)h$ be directed into B at $y_0 = f(x_0)$. Prove that

$$\bar{\partial}(g \circ f)(x_0)h \le \bar{\partial} g(y_0)k = \bar{\partial} g(y_0)(\partial f(x_0)h).$$

5 Let X be a real normed space. Prove that if $p(x) = \frac{1}{2} \|x\|^2$, then p is convex and

$$\partial_* p(x_0) = \{u \in X' : \|u\| = \|x_0\|, u(x_0) = \|u\| \|x_0\|\}.$$

4.3 The tangent cones to the graph and the level surfaces of a function

It has been shown in (1.2.1) that a function f of a real variable has a derivative at an interior point x_0 of its domain if and only if the graph of f has a tangent line not parallel to the y-axis at the point $(x_0, f(x_0))$. Further, the equation of this tangent line is

$$y = f(x_0) + (x - x_0)f'(x_0),$$

i.e. the tangent line is the graph of the approximating function $x \mapsto f(x_0) + (x - x_0)f'(x_0)$. In this section we investigate the analogue of this result for a vector-valued function of a vector variable. We consider also the corresponding problem for the level surfaces of a function. Our results here have applications to constrained maxima and minima which are given in the next two sections.

The definition of tangency which we employ is as follows. Let $E \subseteq X$, and let $x_0 \in X$. We say that a vector $h \in X$ is *tangent* to E at x_0 if for each neighbourhood W of h in X and each positive number ε there exist $k \in W$ and $t \in]0, \varepsilon[$ such that $x_0 + tk \in E$. It is easily verified that if h is tangent to

E at x_0 and $\alpha > 0$, then αh is tangent to E at x_0. The set of vectors tangent to E at x_0 is therefore a cone in X with vertex 0, and we denote this cone by $K_t(E;x_0)$. Clearly $0 \in K_t(E;x_0)$ for all $E \subseteq X$ and all $x_0 \in E$.

Example (a)

If x_0 is an interior point of E, then $K_t(E;x_0) = X$.

Example (b)

If $X = \mathbf{R}^n$, E is the closed unit ball in $\mathbf{R}^n$, and $\|x_0\| = 1$, then $K_t(E;x_0)$ is a half-space.

The following result gives some equivalent formulations of the definition of $K_t(E;x_0)$ in terms of sequences. Here and later, for each non-zero $x \in X$ we use $x^{\wedge}$ to denote the unit vector $x/\|x\|$.

(4.3.1) **Lemma.** *Let $E \subseteq X$, let $x_0 \in E$, and let $h \in X$. Then the following statements are equivalent:*

(i) $h \in K_t(E;x_0)$;

(ii) *there exists a sequence (h_n) in X converging to h and a sequence (t_n) of positive numbers converging to 0 such that $x_0 + t_n h_n \in E$ for all n (or for all sufficiently large n).*

If in addition $h \neq 0$, then each of (i) *and* (ii) *is equivalent to*

(iii) *there exists a sequence (x_n) in $E \backslash \{x_0\}$ converging to x_0 such that $(x_n - x_0)^{\wedge} \to h^{\wedge}$.*

The equivalence of (i) and (ii) is immediate. If now $h \neq 0$, then (ii) implies (iii), for if (h_n) and (t_n) have the properties specified in (ii), and $x_n = x_0 + t_n h_n$, then $x_n \in E$ for all n and $(x_n - x_0)^{\wedge} = (t_n h_n)^{\wedge} = h_n^{\wedge} \to h^{\wedge}$. Conversely, (iii) implies (ii), for if (x_n) has the properties specified in (iii) and $h_n = \|h\| (x_n - x_0)^{\wedge}$, $t_n = \|x_n - x_0\|/\|h\|$, then $h_n \to h, t_n \to 0$, and $x_0 + t_n h_n = x_n \in E$ for all n.

The next two theorems deal with the cone of vectors tangent to the graph of a function; as might be expected from the last lemma, the appropriate form of differentiability to use here is that of Hadamard rather than that of Fréchet.

For any function F from a subset of X into Y we denote the graph of F in $X \times Y$ by $\mathscr{G}(F)$.

(4.3.2) *Let f be a function from a set $A \subseteq X$ into Y, let x_0 be an interior point* of A, let $z_0 = (x_0, f(x_0))$, and let $\mathcal{T}$ be the cone of vectors in $X \times Y$ tangent to $\mathcal{G}(f)$ at z_0.

(i) *If f has a Gâteaux variation at x_0 for each $h \in X$, then $\mathcal{G}(\delta f(x_0)) \subseteq \mathcal{T}$.*

(ii) *If f has a Hadamard variation at x_0 for each $h \in X$, then $\mathcal{G}(\partial f(x_0)) = \mathcal{T}$.*

We may suppose that $x_0 = 0$ and $f(x_0) = 0$ (for if this is not the case we replace f by the function $x \mapsto f(x_0 + x) - f(x_0)$).

To prove (i), let $h \in X$. Then for all $t > 0$, the point $(th, f(th))$ belongs to $\mathcal{G}(f)$, and $(h, f(th)/t) \to (h, \delta f(0)h)$ as $t \to 0+$. Hence $(h, \delta f(0)h) \in \mathcal{T}$, so that $\mathcal{G}(\delta f(0)) \subseteq \mathcal{T}$.

To prove (ii), it is enough, by (i), to show that if $\partial f(0)h$ exists for each $h \in X$, then $\mathcal{T} \subseteq \mathcal{G}(\partial f(0))$. Let $(h, l) \in \mathcal{T}$. Then we can find a sequence $((h_n, l_n))$ in $X \times Y$ converging to (h, l) and a sequence (t_n) of positive numbers converging to 0 such that $(t_n h_n, t_n l_n) \in \mathcal{G}(f)$ for all n. But then $l_n = f(t_n h_n)/t_n \to \partial f(0)h$ as $n \to \infty$, whence $l = \partial f(0)h$, i.e. $(h, l) \in \mathcal{G}(\partial f(0))$.

(4.3.3) (i) *If $\dim Y < \infty$, f is a function from a set $A \subseteq X$ into Y, x_0 is an interior point of A at which f is continuous, and the cone $\mathcal{T}$ of vectors tangent to $\mathcal{G}(f)$ at the point $z_0 = (x_0, f(x_0))$ is contained in the graph of a function $T \in \mathcal{L}(X, Y)$, then f is Hadamard differentiable at x_0 and $\partial f(x_0) = T$. If in addition $\dim X < \infty$, then f is Fréchet differentiable at x_0 and $df(x_0) = T$.*

(ii) *If $\dim Y = \infty$, there exists a function $f_1 : X \to Y$ which is continuous at 0 but not Gâteaux differentiable there, with $f_1(0) = 0$, and such that the cone of vectors tangent to the graph of f_1 at $(0,0)$ is $X \times \{0\}$, i.e. it is the graph of the zero function from X into Y.*

(iii) *If $\dim X = \infty$, $\dim Y < \infty$, there exists a function $f_2 : X \to Y$ which is Hadmard differentiable at 0 but not Fréchet differentiable there, with $f_2(0) = 0$, and such that the cone of vectors tangent to the graph of f_2 at $(0,0)$ is $X \times \{0\}$.*

We recall that if $\dim X < \infty$, then Hadamard and Fréchet differentiability are equivalent (p. 266). Hence to prove (i) it is enough to prove the first part of the result concerning the Hadamard differentiability of f, and in doing this we may again assume that $x_0 = 0$ and $f(x_0) = 0$.

Suppose then that 0 is an interior point of A, that f is continuous at 0 and $f(0) = 0$, and that the cone $\mathcal{T}$ of vectors tangent to $\mathcal{G}(f)$ at $(0,0)$ is contained in $\mathcal{G}(T)$, where $T \in \mathcal{L}(X, Y)$. If f is Hadamard differentiable at 0, then $\mathcal{G}(\partial f(0)) \subseteq \mathcal{G}(T)$ by (4.3.2)(ii) so that $\partial f(0) = T$; and conversely, if $\partial f(0) = T$, then f is obviously Hadamard differentiable at 0. Hence it is enough to show that for each $h \in X$, Th is the Hadamard variation of f at 0 for the increment h.

In the contrary case, there exist a positive number ε, a sequence (h_n) in X converging to a point h, and a sequence (t_n) of positive numbers converging to 0 such that

$$\| f(t_n h_n)/t_n - Th \| \geq \varepsilon \qquad (n = 1, 2, \ldots). \tag{1}$$

If the sequence of points $w_n = f(t_n h_n)/t_n$ is bounded, we can find a subsequence (w_{n_m}) of this sequence converging to a point w (note that a bounded closed set in Y is compact), and since $(h_{n_m}, w_{n_m}) \to (h, w)$ as $m \to \infty$ and $(t_{n_m} h_{n_m}, t_{n_m} w_{n_m}) = (t_{n_m} h_{n_m}, f(t_{n_m} h_{n_m})) \in \mathscr{G}(f)$ for all m, it follows that $(h, w) \in \mathscr{T}$. Hence $(h, w) \in \mathscr{G}(T)$, so that $w = Th$, and this contradicts the condition (1). On the other hand, if the sequence (w_n) is unbounded, we can select a subsequence (w_{r_s}) such that $\| w_{r_s} \| \to \infty$ as $s \to \infty$ and that the sequence $(w_{r_s}^{\wedge})$ converges to a point v (so that also $\| v \| = 1$). But then as n tends to ∞ through the values r_s,

$$(t_n h_n, f(t_n h_n))^{\wedge} = (h_n, w_n)^{\wedge} = \left(\frac{h_n}{\| w_n \|}, w_n^{\wedge} \right)^{\wedge} \to (0, v),$$

and hence $(0, v) \in \mathscr{T}$, by (4.3.1)(iii). Since $(0, v) \notin \mathscr{G}(T)$, this again gives a contradiction, and completes the proof of (i).

Consider now the proof of (ii). Let $\dim Y = \infty$, let (v_n) be a sequence on the unit sphere of Y with no convergent subsequence, let b be a unit vector in X, and let $f_1 : X \to Y$ be given by

$$f_1(b/n) = v_n/n \qquad (n = 1, 2, \ldots), \qquad f_1(x) = 0 \text{ otherwise.}$$

Clearly f_1 is continuous at 0. It is also obvious that f_1 is not Gâteaux differentiable at 0, since the function $t \mapsto f_1(tb)$ from $\mathbf{R}$ into Y does not have a right-hand derivative at 0.

Next, since $(x, f_1(x))^{\wedge} = (b, v_n)^{\wedge}$ if $x = b/n$, $n = 1, 2, \ldots$, it follows that if (x_m) is a sequence in $X \setminus \{0\}$ converging to 0 such that the sequence of points $(x_m, f_1(x_m))^{\wedge}$ tends to a limit, there is only a finite number of m for which x_m is one of the points b_n/n. Hence $(x_m, f_1(x_m))^{\wedge} = (x_m, 0)^{\wedge}$ for all sufficiently large m, and therefore, by (4.3.1)(iii), the cone $\mathscr{T}$ of vectors tangent to the graph of f_1 at $(0, 0)$ is contained in $X \times \{0\}$. On the other hand, if x is a unit vector in X, then

$$(x/(n + \tfrac{1}{2}), f(x/(n + \tfrac{1}{2})))^{\wedge} = (x, 0)^{\wedge} = (x, 0),$$

so that $(x, 0) \in \mathscr{T}$. Hence also $(\alpha x, 0) \in \mathscr{T}$ for all $\alpha > 0$, and therefore $\mathscr{T} = X \times \{0\}$.

It remains now only to prove (iii), and here we can take f_2 to be the function of §4.2, Example (d) (p. 266).

We remark that (4.3.3)(i) shows in particular that the inclusion relation in (4.3.2)(i) cannot be replaced by equality (for otherwise Gâteaux differentiability would imply Hadamard differentiability).

We consider next the cone of vectors tangent to the level surface of a function through a given point. If f is a function from a set $A \subseteq X$ into $Y, x_0 \in A$, and $y_0 = f(x_0)$, then the *level surface of f through the point x_0* is the set $f^{-1}(\{y_0\})$ (this may, of course, consist of the single point x_0). We remark immediately that if f is differentiable at x_0 in one of the three senses defined in Chapter 3 and §§4.1, 2, T is the appropriate differential of f at x_0, and g is the approximating function given by

$$g(x) = y_0 + T(x - x_0) \qquad (x \in X),$$

then the level surface of g through x_0 is the set where $T(x - x_0) = 0$, i.e. it is $x_0 + \ker T$.

We prove first an elementary result.

(4.3.4) *Let f be a function from a set $A \subseteq X$ into Y which is Hadamard differentiable at an (interior) point x_0 of A, and let E be the level surface of f through x_0. Then $K_t(E; x_0) \subseteq \ker \partial f(x_0)$.*

We may again obviously suppose that $x_0 = 0$ and $f(x_0) = 0$. If $h \in K_t(E; x_0)$, we can find a sequence (h_n) in X converging to h and a sequence (t_n) of positive numbers converging to 0 such that $t_n h_n \in E$ for all n, i.e. $f(t_n h_n) = 0$. Since f is Hadamard differentiable at 0, $f(t_n h_n)/t_n \to \partial f(0)h$ as $n \to \infty$, and therefore $\partial f(0)h = 0$, i.e. $h \in \ker \partial f(0)$.

The inclusion relation here cannot be replaced by equality, even if f is Fréchet differentiable at x_0. This can easily be seen by considering the function $f : \mathbf{R} \to \mathbf{R}$ defined by $f(x) = x^2$ at the point $x_0 = 0$, for $df(0) = 0$ but the level surface through 0 is the singleton $\{0\}$. A more sophisticated example is given by defining $g : \mathbf{R}^2 \to \mathbf{R}$ by

$$g(x, y) = x + y^2 \quad (x \geq 0), \qquad g(x, y) = x - y^2 \quad (x < 0),$$

and taking $x_0 = (0, 0)$. Then $dg(x_0)$ exists and, for all $(h_1, h_2) \in \mathbf{R}^2$, $dg(x_0)(h_1, h_2) = h_1$, so that $\ker dg(x_0) = \{0\} \times \mathbf{R}$; but again the level surface through x_0 is the singleton $\{x_0\}$.

In the first of these examples the crucial fact is that $df(x_0)$ is not surjective (see (4.3.7)). In the second, it is that g is not continuous, as we shall now see.

(4.3.5) *Let f be a continuous function from a subset A of the real space X into $\mathbf{R}$ which is Hadamard differentiable at an interior point x_0 of A, with $\partial f(x_0) \neq 0$, and let E be the level surface of f through x_0. Then $K_t(E; x_0) = \ker \partial f(x_0)$.*

By (4.3.4), it is enough to prove that $K_t(E; x_0) \supseteq \ker \partial f(x_0)$, and in doing this we suppose that $x_0 = 0$ and $f(x_0) = 0$. Let $h \in \ker \partial f(0)$ and let $\eta > 0$ be given. Take a point k of X such that $\partial f(0)k > 0$ and $\|k\| < \eta$. Then $\partial f(0)(h + k) > 0$ and therefore $f(t(h + k)) > 0$ for all sufficiently small $t > 0$. Similarly, $\partial f(0)(h - k) < 0$ so that $f(t(h - k)) < 0$ for all sufficiently small $t > 0$. Since f is continuous, it follows that for each sufficiently small $t > 0$ there exists a point k_t on the segment joining k and $-k$ such that $f(t + k_t)) = 0$, i.e. $t(h + k_t) \in E$, and therefore $h \in K_t(E; 0)$.

We shall next prove a result in which equality holds and in which the range of f is a Banach space Y. The proof is a good deal less elementary, and we first require a lemma which is a companion to (3.7.1).

(4.3.6) **Lemma.** *Let* X, Y *be Banach spaces, let* T *be a continuous linear function from* X *into* Y, *let* $H = \ker T$, *and for each* $L \in X/H$ *let* $T_c L = Tx$, *where* x *is any point of* L *(so that, by* (A.2.2)(iii), *$T_c \in \mathscr{LH}(X/H, Y)$). Let also* $M > \|T_c^{-1}\|$, *let* $0 < \varepsilon < 1$, *let* A *be an open set in* X *containing* 0, *and let* $f : A \to Y$ *be a function such that* $f(0) = 0$ *and that*

$$\|f(x) - f(x') - T(x - x')\| \le \frac{\varepsilon}{M} \|x - x'\|$$

for all $x, x' \in A$. *Then if* y *is a point of* Y *such that* A *contains the closed ball* B *in* X *with centre* 0 *and radius* $M\|y\|/(1 - \varepsilon)$, *the equation* $f(x) = y$ *has a solution* x^* *in* B.

The proof is, in its essentials, the same as that of (3.7.1). The main idea is to find a root of the equation $f(x) = y$ using a sequence (x_n) of successive approximations satisfying the recurrence relation

$$T(x_{n+1} - x_n) = y - f(x_n). \tag{2}$$

In (3.7.1), T was invertible, and in this case it would not merely have been easy to define (x_n) inductively from (2), but it was possible to apply the contraction mapping principle and so avoid any explicit mention of (2). In the present proof we use (A.2.3), which shows that, under the hypotheses of this lemma, for each $z \in Y$ we can find $x \in X$ such that $Tx = z$ and that $\|x\| \le M\|z\|$.

Let y and B be as specified in the lemma. We prove first that there exists a sequence (x_n) in B° with $x_0 = 0$ which satisfies both (2) and

$$\|x_{n+1} - x_n\| \le M\varepsilon^n \|y\|. \tag{3}$$

Let $x_0 = 0$. By the result of (A.2.3) mentioned above, with $z = y$, we can choose $x_1 \in X$ such that $Tx_1 = y$ and $\|x_1\| \le M\|y\|$; then $x_1 \in B^\circ$ and (2) and (3) are satisfied with $n = 0$. Suppose then that n is a positive integer such that $x_0, \ldots, x_n$ are defined and lie in $B°$, and that for $r = 0, \ldots, n-1$,

$$T(x_{r+1} - x_r) = y - f(x_r) \quad \text{and} \quad \|x_{r+1} - x_r\| \le M\varepsilon^r \|y\|. \qquad (4)$$

Again by (A.2.3), we can choose x_{n+1} so that

$$T(x_{n+1} - x_n) = y - f(x_n) \quad \text{and} \quad \|x_{n+1} - x_n\| \le M\|y - f(x_n)\|. \qquad (5)$$

Then

$$\begin{aligned}
\|x_{n+1} - x_n\| &\le M\|y - f(x_n)\| \\
&= M\|y - f(x_n) + T(x_n - x_{n-1}) - y + f(x_{n-1})\| \\
&= M\|f(x_n) - f(x_{n-1}) - T(x_n - x_{n-1})\| \\
&\le \varepsilon\|x_n - x_{n-1}\| \le M\varepsilon^n\|y\|,
\end{aligned}$$

so that the conditions (4) are satisfied for $r = n$. Further,

$$\|x_{n+1}\| = \|x_{n+1} - x_0\| \le \|x_{n+1} - x_n\| + \ldots + \|x_1 - x_0\| < M\|y\|/(1-\varepsilon),$$

so that $x_{n+1} \in B°$. It therefore follows by induction that there exists a sequence (x_n) in $B°$ such that $x_0 = 0$ and that (2) and (3) hold for all n. From (3) we now deduce that (x_n) is a Cauchy sequence in $B°$, and hence it converges to a point $x^* \in B$. On making $n \to \infty$ in (2) we deduce that $f(x^*) = y$, and this completes the proof.

(4.3.6. Corollary 1) *Under the conditions of* (4.3.6), $f(A)$ *is open in* Y. *Further, if* $A = X$, *then* $f(A) = Y$.

Let $\bar{y} \in f(A)$, let $\bar{x}$ be a point of A such that $f(\bar{x}) = \bar{y}$, and let

$$g(w) = f(\bar{x} + w) - \bar{y} \qquad (\bar{x} + w \in A).$$

Clearly $g(0) = 0$, and

$$\|g(w) - g(w') - T(w - w')\| \le \frac{\varepsilon}{M}\|w - w'\|$$

for all w, w' in the domain of g. If now A contains the closed ball C in X with centre $\bar{x}$ and radius α, and D is the closed ball in Y with centre $\bar{y}$ and radius $\alpha(1 - \varepsilon)/M$, then for all $y \in D$ the equation $g(w) = y - \bar{y}$ has a solution w such that $\bar{x} + w \in C$, i.e. the equation $f(x) = y$ has a solution $x \in C$. Hence $D \subseteq f(C) \subseteq f(A)$, so that $f(A)$ is open in Y. Further, if $A = X$, then by (4.3.6), $f(A)$ is clearly equal to Y.

(4.3.6. Corollary 2) *If the hypotheses of* (4.3.6) *are satisfied, and* z *is a point of* A *such that* A *contains the closed ball* C *in* X *with centre* z *and radius* $\varepsilon\|z\|/(1 - \varepsilon)$, *then the equation* $f(x) = Tz$ *has a solution in* C.

Let $y = Tz - f(z)$, and let $h(w) = f(z + w) - f(z)$ $(z + w \in A)$. As in Corollary 1, h satisfies the hypotheses of (4.3.6), and

$$\|y\| = \|f(z) - f(0) - T(z - 0)\| \le \varepsilon\|z\|/M,$$

so that the domain of h contains the closed ball D in X with centre 0 and radius $M\|y\|/(1-\varepsilon)$. Hence the equation $h(w)=y$ has a solution w in D, and this gives the result.

(4.3.7) *Let X, Y be Banach spaces, let f be a function from a set $A \subseteq X$ into Y, let x_0 be an interior point of A, and let E be the level surface of f through x_0. Suppose further that f is Hadamard differentiable on a neighbourhood of x_0, that ∂f is continuous at x_0, and that $\partial f(x_0)$ is onto Y. Then $K_t(E;x_0) = \ker \partial f(x_0)$.*†

By (4.3.4) $K_t(E;x_0) \subseteq \ker \partial f(x_0)$, so that it remains to prove the reverse inclusion. Further, we may again suppose that $x_0 = 0$ and $f(x_0) = 0$, and hence it is enough to show that if $T = \partial f(0)$ and h is a non-zero vector in $\ker T$, then there exist a sequence (h_n) in X converging to h and a sequence (t_n) of positive numbers converging to 0 such that $t_n h_n \in E$ for all n.

Let T_c be defined as in (4.3.6), let $M > \|T_c^{-1}\|$, and let $0 < \varepsilon < 1$. Since ∂f is continuous at 0, we can find an open ball A_0 in X with centre 0, contained in A, such that $\|\partial f(x) - T\| \leq \varepsilon/M$ for all $x \in A_0$. By the analogue of (4.1.6) for Hadamard differentiation applied to $f - T$, we then have

$$\|f(x) - f(x') - T(x - x')\| \leq \varepsilon \|x - x'\|/M$$

for all $x, x' \in A_0$, so that the restriction of f to A_0 satisfies the conditions of (4.3.6). It is obvious that for all sufficiently small positive t the ball A_0 contains the closed ball B in X with centre th and radius $\varepsilon\|th\|/(1-\varepsilon)$, and then, by (4.3.6. Corollary 2), there exists $x \in B$ such that $f(x) = T(th) = 0$.

We now apply this result, taking ε to be successively $\tfrac{1}{3}, \tfrac{1}{4}, \dots, 1/n, \dots$, and for $\varepsilon = 1/n$ choosing $t = t_n$ so that $t_n \to 0$ as $n \to \infty$. We thus obtain a sequence (x_n) in E such that $\|x_n - t_n h\| \leq t_n\|h\|/(n-1)$ for all n. If now $h_n = x_n/t_n$, then $t_n h_n = x_n \in E$ for all n and $h_n = x_n/t_n \to h$ as $n \to \infty$, so that $h \in K_t(E;x_0)$.

Exercise 4.3

1 Let X, Y be Banach spaces, let T be a continuous linear function from X onto Y, and let $T_c \in \mathscr{L}\mathscr{H}(X/\ker T, Y)$ be defined as in (4.3.6). Let also S be an element of $\mathscr{L}(X, Y)$ such that $\|S - T\| < 1/\|T_c^{-1}\|$. Prove that S is onto Y, and that if $S_c \in \mathscr{L}\mathscr{H}(X/\ker S, Y)$ is defined as in (4.3.6), then

$$\|S_c^{-1}\| \leq \|T_c^{-1}\|/(1 - \|T_c^{-1}\| \|S - T\|).$$

[This is a generalization of (3.7.1. Corollary 1).
Hint. Choose M such that $\|S - T\| < 1/M < 1/\|T_c^{-1}\|$, and let $\varepsilon = M\|S - T\|$. Then the conditions of (4.3.6) are satisfied with $A = X$ and $f = S$.]

---

† The result of (4.3.7) holds if f is continuous and Gâteaux differentiable on a neighbourhood of x_0, δf is continuous at x_0, and $\delta f(x_0)$ is onto Y. Note that each set of conditions implies that f is Fréchet differentiable at x_0 (4.1.7. Corollary 1).

4.4 Constrained maxima and minima (equality constraints)

In this section and the next, we shall discuss real-valued functions. In order that these may be differentiable, the normed spaces involved will be over the real field. Let X, Y be normed spaces, let $A \subseteq X$, and let $F : A \to \mathbf{R}$ and $G : A \to Y$ be given functions. We say that F has a *local minimum at a point $x_0 \in A$ subject to the condition* $G(x) = 0$ if x_0 belongs to the level surface $S = G^{-1}(\{0\})$ of G and there exists a neighbourhood V of x_0 in X, contained in A, such that $F(x) \geq F(x_0)$ for all $x \in S \cap V$; if strict inequality holds except at x_0, then x_0 is a *strict conditional local minimum*. A *conditional local maximum* of F is defined similarly.

In general, if x_0 is not a local maximum or minimum of F (in the sense defined in §3.6), then as x passes through x_0 crossing successive level surfaces of F, the value of $F(x)$ varies monotonically. It is therefore intuitive that if x_0 is a conditional local minimum or maximum of F on S, then every vector tangent to S at x_0 is also tangent to the level surface $\mathscr{S}$ of F through x_0 (Fig. 4.1). In particular, this would imply that if G satisfies the hypotheses of (4.3.7), and F is Hadamard differentiable at x_0, then

$$\ker \partial G(x_0) = K_t(S; x_0) \subseteq K_t(\mathscr{S}; x_0) = \ker \partial F(x_0). \tag{1}$$

The following result shows that this inclusion relation (1) does in fact hold when the spaces X, Y are complete.

(4.4.1) *Let X, Y be real Banach spaces, let F and G be functions from a set*

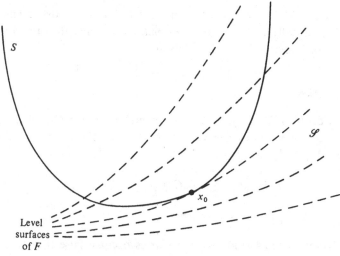

Figure 4.1

$A \subseteq X$ into **R** and Y respectively, and suppose that

(i) F has a local minimum or local maximum at x_0 subject to the condition $G(x) = 0$,

(ii) F is Hadamard differentiable at x_0,

(iii) G is Hadamard differentiable on a neighbourhood of x_0, ∂G is continuous at x_0, and $\partial G(x_0)$ is onto Y. Then ker $\partial G(x_0) \subseteq$ ker $\partial F(x_0)$.

Let $S = G^{-1}(\{0\})$ and suppose that F has a local maximum at x_0 on S. By (4.3.7), ker $\partial G(x_0) = K_t(S; x_0)$, and hence if $h \in$ ker $\partial G(x_0)$ we can find a sequence (h_n) in X converging to h and a sequence (t_n) of positive numbers converging to 0 such that $x_0 + t_n h_n \in S$ for all n. Then $F(x_0 + t_n h_n) \leq F(x_0)$ for all sufficiently large n, whence

$$\partial F(x_0)h = \lim_{n \to \infty} (F(x_0 + t_n h_n) - F(x_0))/t_n \leq 0.$$

But we can repeat this argument with h replaced by $-h$, whence $\partial F(x_0)(-h) \leq 0$, and therefore $\partial F(x_0)h = 0$, as required. The proof when F has a conditional minimum is similar.

(4.4.1. Corollary 1) *If F and G satisfy the hypotheses of the main theorem, there exists $\Lambda \in Y'$ such that $\partial F(x_0) = \Lambda \circ \partial G(x_0)$.*

We apply (A.2.2)(ii). The kernel of the function $T = \partial F(x_0)$ contains the closed subspace $W = $ ker $\partial G(x_0)$. There is therefore $T_c \in \mathscr{L}(X/W, \mathbf{R})$ such that $T = T_c \circ \phi$, where ϕ is the canonical map from X to X/W. Since $\partial G(x_0)$ is onto Y and both X and Y are Banach spaces, there is a linear homeomorphism $\psi : Y \to X/S$ such that $\phi = \psi \circ \partial G(x_0)$. We take $\Lambda = T_c \circ \psi \in \mathscr{L}(Y, \mathbf{R}) = Y'$.

When $X = \mathbf{R}^n$, $Y = \mathbf{R}^m$, the corollary asserts that there exist real numbers $\lambda_1, \ldots, \lambda_m$ (usually called *Lagrange's multipliers*) such that the Jacobian matrices of F and G satisfy the relation

$$[D_j F(x_0)] = [\lambda_1, \ldots, \lambda_m][D_j G_i(x_0)].$$

Thus we have:

(4.4.1. Corollary 2) (Lagrange's multipliers) *If F and G satisfy the hypotheses of the main theorem, where now $X = \mathbf{R}^n$, $Y = \mathbf{R}^m$, there exist real numbers $\lambda_1, \ldots, \lambda_m$ such that*

$$D_j F(x_0) = \sum_{i=1}^{m} \lambda_i D_j G_i(x_0) \qquad (j = 1, \ldots, n), \qquad (2)$$

where $G_1, \ldots, G_m$ are the components of G.

The results of (4.4.1) and its corollaries become false if $\partial G(x_0)$ is not onto Y. For example, if $F, G : \mathbf{R}^2 \to \mathbf{R}$ are given by

$$F(x, y) = y, \quad G(x, y) = (y - x^2)(2y - x^2),$$

then F obviously has a minimum at the origin subject to the condition $G(x, y) = 0$. However, here $\partial G(0, 0) = 0$, and the equations (2), which in this case are

$$0 = \lambda(4x^3 - 6xy) \text{ and } 1 = \lambda(4y - 3x^2),$$

have no solution λ when $x = y = 0$.

In the particular case where G is real-valued, (4.4.1. Corollary 1) shows that there exists $\lambda \in \mathbf{R}$ such that $\partial F(x_0) = \lambda \partial G(x_0)$. This case has an interesting application to the isoperimetric problem of the calculus of variations.

The isoperimetric problem can be stated as follows:

(IP) *Find the function ϕ for which the functional*

$$F(\phi) = \int_a^b f(t, \phi(t), \phi'(t)) dt \tag{3}$$

has a local minimum or maximum, where the admissible functions belong to $C^1([a, b], Y)$, satisfy the endpoint conditions $\phi(a) = c, \phi(b) = d$, and are such that another functional

$$G(\phi) = \int_a^b g(t, \phi(t), \phi'(t)) dt \tag{4}$$

takes a given value l. The name 'isoperimetric' arises from the special case where $c = d = 0$,

$$F(\phi) = \int_a^b \phi(t) dt, \tag{5}$$

$$G(\phi) = \int_a^b (1 + (\phi'(t))^2)^{1/2} dt, \tag{6}$$

since in this case the problem is that of finding the greatest among the areas bounded by the t-axis and a curve $y = \phi(t)$ of fixed length (observe that all such areas have the same perimeter).

To make the problem (IP) precise, let $J = [a, b]$, let Y be a real Banach space, let A be an open set in the metric space $J \times Y \times Y$, and, as in (3.11.1), let $f, g : A \to \mathbf{R}$ be continuous functions whose partial differentials $d_2 f, d_3 f, d_2 g, d_3 g$ are continuous on A. Let also $c, d \in Y$, let $\mathscr{E}$ be the set of $\phi \in C^1(J, Y)$ such that $\phi(a) = c, \phi(b) = d$, and that $(t, \phi(t), \phi'(t)) \in A$ for all $t \in J$, and let $F_{\mathscr{E}}, G_{\mathscr{E}}$ be the restrictions to $\mathscr{E}$ of the functionals defined by (3) and (4). If ϕ is a local minimum or maximum of $F_{\mathscr{E}}$ subject to the condition that $G_{\mathscr{E}}(\phi) = l$, and $\bar{F}, \bar{G}$ are given by

$$\bar{F}(\psi) = F_{\mathscr{E}}(\phi + \psi), \quad \bar{G}(\psi) = G_{\mathscr{E}}(\phi + \psi),$$

where the domain of $\bar{F}$ and $\bar{G}$ is the set of $\psi \in C_0^1(J, Y)$ for which $\phi + \psi \in \mathscr{E}$, then $\psi = 0$ is a local minimum or maximum of $\bar{F}$ subject to the condition that $\bar{G}(\psi) = l$. Since the Fréchet differential of $\bar{F}$ at 0 is given for all $\psi \in C_0^1(J, Y)$ by $d\bar{F}(0)\psi = \int_a^b (d_2 f(\Phi(t))\psi(t) + d_3 f(\Phi(t))\psi'(t))dt$ where $\Phi(t) = (t, \phi(t), \phi'(t))$ (cf. the remarks preceding (3.11.1)) and similarly for $\bar{G}$, we deduce from (4.4.1. Corollary 1) that if $d\bar{G}(0)$ is non-zero then there exists a real number λ such that for all $\psi \in C_0^1(J, Y)$

$$\int_a^b (d_2 f(\Phi(t))\psi(t) + d_3 f(\Phi(t))\psi'(t))dt$$

$$= \lambda \int_a^b (d_2 g(\Phi(t))\psi(t) + d_3 g(\Phi(t))\psi'(t))dt.$$

By combining this with (3.11.4), we thus obtain:

(4.4.2) *Let* $J = [a, b]$, *let* Y *be a real Banach space, let* A *be an open set in the metric space* $J \times Y \times Y$, *and let* $f, g : A \to \mathbf{R}$ *be continuous functions whose partial differentials* $d_2 f, d_3 f, d_2 g, d_3 g$ *are continuous on* A. *If* ϕ *is a solution of the isoperimetric problem* (IP) *which is not an extremal for the functional* G, *and* $\Phi(t) = (t, \phi(t), \phi'(t))$, *there exists a real number* λ *such that the function* $t \mapsto d_3 f(\Phi(t)) - \lambda d_3 g(\Phi(t))$ *has a derivative at each* $t \in J$ *and*

$$\frac{d}{dt}(d_3 f(\Phi(t)) - \lambda d_3 g(\phi(t))) = d_2 f(\Phi(t)) - \lambda d_2 g(\Phi(t)). \tag{7}$$

For example, if F, G are given by (5) and (6), so that

$$f(t, y, y') = y, \qquad g(t, y, y') = (1 + y'^2)^{1/2},$$

the equation (7) is

$$\frac{d}{dt}(-\lambda(1 + (\phi'(t))^2)^{-1/2}\phi'(t)) = 1$$

which is easily solved to show that the largest area is obtained when the curve is an arc of a circle.

A somewhat more difficult example is given by taking

$$F(\phi) = \int_a^b \phi(t)(1 + (\phi'(t))^2)^{1/2} dt, \tag{8}$$

and retaining formula (6) for G. Except for the constant 2π, the functional F then represents the area of the surface obtained by rotating the curve $y = \phi(t)$ about the t-axis (cf. Example (b), §3.11, p. 232). We shall indicate how the solution to this problem may be obtained.

As in §3.11, write $L(t) = (1 + (\phi'(t))^2)^{1/2}$. If we suppress the variable t, (7) applied to the functionals in (8) and (6) gives

$$\frac{d}{dt}(L^{-1}\phi\phi' - \lambda L^{-1}\phi') = L.$$

If we carry out the differentiation and replace ϕ'^2 by $L^2 - 1$, we find

$$-L^{-1} + L^{-3}(\phi - \lambda)\phi'' = 0.$$

On multiplying by $-\phi'$ this equation becomes

$$\frac{d}{dt}(L^{-1}(\phi - \lambda)) = 0.$$

If we now write $\chi(t) = \phi(t) - \lambda$, we see that for some constant α

$$\chi(t) = \alpha(1 + (\chi'(t))^2)^{1/2},$$

and the solution to this problem was given in §3.11 (p.233.)

We now return to the ideas of (4.4.1). In that result we assumed the Hadamard differentiability of F at x_0, but the argument still applies if we assume only that F has a Hadamard variation $\partial F(x_0)h$ for all $h \in X$. More precisely, we have:

(4.4.3) *Let X, Y be real Banach spaces, let $A \subseteq X$, let $x_0 \in A$, and let $F : A \to \mathbf{R}$ and $G : A \to Y$ be functions such that F has a Hadamard variation $\partial F(x_0)h$ for all $h \in X$, G is Hadamard differentiable on a neighbourhood of x_0, ∂G is continuous at x_0, and $\partial G(x_0)$ is onto Y.*
(i) *If F has a local maximum at x_0 subject to the condition that $G(x) = 0$, then*

$$\ker \partial G(x_0) \subseteq \{h \in X : \partial F(x_0)h \le 0, \partial F(x_0)(-h) \le 0\}.$$

In particular, if $\partial F(x_0)$ is sublinear, then

$$\ker \partial G(x_0) \subseteq \{h \in X : \partial F(x_0)h = \partial F(x_0)(-h) = 0\}.$$

(ii) *If F has a local minimum at x_0 subject to the condition that $G(x) = 0$, then*

$$\ker \partial G(x_0) \subseteq \{h \in X : \partial F(x_0)h \ge 0, \partial F(x_0)(-h) \ge 0\}.$$

We note next a simple result concerning a convex function.

(4.4.4) *Let F be a convex function on a convex set $A \subseteq X$, let x_0 be an internal point of A (so that F has a Gâteaux variation $\delta F(x_0)h$ for all $h \in X$), and let $B = \{h \in X : \delta F(x_0)h \ge 0\}$. Then $x_0 + B$ does not meet the set $Q = \{x \in A : F(x) < F(x_0)\}$.*

Suppose on the contrary that there exists $h \in B$ such that $x_0 + h \in Q$. Then on the one hand

$$(F(x_0 + th) - F(x_0))/t \to \delta F(x_0)h \geq 0$$

as $t \to 0 +$, while on the other hand, for $0 < t \leq 1$ we have

$$(F(x_0 + th) - F(x_0))/t \leq F(x_0 + h) - F(x_0) < 0.\dagger$$

If in addition to the hypotheses of (4.4.4) F is continuous, then F has a Hadamard variation $\partial F(x_0)h$ at x_0 for all $h \in X$, $\partial F(x_0) = \delta F(x_0)$, and $F(x_0)$ is sublinear. Hence we have:

(4.4.4. Corollary) *Let X, Y be Banach spaces, let F and G be functions from a convex set $A \subseteq X$ into $\mathbf{R}$ and Y respectively, and suppose that*
  (i) *F has a local maximum or minimum at x_0 subject to the condition that $G(x) = 0$,*
  (ii) *F is continuous and convex,*
  (iii) *G is Hadamard differentiable on a neighbourhood of x_0, ∂G is continuous at x_0, and $\partial G(x_0)$ is onto Y.*
Then $x_0 + \ker \partial G(x_0)$ does not meet the set $\{x \in A : F(x) < F(x_0)\}$.

As an application of (4.4.4. Corollary), we consider the case where $F(x) = \| x \|$, and we prove:

(4.4.5) *Let X be a Banach space, let G be a function from a set $A \subseteq X$ into $\mathbf{R}$, and let $S = G^{-1}(\{0\})$. If x_0 is a local maximum or minimum of the norm function subject to the condition that $G(x) = 0$, G is Hadamard differentiable on a neighbourhood of x_0 and ∂G is continuous at x_0, then $|\partial G(x_0)x_0| = \| \partial G(x_0) \| \, \| x_0 \|$.*
The result is trivial if $x_0 = 0$ or $\partial G(x_0) = 0$, so that we may suppose $x_0 \neq 0$ and $\partial G(x_0) \neq 0$, whence $\partial G(x_0)$ is onto $\mathbf{R}$. Let $T = \partial G(x_0)$, and let $H = \ker T$. Then by (4.4.4. Corollary), $\| x_0 + h \| \geq \| x_0 \|$ for all $h \in H$, whence also $\| x_0\hat{} + h \| \geq 1$ for all $h \in H$, where $x_0\hat{} = x_0/\| x_0 \|$. This implies that the element $L = x_0\hat{} + H$ of the quotient space X/H has norm 1. If now $T_c \in \mathscr{L}\mathscr{H}(X/H, \mathbf{R})$ is defined as in (A.2.2)(iii) then $\| T_c \| = \| T \|$, and since $\pm L$ are the only two points of the unit sphere in X/H (for X/H is one-dimensional) we have also that $\| T_c \| = \| T_c L \|$. Hence

$$\| T \| = \| T_c \| = \| T_c L \| = | Tx_0\hat{} | = | Tx_0 | / \| x_0 \|,$$

as required.

---

† A slight adaptation of this argument shows further that B does not meet the cone of vectors $h \in X$ such that there exists $\lambda > 0$ for which $F(x_0 + \lambda h) < F(x_0)$.

4.5 Constrained maxima and minima (inequality constraints)

We now widen the discussion of §4.4 and consider the minimum of a function F subject to conditions of the form

$$F_i(x) \leq 0 \quad (i = 1, \dots, n), \quad G(x) = 0 \tag{1}$$

(the first n conditions here are known as 'inequality constraints', while the last is an 'equality constraint'). Thus let $F_1, \dots, F_n$ be real-valued functions whose domains are contained in X, let G be a function from a subset of X into Y, let

$$Q_i = \{x \in X : F_i(x) \leq 0\} \quad (i = 1, \dots, n), \quad Q_{n+1} = \{x \in X : G(x) = 0\},$$

$$Q = \bigcap_{i=1}^{n+1} Q_i,$$

and let F be a real-valued function whose domain contains Q (in applications, each of the sets $Q_1, \dots, Q_n$ will have a non-empty interior, while Q_{n+1} will in general have no interior points). We say that a point x_0 of the domain of F is *a local minimum of F subject to the conditions* (1) if $x_0 \in Q$ and there exists a neighbourhood V of x_0 in X such that $F(x) \geq F(x_0)$ for all $x \in Q \cap V$.

We note immediately that if x_0 is a local minimum of F subject to the conditions (1), and

$$Q_0 = \{x \in X : F(x) < F(x_0)\}$$

then there exists a neighbourhood V of x_0 in X with the property that if $x \in \bigcap_{i=1}^{n+1}(Q_i \cap V)$ then $x \notin Q_0$, i.e. such that

$$\bigcap_{i=0}^{n+1}(Q_i \cap V) = \varnothing. \tag{2}$$

We shall see below that this property (2) implies a relation between the cones $K_d(Q_i; x_0)$ $(i = 1, \dots, n)$ and the cone of vectors tangent to Q_{n+1} at x_0. However, we must first investigate further the properties of these cones.

We begin by recalling from §4.2 (p. 258–9) that, for $E \subseteq X$ and $x_0 \in X$, the cone $K_d(E; x_0)$ of vectors directed into E at x_0 consists of those elements $h \in X$ for which there exist a neighbourhood W of h in X and $\varepsilon > 0$ such that $x_0 + tk \in E$ when $k \in W$ and $0 < t < \varepsilon$.

The following elementary properties of this cone are easily checked.

(4.5.1) *Let $E \subseteq X$ and let $x_0 \in X$.*

(i) $K_d(E; x_0)$ *is an open cone with vertex* 0.

(ii) *If* $K_d(E; x_0) \neq \varnothing$ *then* $x_0 \in \bar{E}$ *(so that if* $x_0 \notin \bar{E}$ *then* $K_d(E; x_0) = \varnothing$*).*

(iii) *If* $E \supseteq E'$ *then* $K_d(E; x_0) \supseteq K_d(E'; x_0)$.

(iv) $K_d(E;x_0) = K_d(E^\circ;x_0)$, *so that if* $E^\circ = \varnothing$ *then* $K_d(E,x_0) = \varnothing$.

(v) *If* V *is a neighbourhood of* x_0 *in* X *then* $K_d(E \cap V;x_0) = K_d(E;x_0)$.

(vi) $0 \in K_d(E;x_0)$ *if only if* $x_0 \in E^\circ$, *and then* $K_d(E;x_0) = X$.

(vii) $h \in K_d(E;x_0)$ *if and only if, for each sequence* (h_n) *in* X *converging to* h *and each sequence* (t_n) *of positive numbers converging to* 0, *the points* $x_n + t_n h_n$ *belong to* E *for all sufficiently large* n.†

(viii) *For* $E_i \subseteq X$ $(1 \le i \le m)$, $K_d(\bigcap_{i=1}^m E_i;x_0) = \bigcap_{i=1}^m K_d(E_i;x_0)$.

We now turn to the cone $K_t(E;x_0)$ of vectors tangent to a set E at the point x_0. This was defined in §4.3 (p. 273–4) to be

$\{h \in X :$ for each neighbourhood W of h in X and for each $\varepsilon > 0$

there exist $k \in W$ and $t \in\,]0, \varepsilon[$ such that $x_0 + tk \in E\}$.

Some simple properties of this cone are listed in the next proposition. Here and later, for any set $A \subseteq X$ we denote the complement $X \setminus A$ of A with respect to X by A^c.

(4.5.2) *Let* $E \subseteq X$ *and* $x_0 \in X$.

(i) $K_t(E;x_0) = K_d(E^c;x_0)^c$, *so that* $K_t(E;x_0)$ *is a closed cone with vertex* 0, *and* $0 \in K_t(E;x_0)$ *whenever* $K_t(E;x_0) \ne \varnothing$.

(ii) *If* $K_t(E;x_0) \ne \varnothing$ *then* $x_0 \in \bar{E}$ *(so that if* $x_0 \notin \bar{E}$ *then* $K_t(E;x_0) = \varnothing$).

(iii) *If* $E \supseteq E'$ *then* $K_t(E;x_0) \supseteq K_t(E';x_0)$.

(iv) $K_t(E;x_0) = K_t(\bar{E};x_0)$.

(v) *If* V *is a neighbourhood of* x_0 *in* X *then* $K_t(E \cap V;x_0) = K_t(E;x_0)$.

(vi) $h \in K_t(E;x_0)$ *if and only if there exist a sequence* (h_n) *in* X *converging to* h *and a sequence* (t_n) *of positive numbers converging to* 0 *such that* $x_0 + t_n h_n \in E$ *for all* n.‡

(vii) *For* $E_i \subseteq X$ $(1 \le i \le m)$, $K_t(\bigcup_{i=1}^m E_i;x_0) = \bigcup_{i=1}^m K_t(E_i;x_0)$.

For the special case where E is convex, the relationship between $K_d(E;x_0)$ and $K_t(E;x_0)$ is particularly simple.

(4.5.3) *Let* E *be a non-empty convex set in* X, *let* $x_0 \in \bar{E}$, *let*

$$D = \{h \in X : \text{there exists } \lambda > 0 \text{ for which } x_0 + \lambda h \in E\},$$

$$D_0 = \{h \in X : \text{there exists } \lambda > 0 \text{ for which } x_0 + \lambda h \in E^\circ\},$$

and let $E^\smallfrown$ *be the set of support functionals to* E *at* x_0, *i.e.*

$$E^\smallfrown = \{u \in X' : u(x) \ge u(x_0) \text{ for all } x \in E\}.$$

---

† This condition is given in §4.2.
‡ This is just (4.3.1) (ii).

Then

(i) $K_t(E;x_0) = K_t(\bar{E};x_0) = \bar{D}$ and $K_d(E;x_0) = K_d(E^\circ;x_0) = D_0$,

(ii) $K_t(E;x_0)$ and $K_d(E;x_0)$ are convex,

(iii)   the dual cone $K_t^*(E;x_0)$ of $K_t(E;x_0)$ is equal to $E^\frown$.

If in addition $E^\circ \neq \varnothing$, *then*

(iv) $K_t(E;x_0) = \overline{K_d(E;x_0)}$, *so that the dual cone* $K_d^*(E;x_0)$ *of* $K_d(E;x_0)$
is $E^\frown$.

To prove the first statement in (i), let $h \in K_t(E;x_0)$. Then, given
a neighbourhood W of h, there are $k \in W$ and $\lambda > 0$ with $x_0 + \lambda k \in E$. Thus
$k \in D$, and we conclude $h \in \bar{D}$. On the other hand, if $h \in \bar{D}$, then for each
neighbourhood W of h there exist $k \in W$ and $\lambda > 0$ with $x_0 + \lambda k \in E$.
Since $x_0 \in \bar{E}, x_0 + tk \in \bar{E}$ for $0 < t < \lambda$, whence $h \in K_t(\bar{E};x_0)$. This, together
with (4.5.2)(iv) gives the result.

In proving the second part of (i) we may obviously discard the case
in which $E^\circ = \varnothing$. Let $h \in D_0$, so that there exists $\lambda > 0$ for which $x_0 +
\lambda h \in E^\circ$, and let $W = -\lambda^{-1}x_0 + \lambda^{-1} E^\circ$. Then W is a neighbourhood of h
and if $k \in W$ then $x_0 + \lambda k \in E^\circ$. The convexity of E gives that if $k \in W$ and
$0 < t < \lambda$ then $x_0 + tk \in E^\circ$, whence $h \in K_d(E^\circ;x_0)$. Conversely, if
$h \in K_d(E^\circ;x_0)$ then $x_0 + th \in E^\circ$ for all sufficiently small positive t, so
that $h \in D_0$. This and (4.5.1)(iv) completes the proof of (i).

To prove (ii) it is now enough to show that $\bar{D}$ and D_0 are convex. If we
prove D is convex, then the convexity of $\bar{D}$ will follow, and the same proof
will suffice to show D_0 is convex (for E° is convex when E is). Let $h,k \in D$,
let λ,μ be positive numbers such that $x_0 + \lambda h, x_0 + \mu k \in E$, and let
$v = \min\{\lambda,\mu\}$. Then $x_0 + vh$ and $x_0 + vk$ belong to E, and since E is
convex it follows that

$$x_0 + v(\sigma h + (1 - \sigma)k) = \sigma(x_0 + vh) + (1 - \sigma)(x_0 + vk) \in E$$

for $0 < \sigma < 1$. Hence $\sigma h + (1 - \sigma)k \in D$, so that D is convex.

To prove (iii), we recall (A.5.3) (iv) that the dual of a cone K in X is
identical to the dual of the closure $\bar{K}$ of K, so that it is enough to prove
that $D^* = E^\frown$. Let $u \in D^*$ and let $x \in E$. Then $x - x_0 \in D$, whence $u(x) -
u(x_0) = u(x - x_0) \geq 0$, so that $u \in E^\frown$. Conversely, if $u \in E^\frown$ and $h \in D$, there
exists $\lambda > 0$ such that $x_0 + \lambda h \in E$ and then $u(h) = \lambda^{-1}(u(x_0 + \lambda h) -
u(x_0)) \geq 0$, whence $u \in D^*$.

It remains to prove (iv), and it is enough to show that if $E^\circ \neq \varnothing$ then
$\bar{D}_0 \supseteq D$, for then clearly $\bar{D}_0 = \bar{D}$. Let $h \in D$, let λ be a positive number such
that $x_0 + \lambda h \in E$, let $x_0 + k$ be a point of E°. Then for $0 < \sigma < 1$

$$x_0 + \sigma k + \lambda(1 - \sigma)h = \sigma(x_0 + k) + (1 - \sigma)(x_0 + \lambda h) \in E^\circ,$$

so that $h + \sigma(1 - \sigma)^{-1}\lambda^{-1}k \in D_0$, and therefore $h \in \bar{D}_0$.

The next result contains the kernel of our theory.

(4.5.4) *If* $E_0, \ldots, E_{n+1}$ *are subsets of* X *such that* $\bigcap_{i=0}^{n+1} E_i = \varnothing$, *and* $x_0 \in X$, *then*

$$\left(\bigcap_{i=0}^{n} K_d(E_i ; x_0) \right) \cap K_t(E_{n+1} ; x_0) = \varnothing. \tag{3}$$

If in addition the cones $K_d(E_i ; x_0)$ $(i = 0, \ldots, n)$ *and* $K_t(E_{n+1} ; x_0)$ *are convex and at least one is neither* $\varnothing$ *nor* X *and their dual cones are* $K_d^*(E_i ; x_0)$ *and* $K_t^*(E_{n+1} ; x_0)$ *then there exist* $u_i \in K_d^*(E_i ; x_0)$ $(i = 0, \ldots, n)$ *and* $u_{n+1} \in K_t^*(E_{n+1} ; x_0)$, *not all* 0, *such that*

$$u_0 + u_1 + \ldots + u_n + u_{n+1} = 0. \tag{4}$$

If $\bigcap_{i=0}^{n+1} E_i = \varnothing$ then $E_{n+1} \subseteq \bigcup_{i=0}^{n} E_i^c$, and using (4.5.2) we find

$$K_t(E_{n+1} ; x_0) \subseteq K_t \left(\bigcup_{i=0}^{n} E_i^c ; x_0 \right) = \bigcup_{i=0}^{n} K_t(E_i^c ; x_0)$$

$$= \bigcup_{i=0}^{n} K_d(E_i ; x_0)^c = \left(\bigcap_{i=0}^{n} K_d(E_i ; x_0) \right)^c.$$

This gives (3), and from (3) and (A.5.9) we deduce the existence of $u_0, \ldots, u_{n+1}$, not all 0, with property (4).

When the sets E_i are convex, (4.5.4) has the following partial converse.

(4.5.5) *Let* $E_0, \ldots, E_{n+1}$ *be convex subsets of* X *such that* E_0 *is open and that* $E_1^\circ \cap E_2^\circ \cap \ldots \cap E_n^\circ \cap E_{n+1}$ *is non-empty. Suppose also that there exists* $x_0 \in \bigcap_{i=0}^{n+1} \overline{E_i}$ *such that* (3) *holds. Then the set* $E = \bigcap_{i=0}^{n+1} E_i$ *is empty.*

Suppose on the contrary that $E \neq \varnothing$, let $x \in E$, let $x' \in E_1^\circ \cap \ldots \cap E_n^\circ \cap E_{n+1}$, and for $0 < \sigma < 1$ let $x_\sigma = \sigma x' + (1 - \sigma) x$. Then $x_\sigma \in E_1^\circ \cap \ldots \cap E_n^\circ \cap E_{n+1}$. Moreover, since $x \in E_0$ and E_0 is open, $x_\sigma \in E_0$ for all sufficiently small σ. Choose such a σ and let $h = x_\sigma - x_0$. By (4.5.3)(i), $h \in K_d(E_i ; x_0)$ for $i = 0, \ldots, n$ and $h \in K_t(E_{n+1} ; x_0)$, and this contradicts (3).

We now return to the problem of constrained minima. From (2), (4.5.4), (4.5.1)(v) and (4.5.2)(v) we obtain the following theorem.

(4.5.6) *Let* $F_1, \ldots, F_n$ *be real-valued functions whose domains are contained in the real normed space* X, *let* G *be a function from a subset of* X *into the real normed space* Y, *and let*

$$Q_i = \{ x \in X : F_i(x) \leq 0 \} \ (i = 1, \ldots, n), \qquad Q_{n+1} = \{ x \in X : G(x) = 0 \},$$

$$Q = \bigcap_{i=1}^{n+1} Q_i.$$

Let also F be a real-valued function whose domain contains the set Q, let x_0 be a local minimum of F subject to the condition that $x \in Q$ (so that $x_0 \in Q$) and let

$$Q_0 = \{x \in X : F(x) < F(x_0)\}.$$

Then

$$\left(\bigcap_{i=0}^{n} K_d(Q_i ; x_0) \right) \cap K_t(Q_{n+1} ; x_0) = \varnothing.$$

If in addition the cones $K_d(Q_0 ; x_0), \dots, K_d(Q_n ; x_0), K_t(Q_{n+1} ; x_0)$ are convex and at least one is neither $\varnothing$ nor X, there exist $u_i \in K_d^*(Q_i ; x_0)$ ($i = 0, \dots, n$) and $u_{n+1} \in K_t^*(Q_{n+1} ; x_0)$, not all 0, such that

$$u_0 + u_1 + \dots + u_n + u_{n+1} = 0.$$

In (4.3.7) we saw how to determine $K_t(Q_{n+1} ; x_0)$ in certain cases. We now turn to the corresponding problem for the cones K_d.

(4.5.7) *Let f be a real-valued function with domain contained in the real normed space X, let x_0 be an interior point of its domain and suppose that the upper Hadamard variation $\bar{\partial}f(x_0)h$ exists for all $h \in X$ (see §4.2, p. 268). Let also*

$$Q = \{x \in X : f(x) < f(x_0)\}, \qquad R = \{x \in X : f(x) \le f(x_0)\}.$$

Then

$$\{h \in X : \bar{\partial}f(x_0)h < 0\} \subseteq K_d(Q ; x_0) \subseteq K_d(R ; x_0) \subseteq \{h \in X : \bar{\partial}f(x_0)h \le 0\}. \tag{5}$$

Suppose first that $\bar{\partial}f(x_0)h < 0$. Then for each sequence (h_n) in X converging to h and each sequence (t_n) of positive numbers converging to 0 we have

$$\limsup_{n \to \infty} (f(x_0 + t_n h_n) - f(x_0))/t_n < 0.$$

Hence $f(x_0 + t_n h_n) - f(x_0) < 0$ for all sufficiently large n, whence $x_0 + t_n h_n \in Q$, i.e. $h \in K_d(Q ; x_0)$.

Next let $h \in K_d(R ; x_0)$. Then for each sequence (h_n) in X converging to h and each sequence (t_n) of positive numbers converging to 0 we have $f(x_0 + t_n h_n) - f(x_0) \le 0$ for all sufficiently large n. Hence

$$\limsup_{n \to \infty} (f(x_0 + t_n h_n) - f(x_0))/t_n \le 0,$$

which implies that $\bar{\partial}f(x_0)h \le 0$. Since the central inclusion in (5) is obvious, the proof is complete.

(4.5.7. Corollary) *Suppose that in addition to the hypotheses of (4.5.7)* *the upper variation $\bar{\partial}f(x_0)$ is convex on X and the set $\{h \in X : \bar{\partial}f(x_0)h < 0\}$ is non-empty. Then*

$$K_d^*(Q;x_0) = K_d^*(R;x_0) = \bigcup_{\mu \leq 0} \mu \partial_* f(x_0),$$

where $\partial_* f(x_0)$ is the Hadamard subvariation of f at x_0 (see §4.2, p. 270).

This corollary is simply established by combining (5), (4.2.11. Corollary) and (A.5.6).

It is worth recalling from §4.2 that under the conditions of the corollary, $\partial_* f(x_0)$ is non-empty, and that if f is Hadamard differentiable at x_0, then $\partial_* f(x_0)$ consists of the single element $\partial f(x_0)$.

We can now reinterpret (4.5.6). For simplicity, we take the case when there are only inequality constraints.

(4.5.8) *Let $F_1, \ldots, F_n$ be continuous real-valued functions whose domains are subsets of the real normed space X and let F_0 be a continuous real-valued function whose domain contains $\{x \in X : F_i(x) \leq 0 \ (i = 1, \ldots, n)\}$. Let also x_0 be a local minimum of F_0 subject to the conditions $F_i(x) \leq 0 \ (i = 1, \ldots, n)$, and suppose that $F_0, \ldots, F_n$ possess Hadamard upper variations at x_0, each defined and convex on X. Then there exist non-negative numbers $\lambda_0, \ldots, \lambda_n$, not all 0, such that*

$$\lambda_i F_i(x_0) = 0 \ (i = 1, \ldots, n), \qquad 0 \in \sum_{i=0}^{n} \lambda_i \partial_* F_i(x_0). \tag{6}$$

Moreover, if there exists $h \in X$ such that $\bar{\partial}F_i(x_0)h < 0$ for every integer $i \geq 1$ for which $F_i(x_0) = 0$, then $\lambda_0 \neq 0$.

We begin by discarding some trivial cases. If $0 \in \partial_* F_0(x_0)$, we can take $\lambda_0 = 1$ and $\lambda_i = 0$ for all $i \neq 0$. Again, if there is $j \geq 1$ such that $F_j(x_0) = 0$ and $0 \in \partial_* F_j(x_0)$, we can take $\lambda_j = 1$ and $\lambda_i = 0$ for all $i \neq j$. We can therefore assume that $0 \notin \partial_* F_0(x_0)$ and that for each $i \geq 1$, either $F_i(x_0) \neq 0$ or $0 \notin \partial_* F_i(x_0)$.

Let

$$Q_0 = \{x \in X : F_0(x) < F_0(x_0)\}, \ Q_i = \{x \in X : F_i(x) \leq 0\} \ (i = 1, \ldots, n). \tag{7}$$

Then by (4.5.6) (with $G = 0$) we have $\bigcap_{i=0}^{n} K_d(Q_i;x_0) = \varnothing$. Moreover, since $0 \notin \partial_* F_0(x_0)$, the set $\{h \in X : \bar{\partial}F_0(x_0)h < 0\}$ is non-empty, so that $K_d(Q_0;x_0)$ is non-empty (4.5.7). Further, since $x_0 \notin Q_0$, it follows from (4.5.1)(vi) that $0 \notin K_d(Q_0;x_0)$ so that $K_d(Q_0;x_0) \neq X$. From (4.5.6) we infer the existence of $v_i \in K_d^*(Q_i;x_0) \ (i = 0, \ldots, n)$, not all 0, such that $v_0 + v_1 + \ldots + v_n = 0$. If, for any $i \geq 1$, $F_i(x_0) \neq 0$, so that $F_i(x_0) < 0$ and x_0 is an interior point of Q_i, then by (4.5.1)(vi) $K_d(Q_i;x_0) = X$, so that $K_d^*(Q_i;x_0) = \{0\}$; hence $v_i = 0$, and if we put $\lambda_i = 0$ and take any

$u_i \in \partial_* F_i(x_0)$, we have $-\lambda_i u_i = v_i$. On the other hand, if, for any $i \geq 0$, $0 \notin \partial_* F_i(x_0)$, then $\{h \in X : \bar{\partial} F_i(x_0)h < 0\}$ is non-empty, so by (4.5.7. Corollary), $K_d^*(Q_i; x_0) = \bigcup_{\mu \leq 0} \mu \partial_* F_i(x_0)$; thus we can find $\lambda_i \geq 0$ and $u_i \in \partial_* F_i(x_0)$ with $-\lambda_i u_i = v_i$. We have assumed that for every i one of these possibilities holds, and hence

$$\lambda_0 u_0 + \lambda_1 u_1 + \ldots + \lambda_n u_n = -v_0 - v_1 - \ldots - v_n = 0.$$

Since not every v_i is 0, neither is every λ_i equal to 0, and (6) is established.

Suppose next that the additional hypothesis is satisfied, let $\lambda_0, \ldots, \lambda_n$ be non-negative numbers, not all 0, such that the conditions (6) hold, and let J be the set of integers $i \geq 1$ such that $F_i(x_0) = 0$. If $\lambda_0 = 0$, then $0 \in \sum_{i \in J} \lambda_i \partial_* F_i(x_0)$, and by (4.5.7. Corollary) and (A.5.9) it follows that $\bigcap_{i \in J} K_d(Q_i; x_0) = \varnothing$. An application of (4.5.7) now yields a contradiction to the hypothesis.

For the case in which the F_i are Hadamard differentiable, we have:

(4.5.8. Corollary 1) *Let* $F_1, \ldots, F_n$ *be continuous real-valued functions whose domains are contained in* X, *and let* F_0 *be a continuous real-valued function whose domain contains the set* $\{x \in X : F_i(x) \leq 0, (i = 1, \ldots, n)\}$. *Let also* x_0 *be a local minimum of* F_0 *subject to the conditions* $F_i(x) \leq 0$ $(i = 1, \ldots, n)$, *and suppose that* $F_0, \ldots, F_n$ *are Hadamard differentiable at* x_0. *Then there exist non-negative numbers* $\lambda_0, \ldots, \lambda_n$, *not all 0, such that*

$$\lambda_i F_i(x_0) = 0 \quad (i = 1, \ldots, n), \qquad \sum_{i=0}^n \lambda_i \partial F_i(x_0) = 0.$$

Moreover, if there exists $h \in X$ *such that* $\partial F_i(x_0)h < 0$ *for every integer* i *for which* $F_i(x_0) = 0$, *then* $\lambda_0 \neq 0$.

When the F_i are convex, the additional hypothesis in the last part of (4.5.8) can be put in a more convenient form.

(4.5.8. Corollary 2) *Let* $F_1, \ldots, F_n$ *be continuous convex functions whose domains are subsets of* X *and let* F_0 *be a continuous convex function whose domain contains the set* $\{x \in X : F_i(x) \leq 0, (i = 1, \ldots, n)\}$. *Let* x_0 *be a local minimum of* F_0 *subject to the conditions* $F_i(x) \leq 0$ $(i = 1, \ldots, n)$. *Then there exist non-negative numbers* $\lambda_0, \ldots, \lambda_n$, *not all 0, such that*

$$\lambda_i F_i(x_0) = 0 \quad (i = 1, \ldots, n), \qquad 0 \in \sum_{i=0}^n \lambda_i \partial_* F_i(x_0). \tag{8}$$

Moreover, if there exists $\tilde{x} \in X$ *such that* $F_i(\tilde{x}) < 0$ *for each integer* $i \geq 1$ *for which* $F_i(x_0) = 0$, *then* $\lambda_0 \neq 0$.

Here it is obviously enough to prove the last part, and to do this we have only to note that if F_i is convex then for $0 \le t \le 1$

$$(F_i(x_0 + t(\tilde{x} - x_0)) - F_i(x_0))/t \le F_i(\tilde{x}) - F_i(x_0),$$

so that if $F_i(x_0) = 0$ and $F_i(\tilde{x}) < 0$ then

$$\partial F_i(x_0)(\tilde{x} - x_0) = \delta F_i(x_0)(\tilde{x} - x_0) \le F_i(\tilde{x}) < 0.$$

From (4.5.5) we have the following converse of (4.5.8. Corollary 2).

(4.5.9) *Let* $F_1, \ldots, F_n$ *be continuous convex functions whose domains are subsets of the real normed space* X, *and suppose that there exists* $\tilde{x} \in X$ *such that* $F_i(\tilde{x}) < 0$ *for* $i = 1, \ldots, n$. *Let also* F_0 *be a continuous convex function whose domain contains the set* $Q = \{x \in X : F_i(x) \le 0 \ (i = 1, \ldots, n)\}$, *and let* x_0 *be a point of* Q *with the property that there exist non-negative numbers* $\lambda_0, \ldots, \lambda_n$, *not all 0, such that* (8) *holds. Then* $F_0(x) \ge F_0(x_0)$ *for all* $x \in Q$.

Let $Q_0, \ldots, Q_n$ be defined as in (7). If $Q_0 = \varnothing$ there is nothing to prove. If $Q_0 \ne \varnothing$, then evidently $x_0 \in \overline{Q_0}$; moreover, if $x \in Q_0$, then (exactly as above) $\partial F_0(x - x_0) < 0$, so that by (4.5.7. Corollary) $K_d^*(Q_0 ; x_0) = \bigcup_{\mu \le 0} \mu \partial_* F_0(x_0)$. Next, if $i \ge 1$ and $F_i(x_0) = 0$, then $\partial F_i(x_0)(\tilde{x} - x_0) < 0$, so that $K_d^*(Q_i ; x_0) = \bigcup_{\mu \le 0} \mu \partial_* F_i(x_0)$. On the other hand, if $F_i(x_0) < 0$ then $\lambda_i = 0$ and $K_d(Q_i ; x_0) = X$, so that $\lambda_i \partial_* F_i(x_0) = \{0\} = K_d^*(Q_i ; x_0)$. From (8) and (A.5.9) it follows that $\bigcap_{i=0}^n K_d(Q_i ; x_0) = \varnothing$, whence $\bigcap_{i=0}^n Q_i = \varnothing$ by (4.5.5) and this is the required result.

4.6 Theorems of Lyapunov type for differential equations

In this section we discuss some applications of Lyapunov's 'second method' to questions of stability, uniqueness, and existence of solutions of ordinary differential equations.†

The idea underlying Lyapunov's 'second method' goes back to a theorem first stated by Lagrange, namely that if in a certain position of a conservative mechanical system the potential energy has a strict minimum, then this position is one of stable equilibrium.

Consider, for simplicity, a particle of unit mass moving under a conservative system of forces in $\mathbf{R}^3$, where the potential energy of the particle when it is at the point $x = (x_1, x_2, x_3)$ of $\mathbf{R}^3$ is $U(x)$ (the potential energy

---

† In his work on stability theory, Lyapunov used the term 'second method' to cover methods that did not require a knowledge of the form of the solution of the equation concerned (cf. (4) below).

is independent of the time t, since the system is conservative). We suppose that U has a strict local minimum at 0. Moreover, since U is determined only up to an additive constant, we can suppose that $U(0) = 0$.

The equations of motion of the particle are

$$x_1'' = -\frac{\partial U(x)}{\partial x_1}, \quad x_2'' = -\frac{\partial U(x)}{\partial x_2}, \quad x_3'' = -\frac{\partial U(x)}{\partial x_3},$$

or in vector notation (cf. Exercise 3.5.1),

$$x'' = -\nabla U(x), \tag{1}$$

and since U has a minimum at $0, \nabla U(0) = 0$, so that $x = 0$ is a solution. We wish to show that this solution is stable, and for this purpose we have to consider the pair of first-order equations equivalent to (1), namely

$$x' = y, y' = -\nabla U(x),$$

whose solution is of the form $x = \psi(t), y = \psi'(t)$, where ψ is a solution of (1). Writing $z = \phi(t) = (\psi(t), \psi'(t))$, we have to show that for each $\varepsilon > 0$ we can find $\delta > 0$ such that if $\| \phi(t_0) \| < \delta$ and ϕ reaches to the boundary of $[t_0, \infty[\times \mathbf{R}^6$ on the right, then ϕ is defined and satisfies $\| \phi(t) \| < \varepsilon$ for all $t \geq t_0$.† To do this, we use the fact that the total energy of the system is constant along each solution curve, i.e. that

$$U(\psi(t)) + \tfrac{1}{2} \| \psi'(t) \|^2 = \text{constant}.$$

Let

$$H(z) = H(x, y) = U(x) + \tfrac{1}{2} \| y \|^2 \qquad (x, y \in \mathbf{R}^3).\ddagger$$

Then H is continuous on $\mathbf{R}^6$, $H(0) = 0$, and H has a strict local minimum at 0, so that there exists $\rho > 0$ such that $H(z) > 0$ for $0 < \| z \| \leq \rho$. For $r > 0$ let $E(r)$ be the component of the set $\{ z \in \mathbf{R}^6 : H(z) < r \}$ containing the origin. Clearly $E(r)$ is open. Moreover, $E(r)$ tends to $\{ 0 \}$ as $r \to 0 +$, for if $0 < \varepsilon \leq \rho$, then H has a positive infimum m on the sphere $\{ z : \| z \| = \varepsilon \}$, and hence if $0 < r \leq m$, then $E(r)$ does not meet the sphere, and therefore lies inside it. If now we take such an r and choose $\delta > 0$ so that the ball $\{ z : \| z \| < \delta \}$ is contained in $E(r)$, then if $\| \phi(t_0) \| < \delta, \phi(t) \in E(r)$ for all t in its domain for which $t \geq t_0$, so that $\| \phi(t) \| < \varepsilon$. Since $\varepsilon < \rho$, we infer from (2.4.3)(iv) that ϕ has domain $[t_0, \infty[$, and this is the required result.

What is crucial here is the existence of the (continuous) function H which has a local minimum at 0 and which is constant along each solution curve $z = \phi(t)$. More generally, the argument still applies if the derivative

---

† See Exercises 2.8; the argument actually shows that the zero solution of (1) is uniformly stable.

‡ H is the Hamiltonian of the system (see §3.11, Example (*d*), p. 236).

of H along each solution curve is less than or equal to 0, for this ensures that if the solution starts in $E(r)$ then it remains in $E(r)$.

In the preceding example, the right-hand side of the equation (1) is independent of t. For a general first-order equation

$$y' = f(t, y), \qquad (2)$$

the corresponding approach is to find a function $(t, y) \mapsto V(t, y)$ which is zero for $y = 0$, and whose derivative along any solution curve $y = \phi(t)$ of the equation is less than or equal to 0. Such a function is called a *Lyapunov function for the equation* (2). The Hadamard variation of V enters naturally here, for if V has a Hadamard variation at the point $(t, \phi(t))$ for every increment, then, by (4.2.4),

$$\frac{d}{dt} V(t, \phi(t)) = \partial V(t, \phi(t))(1, \phi'(t)) = \partial V(t, \phi(t))(1, f(t, \phi(t))),$$

so that the required condition is that, for all t,

$$\partial V(t, \phi(t))(1, f(t, \phi(t))) \le 0. \qquad (3)$$

This condition (3) is in turn obviously implied by the condition that, for all (t, y),

$$\partial V(t, y)(1, f(t, y)) \le 0, \qquad (4)$$

and this last condition (4) has the advantage that it does not require any knowledge of the solution ϕ.

More generally, (4) can be replaced by the condition that, for all (t, y),

$$\bar{\partial} V(t, y)(1, f(t, y)) \le 0, \qquad (5)$$

for, by Exercise 4.2.4, this implies that

$$D^+ V(t, \phi(t)) \le \bar{\partial} V(t, \phi(t))(1, \phi'(t)) = \bar{\partial} V(t, \phi(t))(1, f(t, \phi(t))) \le 0, \qquad (6)$$

so that again $t \mapsto V(t, \phi(t))$ is decreasing.

In the theorems below we make a still further generalization, in the spirit of the results of §2.5 and §2.11, by comparing $V(t, \phi(t))$ with the solution of a scalar equation $x' = g(t, x)$ whose behaviour is prescribed. In the first instance we employ the condition

$$\bar{\partial} V(t, y)(1, f(t, y)) \le g(t, V(t, y)),$$

which is the natural extension of (5). Other conditions will be mentioned later.

We prove first a simple differential inequality which underlies the subsequent results on stability.

(4.6.1) *Let* $I = [a, \infty[$, *let* B *be the closed ball in* Y *with centre* 0 *and radius* $\rho > 0$, *let* $f : I \times B \to Y, g : I \times [0, \infty[\to \mathbf{R}$, *and* $V : I \times B \to [0, \infty[$ *be continuous, and let* $t_0 \in I, x_0 \geq 0$. *Suppose further that*

(i) *the maximal solution* Ψ *of the equation* $x' = g(t, x)$ *satisfying* $\Psi(t_0) = x_0$ *and reaching to the boundary of* $I \times [0, \infty[$ *on the right is defined on* $[t_0, \infty[$,

(ii) *for all* (t, y) *for which* $t \in]a, \infty[$, $\|y\| < \rho$, *and* $V(t, y) > \Psi(t)$,

$$\bar{\partial} V(t, y)(1, f(t, y)) \leq g(t, V(t, y)), \tag{7}$$

(iii) ϕ *is a solution of* $y' = f(t, y)$ *on an interval* $[t_0, b[$, *where* $t_0 < b \leq \infty$, *and* $V(t_0, \phi(t_0)) \leq x_0$.
Then for all $t \in [t_0, b[$,

$$V(t, \phi(t)) \leq \Psi(t).$$

Let $\chi(t) = V(t, \phi(t))$ $(t_0 \leq t < b)$. Then $\chi(t_0) \leq x_0$, and, as in (6),

$$D^+ \chi(t) \leq \bar{\partial} V(t, \phi(t))(1, f(t, \phi(t))) \leq g(t, \chi(t)) \tag{8}$$

for all $t \in]t_0, b[$ for which $\chi(t) > \Psi(t)$. The result therefore follows from (1.5.2).

Since $\Psi(t) \geq 0$, the condition (ii) is obviously satisfied if (7) holds whenever $t \in]a, \infty[$ and $0 < \|y\| < \rho$.

We consider now a number of results concerning stability; in these, we require Y to be finite-dimensional. We recall (Exercises 2.8) that if $I = [a, \infty[$, E is a subset of $I \times Y$ open in $I \times Y, f : E \to Y$ is continuous, and the zero function is a solution of $y' = f(t, y)$ on I, then this zero solution is

(S) *stable* if for each $\varepsilon > 0$ and each $t_0 \in I$ there exists $\delta > 0$ such that, if ϕ is a solution of $y' = f(t, y)$ satisfying $\|\phi(t_0)\| < \delta$ and reaching to the boundary of E on the right, then ϕ is defined on $[t_0, \infty[$ and $\|\phi(t)\| < \varepsilon$ for all $t \geq t_0$,

(AS) *asymptotically stable* if it is stable and in addition the δ in (S) can be chosen so that $\phi(t) \to 0$ as $t \to \infty$,

(US) *uniformly stable* if it is stable and the δ in (S) can be chosen to be independent of t_0.

The proof of the first result is modelled on the argument used above for the proof of Lagrange's theorem.

(4.6.2) *Let* Y *be finite-dimensional, let* $I = [a, \infty[$, *let* B *be the closed ball in* Y *with centre* 0 *and radius* $\rho > 0$, *and let* $f : I \times B \to Y$, *g*: $I \times [0, \infty[\to \mathbf{R}$, *and* $V : I \times B \to [0, \infty[$ *be continuous. Suppose further that*

(i) *the equations* $y' = f(t, y)$ *and* $x' = g(t, x)$ *have* $y = 0$ *and* $x = 0$ *as solutions, i.e.* $f(t, 0) = 0$ *and* $g(t, 0) = 0$ *for all* $t \in I$,

(ii) $V(t, 0) = 0$ *for all* $t \in I$, *and for each* $\varepsilon > 0$ *we can find* $r > 0$ *such that,*

if $E(t,r)$ is the component of the set $\{y \in B : V(t,y) < r\}$ that contains 0, then $E(t,r) \subseteq \{y \in Y : \|y\| < \varepsilon\}$ for all $t \in I$,†
(iii) *for all $t \in I^\circ$ and $0 < \|y\| < \rho$,*

$$\partial V(t,y)(1, f(t,y)) \le g(t, V(t,y)).$$

Then if the zero solution of $x' = g(t,x)$ is stable, so is the zero solution of $y' = f(t,y)$.

Suppose that the zero solution of $x' = g(t,x)$ is stable, let $t_0 \in I$, let $0 < \varepsilon < \rho$, and choose $r > 0$ so that $E(t,r) \subseteq \{y \in Y : \|y\| < \varepsilon\}$ for all $t \in I$. We can then find $\eta > 0$ so that each solution ψ of $x' = g(t,x)$ satisfying $\psi(t_0) < \eta$ and reaching to the boundary of $I \times [0, \infty[$ on the right is defined on $[t_0, \infty[$ and satisfies $\psi(t) < r$ for all $t \ge t_0$. Since $y \mapsto V(t_0, y)$ is continuous and $E(t_0, r)$ is open, we can then find $\delta > 0$ such that $V(t_0, y) < \eta$ and $y \in E(t_0, r)$ whenever $\|y\| < \delta$. Let ϕ be a solution of $y' = f(t,y)$ satisfying $\|\phi(t_0)\| < \delta$ and reaching to the boundary of $I \times B$ on the right, and suppose that the domain of ϕ is $[t_0, b[$, where $t_0 < b \le \infty$. Then $V(t_0, \phi(t_0)) < \eta$, and hence the maximal solution Ψ of $x' = g(t,x)$ satisfying $\Psi(t_0) = V(t_0, \phi(t_0))$ is defined on $[t_0, \infty[$, and $\Psi(t) < r$ for all $t \ge t_0$. By (4.6.1), $V(t, \phi(t)) \le \Psi(t) < r$ for all $t \in [t_0, b[$, and since $\phi(t_0) \in E(t_0, r)$ and the set $\{(t,y) : t \ge t_0, y \in E(t,r)\}$ is connected in $I \times Y$, this implies that $\phi(t) \in E(t,r)$ for all $t \in [t_0, b[$, whence $\|\phi(t)\| < \varepsilon$. Since $\varepsilon < \rho$, we infer from (2.4.3)(iv) that $b = \infty$, and hence the zero solution of $y' = f(t,y)$ is stable.

(4.6.3) *Suppose that the conditions of (4.6.2) are satisfied, and that in addition $y \mapsto V(t,y)$ is continuous at 0, uniformly for t in I, and 0 is an interior point of the set $\bigcap_{t \in I} E(t,r)$ for each $r > 0$. Then if the zero solution of $x' = g(t,x)$ is uniformly stable, so is the zero solution of $y' = f(t,y)$.*

If the zero solution of $x' = g(t,x)$ is uniformly stable, the η in the preceding argument can be chosen to be independent of t_0. The additional hypotheses then imply that the δ can also be chosen to be independent of t_0, and this gives the result.

(4.6.4) *Suppose that the conditions of (4.6.2) are satisfied, and that in addition for $0 < \gamma \le \rho$ there exist $\beta(\gamma) > 0$ and $\tau(\gamma) \in I$ such that $V(t,y) \ge \beta(\gamma)$ whenever $t \ge \tau(\gamma)$ and $\gamma \le \|y\| \le \rho$. Then if the zero solution of $x' = g(t,x)$ is asymptotically stable, so is the zero solution of $y' = f(t,y)$.*

---

† Since the sets $E(t,r)$ shrink as r decreases, this is equivalent to the condition that $E(t,r)$ converges to $\{0\}$ as $r \to 0 +$, uniformly for t in I. The condition is strictly weaker than the condition that the component of the set $\{(t,y) \in I \times B : V(t,y) < r\}$ containing $I \times \{0\}$ is contained in the cylinder $\{(t,y) \in I \times Y : \|y\| < \varepsilon\}$ for all sufficiently small r.

If the zero solution of $x' = g(t, x)$ is asymptotically stable, then, exactly as in the proof of (4.6.2), $V(t, \phi(t)) \leq \Psi(t)$ and $\| \phi(t) \| < \varepsilon$ for all $t \geq t_0$, and in addition we can choose the η so that $\Psi(t) \to 0$ as $t \to \infty$. This implies that $\phi(t) \to 0$ as $t \to \infty$, for otherwise we can find $\gamma > 0$ and a sequence (t_n) tending to ∞ such that $\| \phi(t_n) \| \geq \gamma$ for all n. Then $\Psi(t_n) \geq V(t_n, \phi(t_n)) \geq \beta(\gamma) > 0$ for all sufficiently large n, and this gives a contradiction.

The next result, which we deduce from (4.6.2–4), imposes conditions on V which, although more restrictive than those in (4.6.2–4), are easier to verify.

We use $\mathscr{K}$ to denote the set of all real-valued continuous, strictly increasing functions on $[0, \infty[$ which take the value 0 at 0.

(4.6.5) *Suppose that the hypotheses of* (4.6.2) *are satisfied, except that* (ii) *is replaced by*
(ii)$'$ $V(t, 0) = 0$ *for all* $t \in I$, *and there exists* $\kappa \in \mathscr{K}$ *such that* $V(t, y) \geq \kappa(\| y \|)$ *whenever* $t \in I$ *and* $\| y \| \leq \rho$.
Then if the zero solution of $x' = g(t, x)$ *is respectively stable or asymptotically stable, so is the zero solution of* $y' = f(t, y)$. *If in addition there exists* $\mu \in \mathscr{K}$ *such that* $V(t, y) \leq \mu(\| y \|)$ *whenever* $t \in I$ *and* $\| y \| \leq \rho$, *then the uniform stability of the zero solution of* $x' = g(t, x)$ *implies that of the zero solution of* $y' = f(t, y)$.

Here (ii)$'$ implies that the set $E(t, r)$ is contained in the ball $\{ y \in B : \| y \| < \kappa^{-1}(r) \}$ and since κ^{-1} is continuous at 0 and $\kappa^{-1}(0) = 0$, this implies (ii) of (4.6.2). Since (ii)$'$ trivially implies the additional hypothesis in (4.6.4), this proves the statements concerning stability and asymptotic stability. Finally, the existence of μ with the specified properties implies the additional hypotheses in (4.6.3), and this gives the result concerning uniform stability.

The case where V does not depend on t is particularly simple, and since here the condition (ii) of (4.6.2) and the additional hypotheses of (4.6.3, 4) are satisfied if $V(0) = 0$ and $V(y) > 0$ for all non-zero $y \in B$. This result is in fact contained in (4.6.5),† but it is worth while to state it separately.

(4.6.6) *Let* Y *be finite-dimensional, let* $I = [a, \infty[$, *let* B *be the closed*

---

† Take κ to be a continuous, strictly increasing function satisfying $\kappa(r) \leq \inf \{ V(y) : r \leq \| y \| \leq \rho \}$ for $0 \leq r \leq \rho$.

ball in Y with centre 0 and radius $\rho > 0$, *and let* $f : I \times B \to Y$, $g :$
$I \times [0, \infty[\to \mathbf{R}$, *and* $V : B \to \mathbf{R}$ *be continuous. Suppose further that*
(i) *the equations* $y' = f(t, y)$ *and* $x' = g(t, x)$ *have* $y = 0$ *and* $x = 0$ *as*
solutions, i.e. $f(t, 0) = 0$ *and* $g(t, 0) = 0$ *for all* $t \in I$,
(ii) $V(0) = 0$, *and* $V(y) > 0$ *for all non-zero* $y \in B$,
(iii) *for all* $t \in I^\circ$ *and* $0 < \| y \| < \rho$

$$\partial V(y) f(t, y) \le g(t, V(y)).$$

Then if the zero solution of $x' = g(t, x)$ *is respectively stable, asymptotically*
stable, or uniformly stable, so is the zero solution of $y' = f(t, y)$.

We note next some special cases of the condition (iii) of (4.6.2–5), and
also some alternatives to this condition.

(a) *If V is Fréchet differentiable (which requires Y to be real) on the set*
$C = \{(t, y) \in I^\circ \times Y : 0 < \| y \| < \rho\}$, *then the condition* (iii) *is equivalent*
to the condition that for all $(t, y) \in C$

$$\frac{\partial V(t, y)}{\partial t} + d_2 V(t, y) f(t, y) \le g(t, V(t, y)). \tag{9}$$

For instance, if Y is a Hilbert space, and $V(t, y) = e^{-2Mt} \| y \|^2$, then
for all $y, h \in Y$

$$d_2 V(t, y) h = 2 e^{-2Mt} \langle y, h \rangle,$$

so that here (9) with $g = 0$ becomes

$$\langle y, f(t, y) \rangle \le M \| y \|^2.$$

Again, if Y is a Hilbert space, and $V(y) = \| y \|$ (so that V is independent
of t), then

$$\partial V(y) h = d V(y) h = \langle y, h \rangle / \| y \| \qquad (y, h \in Y, y \neq 0),$$

so that here (9) becomes

$$\langle y, f(t, y) \rangle \le \| y \| g(t, \| y \|)$$

(cf. Exercises 2.5.4, 5).

(b) *If* $V(t, y)$ *is locally Lipschitzian in* y *on* C, *then the condition* (iii) *can be*
replaced by the condition that for all $(t, y) \in C$

$$\underline{\delta} V(t, y)(1, f(t, y)) \le g(t, V(t, y)), \tag{10}$$

where $\underline{\delta} V$ *is the lower Gâteaux variation of* V.
  In fact, if ϕ is a solution of $y' = f(t, y)$, then (10) implies that

$$\underline{\delta} V(t, \phi(t))(1, \phi'(t)) \le g(t, V(t, \phi(t))),$$

and therefore that

$$\lim_{\tau \to 0+} \inf (V(t + \tau, \phi(t) + \tau\phi'(t)) - V(t, \phi(t)))/\tau \le g(t, V(t, \phi(t))), \qquad (11)$$

since

$$(V(t + \tau, \phi(t + \tau)) - V(t + \tau, \phi(t) + \tau\phi'(t)))/\tau$$
$$= O(\| \phi(t + \tau) - \phi(t) - \tau\phi'(t) \|/\tau) = o(1).$$

The inequality (11) implies in turn that

$$D_+ V(t, \phi(t)) \le g(t, V(t, \phi(t))),$$

and this can be used in the proof of (4.6.1) instead of (8).

We note in passing that when $V(t, y)$ is locally Lipschitzian in y, then for all $h \in Y$

$$\partial V(t, y)(1, h) = \delta V(t, y)(1, h). \qquad (12)$$

To prove this, let $h \in Y$, let (h_n) be a sequence in Y converging to h, and let $(\tau_n), (k_n)$ be sequences of positive numbers converging to 0 and 1 respectively. Then, by the Lipschitzian property,

$$(V(t + \tau_n k_n, y + \tau_n h_n) - V(t + \tau_n k_n, y + \tau_n k_n h))/(\tau_n k_n) = O(\| h_n - k_n h \|) = o(1).$$

We then have

$$\lim_{n \to \infty} \sup (V(t + \tau_n k_n, y + \tau_n h_n) - V(t, y))/\tau_n$$

$$= \lim_{n \to \infty} k_n \cdot \lim_{n \to \infty} \sup (V(t + \tau_n k_n, y + \tau_n k_n h) - V(t, y))/(\tau_n k_n)$$

$$\le \delta V(t, y)(1, h).$$

Hence $\partial V(t, y)(1, h) \le \delta V(t, y)(1, h)$, and since the reversed inequality is trivial, we obtain (12).

We mention finally a case where V is independent of t.

(c) *If $V : Y \to [0, \infty[$ is continuous and sublinear, then the condition* (iii) *can be replaced by the condition that for all $(t, y) \in C$*

$$V(f(t, y)) \le g(t, V(y))$$

(cf. (2.5.1–3) and (2.5.5)). For, if ϕ is a solution of $y' = f(t, y)$, then, by (1.6.1),

$$D^+ V(\phi(t)) \le V(\phi'(t)) = V(f(t, \phi(t))) \le g(t, V(\phi(t))),$$

and this is the inequality employed in the proof of (4.6.1) for the case where V is independent of t.†

---

† In fact, (c) implies condition (iii) itself, for V is locally Lipschitzian, and therefore, by (12),
$$\partial V(y)f(t, y) = \delta V(y)f(t, y) \le V(f(t, y)) \le g(t, V(y)).$$

We consider next conditions for the uniqueness of solutions of the equation $y' = f(t, y)$. If we were to follow the pattern set by (2.11.2), we should consider

$$\chi(t) = V(t, \phi_1(t) - \phi_2(t)),$$

where ϕ_1, ϕ_2 are solutions of $y' = f(t, y)$ such that $\phi_1(t_0) = \phi_2(t_0)$, and $V(t, y) = 0$ if and only if $y = 0$. However, a further generalization is possible here, in that we can consider

$$\omega(t) = V(t, \phi_1(t), \phi_2(t)),$$

where now $V(t, y, z) = 0$ if and only if $y = z$. Since

$$D^+\omega(t) \leq \bar{\partial}V(t, \phi_1(t), \phi_2(t))(1, \phi_1'(t), \phi_2'(t))$$
$$= \bar{\partial}V(t, \phi_1(t), \phi_2(t))(1, f(t, \phi_1(t)), f(t, \phi_2(t))), \tag{13}$$

the condition we have to use is that

$$\bar{\partial}V(t, y, z)(1, f(t, y), f(t, z)) \leq g(t, V(t, y, z)), \tag{14}$$

where g satisfies an appropriate uniqueness condition.

In (4.6.7) below, we suppose that g satisfies the Iyanaga–Kamke condition (I–K) of §2.11, viz.:

(I–K) *g is a continuous function from the rectangle $]t_0, t_0 + \alpha] \times [0, \beta]$ in $\mathbf{R}^2$ into $[0, \infty[$, satisfying $g(t, 0) = 0$ for all $t \in]t_0, t_0 + \alpha]$, with the property that, for each $t_1 \in]t_0, t_0 + \alpha]$, $\chi = 0$ is the only solution of $x' = g(t, x)$ on $]t_0, t_1]$ such that*

$$\chi(t_0 +) = 0 \quad \text{and that} \quad \lim_{t \to t_0+} \chi(t)/(t - t_0) = 0.$$

We recall (cf. (2.11.1)) that if g satisfies condition (I–K) and $\omega : [t_0, t_0 + \alpha] \to [0, \beta]$ is a continuous function such that $\omega(t_0) = \omega'(t_0) = 0$ and that $D_+\omega(t) \leq g(t, \omega(t))$ for nearly all $t \in]t_0, t_0 + \alpha[$, then $\omega = 0$.

(4.6.7) *Let Y be a Banach space, let $J = [t_0, t_0 + \alpha]$, let E be a subset of $J \times Y$ containing (t_0, y_0), and let $f : E \to Y$ be continuous at (t_0, y_0). Suppose further that*

(i) *$V :]t_0, t_0 + \alpha] \times Y^2 \to [0, \infty[$ is continuous, and for $t_0 < t \leq t_0 + \alpha$, $V(t, y, z) = 0$ if and only if $y = z$,*

(ii) *there exist $L > 0$ and $\lambda \in]0, \alpha]$ such that $V(t, y, z) \leq L\|y - z\|$ whenever $t_0 < t \leq t_0 + \lambda$, $\|y - y_0\| \leq \lambda$, and $\|z - y_0\| \leq \lambda$,*

(iii) *g satisfies condition (I–K) and (14) holds whenever $t \in J^\circ$, (t, y), $(t, z) \in E$, $y \neq z$, and $V(t, y, z) \leq \beta$.*

Then the equation $y' = f(t, y)$ has at most one solution on J taking the value y_0 at t_0.

Let ϕ_1, ϕ_2 be solutions of $y' = f(t, y)$ on J such that $\phi_1(t_0) = \phi_2(t_0) = y_0$,

and let
$$\omega(t) = V(t, \phi_1(t), \phi_2(t)) \quad (t_0 < t \le t_0 + \alpha), \quad \omega(t_0) = 0.$$
Since ϕ_1, ϕ_2 are continuous at t_0, we can find $\mu \in]0, \lambda]$ such that $\|\phi_1(t) - y_0\| \le \lambda$ and $\|\phi_2(t) - y_0\| \le \lambda$ whenever $t_0 \le t \le t_0 + \mu$, whence, by (ii),
$$\omega(t) \le L \|\phi_1(t) - \phi_2(t)\|$$
for $t_0 < t \le t_0 + \mu$. In particular, this shows that $\omega(t) \to 0$ as $t \to t_0 +$, so that ω is continuous at t_0 and therefore is continuous on J. Further, since f is continuous at (t_0, y_0), given $\varepsilon > 0$ we can find $\delta \in]0, \mu]$ such that
$$\|\phi_1'(t) - \phi_2'(t)\| = \|f(t, \phi_1(t)) - f(t, \phi_2(t))\| \le \varepsilon$$
whenever $t_0 \le t \le t_0 + \delta$. Hence for all such t we have
$$\|\phi_1(t) - \phi_2(t)\| \le \varepsilon(t - t_0),$$
and therefore, by (ii) again, $0 \le \omega(t) \le L\varepsilon(t - t_0)$, which in turn implies that $\omega'(t_0) = 0$.

We observe now that if $\phi_1 \ne \phi_2$, we can find $t_1 \in J$ such that $\omega(t_1) > 0$. Hence we can find a subinterval $[\sigma, \tau]$ of $[t_0, t_1]$ such that $\omega(\sigma) = 0$ and that $0 < \omega(t) \le \beta$ for all $t \in]\sigma, \tau]$, and then, as in (13), $D_+ \omega(t) \le g(t, \omega(t))$ for all $t \in]\sigma, \tau[$. If $\sigma = t_0$, then $\omega = 0$ on $[\sigma, \tau]$, by (2.11.1). Similarly, if $\sigma > t_0$, and $\bar{\omega}(t) = 0 \ (t_0 \le t < \sigma), \bar{\omega}(t) = \omega(t) \ (\sigma \le t \le \tau)$, then $\bar{\omega} = 0$, again by (2.11.1). Thus in either case we obtain a contradiction, and this completes the proof.

The condition (14) can be replaced by various alternatives, as in (a)–(c) above. For example, if $V(t, y, z)$ is locally Lipschitzian in (y, z), then (14) can be replaced by the condition that
$$\underline{\delta}V(t, y, z)(1, f(t, y), f(t, z)) \le g(t, V(t, y, z)).$$
We can also subsume condition (ii) of (4.6.7) in further differentiability conditions, as in the following corollary.

(4.6.7. Corollary 1) *Let D be the diagonal $\{(y, y) : y \in Y\}$ in Y^2. Then the result of (4.6.7) continues to hold if the conditions (ii) and (iii) are replaced by*
(ii)' *V is Gâteaux differentiable on $J^\circ \times (Y^2 \backslash D)$ (which implies that Y must be real), and for each bounded set A in Y^2 the partial differentials $\delta_2 V$ and $\delta_3 V$ are bounded on $J^\circ \times (A \backslash D)$,*
(iii)' *g satisfies condition (I–K) and*
$$\delta V(t, y, z)(1, f(t, y), f(t, z)) \le g(t, V(t, y, z))$$

whenever $t \in J^\circ$, (t, y), $(t, z) \in E$, $y \neq z$, *and* $V(t, y, z) \leq \beta$.

Let $R > 0$, and let A be the set of points (t, y, z) such that $t \in J^\circ$, $\|y\| \leq R$, $\|z\| \leq R$. It is enough to prove that if V satisfies (i) and (ii)', then $V(t, y, z)$ is Lipschitzian in (y, z) in A, for this trivially implies the condition (ii) of (4.6.7); also, by an argument similar to that used in the proof of (12), it implies that V is Hadamard differentiable on $J^\circ \times (Y^2 \backslash D)$, so that (iii)' implies (iii).

Let K be the greater of the suprema of $\|\delta_2 V(t, u, v)\|$ and $\|\delta_3 V(t, u, v)\|$ for $(t, u, v) \in A$, $(u, v) \notin D$, and let (t, y, z), $(t, y', z') \in A$. If both (y, z), $(y', z') \in D$, then

$$V(t, y, z) - V(t, y', z') = 0.$$

On the other hand, if at least one of (y, z), (y', z') is not in D, then, by (4.1.3),

$$|V(t, y, z) - V(t, y', z')| \leq |V(t, y, z) - V(t, y', z)| + |V(t, y', z) - V(t, y', z')|$$
$$\leq |\delta_2 V(t, \xi, z)(y - y')| + |\delta_3 V(t, y', \eta)(z - z')|$$

for some ξ, η on the open segments in Y joining y to y' and z to z' such that $(\xi, z), (y', \eta) \notin D$ (note that there is at most one point of D on each of the segments in Y^2 joining (y, z) to (y', z) and (y', z) to (y', z')). Since (t, ξ, z), $(t, y', \eta) \in A$ we therefore have

$$|V(t, y, z) - V(t, y', z')| \leq K(\|y - y'\| + \|z - z'\|),$$

as required.

As a further corollary, we have the following result when $V(t, y, z) = V(t, y - z)$.

(4.6.7. Corollary 2) *Let* $J = [t_0, t_0 + \alpha]$, *let* E *be a subset of* $J \times Y$ *containing* (t_0, y_0), *and let* $f : E \to Y$ *be continuous at* (t_0, y_0). *Suppose further that*

(i) $V :]t_0, t_0 + \alpha] \times Y \to [0, \infty[$ *is continuous, and for* $t_0 < t \leq t_0 + \alpha$, $V(t, y) = 0$ *if and only if* $y = 0$,

(ii) V *is Gâteaux differentiable on* $J^\circ \times (Y \backslash \{0\})$, *and for each bounded set* A *in* Y^2 *the partial differential* $\delta_2 V$ *is bounded on* $J^\circ \times (A \backslash \{0\})$,

(iii) g *satisfies condition (I–K) and*

$$\delta V(t, y - z)(1, f(t, y) - f(t, z)) \leq g(t, V(y - z))$$

whenever $t \in J^\circ$, $(t, y), (t, z) \in E$, $y \neq z$, *and* $V(t, y - z) \leq \beta$.
Then the equation $y' = f(t, y)$ *has at most one solution on* J *taking the value* y_0 *at* t_0.

We consider finally existence theorems corresponding to (4.6.7).

As in (2.11.3), we require that f is continuous, and therefore our results are of interest only when Y is infinite-dimensional.

In the following result, we suppose for simplicity that V is independent of t; the resulting loss in generality is in fact small, since the comparison function g itself depends on t.

(4.6.8) *Let Y be a real Banach space, let $J = [t_0, t_0 + \alpha]$, let B be the closed ball in Y with centre y_0 and radius $\rho > 0$, and let $f : J \times B \to Y$ be a continuous function such that $\| f(t, y) \| \leq M$ for all $(t, y) \in J \times B$. Let also D be the diagonal $\{(y, y) : y \in Y\}$ in Y^2, and suppose that*

(i) *$V : Y^2 \to [0, \infty[$ is continuous, and $V(y, y) = 0$ for all $y \in Y$,*

(ii) *V is Gâteaux differentiable on $Y^2 \backslash D$, and δV is bounded on $A \backslash D$ for each bounded set A in Y^2,*

(iii) *for each $\varepsilon > 0$ there exists $\delta > 0$ such that if $V(y, z) \leq \delta$ then $\| y - z \| \leq \varepsilon$,†*

(iv) *g satisfies condition (I–K) and*

$$\delta V(y, z)(f(t, y), f(t, z)) \leq g(t, V(y, z)) \tag{15}$$

whenever $(t, y), (t, z) \in J^\circ \times B^\circ, y \neq z$, and $V(y, z) \leq \beta$.

Then if $\eta = \min\{\alpha, \rho/M\}$, the equation $y' = f(t, y)$ has a solution on $[t_0, t_0 + \eta]$ taking the value y_0 at t_0. Moreover, if in addition $V(y, z) > 0$ whenever $y \neq z$, this solution is unique.

Let K be the supremum of $\| \delta V(y, z) \|$ for $(y, z) \in B^2 \backslash D$. Then, by (4.1.6), for all $(y, z), (y', z') \in B^2$ we have

$$|V(y, z) - V(y', z')| \leq K(\| y - y' \|^2 + \| z - z' \|^2)^{1/2} \tag{16}$$

(note that the segment in Y^2 joining (y, z) to (y', z') contains at most one point of D unless it lies wholly in D, in which case $V(y, z) = V(y', z') = 0$). In particular, we deduce from (16) that

$$0 \leq V(y, z) = V(y, z) - V(z, z) \leq K \| y - z \| \tag{17}$$

whenever $(y, z) \in B^2$. We observe also that (16) and (ii) together imply that V is Hadamard differentiable on the set specified in (iv) where (15) holds.

Let $I = [t_0, t_0 + \eta]$, and let (ε_n) be a decreasing sequence of positive numbers converging to 0. By (2.3.1), for each positive integer n we can find an ε_n-approximate solution ψ_n of the equation $y' = f(t, y)$ on I, satisfying $\psi_n(t_0) = y_0$, and with the property that for all $s, t \in I$

$$\| \psi_n(s) - \psi_n(t) \| \leq M | s - t | \tag{18}$$

† This is equivalent to the condition that if (y_n), (z_n) are sequences such that $V(y_n, z_n) \to 0$, then $y_n - z_n \to 0$.

(so that $\psi_n(t) \in B^\circ$ whenever $t_0 \leq t < t_0 + \eta$).

For $m > n \geq 1$ let

$$\sigma_{m,n}(t) = V(\psi_m(t), \psi_n(t)) \quad (t_0 < t \leq t_0 + \eta), \quad \sigma_{m,n}(t_0) = 0.$$

Then $\sigma_{m,n}$ is obviously continuous on $I \backslash \{t_0\}$. Further, by (17) we have

$$0 \leq \sigma_{m,n}(t) \leq K \| \psi_m(t) - \psi_n(t) \| \leq 2KM(t - t_0) \qquad (19)$$

for all $t \in I$, and hence $\sigma_{m,n}$ is continuous at t_0 and therefore on I.

Moreover, by (16) and (18), for all $s, t \in I$ we have

$$| \sigma_{m,n}(s) - \sigma_{m,n}(t) | \leq KM\sqrt{2} |s - t|. \qquad (20)$$

Next, since f is continuous at (t_0, y_0), given $\varepsilon > 0$ we can find $\lambda \in \,]0, \eta]$ such that $\| f(t, y) - f(t, z) \| \leq \varepsilon$ whenever $t_0 \leq t \leq t_0 + \lambda$, $\| y - y_0 \| \leq \lambda$, $\| z - z_0 \| \leq \lambda$. Hence if $t_0 \leq t \leq t_0 + \min\{\lambda, \lambda/M\}$, then

$$\| f(t, \psi_m(t)) - f(t, \psi_n(t)) \| \leq \varepsilon,$$

whence also

$$\| \psi_m'(t) - \psi_n'(t) \| \leq \| f(t, \psi_m(t)) - f(t, \psi_n(t)) \| + \| \psi_m'(t) - f(t, \psi_m(t)) \|$$
$$+ \| \psi_n'(t) - f(t, \psi_n(t)) \|$$
$$\leq \varepsilon + 2\varepsilon_n.$$

It follows that for such t we have

$$\| \psi_m(t) - \psi_n(t) \| \leq (\varepsilon + 2\varepsilon_n)(t - t_0),$$

whence, by (19),

$$0 \leq \sigma_{m,n}(t) \leq K(\varepsilon + 2\varepsilon_n)(t - t_0). \qquad (21)$$

Now let $\mu = \min\{\eta, \beta/(2KM)\}$, and let $I_1 = [t_0, t_0 + \mu]$. Then, by (19), $0 \leq \sigma_{m,n}(t) \leq \beta$ for all $t \in I_1$. Hence if t is a point of I_1° for which $\psi_m(t) \neq \psi_n(t)$, (iv) gives

$$\sigma_{m,n}'(t) = \partial V(\psi_m(t), \psi_n(t))(\psi_m'(t), \psi_n'(t))$$
$$= \delta V(\psi_m(t), \psi_n(t))(\psi_m'(t), \psi_n'(t))$$
$$= \delta V(\psi_m(t), \psi_n(t))(f(t, \psi_m(t)), f(t, \psi_n(t)))$$
$$+ \delta V(\psi_m(t), \psi_n(t))(\psi_m'(t) - f(t, \psi_m(t)), \psi_n'(t) - f(t, \psi_n(t)))$$
$$\leq g(t, \sigma_{m,n}(t)) + K(\| \psi_m'(t) - f(t, \psi_m(t)) \|^2 + \| \psi_n'(t) - f(t, \psi_n(t)) \|^2)^{1/2}$$
$$\leq g(t, \sigma_{m,n}(t)) + K\varepsilon_n\sqrt{2}.$$

On the other hand, if $t \in I_1^\circ$ and $\psi_m(t) = \psi_n(t)$, then for all $s \in I_1$ such that $s > t$ we have

$$\sigma_{m,n}(s) - \sigma_{m,n}(t) = \sigma_{m,n}(s) \leq K \| \psi_m(s) - \psi_n(s) \|$$
$$= K \| \psi_m(s) - \psi_m(t) - \psi_n(s) + \psi_n(t) \|,$$

so that

$$D_+ \sigma_{m,n}(t) \leq K \| \psi'_m(t) - \psi'_n(t) \|$$
$$= K \| \psi'_m(t) - f(t, \psi_m(t)) - \psi'_n(t) - f(t, \psi_n(t)) \|$$
$$\leq 2K\varepsilon_n.$$

Hence for all $t \in I_1^\circ$ we have

$$D_+ \sigma_{m,n}(t) \leq g(t, \sigma_{m,n}(t)) + 2K\varepsilon_n. \tag{22}$$

For each positive integer n let $\omega_n = \sup\limits_{m>n} \sigma_{m,n}$. Then $\omega_n(t_0) = 0$, and, by (20), (22) and (2.11.5),

$$|\omega_n(s) - \omega_n(t)| \leq KM\sqrt{2}|s - t|$$

for all $s, t \in I_1$, and

$$D_+ \omega_n(t) \leq g(t, \omega_n(t)) + 2K\varepsilon_n \tag{23}$$

for all $t \in I_1^\circ$. Moreover, by (21),

$$0 \leq \omega_n(t) \leq K(\varepsilon + 2\varepsilon_n)(t - t_0) \tag{24}$$

whenever $t_0 \leq t \leq \min\{\lambda, \lambda/M\}$. The sequence (ω_n) is therefore equicontinuous and uniformly bounded on I_1, and hence (A.1.2) it has a subsequence (ω_{n_r}) converging uniformly on I_1 to a function ω, and clearly $\omega(t_0) = 0$. By (23) and (2.11.4),

$$D_+ \omega(t) \leq g(t, \omega(t)) + 2K\varepsilon_n$$

for all $t \in I_1^\circ$, and therefore also

$$D_+ \omega(t) \leq g(t, \omega(t)).$$

Further, by (24), $0 \leq \omega(t) \leq K\varepsilon(t - t_0)$ whenever $t_0 \leq t \leq t_0 + \min\{\lambda, \lambda/M\}$. Hence $\omega'(t_0) = 0$, and therefore $\omega = 0$, by (2.11.1).

Now let $\varepsilon' > 0$, and choose $\delta > 0$ so that if $V(y, z) \leq \delta$ then $\| y - z \| \leq \varepsilon'$. Since $\omega = 0$, we can find an integer N such that $\sigma_{m,n_r}(t) \leq \delta$ whenever $m > n_r, r \geq N$, and $t \in I_1$, and hence we have also $\| \psi_m(t) - \psi_{n_r}(t) \| \leq \varepsilon'$. This implies in turn that $\| \psi_m(t) - \psi_n(t) \| \leq 2\varepsilon'$ whenever $m, n > r_N$ and $t \in I_1$, i.e. (ψ_n) converges uniformly on I_1. Hence $\phi = \lim \psi_n$ is a solution of $y' = f(t, y)$ on I_1 such that $\phi(t_0) = y_0$. If $\mu = \eta$, the proof is now complete. On the other hand, if $\mu < \eta$, we can repeat the argument starting from the point $(t_0 + \mu, \phi(t_0 + \mu))$, and after a finite number of such repetitions we obtain the required solution on I.

Uniqueness is immediate from (4.6.7. Corollary 1).

We conclude by stating without proof an extension of (4.6.8) in which V may depend on t.

(4.6.9) *Let* Y *be a real Banach space, let* $J = [t_0, t_0 + \alpha]$, *let* B *be the closed ball in* Y *with centre* y_0 *and radius* $\rho > 0$, *and let* $f: J \times B \to Y$ *be a continuous function such that* $\|f(t, y)\| \leq M$ *for all* $(t, y) \in J \times B$. *Let also* D *be the diagonal* $\{(y, y) : y \in Y\}$ *in* Y^2, *and suppose that*

   (i) $V:]t_0, t_0 + \alpha] \times Y^2 \to [0, \infty[$ *is continuous, and* $V(t, y, y) = 0$ *for all* $t \in]t_0, t_0 + \alpha]$ *and* $y \in Y$,

   (ii) V *is Gâteaux differentiable on* $J° \times (Y^2 \backslash D)$, *and for each bounded set* A *in* Y^2 *and each* $\tau \in]t_0, t_0 + \alpha[$, δV *is bounded on* $[\tau, t_0 + \alpha[\times (A \backslash D)$,

   (iii) *there exist* $L > 0$ *and* $\lambda \in]0, \alpha]$ *such that* $V(t, y, z) \leq L\|y - z\|$ *whenever* $t_0 < t \leq t_0 + \lambda$, $\|y - y_0\| \leq \lambda$, *and* $\|z - y_0\| \leq \lambda$,

   (iv) *for each* $\varepsilon > 0$ *there exists* $\delta > 0$ *such that if* $V(t, u, v) \leq \delta$ *then* $\|u - v\| \leq \varepsilon$,

   (v) *g satisfies condition* (I–K) *and*

$$\delta V(t, y, z)(1, f(t, y), f(t, z)) \leq g(t, V(t, y, z))$$

whenever $(t, y), (t, z) \in J° \times B, y \neq z$, *and* $V(t, y, z) \leq \beta$.
Then if $\eta = \min \{\alpha, \rho/M\}$, *the equation* $y' = f(t, y)$ *has a solution on* $[t_0, t_0 + \eta]$ *taking the value* y_0 *at* t_0. *Moreover, if in addition* $V(t, y, z) > 0$ *whenever* $t \in J°$ *and* $y \neq z$, *this solution is unique.*

4.7 Historical note on differentials

The motivation for differentiating functions of a vector variable came from the calculus of variations. If a good way of finding maxima or minima of ordinary functions was through the differential calculus, then perhaps a similar calculus would offer the same advantages for functions of functions. Volterra began the study from this viewpoint in 1887 (though soon, using a geometrical perspective, he came to call the subject 'functions of lines'). Work in this field, which is still continuing, formed a substantial part of the inspiration for the formation of functional analysis.

   In his paper of 1887, Volterra (1887a) considers a function f defined on a subset of the space $C[a, b]$ of continuous real-valued functions defined on the interval $[a, b]$, and in effect calculates the Gâteaux variation of f at ϕ for the increment ψ. If we assume that f is Fréchet differentiable at ϕ (which Volterra's conditions imply), this variation is simply $df(\phi)\psi$. Now $df(\phi)$ belongs to the dual space of $C[a, b]$ and so by the Riesz representation theorem corresponds to a measure; we assume that this measure is absolutely continuous, and denote it by $f'(\phi, s)ds$. The variation then appears in the form

$$df(\phi)\psi = \int_a^b f'(\phi, s)\psi(s)ds$$

which is the formula given by Volterra. Of course, Volterra lacked the advantages conferred by measure theory. His strategy was first to proceed locally: fix $s \in [a,b]$, let $\varepsilon > 0$, take $\psi \geq 0$ with support in $[s - \varepsilon, s + \varepsilon]$, and define $f'(\phi, s)$ to be the limit of the expression

$$f(\phi + \psi) \bigg/ \int_{s-\varepsilon}^{s+\varepsilon} \psi(t)dt$$

as sup $\psi \to 0$ and $\varepsilon \to 0$. He then proceeded to obtain his result for 'global' variations ψ.

Volterra (1887a) also gave a Taylor expansion for a function of functions. This necessitated producing a higher variation and this appeared in the form

$$\int_a^b \cdots \int_a^b f^{(n)}(\phi, s_1, \ldots, s_n) \psi(s_1) \ldots \psi(s_n) ds_1 \ldots ds_n.$$

(Here, as with f' above, $f^{(n)}$ is not an ordinary derivative.) Summaries of this part of his work are given in Volterra's book (1913).

In 1911, Fréchet took a decisive step by declaring that a real-valued function f of two real variables was differentiable at a point (a,b) if and only if its graph had a tangent plane at that point. Later that year he conceded that the analytic expression asserting this,

$$f(a + h, b + k) - f(a,b) = D_1 f(a,b)h + D_2 f(a,b)k + \varepsilon(h,k)(h^2 + k^2)^{1/2} \quad (1)$$

where $\varepsilon(h,k) \to 0$ as $h, k \to 0$, had been given by Young (1909a). Young, as stated in his book (1910), was aiming at 'rigidity of proof and novelty of treatment', a target which he must be judged to have hit, but he himself concedes priority in using (1) to Stolz (1893).

In his note of 1911, Fréchet points out that the formula (1) can be generalized to infinite-dimensional spaces, provided a linear functional is used to replace the first two terms on the right-hand side and a metric is given on the space to provide the possibility of convergence.† He also asserts that the elementary properties of differentials can easily be transferred to the infinite-dimensional case.

Two years later, a new differential appeared in a note by Gâteaux (1913). Unfortunately, Gâteaux was killed very early in the Great War, and it was not until 1922 that a full version of his work appeared. In this latter paper, the importance of the linearity of the variation is stressed, and the conclusion of (4.1.4) is obtained under stronger hypotheses

---

† Perhaps Fréchet allowed his enthusiasm to carry him away here. As Nashed (1971, p. 116) points out, this definition is not what is required in a linear metric space, even if the dimension is 1. However, it was probably a normed space which Fréchet had in mind.

(joint continuity in x and h). Gâteaux's papers were prepared for publication by Lévy, who in his book (1922) also emphasizes the value of the linearity of the variation and points out that, even with linearity imposed as an additional requirement, the Gâteaux differential is more general than that of Fréchet.

Until 1925, the functions involved were all scalar-valued (and the word 'functional' was used to describe such a mapping). In that year, Fréchet recognized that his definition of 1911 needed very little modification to apply to functions between normed spaces. The announcement of this in Fréchet (1925a) is discussed at length in Fréchet (1925b).

With Hildebrandt and Graves (1927), calculus in normed spaces really becomes a subject of its own. It is interesting that they consider it worth while to begin with a list of axioms for a normed space (the algebraic as well as the analytic); the basic concepts were not yet common knowledge. They work with the Fréchet differential and prove its uniqueness, the result of (3.1.1), that a function with a bounded Fréchet differential in an open set satisfies a Lipschitz condition (3.2.2), they define partial differentials on a product of normed spaces and show that the existence and continuity of the partial differentials implies the existence and continuity of the differential (3.3.3), their higher order differentials are multilinear (§3.5) and they give a complicated form of the chain rule. The highlight of the paper is a local implicit function theorem which they establish using the contraction mapping principle (a result which they also prove). They establish that the implicitly defined function is of class C^r if the requisite conditions are satisfied (cf. (3.8.1)). (The finite-dimensional version of the latter result is due to Young (1909b).)

In a later paper in the same volume of the *Transactions of the American Mathematical Society*, Graves (1927a) gives a version of Taylor's theorem with an integral form of the remainder (cf. Exercise 3.6.3). He generally works with the Gâteaux differential but does also prove that the higher order Fréchet differential is symmetric (3.5.7).

Perhaps the oldest notion of derivative for a function of a vector variable is that of the gradient of a real-valued function defined on a Euclidean space. This idea was placed in an abstract (Hilbert space) setting by Golomb (1935), who essentially gives the definition of Exercise 3.5.1, although he does not mention the Fréchet differential explicitly. An earlier gradient in function spaces had been described by Courant and Hilbert (1930) in their chapter on the calculus of variations. (This material was not in the first German edition of the book but is translated in the first English edition of 1953.) Unfortunately, the text is not very precise about the domains of definition of its functionals or the conditions under which the gradient is defined, but it is clear that the underlying concept of differential is of Hadamard rather than Fréchet type.

The rigorous definition of the Hadamard differential appears in the work of Fréchet (1937). Fréchet complains that the Gâteaux differential, even when required to be linear as advocated by Lévy, is seriously defective in that it fails to satisfy the chain rule. He therefore proposes a new differential for which the condition of (4.2.8)(iii) is taken as definition, i.e. that the chain rule should hold for compositions with functions whose domains are intervals. This idea was used by Hadamard (1923) to make sense of the formula

$$dz = \frac{\partial z}{\partial x}dx + \frac{\partial z}{\partial y}dy$$

and so Fréchet names this new differential after him.

Fréchet proves that for finite-dimensional spaces, the Fréchet and Hadamard differentials coincide (see above, p. 266). He also considers the example $M(x) = \max_{0 \leq t \leq 1} |x(t)|$ for x in $C[0,1]$ to show that these two differentials may be distinct in infinite dimensions. Perhaps a little strangely, Fréchet does not go on to prove that the Hadamard differential satisfies the general chain rule (4.2.4). This task is left to Ky Fan (1942) who also shows that a Hadamard differentiable function is continuous (4.2.6).

In the second version of his book, Lévy (1951) accepts Fréchet's criticism of the Gâteaux differential. He points out that the weaker the definition of differential, the fewer will be its properties, and he gives a long discussion of the merits of various possible definitions (pp. 37–40). His conclusion is that, despite the drawback that it depends on the norm in the space, the Fréchet differential is preferable because it is simpler and more natural.

Lévy's conclusion seems to be shared by most 'pure' analysts, and it is the Fréchet differential which forms the subject of textbooks on calculus (see Cartan (1967) or Dieudonné (1960)). Those who work in the calculus of variations, however, take a different view. Vainberg (1956) finds the Gâteaux differential valuable and discusses its relationship with the Fréchet differential; he proves (4.1.7. Corollary 1) and gives an example of a function everywhere Gâteaux differentiable but nowhere Fréchet differentiable. Nashed (1971) finds all three differentials of importance and includes many others. He gives a detailed study of the connections between various differentials and has extensive information on the history of the subject. Most of the results on Gâteaux and Hadamard variations which have not been ascribed to their originators here may be found in Nashed's paper, though it should be mentioned that Sova (1964) was responsible for the characterizations of Hadamard differentiability in (4.2.8).

The subject is now developing in the direction of greater abstraction, principally to differentiation in linear topological spaces. The Gâteaux differential has a great advantage here, since it is independent of the topology placed on the spaces involved. However, all the differentials mentioned have several generalizations to topological vector spaces: see, for example, the paper by Nashed (1971) already cited, or Averbukh and Smolyanov (1967).

APPENDIX

Here we collect together some results which are needed in the main body of the work. Where they are standard, they are given without proof; where they are less familiar or differ in detail from standard results, short proofs are given. References can be found in the notes on the appendix (p. 338).

Except where noted, all vector spaces are over a field **K** which can be either the field **R** of real numbers or the field **C** of complex numbers.

A.1 Metric and topological spaces

(A.1.1) (The Tietze extension theorem) *Let X be a metric space, let A be a closed subspace, and let $f : A \to \mathbf{R}$ be a continuous bounded function. Then there exists a continuous function $g : X \to \mathbf{R}$ such that g coincides with f on A and*

$$\inf_{x \in X} g(x) = \inf_{x \in A} f(x), \qquad \sup_{x \in A} f(x) = \sup_{x \in X} g(x).$$

(A.1.1. Corollary) *Let X, A be as in (A.1.1). Let $f : A \to \mathbf{R}^n$ (where n is a positive integer) be continuous and bounded. Then there is a continuous and bounded $g : X \to \mathbf{R}^n$ which coincides with f on A.*

To see this, simply write f in terms of n real-valued functions as $f = (f_1, \ldots, f_n)$ and extend each f_i to X by (A.1.1).

For the statement of our second theorem, we recall that a set F of functions defined on a metric space X (with metric ρ) with values in a normed space Y is *equicontinuous* at the point x_0 of X if, for each $\varepsilon > 0$ we can find $\delta > 0$ such that $\| f(x) - f(x_0) \| < \varepsilon$, for all f in F, whenever $\rho(x, x_0) < \delta$. The set F is *equicontinuous* on X if it is equicontinuous at each point of X. There is a corresponding notion of uniform equicontinuity. As is the case with ordinary continuity, when the space X is compact, uniform equicontinuity is equivalent to equicontinuity on X.

(A.1.2) (The Ascoli–Arzelà theorem) *Let X be a compact metric space, let Y be a finite-dimensional normed space and let (f_n) be an equicontinuous and uniformly bounded sequence of functions from X into Y. Then there is a subsequence of (f_n) which converges uniformly on X.*

This theorem is usually stated as a criterion for compactness. The vector space $C(X, Y)$ of continuous bounded functions on X with values in Y becomes a Banach space when provided with the norm

$$\|f\| = \sup \{ \|f(x)\| : x \in X \}.$$

An equivalent formulation of the Ascoli–Arzelà theorem is that if F is equicontinuous and bounded in $C(X, Y)$, then it has compact closure.

The next result, though well known, is perhaps not quite so easy to find in the literature.

(A.1.3) *Let X, Z be topological spaces, let X be compact, let Y be a metric space, and let $f : X \times Z \to Y$. Then f is continuous if and only if both $x \mapsto f(x, z)$ is continuous on X for each $z \in Z$ and also, for each $z_0 \in Z, f(x, z) \to f(x, z_0)$ uniformly for $x \in X$ as $z \to z_0$.*

Let ρ be the metric on Y. If f is continuous, then obviously $x \mapsto f(x, z)$ is continuous for each z. Also, given $z_0 \in Z$ and $\varepsilon > 0$, for each $x \in X$ we can find neighbourhoods $U(x)$ of x and $V(x)$ of z_0 such that $\rho(f(x', z), f(x, z_0)) \le \varepsilon/2$ for $x' \in U(x)$ and $z \in V(x)$. Since X is compact we can find $x_1, \ldots, x_n$ such that $U(x_1), \ldots, U(x_n)$ cover X; put $V = V(x_1) \cap \ldots \cap V(x_n)$. Then V is a neighbourhood of z_0, and if $z \in V$, for all $x \in X$,

$$\rho(f(x, z), f(x, z_0)) \le \min_i \{ \rho(f(x, z), f(x_i, z_0)) + \rho(f(x_i, z_0), f(x, z_0)) \}$$
$$\le \varepsilon$$

since each x belongs to $U(x_i)$ for some i. Thus, $f(x, z) \to f(x, z_0)$ uniformly for $x \in X$.

To see the converse, observe that for $(x, z), (x_0, z_0) \in X \times Z, \rho(f(x, z), f(x_0, z_0)) \le \rho(f(x, z), f(x, z_0)) + \rho(f(x, z_0), f(x_0, z_0))$. Under the two given conditions each of the terms on the right tends to zero as (x, z) tends to (x_0, z_0).

We need a further result in which compactness plays a vital role. The *uniform norm* on the space $C(X)$ of continuous bounded functions on X is defined for $f \in C(X)$ by

$$\|f\| = \sup_{x \in X} |f(x)|.$$

(A.1.4) (The Stone–Weierstrass theorem) *Let X be a compact topological space and let $C(X)$ denote the real vector space of all continuous real-valued functions on X. Let F be a vector subspace of $C(X)$ which contains the constant functions, which separates the points of X (i.e. given $x_1, x_2 \in X$ there is $f \in F$ such that $f(x_1) \neq f(x_2)$), and which is closed under pointwise multiplication (i.e. if $f, g \in F$, the function fg defined by $fg(x) = f(x)g(x)$ for each $x \in X$ is also in F). Then F is uniformly dense in $C(X)$. The conclusion also holds for complex-valued functions provided that F satisfies the additional condition that, for each $f \in F$, the complex conjugate $\bar{f}$ also belongs to F.*

We shall apply this theorem to show that under certain conditions, Lipschitzian functions are dense. Let X, Y and Z be normed spaces, let $E \subseteq X$ and let $g : E \to Z$. Then g is *Lipschitzian* if there is a constant K such that

$$\| g(x_1) - g(x_2) \| \leq K \| x_1 - x_2 \|$$

for all x_1, x_2 in E; if we wish to call attention to the constant K involved, we say that g is *K-Lipschitzian*. If $E \subseteq X \times Y$ and $f : E \to Z$, we say f is *Lipschitzian in y* if there is a constant K such that

$$\| f(x, y_1) - f(x, y_2) \| \leq K \| y_1 - y_2 \|$$

for all $(x, y_1), (x, y_2)$ in E. (Note that the latter definition requires that the same value of K can be taken for each relevant x.)

(A.1.5) *Let Y and Z be finite-dimensional normed spaces. Let B be a closed ball in Y and let J be a compact subset of $\mathbf{R}$. Let $h : J \times B \to Z$ be continuous. Then for each $\varepsilon > 0$ there is a function $f : J \times B \to Z$, Lipschitzian in y, satisfying*

(i) $\| h(t, y) - f(t, y) \| < \varepsilon$ *for all* $(t, y) \in J \times B$;

(ii) $\sup \{ \| f(t, y) \| : (t, y) \in J \times B \} \leq \sup \{ \| h(t, y) \| : (t, y) \in J \times B \}$.

To prove this, we first consider the case when $Z = \mathbf{K}$. The set $J \times B$ is compact, and the space F of functions Lipschitzian in y on $J \times B$ satisfies the conditions of the Stone–Weierstrass theorem (A.1.4). Thus in this case, (i) holds. To obtain (i) in general, when $Z = \mathbf{K}^m$ for some integer m, it is enough to approximate each of the coordinate functions separately, i.e. to apply m times the case $Z = \mathbf{K}$. To see how to modify f so that (ii) also holds, write $M = \sup_{J \times B} \| h(t, y) \|$. Define $u : Z \to Z$ by

$$u(z) = \begin{cases} z & \text{if } \| z \| \leq M \\ Mz / \| z \| & \text{if } \| z \| \geq M. \end{cases}$$

Then there is a constant $K\dagger$ for which

$$\| u(z_1) - u(z_2) \| \leq K \| z_1 - z_2 \|,$$

so that, for $(t, y) \in J \times B$,

$$\| u \circ h(t, y) - u \circ f(t, y) \| < K\varepsilon$$

and the result follows since $u \circ h = h$ while $\sup\limits_{J \times B} \| u \circ f(t, y) \| \leq M$.

Our final result in this section is the Banach fixed point theorem or the contraction mapping principle.

(A.1.6) (Contraction mapping principle) *Let X be a complete metric space and let ρ be its metric. Let $f : X \to X$ be a contraction (i.e. there is k with $0 \leq k < 1$ such that $\rho(f(x), f(x')) \leq k\rho(x, x')$ for $x, x' \in X$). Then f has a unique fixed point (i.e. there is a unique $x \in X$ with $f(x) = x$).*

A.2 Normed spaces

For X, Y normed spaces, we denote by $\mathscr{L}(X, Y)$ the space of all continuous linear mappings from X to Y with norm

$$\| T \| = \sup\{ \| Tx \| : x \in X, \| x \| \leq 1 \}.$$

The space $X' = \mathscr{L}(X, \mathbf{K})$ where $\mathbf{K}$ is the field of scalars, is the *dual* of X. If $T \in \mathscr{L}(X, Y)$ is bijective, then the inverse map T^{-1} is linear. If T^{-1} is continuous, then T is a homeomorphism. We shall denote the set of all linear homeomorphisms from X to Y by $\mathscr{L}\mathscr{H}(X, Y)$.

If X and Y are finite-dimensional, each linear map from X to Y can be represented by a matrix with respect to any given bases in these spaces, and is automatically continuous. In general, discontinuous linear maps exist; the following theorem gives an important criterion for continuity.

(A.2.1) (The closed graph theorem) *Let X, Y be Banach spaces. Let $T : X \to Y$ be linear and suppose the graph of T,*

$$\{(x, y) : y = Tx\},$$

is a closed subset of $X \times Y$. Then T is continuous.

The following corollary is an almost immediate consequence.

---

† When Z has the Euclidean norm, this result is easily seen to hold with $K = 1$ from elementary geometrical considerations, since the space spanned by the real multiples of z_1 and z_2 is isomorphic with the Euclidean plane. The general inequality follows since all norms on Z are equivalent. A detailed argument is given in (2.10.2).

(A.2.1. Corollary) *Let X, Y be Banach spaces. Let $T: X \to Y$ be linear, bijective and continuous. Then T^{-1} is continuous (or equivalently, T is a linear homeomorphism).*

A situation to which this corollary applies is given in the last part of the next theorem.

(A.2.2) (i) *Let X be a normed space and let S be a closed subspace. For $x \in X$, write $\| x + S \| = \inf\{ \| x + y \| : y \in S \}$. Then the space X/S of cosets of S becomes a normed space, the mapping $\phi: X \to X/S$ defined by $\phi(x) = x + S$ is continuous, and $\| \phi \| = 1$. If X is complete, then X/S is complete.*

(ii) *Suppose in addition that Y is also a normed space, that $T: X \to Y$ is linear and continuous, and that $S \subseteq \ker T$. Then the mapping $T_0: X/S \to Y$ defined by $T_0(x + S) = T(x)$ is well defined, linear and continuous, $\| T_0 \| = \| T \|$, and $T = T_0 \circ \phi$.*

(iii) *Suppose now $S = \ker T$ and write T_c for T_0. If in addition both X and Y are Banach spaces and T is surjective, then $T_c: X/\ker T \to Y$ is a linear homeomorphism.*

When $T: X \to Y$ is not bijective, it does not have an inverse. The following replacement is useful in some situations.

(A.2.3) *Let X, Y be Banach spaces. Let $T: X \to Y$ be linear, surjective and continuous. Let T_c be as in (A.2.2) and take $k > \| T_c^{-1} \|$. Then for each $y \in Y$ we can find $x \in X$ with $T(x) = y$ and $\| x \| \leq k \| y \|$.*

Using the notation of (A.2.2), if $y \neq 0, \| T_c^{-1}(y) \| < k \| y \|$, so we can find x in the coset $T_c^{-1}(y)$ with $\| x \| < k \| y \|$. Obviously $\phi(x) = T_c^{-1}(y)$, so $T(x) = y$.

Another result which, like the closed graph theorem, has a proof depending on the theory of Baire category, is the uniform boundedness principle (or Banach–Steinhaus theorem).

(A.2.4) (The uniform boundedness principle) *Let Γ be a set of continuous linear mappings from a Banach space X to a normed space Y, and suppose that for each $x \in X$, $\{ \| Tx \| : T \in \Gamma \}$ is bounded. Then $\{ \| T \| : T \in \Gamma \}$ is bounded.*

Let $X_1, \ldots, X_n$ be normed spaces. The *product* $X_1 \times \ldots \times X_n$ can be made into a normed space in many equivalent ways; when the norm is

not explicitly mentioned, the product will be understood to have the
Euclidean norm, defined for $x = (x_1, \dots, x_n) \in X_1 \times \dots \times X_n$ by

$$\|x\| = (\|x_1\|^2 + \dots + \|x_n\|^2)^{1/2}.$$

This remark holds in particular when $X_i = \mathbf{K}$ for $1 \le i \le n$.

Now let Y be another normed space. We denote by $\mathscr{L}(X_1, \dots, X_n; Y)$ the space of continuous multilinear mappings from $X_1 \times \dots \times X_n$ to Y. If $X_1 = \dots = X_n = X$, we write $\mathscr{L}_n(X; Y)$ for $\mathscr{L}(X \times \dots \times X; Y)$.

(A.2.5) *Let* $f : X_1 \times \dots \times X_n \to Y$ *be multilinear. The following assertions are equivalent*:

  (i) f *is continuous*;

  (ii) f *is continuous at one point*;

  (iii) $\|f\| = \sup\{\|f(x_1, \dots, x_n)\| : \|x_1\| \le 1, \dots, \|x_n\| \le 1\}$ *is finite*;

  (iv) $E = \{k \ge 0 : \|f(x_1, \dots, x_n)\| \le k\|x_1\| \dots \|x_n\| \text{ for all } x_1, \dots, x_n\} \ne \varnothing$.

If these conditions hold, then $\|f\| \in E$ *and* $\|f\| = \inf E$.

The norm defined in (iii) of this theorem makes $\mathscr{L}(X_1, \dots, X_n; Y)$ into a normed space, and it is complete if Y is complete.

Now let Z be another normed space. For $f \in \mathscr{L}(X_1, \dots, X_n, Z; Y)$ and $x = (x_1, \dots, x_n) \in X_1 \times \dots \times X_n$, write

$$f_x(z) = f(x_1, \dots, x_n, z) \qquad (z \in Z).$$

Then clearly $f_x \in \mathscr{L}(Z, Y)$. The map $x \mapsto f_x$ is multilinear, and its norm is

$$\sup\{\|f_x\| : \|x_1\| \le 1, \dots, \|x_n\| \le 1\} = \|f\|,$$

since $\|f_x\| = \sup\{\|f_x(z)\| : \|z\| \le 1\}$. The correspondence which assigns to $f \in \mathscr{L}(X_1, \dots, X_n, Z; Y)$ the map $x \mapsto f_x$ in $\mathscr{L}(X_1, \dots, X_n; \mathscr{L}(Z, Y))$ is therefore an isometry. It is obviously linear. By retracing the steps of the argument we see that it has an inverse. This gives the next theorem.

(A.2.6) *The mapping described above is an isometric isomorphism between* $\mathscr{L}(X_1, \dots, X_n, Z; Y)$ *and* $\mathscr{L}(X_1, \dots, X_n; \mathscr{L}(Z, Y))$. *By induction,* $\mathscr{L}(X_1, \dots, X_n; Y)$ *is isomorphic with* $\mathscr{L}(X_1, \mathscr{L}(X_2, \dots, \mathscr{L}(X_n, Y) \dots))$.

Many of the more important normed spaces are inner product spaces. An *inner product* on X is a mapping $(x, x') \to \langle x, x' \rangle$ of $X \times X$ into $\mathbf{C}$ which satisfies $\langle x, x \rangle \ge 0$ and $\langle x, x \rangle = 0$ if and only if $x = 0$; $\langle x', x \rangle = \overline{\langle x, x' \rangle}$; and which is linear in x. A norm $\|\ \|$ can be defined on an inner product space by writing $\|x\|^2 = \langle x, x \rangle$. A complete inner product space is called a *Hilbert space*. Thus, finite-dimensional inner product spaces are Hilbert spaces.

If X is a Hilbert space, an element T of $\mathscr{L}(X,X)$ has an *adjoint* T^* which is determined by the formula $\langle Tx, x' \rangle = \langle x, T^*x' \rangle$ for all x, x' in X; T is *self-adjoint* if $T = T^*$. The *spectrum* of T is the set

$$\{\lambda \in \mathbf{C} : T - \lambda 1_X \text{ has no inverse in } \mathscr{L}(X,X)\}$$

where 1_X is the identity on X. If X is finite-dimensional, this is just the set of eigenvalues; in general it is a compact subset of $\mathbf{C}$. The spectrum of a self-adjoint operator is real, and therefore has a maximal element.

(A.2.7) *If T is a self-adjoint operator on a Hilbert space X, the maximal element of the spectrum of T is*

$$\sup \{\langle Tx, x \rangle : x \in X, \|x\| \leq 1\}.$$

A.3 Convex sets and functions

A set E in a normed space X is *convex* if, whenever $x, y \in E$ and $0 \leq \lambda \leq 1$, then $\lambda x + (1 - \lambda)y \in E$. A point $x \in E$ is *internal* to E (or E is *radial* or *absorbing* at x) if, for every $h \in X$ there is $\lambda > 0$ such that $x + \lambda h \in E$.

(A.3.1) *Let X be a normed space and let E be a convex subset of X. Then*
 (i) *the interior E°, the closure $\bar{E}$ and the set of internal points of E are convex,*
 (ii) *if $x \in E^\circ, y \in E$, then $\lambda x + (1 - \lambda)y \in E^\circ$ for $0 < \lambda \leq 1$,*
 (iii) *if $E^\circ \neq \varnothing$, then $\overline{E^\circ} = \bar{E}$.*
  To see (ii), observe that if U is open and $x \in U \subseteq E$, then for $0 < \lambda < 1$, the set $\lambda U + (1 - \lambda)y$ is open, contains $\lambda x + (1 - \lambda)y$ and is contained in E. The rest is easy.

It is obvious that an interior point of a set is internal. There are situations in which the converse is true.

(A.3.2) *Let X be a Banach space and let $E \subseteq X$ be convex and closed. Then*
 (i) *every point internal to E belongs to E°,*
 (ii) *if X is a finite-dimensional space and H is the minimal closed linear variety containing E, then E has an interior point relative to H.*
  Let x be internal to E so that 0 is internal to the translate $E - x$. Then $X = \bigcup_{n=1}^\infty n(E - x)$. Since each set $n(E - x)$ is closed and X is a complete metric space, one of these sets has a non-empty interior by the Baire category theorem. Hence E has non-empty interior. Let $z \in E^\circ$. Since x is internal, for some $\alpha > 0$, $x + \alpha(x - z) \in E$. Then, using (A.3.1)(ii)

with $\lambda = \alpha/(1 + \alpha)$ we see that

$$x = \lambda z + (1 - \lambda)(x + \alpha(x - z)) \in E^\circ.$$

We now turn to (ii). Let $x \in E$. Then $H - x$ is the smallest linear subspace of X containing $E - x$. If m is its dimension, there must exist m linearly independent points $\{x_1, \ldots, x_m\}$ in $E - x$. The convex hull of $\{0, x_1, \ldots, x_m\}$ is a non-degenerate simplex in $H - x$ which is easily seen to have internal points. Since a finite-dimensional space is complete, $E - x$ has an interior point relative to $H - x$ by part (i).

The (*closed*) *convex hull* of a set A is the smallest (closed) convex set which contains A. The convex hull of A will be denoted by co A; the closed convex hull of A is just the closure $\overline{\text{co}} \, A$. An *extremal point* of a convex set E is a point $x \in E$ such that if $x = \lambda x_1 + (1 - \lambda)x_2$ with $x_1, x_2 \in E$ and $0 < \lambda < 1$, then $x = x_1 = x_2$.

(A.3.3) (The Krein–Milman theorem) *Let A be the set of extremal points of the compact convex set E in a normed space X. Then $E = \overline{\text{co}} \, A$.*

In general, it is difficult to find the extremal points of any given set, so that the following result is useful.

(A.3.4) *Let A be a closed set in a normed space X for which $E = \overline{\text{co}} \, A$ is compact. Then A contains the extremal points of E.*

Under certain conditions, the question of when a closed convex hull is compact can be answered. The convex hull of a totally bounded set in a normed space is again totally bounded, and the closure of a totally bounded set in a complete space is compact.

(A.3.5) *Let A be a compact subset of X.*
(i) *If X is a Banach space, then $\overline{\text{co}} \, A$ is compact.*
(ii) *If X is finite-dimensional, then co A is compact.*

To see why (ii) holds, recall that if the dimension of X is n, then each point of co A belongs to the convex hull of a subset of A consisting of no more than $(n + 1)$ points. Thus, co A is the image of the compact set

$$\{(\lambda_1, \ldots, \lambda_{n+1}) \in \mathbf{R}^{n+1} : \lambda_1 \geq 0, \ldots, \lambda_{n+1} \geq 0, \textstyle\sum \lambda_i = 1\} \times A^{n+1}$$

under the continuous mapping

$$((\lambda_1, \ldots, \lambda_{n+1}), x_1, \ldots, x_{n+1}) \mapsto \lambda_1 x_1 + \ldots + \lambda_{n+1} x_{n+1},$$

and it is therefore compact.

We now turn to consider convex functions. A real-valued† function f whose domain is a convex set E is called *convex* if, for all $x, y \in E$ and λ with $0 \le \lambda \le 1$,

$$f(\lambda x + (1 - \lambda)y) \le \lambda f(x) + (1 - \lambda)f(y).$$

Rearranging this inequality gives

$$\lambda(f(\lambda x + (1 - \lambda)y) - f(x)) \le (1 - \lambda)(f(y) - f(\lambda x + (1 - \lambda)y)).$$

In particular, we obtain the following proposition by taking

$$x = x_0 + sh, y = x_0 + uh, \lambda = (u - t)/(u - s).$$

(A.3.6) *Let f be a convex function defined on the convex set E. Let $s < t < u$ be real numbers and let $x_0, h \in X$ be such that $x_0 + sh, x_0 + uh \in E$. Then*

$$\frac{f(x_0 + th) - f(x_0 + sh)}{t - s} \le \frac{f(x_0 + uh) - f(x_0 + th)}{u - t}.$$

The function f is *quasi-convex* if, for each $\alpha \in \mathbf{R}$, $\{x \in E : f(x) \le \alpha\}$ is convex. Evidently, a convex function is quasi-convex, but the converse is not true. Many results which hold for convex functions (including some proved below) are also true for quasi-convex functions.

(A.3.7) *The following statements are equivalent for a real-valued function defined on a convex set E.*
 (i) f *is quasi-convex*;
 (ii) *for each $\alpha \in \mathbf{R}$, $\{x \in E : f(x) < \alpha\}$ is convex*;
 (iii) *for $0 \le \lambda \le 1, x, y \in E, f(\lambda x + (1 - \lambda)y) \le \max\{f(x), f(y)\}$.*
The proof is straightforward.

Part (iii) of the last theorem motivates the following definition. A function f defined on a convex set E is *strictly quasi-convex* if $x, y \in E, x \ne y$ and $0 < \lambda < 1$ imply $f(\lambda x + (1 - \lambda)y) < \max\{f(x), f(y)\}$.

A real-valued function f defined on a vector space X is *sublinear* if it is both *positive homogeneous* (i.e. $f(\lambda x) = \lambda f(x)$ for $\lambda \ge 0$ and $x \in X$) and *subadditive* (i.e. $f(x + y) \le f(x) + f(y)$ for $x, y \in X$).

(A.3.8) *A sublinear function is convex. A function which is both convex and positive homogeneous and whose domain is a vector space is sublinear.*

It need only be shown that if f is positive homogeneous and convex, then it is subadditive. Now, for any x, y,

$$f(x + y) = f(\tfrac{1}{2} \cdot 2x + \tfrac{1}{2} \cdot 2y) \le \tfrac{1}{2}f(2x) + \tfrac{1}{2}f(2y) = f(x) + f(y).$$

† It is usual to allow convex functions to take the value $+ \infty$, and often $- \infty$ is permitted as well. For our purposes, the restriction to finite values is convenient.

Certain convex sets are naturally associated with sublinear functionals. Let E be a convex set which has 0 as an internal point. Then the *Minkowski functional* (or *gauge*) p of E is defined at $x \in X$ by

$$p(x) = \inf\{\lambda : x \in \lambda E\}.$$

(A.3.9) *Let p be the Minkowski functional of the convex set E. Then p is sublinear, $\{x : p(x) < 1\}$ consists of all internal points of E, and $E \subseteq \{x : p(x) \leq 1\}$.*

We next consider continuity properties of convex functions.

(A.3.10) *Let f be convex and bounded above on a convex symmetric neighbourhood N of x (so that $x + y \in N$ implies $x - y \in N$). Then f is also bounded below on N.*

This follows immediately from the inequality

$$\tfrac{1}{2}f(x + y) \geq f(x) - \tfrac{1}{2}f(x - y).$$

(A.3.11) *Let f be convex on a convex set E and let $x \in E^\circ$. If f is bounded above on a neighbourhood of x (and so in particular if f is continuous at x) then f is Lipschitzian, and a fortiori uniformly continuous, in some neighbourhood of x.*

By the previous result, there is an open ball $B(x_1, 2\varepsilon)$ in which f is bounded, say $b_1 \leq f(y) \leq b_2$ for $y \in B(x, 2\varepsilon)$. Let $y, z \in B(x, \varepsilon)$. For $y \neq z$, write $\lambda = \varepsilon / \|y - z\|$ and $w = (1 + \lambda)z - \lambda y$. Then $w \in B(x, 2\varepsilon)$ and $z = \lambda y/(1 + \lambda) + w/(1 + \lambda)$. Therefore

$$f(z) \leq \frac{\lambda}{1 + \lambda}f(y) + \frac{1}{1 + \lambda}f(w) = f(y) + \frac{1}{1 + \lambda}(f(w) - f(y))$$

and so

$$f(z) - f(y) \leq \frac{1}{\lambda}(f(w) - f(y)) \leq \frac{b_2 - b_1}{\varepsilon}\|y - z\|.$$

The result follows.

In finite dimensions, the boundedness is automatic.

(A.3.11. Corollary) *Let X be finite-dimensional, let f be convex on the convex set E, and let $x \in E^\circ$. Then f is Lipschitzian on some neighbourhood of x.*

From the theorem, we have only to show that f is bounded above on some neighbourhood of x. Take a finite subset $\{x_1, \ldots, x_n\}$ of E whose convex hull N is a neighbourhood of x. If $y \in N$, then $y = \sum_{i=1}^{n} \lambda_i x_i$ where $\sum_{i=1}^{n} \lambda_i = 1$ and $\lambda_i \geq 0$ $(1 \leq i \leq n)$, so that

$$f(y) \leq \sum_{i=1}^{n} \lambda_i f(x_i) \leq \max_{1 \leq i \leq n} f(x_i).$$

The last result of this section is not as immediate a consequence of the above as appears at first sight. It would be if it were about upper, rather than lower, semicontinuous functions.

(A.3.12) *Let X be a Banach space. Let f be convex and lower semicontinuous on a closed convex set E which has an internal point. Then E° is non-empty and f is continuous on E°.*

From (A.3.2), E° is the set of internal points of E. Let $x \in E^{\circ}$ and put $A = \{y \in E : f(y) \leq f(x) + 1\}$. Then A is closed since f is lower semicontinuous and E is closed; A is convex because f is convex; and x is internal to A since, for each $h \in X$, the restriction of f to the (one-dimensional) segment $\{x + \lambda h : x + \lambda h \in E, \lambda \in R\}$ is continuous by (A.3.11. Corollary). By (A.3.2) again, $x \in A^{\circ}$, so that f is bounded above in a neighbourhood of x. The result now follows from (A.3.11).

A.4 The Hahn–Banach theorem

In this section, we take $\mathbf{K} = \mathbf{R}$, or in other words, all vector spaces are real. (This is merely for simplicity of exposition; there are forms of these theorems valid for complex spaces.)

The results described here are all forms of, or consequences of, the Hahn–Banach theorem. We begin with the analytic form. (Recall that a *functional* is a mapping into the field of scalars.)

(A.4.1) *Let p be a sublinear functional defined on the real normed space X. Let u be a linear functional defined on a subspace S of X for which $u(x) \leq p(x)$ for $x \in S$. Then there is a linear functional f on X with $f(x) \leq p(x)$ for $x \in X$ whose restriction to S is u.*

The special case in which, for some fixed $x \in X$, $S = \{\lambda x : \lambda \in \mathbf{K}\}$ and u is defined by $u(\lambda x) = \lambda p(x)$, is worthy of note.

(A.4.1. Corollary 1) *Let p be a continuous sublinear functional on the real normed space X and let $x \in X$. Then there is a continuous linear functional f on X with $f \leq p$ and $f(x) = p(x)$.*

The particular case of the corollary in which p is just the norm yields a linear functional f with $\|f\| = 1$ and $f(x) = \|x\|$.

Another way of formulating this corollary is as follows.

(A.4.1. Corollary 2) *If p is continuous and sublinear and $x \in X$, then*

$$p(x) = \sup\{f(x) : f \in X', f \le p\}.$$

The geometric forms of the Hahn–Banach theorem are concerned with convex sets.

(A.4.2) *Let A, B be disjoint convex sets in the real normed space X.*

(i) *If A is open, there exist $f \in X'$ and $\alpha \in \mathbf{R}$ such that $f(x) > \alpha$ for $x \in A$ and $f(x) \le \alpha$ for $x \in B$.*

(ii) *If A is closed and B is compact, there exist $f \in X'$ and $\alpha \in \mathbf{R}$ such that $f(x) > \alpha$ for $x \in A$ and $f(x) < \alpha$ for $x \in B$.*

A *closed half-space* is any set of the form $\{x : f(x) \ge \alpha\}$ for $f \in X', f \ne 0$, $\alpha \in \mathbf{R}$. An *open half-space* has the form $\{x : f(x) > \alpha\}$. A *hyperplane* is a set which can be written as $\{x : f(x) = \alpha\}$ for some non-zero linear functional f on X and some $\alpha \in \mathbf{R}$; in this case, $f(x) = \alpha$ is called an *equation of the hyperplane*. Of course, when $\alpha = 0$, the hyperplane is the subspace $\ker f$. The hyperplane is a closed set if and only if f is continuous. A hyperplane H is a *hyperplane of support* for a convex set E if $H \cap E \ne \varnothing$ and E lies on one side of H (i.e. if H has equation $f(x) = \alpha$, then either $f(x) \ge \alpha$ for all $x \in E$, or $f(x) \le \alpha$ for all $x \in E$).

(A.4.3) *The closed convex hull of a set E is the intersection of the closed half-spaces which contain E.*

This follows easily from (A.4.2)(ii) on taking $A = \overline{\text{co}}\, E$ and taking $B = \{x\}$ for each x not belonging to A.

(A.4.4) *If E is convex, $E^\circ \ne \varnothing$, and $x \in E \backslash E^\circ$, then there is a closed hyperplane of support H of E with $x \in H$.*

First apply (A.4.2)(i) with $A = E^\circ, B = \{x\}$. Since $f(y) > \alpha$ for $y \in E^\circ$ and $E \subseteq \overline{E^\circ}$ (by (A.3.1) (iii)), we have $f(y) \ge \alpha$ for $y \in \bar{E}$. Thus $x \in H = \{y : f(y) = \alpha\}$, and so H is a hyperplane of support.

A.5 Cones and their duality

In this section, again all normed spaces are over the real field.

A set K in a vector space X is a *cone with vertex a* if it is a union of half-lines with one endpoint at a, or equivalently, if for each $\lambda > 0$, $\lambda(K - a) \subseteq K - a$. Thus, cones are not necessarily convex, and they may contain whole lines (for example, half-spaces are cones).

Our first result follows easily from (A.4.2)(i), since if $f(x) > \alpha$ for all x in some cone with vertex 0, it must follow that $\alpha \le 0$.

(A.5.1) *Let K be an open convex cone with vertex 0, and let B be convex and disjoint from K, in the real normed space X. Then there is a continuous linear functional f with $f(x) > 0$ for $x \in K$ and $f(x) \le 0$ for $x \in B$.*

With the usual interpretation of a cone as a set of 'positive' elements, the next theorem gives conditions under which a positive linear functional defined on a subspace has a positive extension to the whole space. It is in fact another form of the Hahn–Banach theorem.

(A.5.2) *Let K be a convex cone with vertex 0 in the real normed space X. Let S be a linear subspace of X with $S \cap K^\circ \ne \varnothing$ and let $u : S \to \mathbf{R}$ be linear with $u(S \cap K) \ge 0$. Then there is a continuous linear functional f on X with $f(K) \ge 0$ whose restriction to S coincides with u.*

We turn next to duality properties of cones – properties of pairings between cones in X and in X'. For any sets $E \subseteq X, F \subseteq X'$, we write†

$$E^* = \{ f \in X' : f(x) \ge 0 \, (x \in E) \},$$
$$F_* = \{ x \in X : f(x) \ge 0 \, (f \in F) \}.$$

We call E^* and F_* the cones *dual* to E and F respectively.

The following properties are easily checked.

(A.5.3) *Let $E, E_1, E_2, E_i \subseteq X$ (for $i \in I$) where X is a real normed space. Then*
  (i) *E^* is a closed convex cone with vertex 0,*
  (ii) *if $E_1 \subseteq E_2$, then $E_2^* \subseteq E_1^*$,*
  (iii) *$E \subseteq (E^*)_*$,*
  (iv) *$(\bar{E})^* = E^*$,*
  (v) *$(\bigcup_{i \in I} E_i)^* = \bigcap_{i \in I} E_i^*$.*
Corresponding results hold for subsets of X'.

An important question asks which cones are the duals of their duals. Its solution for cones in X is an application of the theory of polar sets.

---

† If E and F are themselves cones, then E^*, F_* are the polars of E and F respectively, and much of the theory can be recovered from the theory of polars of convex sets.

(A.5.4) *Let K be a closed convex cone with vertex 0 in the real normed space X. Then $(K^*)_* = K$.*

The solution to this question for cones in K' is not so simple, and requires the introduction of a new topology. It is easiest to do this in terms of nets. We say that a net $(f_i : i \in I)$ in X' converges to f in the *weak* topology* in X' if, for each $x \in X, f_i(x) \to f(x)$. Obviously if $f_i \to f$ in the norm of X', then $f_i \to f$ in the weak* topology, so that every weak*-closed set is norm-closed. It is also easy to see from the definition that for each $E \subseteq X$, E^* is weak*-closed.

(A.5.5) *Let $K \subseteq X'$ be a weak*-closed cone with vertex 0. Then $(K_*)^* = K$.*
Again, the proof is a standard part of the theory of polar sets.

We shall use this result to determine the dual of a particular cone.

(A.5.6) *Let p be a continuous sublinear functional on the real normed space X, and suppose that $K_0 = \{x : p(x) < 0\} \neq \emptyset$. Write*
$$K = \{x : p(x) \leq 0\}, \ A = \{f \in X' : f \leq p\}.$$
Then $K^* = K_0^* = \bigcup_{\mu \leq 0} \mu A$.
We first prove that $K = (-A)_*$. It is easy to see that $K \subseteq (-A)_*$. To obtain the reverse inequality, given $x \notin K$ use (A.4.1. Corollary 1) to find $f \in X'$ with $f(x) = p(x) > 0$ and $f \leq p$; then $(-f) \in (-A)$ and $-f(x) < 0$, so $x \notin (-A)_*$.

It now follows easily that $K = (\bigcup_{\mu \leq 0} \mu A)_*$. The result can be deduced from (A.5.5) once it is shown that the cone $\bigcup_{\mu \leq 0} \mu A$ is weak*-closed. Since the map $f \mapsto (-f)$ is obviously a weak*-homeomorphism of X', it is enough to prove that $\bigcup_{\lambda \geq 0} \lambda A$ is weak*-closed. Let the net (f_i) in $\bigcup_{\lambda \geq 0} \lambda A$ converge to f in the weak* sense. For each i, there is $\lambda_i \geq 0$ such that $f_i(x) \leq \lambda_i p(x)$ for all $x \in X$. Therefore,
$$f(x) = \lim_i f_i(x) \leq (\limsup_i \lambda_i) p(x)$$
for all $x \in X$. A particular choice of x such that $p(x) < 0$ shows that $\lambda = \limsup_i \lambda_i < \infty$, and we then conclude that $f \in \lambda A \subseteq \bigcup_{\lambda \geq 0} \lambda A$.

Since $\bar{K}_0 = K$, that $K_0^* = K^*$ follows from (A.5.3)(iv).

We shall need to know the form of the dual cone of a finite intersection of cones under certain conditions. We first give a lemma on intersections of convex sets.

(A.5.7) *Let E_i ($i \in I$) be convex sets in the normed space X with $\bigcap_{i \in I} E_i^\circ \neq \varnothing$. Then*

$$\overline{\bigcap_{i \in I} E_i^\circ} = \overline{\bigcap_{i \in I} E_i} = \bigcap_{i \in I} \overline{E_i}.$$

It is clear that

$$\overline{\bigcap_{i \in I} E_i^\circ} \subseteq \overline{\bigcap_{i \in I} E_i} \subseteq \bigcap_{i \in I} \overline{E_i}.$$

Let $x \in \overline{E_i}$ for every i. Take $y \in E_i^\circ$ for every i. By (A.3.1)(ii), we see that $\lambda y + (1 - \lambda)x \in E_i^\circ$ for every i and every λ with $0 < \lambda \leq 1$. On letting $\lambda \to 0$ we conclude that $x \in \overline{\bigcap_{i \in I} E_i^\circ}$.

(A.5.8)*Let $K_1, \ldots, K_n$ be convex cones with vertex 0 in the real normed space X and suppose $\bigcap_{i=1}^n K_i^\circ \neq \varnothing$. Then*

$$\left(\bigcap_{i=1}^n K_i \right)^* = K_1^* + \ldots + K_n^*.$$

By (A.3.1)(iii), (A.5.3)(iv) and (A.5.7), we may replace the cones K_i by their interiors, i.e. we may assume each K_i is open.

One inclusion is easy. Since $\bigcap_{i=1}^n K_i \subseteq K_j$ for each j, we see from (A.5.3)(ii) that $K_j^* \subseteq (\bigcap_{i=1}^n K_i)^*$ for each j. Since these sets are convex cones, we conclude that $K_1^* + \ldots + K_n^* \subseteq (\bigcap_{i=1}^n K_i)^*$.

To obtain the reverse inequality, consider the open cone $K = K_1 \times \ldots \times K_n$ in X^n. Let $S = \{(x, x, \ldots, x) : x \in X\}$ be the diagonal of X^n. Let $f \in X'$ and define u on S by $u(x, x, \ldots, x) = f(x)$ for $x \in X$. Since $(x, x, \ldots, x) \in S \cap K$ if and only if $x \in \bigcap_{i=1}^n K_i$, we see that if $f \in (\bigcap_{i=1}^n K_i)^*$, then u is positive on $S \cap K$. Using (A.5.2) we find $g \in (X^n)'$ with $g(K) \geq 0$ and $g(x, x, \ldots, x) = u(x, x, \ldots, x) = f(x)$ for $x \in X$. Let $g_1, g_2, \ldots, g_n \in X'$ be the coordinate functions for g, so that

$$g(x_1, \ldots, x_n) = g_1(x_1) + \ldots + g_n(x_n).$$

Since $g(\bar{K}) \geq 0$, $g_i(K_i) \geq 0$ for each i, so that $g_i \in K_i^*$ for each i. Moreover, for each $x \in X$,

$$f(x) = g(x, x, \ldots, x) = g_1(x) + \ldots + g_n(x),$$

whence $f \in K_1^* + \ldots + K_n^*$. This completes the proof.

(A.5.9) *Let $K_1, \ldots, K_n, K_{n+1}$ be convex cones with vertex 0 in the real normed space X with $K_1, \ldots, K_n$ open. Then $\bigcap_{i=1}^{n+1} K_i = \varnothing$ if and only if there exist $f_i \in K_i^*$, not all zero, such that*

$$f_1 + \ldots + f_n + f_{n+1} = 0.$$

First, suppose $\bigcap_{i=1}^{n+1} K_i = \varnothing$. If any K_i is empty, the result is trivial. Suppose this is not so, and let m be the first integer for which $\bigcap_{i=1}^{m+1} K_i = \varnothing$.

By (A.5.1) there is $f_{m+1} \in X'$ with $f_{m+1}(\bigcap_{i=1}^{m} K_i) < 0, f_{m+1}(K_{m+1}) \geq 0$. Thus, $f_{m+1} \neq 0$,

$$-f_{m+1} \in \left(\bigcap_{i=1}^{m} K_i \right)^* = K_1^* + \dots + K_m^*$$

by (A.5.8), and $f_{m+1} \in K_{m+1}^*$. This proves necessity.

To establish the converse, take m with $1 \leq m \leq n$,

$$f_1 + \dots + f_{m+1} = 0$$

$(f_i \in K_i^*)$ and $f_{m+1} \neq 0$. Then

$$-f_{m+1} = f_1 + \dots + f_m \in K_1^* + \dots + K_m^* = \left(\bigcap_{i=1}^{m} K_i \right)^*,$$

using (A.5.8), so that $f_{m+1}(\bigcap_{i=1}^{m} K_i) < 0$ (the intersection is an open set and f_{m+1}, being a continuous linear mapping onto $\mathbf{R}$, is open) and $f_{m+1}(K_{m+1}) \geq 0$. Hence $\bigcap_{i=1}^{m+1} K_i = \varnothing$.

A.6 Measurable vector-valued functions

The purpose of this section is to describe a particular result ((A.6.2) below) concerning the measurability of certain functions with values in a finite-dimensional space.

We recall briefly some concepts from measure theory. For simplicity, we consider only Banach spaces which are separable (i.e. which have a countable dense subset). A *Borel set* in a Banach space Z is an element of the σ-algebra generated by the open sets of Z; when Z is separable, this same σ-algebra is generated by the closed balls of Z. A set $A \subseteq \mathbf{R}$ has *Lebesgue measure zero* if, for each $\varepsilon > 0$ there is a sequence I_n of intervals with $A \subseteq \bigcup_n I_n$ and $\sum_n |I_n| < \varepsilon$, where $|I_n|$ denotes the length of I_n. A property which holds for all values of a variable except those in a set of measure zero is said to obtain *almost everywhere*. A *Lebesgue measurable set* in $\mathbf{R}$ is one which differs from a Borel set of $\mathbf{R}$ by a set of measure zero. The *Lebesgue measure on* $\mathbf{R}$ is the unique measure on the σ-algebra of Lebesgue measurable sets in $\mathbf{R}$ which extends length on intervals. If $E \subseteq \mathbf{R}, Z$ is a separable Banach space and $f : E \to Z$ then f is *(Lebesgue) measurable* if $f^{-1}(S)$ is Lebesgue measurable for each Borel set S in Z; in fact it is enough that this should hold when S is any closed ball in Z. If $f : E \to [0, \infty[$ is measurable, then

$$\int_E f(t)dt = \inf \left\{ \int_E g(t)dt : f \leq g \text{ and } g \text{ is lower semicontinuous} \right\},$$

where we allow the integrals to take infinite values.

In what follows, Y will be a finite-dimensional normed space. The vector space Z of continuous functions from Y to Y is unfortunately not a Banach space. However, for each $n > 0$ the ball $B_n = \{y \in Y : \|y\| \le n\}$ is compact since Y is finite-dimensional, and the space $Z_n = C(B_n, Y)$ becomes a Banach space if we write, for $z \in Z_n$,

$$\|z\| = \sup \{\|z(y)\| : y \in B_n\}.$$

The space Z_n is separable for each n. Indeed, in the space $C(B_n, \mathbf{R})$ the countable set consisting of the polynomials in the coordinates of Y with rational coefficients is dense by the Stone–Weierstrass theorem (A.1.4). The result for Z_n follows by treating Y as $\mathbf{R}^k$ and Z_n as the direct sum of k copies of $C(B_n, \mathbf{R})$ (cf. (A.1 5)).

Let $I \subseteq \mathbf{R}$ be a compact interval. To each function $f : I \times B_n \to Y$ which is continuous in its second variable there corresponds a mapping $u : I \to Z_n$ via the formula

$$u(t) = f(t, \cdot) \qquad (t \in I).$$

Moreover, f is continuous (as a function of two variables) if and only if u is continuous (use (A.1.3) to see this). There is a more subtle connection between the measurability of f and of u.

(A.6.1) **Lemma.** *With the above notation, suppose that for each $t \in I$, $y \mapsto f(t, y)$ is continuous on B_n and that for each $y \in B_n$, $t \mapsto f(t, y)$ is measurable on I. Then u is measurable.*

The compact metric space B_n has a dense subsequence (y_k), and so for $z \in Z_n$, $\|z\| = \sup_k \|z(y_k)\|$. Let S be the closed ball with centre z_0 and radius $\varepsilon > 0$ in Z_n. Then

$$u^{-1}(S) = \{t : \|u(t) - z_0\| \le \varepsilon\} = \{t : \sup_k \|u(t)(y_k) - z_0(y_k)\| \le \varepsilon\}$$

$$= \bigcap_k \{t : \|f(t, y_k) - z_0(y_k)\| \le \varepsilon\}$$

which, by hypothesis, is the intersection of a sequence of measurable sets and thus measurable.

This brings us to the theorem we were seeking.

(A.6.2) *Let $I \subseteq \mathbf{R}$ be a compact interval, let Y be finite-dimensional, and let $f : I \times Y \to Y$ have the properties that for each $t \in I$, $y \mapsto f(t, y)$ is continuous and for each $y \in Y$, $t \mapsto f(t, y)$ is measurable. Then there are a sequence (f_n) of continuous functions from $I \times Y$ into Y and a set A of Lebesgue measure zero such that, for each compact set $K \subseteq Y$,*

$$\sup_{y \in K} \| f_n(t, y) - f(t, y) \| \to 0 \tag{1}$$

for all $t \notin A$.

We begin by defining for each positive integer n, functions $v_n : I \to Z_n$ by writing, for $t \in I$,

$$v_n(t)(y) = f(t, y) \qquad (y \in Y, \|y\| \leq n).$$

By (A.6.1), each v_n is measurable. Theorems about the approximation of measurable real-valued functions by simple functions are usually also valid for functions with values in a separable Banach space; here, the function v_n takes values in the separable Banach space Z_n, and we can find a simple function $\sum_{i=1}^{r} a_i \chi_{K_i}$ (where a_i is an element of Z_n and χ_{K_i} is the characteristic function of a compact set $K_i \subseteq I$) such that $|I \backslash \bigcup_i K_i| <$ $1/n$ and for every $t \in \bigcup_i K_i$

$$\sup_{y \in B_n} \left\| \sum_{i=1}^{r} a_i \chi_{K_i}(t)(y) - v_n(t)(y) \right\| < \frac{1}{n}$$

(the supremum on the left here is the norm in Z_n). Write $F_n = \bigcup_i K_i$. The complement of the closed set F_n is the union of a sequence of intervals open in I; let $w_n : I \to Z_n$ be a function which is continuous, coincides with $\sum_i a_i \chi_{K_i}$ on F_n and is linear on each interval of $I \backslash F_n$.

Now define $f_n : I \times Y \to Y$ by

$$f_n(t, y) = \begin{cases} w_n(t)(y) & \text{if } \|y\| \leq n \\ w_n(t)\left(\dfrac{ny}{\|y\|} \right) & \text{if } \|y\| > n \end{cases}$$

for $(t, y) \in I \times Y$. The function f_n is continuous. Moreover, if K is a compact subset of Y, for all sufficiently large n we have $K \subseteq B_n$ and so, for $t \in F_n$,

$$\sup_{y \in K} \| f_n(t, y) - f(t, y) \| < 1/n.$$

Write $A = I \backslash \bigcup_n \bigcap_{m>n} F_m$. Then A has measure zero, and for each compact $K \subseteq I$, (1) holds for $t \notin A$.

The remaining results are simple variations on the above theme.

(A.6.2. Corollary 1) *Suppose that the hypotheses of (A.6.2) are satisfied and that in addition there is an increasing sequence (μ_n) of continuous functions mapping I into $[0, \infty[$ and a continuous function h mapping $[0, \infty[$ into $[0, \infty[$ such that, for all $(t, y) \in I \times Y$,*

$$\| f(t, y) \| \leq \lim_{n} \mu_n(t) h(\|y\|).$$

Then the functions (f_n) of (A.6.2) *can be chosen to satisfy the additional
condition*

$$\| f_n(t, y) \| \le \mu_n(t) h(\| y \|) \qquad (2)$$

for each $(t, y) \in I \times Y$ *and each n.*

This is easily achieved by replacing $f_n(t, y)$ by

$$f_n(t, y) \frac{\mu_n(t) h(\| y \|)}{\| f_n(t, y) \|}$$

for all values of (t, y) for which (2) is violated.

(A.6.2. Corollary 2) *Let f be as in* (A.6.2) *and let* $\phi : I \to Y$ *be continuous.
Then* $t \mapsto f(t, \phi(t))$ *is measurable.*

The function $t \mapsto f(t, \phi(t))$ is the limit almost everywhere of the sequence
of continuous functions $t \mapsto f_n(t, \phi(t))$ and is therefore measurable.

NOTES

Chapter one

The results of §§1.1–2 are essentially classical, and references can be found in Boyer (1959) and the encyclopaedia article of Voss and Molk (1909). The treatment of convexity in §1.1, Example (g), and Exercise 1.7.3 follows that of Bourbaki (1949, pp. 41–8). Exercise (1.1.1) is a theorem of Peano (1892a). The result of Exercise 1.1.2 is due to Hukuhara (1940a); the solution is an adaptation of an argument used by Stein and Peck (1955) for the case where p is the norm function.

References to (1.3.1–6) and (1.3.10. Corollary 1) are given in §1.11. The argument of (1.3.8) is due to Aziz and Diaz (1963b), who use it to obtain the analogue of (1.3.8) for the weak derivative. The result of (1.3.9) was obtained by Darboux (1875, p. 109); the proofs in the text are unlikely to be new, but I have not seen them in print before. The counterexample to an n-dimensional extension of Darboux's theorem in Exercise 1.3.2 is given by Bourbaki (1949, p.29); the related Exercises 1.3.4, 5 on the connectedness of the set of tangent directions are new. Theorem (1.3.11) is due essentially to Valiron (1927). For (1.3.12) see Flett (1957); Exercise 1.3.3, which depends on (1.3.12), is essentially in Flett (1955).

The results of (1.4.1–4) have been discussed in §1.11. Theorem (1.4.5) is essentially due to Perron (1915); the proof in the text is that of Ważewski (1960c). For results using other types of derivative see Bruckner and Leonard (1966).

The history of (1.5.1) is discussed in §2.14. Theorem (1.5.2) is due to Ważewski (1951a).

Lemma (1.6.1) is due to Hukuhara (1940a). Theorem (1.6.2) and its generalization (1.10.5) are special cases of a result of McLeod (1965); the proof in the text is that of Flett (1972a).

The extensions of Darboux's increment inequality (see §1.11) given in (1.6.3) and (1.10.6) are special cases of a theorem of Flett (1972a); for $\psi(t) = t$ these results were obtained by McLeod (1965), though his formulation and proof are different. Both McLeod and Flett consider also results of this type for topological linear spaces. The method of proof of (1.6.6) is due to Mlak (see Szarski, 1965, p. 222), but the actual result here is new, as are those of Exercises 1.6.3, 4 and 1.10.3. As mentioned in §1.11, earlier results of the Darboux type were obtained by Mie (1893), Halperin (1954), and Aziz and Diaz (1963a, b) (see also Aziz, Diaz, and Mlak, 1966).

It has already been remarked in §1.11 that the extension (1.6.4) of the Weierstrass–Peano theorem is a special case of a result of Ważewski (1949, 1951b) (Ważewski's result also contains (1.6.2. Corollary 1) which was obtained independently by

Bourbaki (1949, p. 22)). Ważewski's argument was developed by Alexiewicz (1951) for the weak derivative, and by Mlak (1957); the latter's result is given in Exercise 1.6.6. The stronger result of (1.10.7) is due to McLeod (1965), as is also (1.6.5). Our results of the Weierstrass–Peano type (see §1.11) have been given by Hukuhara (1959, p. 112), Frölicher and Bucher (1966, p. 50), Averbukh and Smolyanov (1967, p. 217), and Flett (1972a). Hukuhara, Averbukh and Smolyanov, McLeod and Flett consider also results of this type for linear topological spaces.

Exercise 1.6.1 is the result underlying the Nagumo–Perron uniqueness criterion for solutions of ordinary differential equations (see §2.14, p. 162)

The results of (1.7.1, 2) are essentially classical. Theorem (1.7.3) is due to Wintner (1946a) (see also Hukuhara, 1959, p. 128); the proof here is new. The version of l'Hospital's theorem in Exercise 1.7.4, which contains (1.7.4,5), is due to Ważewski (1955); again the proof is new.

The results of Exercise 1.7.5 on rectifiable paths are essentially due to Jordan (1887). The results of Exercise 1.7.8 on paths with continuously turning tangents were proved by Valiron (1927) for the case where $Y = \mathbf{R}^3$; the proof here via Exercises 1.7.6, 7 and Exercise 1.6.6 is new.

The result of (1.8.1) is similar to a result of Flett (1972a) for topological linear spaces, and generalizes a little-known theorem of Peano (1889), namely that if $\phi : [a, b] \to \mathbf{R}$ is a continuous function possessing derivatives of orders $1, \ldots, n-1$ on $[a, b[$, there is at least one $\xi \in]a, b[$ for which

$$\phi(b) = \phi(a) + (b - a)\phi'(a) + \ldots + \frac{(b-a)^{n-1}}{(n-1)!}\phi^{(n-1)}(a)$$
$$+ \frac{(b-a)^n}{n!}\left(\frac{\phi^{(n-1)}(\xi) - \phi^{(n-1)}(a)}{\xi - a}\right).$$

The result of (1.8.2) for real-valued ϕ is stated without proof by Peano in Genocchi (1884); Peano subsequently gave three proofs, via l'Hospital's theorem (1887b, p. 49), via the formula above (1889), and by induction (1891b). Peano also extended the result to functions with values in $\mathbf{R}^n$ (1887b, p. 49). The infinite-dimensional version of (1.8.2) is due to Bourbaki (1949, p. 33), who uses an induction argument similar to one used by Young (1910, pp. 27–30) to prove the analogous result for functions of several real variables. The proof of (1.8.2) in the text is that of Flett (1972a), who gives also versions of (1.8.2, 3) for topological linear spaces. For further references concerning other versions of Taylor's theorem see Pringsheim (1900). Theorem (1.8.4) is due to Hadamard (1914).

The discussion of regulated functions and the integral in §1.9 follows that of Bourbaki (1949, ch. 2) and Dieudonné (1960, pp. 139, 159). Exercise 1.9.5 is implicit in Wintner (1946d) (see the notes on (2.5.4)).

The continuous case of (1.10.1) and the associated lemma (1.10.2) were proved by la Vallée Poussin (1909, pp. 80–1).

Lemma (1.10.3) is due to Banach (1925); the condition that $\phi(E)$ has measure 0 whenever E has measure 0 appears in the work of Levi (1906a) on differentiation, and was named 'condition (N)' by Lusin in 1915 (see Saks, 1937, p. 224). Lemma (1.10.3) has a partial converse, also due to Banach, namely that if ϕ is a continuous function of bounded variation satisfying condition (N), then ϕ is absolutely continuous. Theorem (1.10.4) is due to Levi (1906b); the proof in the text is given by Saks

(1937, p. 225). Note that the proof fails if we replace the hypothesis that ϕ is absolutely continuous by the hypothesis that ϕ is continuous and satisfies condition (N), for this latter hypothesis does not imply that ψ satisfies condition (N) (see Saks, 1937, p. 226). For further results in the direction of (1.10.4) see Saks (1937, ch. 9).

The argument used in the proof of (1.10.8) is familiar (see, for example, Carathéodory (1927, p. 597), but I have not seen this particular formulation of the result in print before.

Chapter two

The history of the theory of ordinary differential equations up to 1910 is treated in detail in the encyclopaedia articles by Painlevé (1910) and Vessiot (1910); Müller (1928b) gives a good account of existence and uniqueness theorems. Comprehensive accounts of the modern theory for finite-dimensional Y are given in the books of Coddington and Levinson (1955), Hartman (1964) and Reid (1971); the infinite-dimensional case is considered by Bourbaki (1951) (for applications of the infinite-dimensional case see Saaty (1967, ch. 8) and Barbashin (1970, ch. 3)). For the general theory of differential and integral inequalities and their application to differential equations the reader should consult the books of Szarski (1965), Lakshmikantham and Leela (1969) and W. Walter (1970).

The finite-dimensional case of (2.3.1) has already been mentioned in §2.14. The infinite-dimensional case is proved by Bourbaki (1951, p. 6, prop. 3), using Zorn's lemma; the proof in the text follows that of Cartan (1967, p. 112; 1971, p. 102). Theorem (2.3.2) is taken from Hartman (1964, p. 4). The result of Exercise 2.3.1 is due to Müller (1927).

Theorems (2.4.1–5) and Exercises 2.4.1–3 have been discussed in §2.14. The example at the end of §2.4 (p. 84) to show that the continuity of f does not imply existence when dim $Y = \infty$ is similar to an example of Bourbaki (1951, p. 25, ex. 17). Exercise 2.4.4 is due to Perron (1915).

Comparison theorems of the type discussed in §2.5 seem to have been considered first by Wintner (1945) (summaries can be found in Wintner (1957b) and Brauer (1965)). The results of (2.5.1,2) and Exercise 2.5.4 can hardly be new, but I have not seen them in print before; Stokes (1959) uses the Tychonoff fixed point theorem to obtain the special case of (2.5.1) where $x \mapsto g(t, x)$ is increasing. The existence part of (2.5.3) and the upper estimate given there are in essence due to Conti (1956) and Wintner (1957b); the lower estimate is given by Lakshmikantham (1962b). The existence part of (2.5.4) is due to Wintner (1946c); the two estimates, obtained by the method of Exercise 1.9.5, are implicit in Wintner (1946d), and were made explicit by Langenhop (1960); the upper bound implies an integral inequality of La Salle (1949) and Bihari (1956). The bounds can also be obtained from (2.5.3) and Exercise 2.5.1 (see Lakshmikantham, 1962b, and Brauer, 1963). The results of (2.5.4. Corollary) are due to Wintner (1946c, d). Theorem (2.5.5)(i) is due to Brauer (1965); (2.5.5)(ii) is new; Brauer (1965) and Lakshmikantham and Leela (1969, i, p. 88) wrongly state that the solution ϕ in (ii) can be taken with domain I (cf. Exercise 2.5.8). Exercises 2.5.1–3 are given by many authors, though case (ii) in Exercise 2.5.1 is rarely noted. Exercises 2.5.5, 6 are essentially due to Wintner (1957b, 1946d). Exercise 2.5.7 is a generalization of a result of Wintner (1957b) and the indicated solution of the last part may be new.

The results of §2.6 and the finite-dimensional cases of those of §2.7 are discussed in §2.14. The results of §2.7 for dim $Y = \infty$ are given by Bourbaki (1951), who uses a special case of (2.6.2) in place of (2.6.1). The alternative treatment in Exercise 2.7.2 is attributed by W. Walter (1970, p. 62) to Morgenstern. Exercises 2.6.1 and 2.6.3 are obvious generalizations of (2.6.1, 2); Exercise 2.6.4 is due to Baiada (1947); Exercises 2.6.2 and 2.7.3 are new.

The treatment of linear equations in §2.8 is broadly that of Bourbaki (1951); a detailed history of the subject can be found in Vessiot (1910, pp. 108–31).

The result of (2.8.2) for an nth-order linear equation (i.e. (2.8.8) and the remark following it) was proved by Hesse in 1857, while the remarks following (2.8.2) concerning the form of the general solution, again for an nth-order equation, go back to Lagrange and d'Alembert c. 1765 (see Vessiot, 1910, pp. 109–10, 112); the proof of (2.8.2) in the text is that of Bourbaki (1951).

Fundamental systems of solutions of an nth-order equation (cf. the remarks following the proof of (2.8.6) were first considered by Lagrange c.1765; the term 'fundamental system' which we have changed into 'fundamental kernel', was introduced by Fuchs in 1866 (see Vessiot, 1910, p. 110). The equation $X' = A(t) \circ X$ (in matrix form) was first considered by Volterra (1887b). A proof of the finite-dimensional case of (2.8.5) is given by Winter (1946b); the proof in the text is taken from Bellman (1970, pp. 115–6).

Euler and Daniel Bernoulli applied the method of variation of parameters (see (2.8.6)) to particular equations c 1740; the general method was developed by Lagrange c. 1774–5 (see Vessiot, 1910, p. 113). The identity §2.8(8) is due to Liouville (1838); the special case for a second-order equation was obtained by Abel in 1827 (see Vessiot, 1910, p. 117).[†]

The proof of the inequality §2.8 (11) in the text is due to Kitamura (1943); the special case where $b(t) = 0$ was obtained by Tôyama (1940) by a different method. The logarithmic norm was discovered by Lozinskii (1958) and independently by Dahlquist (1959); the name is Lozinskii's. The special cases of (2.8.7) and Exercise 2.8.6 where Y is a Hilbert space (so that $\mu(X)$ is the greatest element of the spectrum of the associated self-adjoint transformation W) were obtained by Ważewski (1948) and Wintner (1950) for finite-dimensional Y and by Wintner (1957a) for infinite-dimensional Y. The corollary improves a result of Ważewski and Wintner (Ważewski, 1948; Wintner, 1950).

Exercise 2.8.3 is due to Bôcher, Dunkel and Wintner (see Wintner, 1957a); Exercises 2.8.4 and 2.8.7 are due to Wintner (1946b).

For the general theory of stability of solutions of differential equations see Bellman (1953) and Cesari (1959); Conti (1965) gives a summary of the theory for linear equations. Exercises 2.8.8–10 are to be found, more or less, in Cesari (1959), pp. 44–5). Exercise 2.8.11 is due to Wintner (1957a). Exercises 2.8.13, 14 are generalizations of the Dini–Hukuhara theorem, the history of which can be found in Cesari (1959, pp. 36–7); Exercise 2.8.13 is the analogue for non-linear equations of a theorem of Bellman (1958) for linear equations; Exercise 2.8.14 is given by Cesari (1959, p. 95). Exercise 2.8.15 is due to Conti (see Cesari, 1959, p. 47).

[†] As Hartman points out, Vessiot incorrectly attributes §2.8(8) to Jacobi, and only the special case §2.8(20) to Liouville.

The operator exp(tA) (§2.9) was defined by Peano (1887a). The proof of (2.9.1) follows the treatment of Hartman (1964, p. 58). Exercise 2.9.3 is due to Perron (1928b).

It has already been remarked in §2.14 that Moigno proved the continuity of the local solution with respect to y_0 in the Lipschitz case. The continuity with respect to (t_0, y_0) of solutions in the Lipschitz case having maximal domain of existence (cf. (2.10.1)) was proved by Bliss (1905). A result of the form of (2.10.4) for solutions reaching to the boundary in the continuous case seems first to have been obtained by Yosie (1926) and Kamke (1928; 1929; 1930b, pp. 86–9, 142–51). Theorem (2.10.3) is also essentially due to Kamke (1932). The proofs of (2.10.1, 4) given in the text may be new. The extension lemma (2.10.2) is given by Graves (1927b); it was rediscovered in the 1960s (see the references in Flett (1974) who provides a generalization). Theorem (2.10.5) (or, more accurately (2.10.5. Corollary)) is a theorem of Kneser (1923); the proof in the text is due to Müller (1928a) (for further references see Pugh (1964) and Sell (1965)). Exercise 2.10.1 is due to Montel (1926).

References to uniqueness theorems and, in particular, to conditions (I–K) and (P) of §2.11 and to (2.11.1, 2) and Exercises 2.11.1, 2 have already been given in §2.14. Exercises 2.11.3, 4 improve a result of W. Walter (1970, p. 83), and Exercise 2.11.7 is new. The special case $c = 1$ of Exercise 2.11.8 is given by Peano (1890); Nagumo (1926) mentions the general case, but proves only the 'if' in part (iii).

Theorem (2.11.3) with condition (6) is due to Ważewski (1960c), and with condition (7) to Flett (1972b); the proof of both parts given here is that of Flett. Lemma (2.11.4) is implicit in Perron (1915); (2.11.5) is due to Flett (1972b).

Theorem (2.12.1) is due to Ważewski (1960a); the result implicit in the proof of (2.12.2) is due to Olech (1960) (see also W. Walter, 1964). Exercise 2.12.1 is due to Müller (1926).

Theorem (2.13.1) is discussed in §2.14.

Chapter three

The results of §§3.1–3.6 in the main have been discussed in the section (§4.7) on the history of the idea of differential. Perhaps attention should be called to the form of Taylor's theorem given in (3.6.3), which is from Flett (1972a). The theory of unconstrained maxima and minima for functions with domains in Banach spaces is well known; it can be found in Cartan (1967), while Vainberg (1956) gives a slightly weaker condition than that of (3.6.5), though only for reflexive spaces.

It was pointed out in §4.7 that' the standard implicit function theorem (3.8.1) was due to Hildebrandt and Graves (1927). The more general, and less well known, theorem (3.8.3) was in fact proved by Graves in the same year (Graves, 1927b).

The function G of (3.9.1) has been widely studied under the name of the superposition operator or Nemytskii operator (the latter being used more frequently in the Russian literature), while the integral operator of (3.9.2) is usually treated as a special case of the Hammerstein operator. These mappings are often considered between L_p-spaces, where there are substantial technical difficulties, although the form of their differentials is similar (see Vainberg, 1956; or Krasnosel'skii *et al.*, 1966), but the simple case considered in the text may be found as an example in Rall (1969) and in the second edition of Liusternik and Sobolev (1965).

The results of §3.10 on the dependence of the solutions of differential equations on parameters are in principle well known (Cartan, 1967; Dieudonné, 1960) although some details—in particular, the variation of the maximal interval of definition of the solution—are less well established. Results of the latter kind for finite-dimensional spaces are in Hartman (1964).

The reader interested in the long history of the calculus of variations can consult the German encyclopaedia articles of Kneser (1900) and Zermelo and Hahn (1904) or that in the French encyclopaedia by Lecat (1913). For work in the first few decades of this century, Carathéodory (1935) provides an extensive bibliography with comments; the English translation contains a few additional references up to 1964. In the present book, the aim has been simply to establish the most basic results of the theory in Banach spaces. This was done by Goldstine (1942) who proved theorems (3.11.1), (3.11.2), (3.11.6) and (3.11.8) among others.

The Kantorovich theorem (3.12.1), extending Newton's method to Banach spaces, is of course due to Kantorovich (1948). He, in fact, used the boundedness of the second derivative in place of the assumption that df is Lipschitzian; Fenyö (1954) first made the latter modification. Also, Kantorovich's error estimate was weaker; the present ones (which are sharp) were given by Dennis (1969). The proof in the text follows Ortega (1968) and Tapia (1971). Bartle (1955) observed that considerable flexibility could be allowed in replacing $df(x_n)$ by operators T_n, as in (3.12.3), provided these satisfied certain conditions. Rheinboldt (1968) considers a more general process, but the applications he gives to the Newton–Kantorovich situation are all covered by (3.12.3).

Chapter four

The basic results about Gâteaux and Hadamard differentials are covered in §4.7. The notion of subvariation was introduced by Moreau (1963) and independently by Minty (1964), who use (4.2.12)(iii) as the defining condition. The idea of relating the upper Hadamard variation to the subvariation is new. For the use of the subvariation in convexity theory, see Moreau (1967).

The definition of tangent cone was given by Bouligand (1932). The results of §4.3 are improvements of theorems in Flett (1967) as the use of the Hadamard differential allows more precise characterizations to be given.

Theorem (4.4.1) and its first corollary are from Flett (1967), again with the improvement from Fréchet to Hadamard differentiability. The theory of Lagrange multipliers (4.4.1. Corollary 2) has a long history. The first results for infinite-dimensional spaces were obtained in special situations in the calculus of variations; the first general theorems in this area are due to Liusternik (1934) and Goldstine (1938). The development in the text follows Flett (1966) who discussed the finite-dimensional case.

The exposition of §4.5 follows in a general way that of Girsanov (1970). The inspiration for the theory comes from the work of Kuhn and Tucker (1951), although their results were anticipated by John (1948). They were concerned with the finite-dimensional case (a good exposition of which may be found in Rockafellar (1970)). Ritter (1967) established the basic results ((4.5.8. Corollary 2) and (4.5.9)) for Fréchet differentiable functions on Banach spaces. The general theory developed here is the invention of Dubovitskii and Milyutin (1963) who discovered the fundamental theorem (4.5.4). The application of subvariations in this context was first made by Laurent (1969), who proved results similar to (4.5.8) in special cases.

The theorem of Lagrange mentioned at the beginning of §4.6 can be found in his *Mécanique Analytique* (1811). Lyapunov (1892) first used comparison functions in the way they are employed here; he proved the basic theorem of the form of (4.6.5), although of course his version is much less sophisticated. Corduneanu (1960) and Brauer (1961) have versions of (4.6.5) with the comparison function g taken to be zero; Antosiewicz (1962) introduces a non-zero comparison function, while Lakshmikantham (1962a) has an additional innovation in that he uses upper derivatives. The first step in the proof, (4.6.1), is in Conti (1956) for finite-dimensional spaces. The approach via the sets $E(t,r)$ and the theorem (4.6.2) appear to be new. The existence theorems (4.6.8,9) and the uniqueness theorem (4.6.7) are essentially Murakami's (1966); the new feature is the introduction of a non-zero comparison function. Murakami's results can also be found in the book by Lakshmikantham and Leela (1969).

Appendix

The material in the appendix is on the whole standard in functional analysis, and the reader who browses through a few books in the field will find most of it. In particular, it is nearly all in the encyclopaedic work of Dunford and Schwartz (1958)–often in much greater generality. However, we shall give more detailed references for the reader who prefers this.

The elementary theory of metric and of normed spaces is in Dieudonné (1960, ch. 3,5), where (A.1.1–4) and (A.2.5,6) can be found. Liusternik and Sobolev (1965, §1.7) give (A.1.6). For (A.2.1–4) see Dunford and Schwartz (1958, ch. 2, §2) or Taylor (1958, §4.2). Taylor also gives (A.2.7) in his §6.2; this result may also be found in Liusternik and Sobolev (1965, §7.1).

General theorems on convexity are given in Rockafellar (1970); he restricts his discussion to finite dimensions, but the proofs are usually the same. The Krein–Milman theorem (A.3.3) and the related results (A.3.4) and (A.3.5)(i) are in Dunford and Schwartz (1958, ch. 5, §8).

For the various forms of the Hahn–Banach theorem in §A.4 and (A.5.1,2) see Bourbaki (1955, ch. 2) or Kelley and Namioka (1963, ch. 1, §3). The theory of polar sets in Bourbaki (1955, ch. 4) gives (A.5.3–5). Theorems (A.5.8,9) are in Girsanov (1970, Lecture 5).

For §A.6 suitable references are Dinculeanu (1967, §6) and Bourbaki (1965, particularly ch. 4, §3 and §5).

BIBLIOGRAPHY

Alexiewicz, A. (1951). On a theorem of Ważewski. *Ann. Soc. Polonaise Math.*, 24 (1951), 129–31.

Alexiewicz, A. and Orlicz, W. (1955). On a theorem of Carathéodory. *Ann. Polonici Math.* 1 (1955), 414–17.

Ampère, A. M. (1806). Recherches sur quelques points de la théorie des fonctions dérivées qui conduisent à une nouvelle démonstration de la série de Taylor, et à l'expression finie des termes qu' on néglige lorsqu' on arrête cette série à un terme quelconque. *J. de l'École Polytechnique*, 14 (1806), 148–81.

Antosiewicz, H. A. (1962). An inequality for approximate solutions of ordinary differential equations. *Math. Z.* 78 (1962), 44–52.

Arzelà, C. (1895). Sull'integrabilità delle equazioni differenziali ordinarie. *Memorie Inst. Bologna* (5), 5 (1895–6), 257–70.

– (1896). Sull'esistenza degli integrali nelle equazioni differenziali ordinarie. *Memorie Inst. Bologna* (5), 6 (1896–7), 131–40.

Averbukh, V. I. and Smolyanov. O. G. (1967). The theory of differentiation in linear topological spaces. *Uspehi Mat. Nauk.* 22 No. 6 (1967), 201–60, Russian; *Russian Math. Surveys*, 22 No. 6 (1967), 201–58, English translation.

Aziz, A. K. and Diaz, J. B. (1963a). On a mean-value theorem of the differential calculus of vector-valued functions, and uniqueness theorems for ordinary differential equations in a linear normed space. *Contrib. to Diff. Eqns.* 1 (1963), 251–69.

– (1963b). On a mean-value theorem of the weak differential calculus of vector-valued functions. *Contrib. to Diff. Eqns.* 1 (1963), 271–3.

Aziz, A. K., Diaz, J. B. and Mlak, W. (1966). On a mean value theorem for vector-valued functions, with applications to uniqueness theorems for 'right-hand-derivative' equations. *J. Math. Anal. Appl.* 16 (1966), 302–7.

Baiada, E. (1947). Confronto e dipendenza dei parametri degli integrali delle equazioni differenziali. *Atti Accad. Naz. Lincei Rend.* (8), 3 (1947), 258–71.

Banach, S. (1925). Sur les lignes rectifiables et les surfaces dont l'aire est fini. *Fund. Math.* 7 (1925), 225–36.

Barbashin, E. A. (1970). *Introduction to the Theory of Stability* (Wollers-Noordhoff, Groningen, 1970).

Bartle, R. G. (1955). Newton's method in Banach spaces. *Proc. Amer. Math. Soc.* 6 (1955), 827–31.

Bellman, R. (1943). The stability of solutions of differential equations. *Duke Math. J.* 10 (1943), 643–7.

– (1953). *Stability Theory of Differential Equations* (McGraw-Hill, New York, 1953).

– (1958). On a generalization of a result of Wintner. *Quart. Appl. Math.* 16 (1958), 431–2.

– (1970). *Methods of Non-linear Analysis*, vol. 1 (Academic Press, New York, 1970).

Bendixson, I. (1893). Sur le calcul des intégrales d'un systéme d'équations différentielles par des approximations successifes. *Öfversigt Vetensk. Akad. Förhandl. (Stockholm)*, 50 (1893), 599–612.

– (1897) *Öfversigt Vetensk. Akad. Förhandl. (Stockholm)*, 54 (1897) 617.

Bieberbach, L. (1934). *Lehrbuch der Funktionentheorie*, vol. 1 (Teubner, Leipzig, 1934).

Bihari, I. (1956). A generalization of a lemma of Bellman and its application to uniqueness problems of differential equations. *Acta Math. Acad. Sci. Hungar.* 7 (1956), 81–94.

Bliss, G. A. (1905). The solutions of differential equations of the first order as functions of their initial values. *Ann. of Math.* 6 (1904–5), 49–68.

Boas, R. P. (1960). *A Primer of Real Functions* (Wiley, New Jersey, 1960).

Bolzano, B. (1817). Rein analytischer Beweis des Lehrsatzes, dass zwischen je zwey Werthen, die ein entgegengesetzes Resultat gewähren, wenigstens eine reelle Wurzel der Gleichung liege. *Abh. Gesell. Wiss. Prague*, (3) 5 (1814–7), 1–60 (separate pagination); reprinted in *Ostwald's Klassiker*, No. 153, ed. P. E. B. Jourdain (Engelmann, Leipzig, 1905).

Bompiani, E. (1925). Un teorema di confronto ed un teorema di unicità per l'equazione differenziale $y' = f(x, y)$. *Atti Accad. Naz. Lincei Rend.* (6), 1 (1925), 298–302.

Bouligand, G. (1932). *Introduction à la Géométrie Infinitésimale Directe* (Paris, 1932).

Bourbaki, N. (1949). *Éléments de Mathématique*, vol. 4, *Fonctions d'une Variable Réelle*, chs. 1–3 (Hermann, Paris, 1949).

– (1951). *Éléments de Mathématique*, vol. 4, *Fonctions d'une Variable Réelle*, chs. 4–7 (Hermann, Paris, 1951).

– (1955). *Éléments de Mathématique*, vol. 5 *Espaces Vectoriels Topologiques* (Hermann, Paris, 1955).

– (1965), *Éléments de Mathématique*, vol. 6 *Intégration*, chs. 1–4 (Hermann, Paris, 1965).

Boyer, C. B. (1959). *The History of the Calculus and its Conceptual Development.* (Dover, New York, 1959).

Brauer, F. (1961). Global behavior of solutions of ordinary differential equations. *J. Math. Anal. Appl.* 2 (1961), 145–58.

– (1963). Bounds for solutions of ordinary differential equations. *Proc. Amer. Math. Soc.* 14 (1963), 36–43.

– (1965). The use of comparison theorems for ordinary differential equations. *Proc. NATO Advanced Study Inst.*, Padua, 1965, pp. 29–50 (Oderisi, Gubbio, Italy, 1965).

Bruckner, A. M. and Leonard, J. L. (1966). Derivatives. *Amer. Math. Monthly*, 73 (1966), 24–56.

Carathéodory, C. (1918, 1927). *Vorlesungen über reele Funktionen* (Teubner, Leipzig, 1st edn 1918, 2nd edn 1927).

– (1935). *Variationsrechnung und partielle Differentialgleichungen erster Ordnung* (Teubner, Berlin, 1935); (Holden-Day, San Francisco, 1967), English translation.

Cartan, H. (1967). *Calcul Différentiel* (Hermann, Paris, 1967); (Kershaw, London, 1971), English translation.

Cauchy, A. (1821). *Cours d'Analyse de l'École Royale Polytechnique; Première Partie, Analyse Algébrique* (Paris, 1821); *Oeuvres Complètes*, 2nd series, vol. 3 (Paris, 1897).

– (1823). *Résumé des Leçons données à l'École Royale Polytechnique sur le Calcul Infinitésimal*, vol. 1 (Paris, 1823); *Oeuvres Complètes*, 2nd series, vol. 4 (Paris, 1899).

– (1829). *Leçons sur le Calcul Différentiel* (Paris, 1829); *Oeuvres Complètes*, 2nd series, vol. 4 (Paris, 1899).

– (1840). *Mémoire sur l'Intégration des équations Différentielles, Exercices d'Analyse et de Physique Mathématique*, vol. 1 (Paris, 1840); *Oeuvres Complètes*, 2nd series, vol. 11 (Paris, 1913). The memoir was originally lithographed in Prague in 1835.

Cesari, L. (1959). *Asymptotic Behavior and Stability Problems in Ordinary Differential Equations* (Ergebnisse der Mathematik und ihrer Grenzgebiete, New Series, 16, Springer, Berlin, 1959).

Coddington, E. A. and Levinson, N. (1955). *Theory of Ordinary Differential Equations* (McGraw-Hill, New York, 1955).

Conti, R. (1956). Sulla prolungabilità delle soluzioni di un sistema di equazioni differenziali ordinarie. *Boll. Un. Mat. Ital.* 11 (1956), 510–14.

– (1965). On linear affine and linear differential equations. *Proc. NATO Advanced Study Institute*, Padua, 1965 pp. 1–18. (Oderisi, Gubbio, Italy, 1965).

Corduneanu, C. (1960). The application of differential inequalities to the theory of stability. *An. Sti. Univ. 'Al. I. Cuza' Iasi*, Sect. 1, 6 (1960), 47–58, Russian.

– (1964). Sur les inégalités différentielles. *Mathematika*, 6 (29) (1964), 31–3.

Coriolis, G. (1837). Sur le degré d'approximation qu'on obtient pour les valeurs numériques d'une variable qui satisfait à une équation différentielle, en employant pour calculer ces valeurs diverses équations aux différences plus ou moins approchées. *J. Math. Pures Appl.* (1), 2 (1837), 229–44.

Courant, R. and Hilbert, D. (1930). *Methoden der mathematischen Physik*, vol. 1, Grundlehren der mathematischen Wissenschaften 11, 2nd edn (Springer, Berlin, 1930); *Methods of Mathematical Physics*, vol. 1 (Interscience, New York, 1953) modified English translation.

– (1953). *Methods of Mathematical Physics*, vol. 1 (Interscience, New York, 1953).

Dahlquist, G. (1959). Stability and error bounds in the numerical integration of ordinary differential equations. *Kungl. Tekn. Högsk. Handl. Stockholm*, 130 (1959), 87 pp.

Darboux, G. (1875). Mémoire sur les fonctions discontinues. *Ann. de l'École Norm. Sup.* (2) 4 (1875), 57–112.

– (1876). Sur les développements en série des fonctions d'une seule variable. *J. de Math.* (3) 2 (1876), 291–312.

Dennis, J. E. (1969). On the Kantorovich hypothesis for Newton's method. *SIAM J. Numer. Anal.* 6 (1969), 493–507.

Dieudonné, J. (1960). *Foundations of Modern Analysis* (Academic Press, New York, 1960).

Dinculeanu, N. (1967). *Vector Measures* (Pergamon, Oxford, 1967).

Dini, U. (1878). *Fundamenti per la Teorica delle Funzioni di Variabili Reali* (Pisa, 1878).

Dubovitskii, A. Ya. and Milyutin, A. A. (1963). The extremum problem in the presence of constraints. *Dokl. Akad. Nauk. SSSR*, 149 (1963), 759–62, Russian; *Sov. Math. Dokl.* 4 (1963), 452–5, English translation.

Dunford, N. and Schwartz, J. T. (1958). *Linear Operators*, Part 1, *General Theory* (Interscience, New York, 1958).

Fan, Ky (1942). Sur quelques notions fondamentals de l'analyse générale. *J. Math. Pures Appl.* 21 (1942), 289–368.

Fenyö, I. (1954). Über die Lösung der im Banachschen Raum definierten nichtlinearen Gleichungen. *Acta Math. Acad. Sci. Hungar.* 5 (1954), 85–93.

Flett, T. M. (1955). Some remarks on schlicht functions and harmonic functions of uniformly bounded variation. *Quart. J. Math.* 6 (1955), 59–72.

– (1957). The definition of a tangent to a curve. *Edinburgh Math. Notes*, 41 (1957), 1–9.

– (1966). *Mathematical Analysis* (McGraw-Hill, London, 1966).

– (1967). On differentiation in normed spaces. *J. London Math. Soc.* 42 (1967), 523–33.

– (1972a). Mean value theorems for vector-valued functions. *Tôhoku Math. J.* (2) 24 (1972), 141–51.

– (1972b). Some applications of Zygmund's lemma to non-linear differential equations in Banach and Hilbert spaces. *Studia Math.* 44 (1972), 335–44 and addendum 649–50.

– (1974). Extensions of Lipschitz functions. *J. London Math. Soc.* (2) 7 (1974), 604–8.

Fréchet, M. (1911). Sur la notion de différentielle. *C. R. Acad. Sci. Paris*, 152 (1911), 845–7 and 1050–1

– (1925a). La notion de différentielle dans l'analyse générale. *C. R. Acad. Sci. Paris*, 180 (1925), 806–9.

– (1925b). La notion de différentielle dans l'analyse générale. *Ann. Sci. École Norm. Sup.* (3) 42 (1925), 293–323.

– (1937). Sur la notion de différentielle. *J. Math. Pures Appl.* 16 (1937) 233–50.

Frölicher, A. and Bucher, W. (1966). *Calculus in Vector Spaces without Norm, Lecture Notes in Mathematics* 30 (Springer, Berlin, 1966).

Fukuhara, M. (1928a). Sur le théorème d'existence des intégrales des équations différentielles du premier ordre. *Jap. J. Math.* 5 (1928), 239–51.

– (1928b). Sur les systèmes des équations différentielles ordinaires. *Jap. J. Math.* 5 (1928), 345–50.

– (1928c). Sur les systèmes des équations différentielles ordinaires. *Proc. Imp. Acad. Japan*, 4 (1928), 448–9.

Gâteaux, M. R. (1913). Sur les fonctionnelles continues et les fonctionnelles analytiques. *C. R. Acad. Sci. Paris*, 157 (1913), 325–7.

– (1922). Sur diverses questions de calcul fonctionnel. Premier Mémoire: Sur les fonctionnelles continues et les fonctionnelles analytiques. *Bull. Soc. Math. France*, 50 (1922), 1–21.

Genocchi, A. (1884). *Calcolo Differenziale e Principii di Calcolo Integrale*, Pubblicato con Aggiunte da G. Peano (Turin, 1884) (see G. Peano, *Opere Scelte*, vol.1, p. 60 (Rome, 1957)).

Ghizzetti, A. (1969). *Theory and Application of Monotone Operators* (Oderisi, Gubbio, Italy, 1969).

Gilbert, Ph. (1884). Lettre de M. le prof. Ph. Gilbert. *Nouvelles Ann. de Math.* Serie 3, A (1884), 153–5 (reprinted in G. Peano, *Opere Scelte*, vol. 1, pp. 41–2 (Rome, 1957)).

Girsanov, I. V. (1970). *Lectures on Mathematical Theory of Extremum Problems* (Moscow University, 1970), Russian; *Lecture Notes in Economics and Mathematical Systems*, 67 (Springer, Berlin, 1972), English translation.

Giuliano, L. (1940). Sull'unicità delle soluzioni dei sistemi di equazioni differenziali ordinarie *Boll. Un. Mat. Ital.* (2) 2 (1940), 221–7.

Goldstine, H. H. (1938). A multiplier rule in abstract spaces. *Bull. Amer. Math. Soc.* 44 (1938), 388–94.

– (1942). The calculus of variations in abstract spaces. *Duke Math. J.* 9 (1942), 811–22.

Gollwitzer, H. E. (1969). A note on a functional inequality. *Proc. Amer. Math. Soc.* 23 (1969), 642–7.

Golomb, M. (1935). Zur Theorie der nichtlinearen Integralgleichungen,

Integralgleichungssysteme und allgemeinen Funktionalgleichungen. *Math. Z.* 39 (1935), 45–75.

Grassmann, H. (1847). *Geometrische Analyse* (Leipzig, 1847); *Gesammelte mathematische und physikalische Werke*, vol. 1, pp. 320–99 (Leipzig, 1894).

– (1862). *Die Ausdehnungslehre*, 2nd edn (Berlin, 1862); *Gesammelte Mathematische und Physikalische Werke*, vol. 1, part 2, pp. 1–383 (Leipzig, 1896).

Grattan-Guinness, I. (1970). *The Development of the Foundations of Mathematical Analysis from Euler to Riemann* (MIT, Cambridge, Mass., 1970).

Graves, L. M. (1927a). Riemann integration and Taylor's theorem in general analysis. *Trans. Amer. Math. Soc.* 29 (1927), 163–77.

– (1927b). Implicit functions and differential equations in general analysis. *Trans. Amer. Math. Soc.* 29 (1927), 514–52.

Gronwall, T. H. (1919). Note on the derivatives with respect to a parameter of the solutions of a system of differential equations. *Ann. of Math.* 20 (1919), 292–6.

Grunsky, H. (1961). Ein konstruktiver Beweis für die Lösbarkeit der Differential-gleichung $y' = f(x,y)$ bei stetigem $f(x, y)$. *Jber. Deutsch. Math.-Verein*, 63 (1961), 78–84.

Hadamard, J. (1914). Sur le module maximum d'une fonction et de ses dérivées. *Bull. Soc. Math. France*, C. R. des Séances (1914), 68–72.

– (1923). La notion de différentielle dans l'enseignement. *Scripta Univ. Ab. Bib. Hierosolymitanarum, Jerusalem*, 1 (1923), 3.

Halperin, I. (1954). A fundamental theorem of the calculus. *Amer. Math. Monthly*, 61 (1954), 122–3.

Hartman, P. (1964). *Ordinary Differential Equations* (Wiley, New York, 1964).

Heine, E. (1870). Ueber trigonometrische Reihen. *J. für Math.* 71 (1870), 353–65.

– (1872). Die Elemente der Funktionenlehre. *J. für Math.* 74 (1872), 172–88.

Hermite, C. (1891). *Cours de M. Hermite rédigé en 1882 par M. Andoyer*, 3rd edn (Paris, 1891).

Hildebrandt, T. H. and Graves, L. M. (1927). Implicit functions and their differentials in general analysis. *Trans. Amer. Math. Soc.* 29 (1927), 127–53.

Hukuhara, M. (1940a). Sur la fonction $S(x)$ de M. E. Kamke. *Japan. J. Math.* 17 (1940), 289–98.

– (1940b, 1941). Théorèmes fondamentaux de la théorie des équations différentielles ordinaires, I, II. *Mem. Fac. Sci. Kyūsyū Univ.* A 1 (1940), 111–27; 2 (1941), 1–25.

– (1959). Théorèmes fondamentaux de la théorie des équations différéntielles ordinaires dans l'espace vectoriel topologique. *J. Fac. Sci. Univ. Tokyo, Sect. I Math. Astr. Phys. Chem.* 8 (1959–60), 111–38.

Iyanaga, S. (1928). Uber die Unitätsbedingungen der Lösung der Differentialgleichung $dy/dx = f(x,y)$. *Japan. J. Math.* 5 (1928), 253–7.

John, F. (1948). Extremum problems with inequalities as subsidiary conditions. *Studies and Essays, Courant Anniversary Volume* (Interscience, New York, 1948).

Jordan, C. (1882). *Cours d'Analyse*, 1st edn, vol. 1 (Paris, 1882).

– (1887). *Cours d'Analyse*, 1st edn, vol. 3 (Paris, 1887).

Kamke, E. (1928). Zur Theorie der Differentialgleichung $y' = f(x,y)$. *Acta Math.* 52 (1928), 327–39.

– (1929). Zur Theorie der Systeme gewöhnlicher Differentialgleichungen. *J. für Math.* 161 (1929), 194–8.

344   *Bibliography*

- (1930a). Über die eindeutige Bestimmtheit der Integrale von Differentialgleichungen. *Math. Z.* 32 (1930), 101–7.
- (1930b). *Differentialgleichungen reeller Funktionen* (Leipzig, 1930).
- (1930c). Über die eindeutige Bestimmtheit der Integrale von Differentialgleichungen II. *Sitz.-ber. Heidelberg Akad. Wiss.*, Math.-Naturw. Kl., 17 Abhandl. (1930).
- (1932). Zur Theorie der Systeme gewöhnlicher Differentialgleichungen II. *Acta Math.* 58 (1932), 57–85.

Kantorovich, L. V. (1948). Functional analysis and applied mathematics. *Uspekhi Mat. Nauk.* 3, 6 (28) (1948), 89–185, Russian; *Nat. Bur. Standards*, Washington DC. (1953), English translation.

Kelley, J. L. and Namioka, I. (1963). *Linear Topological Spaces* (Van Nostrand, Princeton, 1963).

Kitamura, T. (1943). Some inequalities on a system of solutions of linear simultaneous differential equations. *Tôhoku Math. J.* 49 (1943), 308–11.

Kneser, A. (1900). Variationsrechnung. *Encyklopädie der math. Wiss.* (Teubner, Leipzig, 1899–1916).

Kneser, H. (1923). Über die Lösungen eines Systems gewöhnlicher Differential- gleichungen das der Lipschitzschen Bedingung nicht genügt. *Sitz.-ber. der Preussischen Akad. der Wiss. Phys.-Math. Kl.* (1923), 171–4.

Kowalewski, G. (1900). Einige Bemerkungen zur Theorie der stetigen Funktionen einer reellen Veränderlichen. *Berichte der Sächsischen Gesellschaft (Akademie) der Wissenschaften zu Leipzig:* Math. Phys. Kl. 52 (1900), 214–19.

Krasnosel'skii, M. A., Zabreiko, P. P., Pustyl'nik E. I. and Sobolevskii, P. E. (1966). *Integral Operators in Spaces of Summable Functions* (Nauka, Moscow, 1966), Russian; (Noordhoff International, Leiden, 1976), English translation.

Kuhn, H. W. and Tucker, A. W. (1951). *Nonlinear Programming.* Proceedings of the Second Berkeley Symposium on Mathematical Statistics and Probability (Univ. California Press, Berkeley, 1951).

Lagrange, J. L. (1797). *Théorie des Fonctions Analytiques*, 1st edn (Paris, 1797).
- (1801). *Leçons sur le Calcul des Fonctions*, 1st edn (Recueil des leçons de l'École Normale, Paris, 1801); reprinted in *J. de l'École Polytechnique*, vol. 12 (1804). 2nd edn (Recueil des leçons de l'École Normale, Paris, 1806); reprinted in *Oeuvres*, vol. 10 (Paris, 1867–92).
- (1811). *Mécanique Analytique*, vol. 1, 2nd edn (Courier, Paris, 1811).

Lakshmikantham, V. (1962a). Differential systems and extensions of Lyapunov's methods. *Michigan Math. J.* 9 (1962), 311–20.
- (1962b). Upper and lower bounds of the norm of solutions of differential equations. *Proc. Amer. Math. Soc.* 13 (1962), 615–16.

Lakshmikantham, V. and Leela, S. (1969). *Differential and Integral Inequalities* (Academic Press, New York, 1969).

Langenhop, C. E. (1960). Bounds on the norm of a solution of a general differential equation. *Proc. Amer. Math. Soc.* 11 (1960), 795–9.

La Salle, J. (1949). Uniqueness theorems and successive approximations. *Ann. Math.* 50 (1949), 722–30.

Laurent, P-J. (1969). Cônes de déplacements admissables et approximation convexe dans un espace normé. *Theory and Application of Monotone Operators*, ed. A. Ghizzetti, pp. 265–74, (Oderisi, Gubbio, Italy, 1969).

Lavrentieff, M. (1925). Sur une équation différentielle du premier ordre. *Math. Z.* 23 (1925), 197–209.

Lebesgue, H. (1904). *Leçons sur l'Intégration et la Recherche des Fonctions Primitives* (Paris, 1904).

Lecat, M. (1913). Calcul des Variations, *Encylopédie des Sci. Math. Pures et Appl.* vol. 2.6.1, pp. 1–288 (Paris, 1913).

Levi, B. (1906a). Ricerche sulle funzioni derivate. *Atti R. Accad. Lincei Rend.* (5) 15 I (1906), 551–8.

– (1906b). Ricerche sopra le funzioni derivate. *Atti R. Accad. Lincei Rend.* (5) 15 II (1906), 674–84.

Levinson, N. (1949). On stability of nonlinear systems of differential equations. *Colloq. Math.* 2 (1949), 40–5.

Lévy, P. (1922). *Leçons d'Analyse Fonctionnelle* (Gauthier-Villars, Paris, 1922).

– (1934). Sur une géneralisation du théorème de Rolle. *C. R. Acad. Sci. Paris*, 198 (1934), 424–5.

– (1951), *Problèmes Concrets d'Analyse Fonctionnelle* (Gauthier-Villars, Paris, 1951).

Lindelöf, E. (1894). Sur l'application des méthodes d'approximations successives à l'étude des intégrales réelles des équations différentielles ordinaires. *J. Math. Pures Appl.* (4) 10 (1894), 117–28.

Liouville, J. (1838). Sur la théorie de la variation des constants arbitraires. *J. Math. Pures Appl.* (1) 3 (1838), 342–9.

Lipschitz, R. (1868). Disamina della possibilità d'integrare completamente un dato sistema di equazioni differenziali ordinarie. *Ann. Mat. Pura Appl.* (2) 2 (1868–9), 288–302.

– (1876). Sur la possibilité d'intégrer complètement un système donné d'équations différentielles. *Bull. Sci. Math.* (1) 10 (1876), 149–59.

Liusternik, L. A. (1934). On conditional extrema of functionals. *Mat. Sb.* 41 (1934), 390–401, Russian.

Liusternik, L. A. and Sobolev, V. I. (1965). *Elements of Functional Analysis*, 2nd edn (Nauka, Moscow, 1965), Russian; (Wiley, New York, 1974), English translation.

Lozinskii, S. M. (1958). Error estimates for the numerical integration of ordinary differential equations I. *Izv. Vyss. Ucebn. Zaved. Matematika*, 5 (1958), 52–90, Russian.

Lüroth, J. (1873). Bemerkung uber gleichmässige Stetigkeit. *Math. Ann.* 6 (1873), 319–20.

Lyapunov, A. M. (Liapounoff, A. M.) (1892). *Problème générale de la stabilité du mouvement* (Kharkov, 1892), Russian; *Ann. Fac. Sci. Univ. Toulouse* (2) 9 (1907), 203–475, French translation; the latter reprinted in *Ann. Math. Studies*, 17 (Princeton, 1947).

McLeod, R. M. (1965). Mean value theorems for vector valued functions. *Proc. Edinburgh Math. Soc.* (2) 14 (1964–5), 197–209.

McShane, E. J. (1939). On the uniqueness of the solutions of differential equations. *Bull. Amer. Math. Soc.* 45 (1939), 755–7.

Mie, G. (1893). Beweis der Integrirbarkeit gewöhnlicher Differentialgleichungssysteme nach Peano. *Math. Ann.* 43 (1893), 553–68.

Minty, G. J. (1964). On the monotonicity of the gradient of a convex function. *Pacific J. Math.* 14 (1964), 243–7.

Mlak, W. (1957). Note on the mean value theorem. *Ann. Polon. Math.* 3 (1957), 29–31.

Moigno, F. N. M. (1840, 1844). *Leçons de Calcul Différentiel et de Calcul Intégral*,

Rédigées d'après les Méthodes de M. A.-L. Cauchy, 2 vols. (Bachelier, Paris, 1840, 1844).

Montel, P. (1907). Sur les suites infinies des fonctions. *Ann. Sci. École Norm. Sup.* (3) 24 (1907), 233–334.

– (1926). Sur l'intégrale supérieure et l'intégrale inférieure d'une équation différentielle. *Bull. Sci. Math.* (2) 50 (1926), 205–17.

Moreau, J. J. (1963). Propriétés des applications 'prox'. *C. R. Acad. Sci. Paris*, 256 (1963), 1069–71.

– (1967). Sous-différentiabilité. In *Proceedings of the Colloquium on Convexity*, Copenhagen 1965, ed. W. Fenchel, pp. 185–201 (Copenhagen Mat. Inst., 1967).

Müller, M. (1926). Über das Fundamental theorem in der Theorie der gewöhnlichen Differentialgleichungen. *Math. Z.* 26 (1926), 619–45.

– (1927). Über die Eindeutigkeit der Integrale eines Systems gewöhnlicher Differentialgleichungen und die Konvergenz einer Gattung von Verfahren zur Approximation dieser Integrale. *Sitz.-ber. Heidelberg Akad. Wiss.*, Math.-Naturw. Kl. (1927), 9, pp. 1–38.

– (1928a). Beweis eines Satzes des Herrn H. Kneser über die Gesamtheit der Lösungen, die ein System gewöhnlicher Differentialgleichungen durch einen Punkt schickt. *Math. Z.* 28 (1928), 349–55.

– (1928b). Neuere Untersuchungen über den Fundamentalsatz in der Theorie der gewöhnlichen Differentialgleichungen. *Jber. Deutsch. Math.-Verein.* 37 (1928), 33–48.

Murakami, H. (1966). On non-linear ordinary and evolution equations. *Funkcial. Ekvas.* 9 (1966), 151–62.

Nagumo, M. (1926). Eine hinreichende Bedingung für die Unität der Lösung von Differentialgleichungen erster Ordnung. *Japan. J. Math.* 3 (1926), 107–12.

Nashed, M. Z. (1971). Differentiability and related properties of nonlinear operators: some aspects of the role of differentials in nonlinear functional analysis. *Nonlinear Functional Analysis and Applications*, ed. L. B. Rall, pp. 103–309 (Academic Press, New York, 1971).

Newman, M. H. A. (1951). *Elements of the Topology of Plane Sets of Points*, 2nd edn (Cambridge, 1951).

Olech, C. (1960). Remarks concerning criteria for uniqueness of solutions of ordinary differential equations. *Bull. Acad. Polon. Sci. Ser. Sci. Math. Astr. Phys.* 8 (1960), 661–6.

Ortega, J. M. (1968). The Newton-Kantorovich theorem. *Amer. Math. Monthly*, 75 (1968), 658–60.

Osgood, W. F. (1898). Beweis der Existenz einer Lösung der Differentialgleichung $dy/dx = f(x, y)$ ohne Hinzunahme der Cauchy–Lipschitz'schen Bedingung. *Monatsh. Math. Phys.* 9 (1898), 331–45.

Painlevé, P. (1910). Existence de l'intégrale générale. Détermination d'une intégrale particulière par ses valeurs initiales. *Encyclopédie des Sci. Math. Pures Appl.*, vol. 2.3.1, pp. 1–57 (Paris, 1910).

Peano, G. (1884a). Extrait d'une lettre de M. le Dr J. Peano. *Nouvelles Ann. Math.* Serie 3, A (1884), 45–7; *Opere Scelte*, vol. 1 (Rome, 1957), pp. 40–1.

– (1884b). Lettre de M. le Dr J. Peano. *Nouvelles Ann. Math.*, Serie 3, A (1884), 252–6; *Opere Scelte*, vol. 1 (Rome, 1957), pp. 43–6.

– (1886). Sull'integrabilità delle equazioni differenziali di primo ordine. *Atti Accad.*

Torino, 21 A (1886), 677–85; *Opere Scelte,* vol. 1, pp. 74–81 (Rome, 1957).

- (1887a). Integrazione per serie delle equazioni differenziali lineari. *Atti Accad. Torino,* 22 A (1887), 437–46; *Opere Scelte,* vol. 1, pp. 82–90 (Rome, 1957). A translation into French appeared as Intégration par séries des équations différentielles linéaires. *Math. Ann.* 32 (1888), 450–6.

- (1887b). *Applicazioni Geometriche del Calcolo Infinitesimale* (Bocca, Turin, 1887).

- (1888). *Calcolo Geometrico Secondo l'Ausdehnungslehre di H. Grassmann, Preceduto dalle Operazioni della Logica Deduttivo* (Bocca, Turin, 1888).

- (1889). Une nouvelle forme du reste dans la formule de Taylor. *Mathesis* 9 (1889), 182–3; *Opere Scelte,* vol. 1 (Rome, 1957) pp. 95–6.

- (1890). Démonstration de l'intégrabilité des équations différentielles ordinaires. *Math. Ann.* 37 (1890), 182–228; *Opere Scelte,* vol. 1, pp. 119–70 (Rome, 1957).

- (1891a). *Gli elementi di calcolo geometrica* (Candeletti, Turin, 1891); *Opere Scelte,* vol. 3, pp. 41–71 (Rome, 1959).

- (1891b). Sulla formola di Taylor. *Atti Accad. Sci. Torino,* 27 A (1891), 40–6; *Opere Scelte,* vol. 1, pp. 204–9 (Rome, 1957).

- (1892a). Sur la définition de la derivée. *Mathesis* (2) 2 (1892), 12–14; *Opere Scelte,* vol. 1, pp. 210–12 (Rome, 1957).

- (1892b). Sur le théorème général relatif à l'existence des intégrales des équations différentielles ordinaires. *Nouvelles Ann. Math.* (3) 11 (1892), 79–82; *Opere Scelte,* vol. 1, pp. 215–17 (Rome, 1957).

- (1896). Saggio di calcolo geometrico. *Atti Accad. Sci. Torino,* 31 A (1895–6), 952–75; *Opere Scelte,* vol. 3, pp. 167–86 (Rome, 1959).

Perron, O. (1915). Ein neuer Existenzbeweis für die Integrale der Differentialgleichung $y' = f(x, y)$. *Math. Ann.* 76 (1915), 471–84.

- (1926) Über Ein- und Mehrdeutigkeit des Integrales eines Systems von Differentialgleichungen. *Math. Ann.* 95 (1926), 98–101.

- (1928a). Eine hinreichende Bedingung für die Unität der Lösung von Differentialgleichungen erster Ordnung. *Math. Z.* 28 (1928), 216–19.

- (1928b). Über Stabilität und asymptotisches Verhalten der Integrale von Differentialgleichungssysteme. *Math. Z.* 29 (1928), 129–60.

Picard, E. (1890). Mémoire sur la théorie des équations aux derivées partielles et la méthode des approximations successives, *J. Math. Pures Appl.* (4) 6 (1890), 145–210.

- (1891). Sur le théorème générale relatif à l'existence des intégrales des équations différentielles ordinaires. *Bull. Soc. Math. France,* 19 (1890–91), 61–4.

- (1893). *Traité d'Analyse,* vol. 2 (Gauthier-Villars, Paris, 1893).

Pringsheim, A. (1900). Zur Geschichte des Tayloschen Lehrsatzes. *Bibl. Math.* (3) 1 (1900), 433.

- (1916).Grundlagen der allgemeinen Functionenlehre. *Encyklopädie Math. Wiss.* vol. 2.1.1, pp. 1–53 (Leipzig, 1899–1916).

Pugh, C. C. (1964). Cross-sections of solution funnels. *Bull. Amer. Math. Soc.* 70 (1964), 580–3.

Rall, L. B. (1969). *Computational Solution of Nonlinear Operator Equations* (Wiley, New York, 1969).

- (1971). *Nonlinear Functional Analysis and Applications* (Academic Press, New York, 1971).

Reid, W. T. (1971). *Ordinary Differential Equations* (Wiley, New York, 1971).

Rheinboldt, W. C. (1968). A unified convergence theory for a class of iterative processes. *SIAM J. Numer. Anal.* 5 (1968), 42–63.

Ritter, K. (1967). Duality for nonlinear programming in a Banach space. *SIAM J. Appl. Math.* 15 (1967), 294–302.

Rockafellar, R. T. (1970). *Convex Analysis* (Princeton, 1970).

Rosenblatt, A. (1909). Über die Existenz von Integralen gewöhnlicher Differentialgleichungen. *Ark. Mat. Astr. Fys.* 5 (1909), No. 2.

Saaty, T. (1967). *Modern Nonlinear Equations* (McGraw-Hill, New York, 1967).

Saks, S. (1937). *Theory of the Integral* (Hafner, New York, 1937).

Scheeffer, L. (1884a). Allgemeine Untersuchungen über Rectification der Curven. *Acta Math.* 5 (1884), 49–82.

– (1884b). Zur Theorie der stetigen Funktionen einer reellen Veränderlichen. *Acta Math.* 5 (1884), 183–94 and 279–96.

Sell, G. R. (1965). On the fundamental theory of ordinary differential equations. *J. Differential Equations*, 1 (1965), 370–92.

Serret, J. -A. (1868). *Cours de Calcul Différentiel et Intégral*, 2 vols. (Paris, 1868).

Sova M. (1964). General theory of differentiability in linear topological spaces. *Czech. Math. J.* 14 (1964), 485–508, Russian.

Stein, P. and Peck, J. E. L. (1955). The differentiability of the norm of a linear operator. *J. London Math. Soc.* 30 (1955), 496–501.

Stokes, A. P. (1959). The application of a fixed-point theorem to a variety of non-linear problems. *Proc. Nat. Acad. Sci. USA*, 45 (1959), 231–5.

Stolz, O. (1893). *Grundzüge der Differential- und Integralrechnung*, vol. 1 (Teubner, Leipzig. 1893).

Szarski, J. (1965). *Differential Inequalities* (P.W.N., Warsaw, 1965).

Tapia, R. A. (1971). The Kantorovich theorem for Newton's method. *Amer. Math. Monthly*, 78 (1971), 389–92.

Taylor, A. E. (1958). *Introduction to Functional Analysis* (Wiley, New York, 1958).

Tonelli, L. (1925). Sull'unicità della soluzione di un'equazione differenziale ordinaria. *Atti Accad. Naz. Lincea Rend.* (6) 1 (1925), 272–7.

Tôyama, H. (1940). Some inequalities in the theory of linear differential equations. *Tôhoku Math. J.* 47 (1940), 210–16.

Vacca, G. (1901). *Bibl. Math.* (3) 2 (1901), 148–9.

Vainberg, M. M. (1956). *Variational Methods for the Study of Nonlinear Operators* (Gostekhizdat, Moscow, 1956), Russian; (Holden-Day, San Francisco, 1964), English translation.

Valiron, G. (1927). Sur les courbes qui admettent une tangente en chaque point. *Nouvelles Ann. Math.* (6) 2 (1927), 46–51.

Vallée Poussin, C. J. de la (1893). Mémoire sur l'intégration des équations différentielles, Mémoires Couronnés et Autres Mémoires. *Acad. Roy. des Sciences, des Lettres et des Beaux-Arts de Belgique*, 47 (1892–3), 82 pp.

– (1909, 1912). *Cours d'Analyse Infinitésimale*, 2nd edn 2 vols. (Paris, 1909, 1912).

Vessiot, E. (1910). Méthodes d'intégration élémentaires. Étude des équations différentielles ordinaires au point de vue formel. *Encyclopédie des Sci. Math. Pures Appl.*, vol. 2.1.3, pp. 58–170 (Paris, 1910).

Viswanatham, B. (1952). On the asymptotic behaviour of solutions of non-linear differential equations. *Proc. Indian Acad. Sci. Sect.* A 36 (1952), 335–42. A proof of the main theorem is reproduced in 'A generalisation of Bellman's lemma.' *Proc. Amer. Math. Soc.* 14 (1963), 15–18.

Volterra, V. (1882). Sui principii del calcolo integrale, *Giorn. Mat.* 19 (1882), 333–72; *Opere Matematiche*, vol. 1, pp. 16–48 (Accad. Nazionale die Lincei, Roma, 1954).

- (1887a). Sopra le funzioni che dipendono da altre funzioni, Nota 1. *Rend. Lincei* Ser. 4, 3 (1887), 97–105; *Opere Matematiche*, vol. 1, pp. 294–302 (Accad. Nazionale die Lincei, Roma, 1954).

- (1887b). Sulle equazioni differenziali lineari, *Rend. Lincei*, Ser. 4a, 3 (1887), 393–6; *Opere Matematiche*, vol. 1, pp. 291–3 (Accad. Nazionale dei Lincei, Roma, 1954).

- (1913). *Lecons sur les Fonctions de Lignes* (Gauthier-Villars, Paris, 1913).

Voss, A. and Molk, J. (1909). Calcul Differentiel. *Encyclopédie des Sci. Math. Pures Appl.* vol. 2.1.2, pp. 242–336 (Paris, 1909).

Walter, J. (1973). On elementary proofs of Peano's existence theorems. *Amer. Math. Monthly*, 80 (1973), 281–5.

Walter, W. (1964). Bemerkungen zu verschiedenen Eindeutigkeitskriterien für gewöhnliche differentialgleichungen. *Math. Z.* 84 (1964), 222–7.

- (1970). *Differential and Integral Inequalities*, Ergebnisse der Math. vol. 55 (Springer, Berlin, 1970).

- (1971). There is an elementary proof of Peano's existence theorem. *Amer. Math. Monthly*, 8 (1971), 170–3.

Ważewski, T. (1948). Sur la limitation des intégrales des systèmes d'équations différentielles linéaires ordinaires, *Studia Math.* 10 (1948), 48–59.

- (1949). Une généralisation des théorèmes sur les accroissements finis au cas des espaces abstraits. Applications. *Bull. Acad. Polon. Sci.* A (1949), 183–5.

- (1951a). Cértaines propositions de caractère 'épidermique' relatives aux inégalités différentielles. *Ann. Soc. Polon. Math.* 24 (1951), 1–12.

- (1951b). Une généralisation des théorèmes sur les accroissements finis au cas des espaces de Banach et application à la généralisation du théorème de l'Hôpital. *Ann. Soc. Polon. Math.* 24 (1951), 132–47.

- (1955). Une modification du théorème de l'Hôpital liée au problème du prolongement des intégrales des équations différentielles. *Ann. Polon. Math.* 1 (1955), 1–12.

- (1960a). Sur une extension du procédé de I. Jungermann pour établir la convergence des approximations successives au cas des équations différentielles ordinaires. *Bull. Acad. Polon. Sci. Ser. Sci. Math. Astr. Phys.* 8 (1960), 213–16.

- (1960b). Sur la dérivabilité de la limite d'une suite de fonctions possédant une dérivée approximative unilatérale (cas de l'espace de Banach). *Bull. Acad. Polon. Sci. Ser. Sci. Math. Astr. Phys.* 8 (1960), 295–9.

- (1960c). Sur l'existence et l'unicité des intégrales des équations différentielles ordinaires au cas de l'espace de Banach. *Bull. Acad. Polon. Sci. Ser. Sci. Math. Astr. Phys.* 8 (1960), 301–5.

Weyl, H. (1946). Comment on the preceding paper (of N. Levinson). *Amer. J. Math.* 68 (1946), 7–12.

Wintner, A. (1945). The non-local existence problem of ordinary differential equations. *Amer. J. Math.* 67 (1945), 277–84.

- (1946a). The infinities in the non-local existence problem of ordinary differential equations. *Amer. J. Math.* 68 (1946), 173–8.

- (1946b). Linear variation of constants. *Amer. J. Math.* 68 (1946), 185–213.

- (1946c). An abelian lemma concerning asymptotic equilibrium. *Amer. J. Math.* 68 (1946), 451–4.
- (1946d). Asymptotic integration constants. *Amer. J. Math.* 68 (1946), 553–9.
- (1950). On free vibrations with amplitudinal limits. *Quart. Appl. Math.* 8 (1950), 102–3.
- (1957a). Bounded matrices and linear differential equations. *Amer. J. Math.* 79 (1957), 139–51.
- (1957b). Ordinary differential equations and Laplace transforms, *Amer. J. Math.* 79 (1957), 265–94.

Yosie, T. (1926). Über benachbarte Differentialgleichungen erster Ordnung. *Proc Physico-Math. Soc. Japan*, (3) 8 (1926), 16–20.

Young, W. H. (1909a). On differentials. *Proc. London Math. Soc.* (2) 7 (1909), 157–80.
- (1909b). Implicit functions and their differentials. *Proc. London Math. Soc.* (2) 7 (1909), 398–421.
- (1910). *The Fundamental Theorems of the Differential Calculus.* Cambridge Tracts in Mathematics and Mathematical Physics, 11 (Cambridge, 1910).

Zermelo, E. and Hahn, H. (1904). Weiterentwicklung der Variationsrechnung in den letzten Jahren. *Encyklopädie der math. Wiss.* vol. 2.1.1, pp. 626–41 (Teubner, Leipzig, 1899–1916).

Zoretti, L. (1912). Recherches contemporaines sur la théorie des fonctions. Les ensembles de points. *Encyclopédie des Sci. Math. Pures Appl.* vol. 2.1.2, pp. 113–70 (Paris, 1912).

MATHEMATICAL NOTATION

Principal symbols, brief description and, if relevant, the page on which a definition can be found.

1 Derivatives and differentials

ϕ'	derivative of a function of a real variable; p. 3.
$\dfrac{d\phi}{dt}$	alternative for ϕ'; p. 4.
ϕ'_+, ϕ'_-	right- (resp. left-) hand derivative; p. 6.
ϕ'_w	weak derivative; p. 35.
$\phi^{(r)}$	rth derivative; p. 41.
$D^+\phi, D_+\phi,$	
$D^-\phi, D_-\phi$	Dini derivative; p. 21.
$df(x_0)$	Fréchet differential of f at x_0; p. 167.
df	Fréchet differential of f; p. 167.
$d_j f(x_0), d_j f$	partial Fréchet differential; p. 178.
$D_j f(x_0), D_j f$	partial derivatives of f; p. 181.
$\partial f/\partial x_j$	alternative for $D_j f$; p. 182.
$[D_j f_i(x_0)]$	Jacobian (matrix); p. 183.
$\dfrac{\partial(f_1, \ldots, f_n)}{\partial(x_1, \ldots, x_n)}$	Jacobian (matrix); p. 183.
$\delta f(x_0)h$	Gâteaux variation of f at x_0 for h; p. 251.
$\delta f(x_0)$	Gâteaux variation; p. 251. Gâteaux differential; p. 255.
δf	Gâteaux differential; p. 256.
$\overline{\delta} f(x_0)h,$	
$\underline{\delta} f(x_0)h$	upper (resp. lower) Gâteaux variations; p. 258.
$\partial f(x_0)h$	Hadamard variation of f at x_0 for h; p. 259.
$\partial f(x_0)$	Hadamard variation; p. 259. Hadamard differential; p. 264.
∂f	Hadamard differential; p. 264.
$\overline{\partial} f(x_0)h,$	
$\underline{\partial} f(x_0)h$	upper (resp. lower) Hadamard variations; p. 268.
$\partial_* f(x_0)$	Hadamard subvariation; p. 270.
∇	gradient in Hilbert space; p. 196.

2 Mathematical symbols

$]a,b[,[a,b]$	open (resp. closed) interval
E^{o}	interior of set.
$\bar{E}$	closure of set.
$\langle x,y \rangle$	inner product; p. 318.
$\lvert x \rvert$	modulus of scalar x.
$\lVert x \rVert$	norm of x.
X'	dual of normed space X; p. 316.
X/S	quotient of normed space X by S; p. 317.
$X_1 \times \ldots \times X_n$	product of normed spaces; p. 317–8
E^*, E_*	cones dual to set E; p. 325.
T^*	adjoint of operator T; p. 319.
$E^{\,\char`^}$	support functionals to convex set E; p. 288.
H	Hamiltonian; p. 235.
$x^{\char`^}$	unit vector $x/\lVert x \rVert$; p. 274.
$\begin{bmatrix} a\,b \\ c\,d \end{bmatrix}$	matrix.
$\mathbf{V}$	matrix of linear map V.
$\mathrm{tr}\,\mathbf{V}$	trace of matrix $\mathbf{V}$.
$(h)^r$	$(h,h,\ldots,h)$ with r occurrences of h; p. 198.
$\alpha!$	if $\alpha=(\alpha_1,\ldots,\alpha_n)$, then $\alpha!=\alpha_1!\alpha_2!\ldots\alpha_n!$; p. 185.

3 Alphabetical list

$\mathbf{C}$	complex numbers.
$C[a,b]$	continuous scalar-valued functions on $[a,b]$.
$\mathrm{co}\,E, \overline{\mathrm{co}}\,\hat{E}$	convex (resp. closed convex) hull of E; p. 320.
C^r	r times continuously differentiable; p. 186.
C^∞	C^r for every r; p. 186.
$C(X,Y),$	
$\quad C^r(X,Y)$	continuous (resp. C^r) functions from X to Y.
d, ∂, δ, D	see *Derivatives* above.
$\mathscr{D}(\phi)$	domain of function ϕ.
$\det$	determinant.
$\mathscr{G}(\phi)$	graph of function ϕ.
$H, \hat{H}$	Hamiltonians; pp. 236, 235.
I	usually an interval in the real line.
J	as I.
$\mathbf{K}$	the scalar field, either $\mathbf{R}$ or $\mathbf{C}$.
$\ker$	kernel of a linear mapping.
$K_d(E;x_0)$	cone of vectors directed into E at x_0; p. 259.
$K_t(E;x_0)$	cone of vectors tangent to E at x_0; p. 288.
$\mathscr{L}\mathscr{H}(X,Y)$	linear homeomorphisms between normed spaces X, Y; p. 316.
$\mathscr{L}_n(X;Y)$	continuous multilinear maps from $X \times \ldots \times X$ to Y; p. 318.
$\mathscr{L}(X,Y)$	continuous linear maps from X to Y; p. 316.

$\mathcal{L}(X_1, ..., X_n; Y)$ continuous multilinear maps from $X_1 \times ... \times X_n$ to Y; p. 318.

$O(\phi(t))$ for small t, there is a constant k such that $|\phi(t)| \le k|t|$.

$o(\phi(t))$ $\phi(t)/t \to 0$ as $t \to 0$.

R   real numbers.

tr **V**   trace of matrix **V**.

X, Y, Z usually, but not invariably, normed spaces.

AUTHOR INDEX

Abel 108, 135
d' Alembert 335
Alexiewicz 160, 333
Ampère 57–9
Antosiewicz 338
Arzelà 159
Averbukh 312, 333
Aziz 67, 332

Baiada 335
Banach 333
Barbashin 334
Bartle 337
Bellman 95, 164, 335
Bendixson 155, 161
Bernoulli, D. 335
Bernoulli, J. 225
Bieberbach 65
Bihari 334
Bliss 159, 336
Bôcher 335
Bolzano 54, 56–7
Bompiani 162
Bonnet 62
Borel 63
Bouligand 337
Bourbaki 332–5, 338
Boyer 332
Brauer 334, 338
Bruckner 332
Bucher 333

Caqué 155
Carathéodory 160–1, 334, 337
Cartan 311, 334, 336–7
Cauchy 54, 56–62, 148–54
Cavalieri 54
Cesari 335
Coddington 160, 334
Conti 334–5, 338
Corduneanu 158, 338
Coriolis 148–9
Courant 310

Dahlquist 335

Darboux 65, 67
Dedekind 56
Dennis 337
Diaz 67, 332
Dieudonné 311, 333, 337–8
Dinculeanu 338
Dini 64
Dubovitskii 337
Dunford 338
Dunkel 335

Euler 335

Fan, Ky 311
Fenyö 337
Flett 332–3, 336–7
Fréchet 309–10, 311
Frölicher 333
Fuchs 155, 335
Fukuhara 158, 161

Gâteaux 309–10
Genocchi 333
Gilbert 62
Girsanov 337–8
Giuliano 162
Goldstine 337
Gollwitzer 164
Golomb 310
Grassmann 66
Graves 310, 336
Gronwall 95, 163
Grunsky 158

Hadamard 311, 333
Hahn 337
Halperin 67, 332
Hamilton 237
Hartman 160, 164, 334–7
Heine 63, 151
Hermite 65
Hesse 335
Hilbert 310
Hildebrandt 310, 336
Hukuhara 157, 162, 332–3

SUBJECT INDEX

An asterisk indicates that the reference occurs in an historical note.